Phytochemicals in Vegetables: A Valuable Source of Bioactive Compounds

Edited by

Spyridon A. Petropoulos

University of Thessaly, Department of Agriculture, Crop Production and Rural Environment Greece

Isabel C.F.R. Ferreira

Mountain Research Centre (CIMO), Polytechnic Institute of Bragança, Portugal

&

Lillian Barros

Mountain Research Centre (CIMO), Polytechnic Institute of Bragança, Portugal

General:

1. Any dispute or claim arising out of or in connection with this License Agreement or the Work (including non-contractual disputes or claims) will be governed by and construed in accordance with the laws of the U.A.E. as applied in the Emirate of Dubai. Each party agrees that the courts of the Emirate of Dubai shall have exclusive jurisdiction to settle any dispute or claim arising out of or in connection with this License Agreement or the Work (including non-contractual disputes or claims).
2. Your rights under this License Agreement will automatically terminate without notice and without the need for a court order if at any point you breach any terms of this License Agreement. In no event will any delay or failure by Bentham Science Publishers in enforcing your compliance with this License Agreement constitute a waiver of any of its rights.
3. You acknowledge that you have read this License Agreement, and agree to be bound by its terms and conditions. To the extent that any other terms and conditions presented on any website of Bentham Science Publishers conflict with, or are inconsistent with, the terms and conditions set out in this License Agreement, you acknowledge that the terms and conditions set out in this License Agreement shall prevail.

Bentham Science Publishers Ltd.
Executive Suite Y - 2
PO Box 7917, Saif Zone
Sharjah, U.A.E.
Email: subscriptions@benthamscience.org

**BENTHAM
SCIENCE**

CONTENTS

FOREWORD

Vegetables play a crucial role in the human diet, being relevant contributors to the intake of micronutrients (*i.e.*, vitamins and minerals) and dietary fiber and prebiotics, as well as occasionally of digestible carbohydrates and proteins (*e.g.*, tubers and pulses). Furthermore, beyond their nutrient composition, vegetables contain a range of non-essential bioactive compounds (*i.e.*, phytochemicals), among which carotenoids and polyphenols, including flavonoids, phenolic acids, stilbenes, lignans or tannins, are prominent, with others such as glucosinolates (in Brassicaceae), cysteine sulfoxides (in *Allium* species) or betalains (in beets) having more limited distribution.

Phytochemicals have attracted much attention in recent times as they may provide additional health benefits to the consumption of vegetables and other plant foodstuffs. The dietary intake of these compounds has been related with the prevention of some chronic and degenerative diseases that constitute major causes of death and incapacity in developed countries, such as cardiovascular diseases, type II diabetes, some types of cancers or neurodegenerative disorders like Alzheimer's and Parkinson's diseases. Nowadays it is considered that phytochemicals contribute, at least in part, for the protective effects of fruit and vegetable-rich diets, so that the study of their role in human nutrition has become a central issue in food research.

Consumers more demand for healthy and nutritious natural foods, while they are increasingly reluctant to chemical additives. These are requirements that fresh or minimally processed plant foods like vegetables can meet. Nevertheless, time constraints in developed countries have led to a decreasing tendency in the preparation of daily meals based on fresh ingredients. In this context, phytochemicals-rich foods are of great interest for both consumers and food industry that can use them as sources of bioactive ingredients for functional foods, nutraceuticals or dietary supplements. Moreover, owing to their properties, some phytochemicals might be used as natural additives, like antioxidants, preservatives, colorants or taste enhancers. Last but not the least, their bioactivity makes them also interesting to pharmaceutical and cosmetic industries for the development of drugs or cosmeceuticals.

Acknowledged experts in their fields have collaborated in the preparation of this book under the coordination of Prof. Spyridon A. Petropoulos, Prof. Isabel C.F.R. Ferreira and Dr. Lillian Barros. Throughout 12 chapters, a comprehensive overview is provided on the main groups of cultivated edible vegetables, as well as on some particular less used or locally employed native species that might be promoted for larger use in human nutrition. The coverage is ample, while the main focus is put into the interest of vegetables as phytochemicals sources, aspects such as plant description, chemical composition, influence of breeding, post-harvest or processing on bioactive compounds, health effects, bioaccessibility or bioavailability are also dealt with. No doubt that the book will be very useful for academic and industrial scientists, but also for students and consumers concerned about their health or who wish to delve into the knowledge of vegetables, their nutrient and phytochemical composition and their undoubted relevance in the human diet.

Celestino Santos-Buelga
Food Science, Faculty of Pharmacy,
University of Salamanca,
Spain

PREFACE

The present e-book aims at presenting the phytochemicals content of the main cultivated vegetables, as well as their health and therapeutic effects based on ¬*in vitro* and *in vivo*, animal and clinical studies. The importance of vegetables on human health is mostly attributed to their nutritional value; however, not always nutrients are the sole responsible compounds for such properties and several other compounds can also contribute to health-promoting effects. These compounds have been identified as secondary metabolites and plants usually synthesize them for their own protection from pests and diseases or their biosynthesis is triggered under specific environmental conditions.

Book structure has been arranged in individual chapters, each one of them dealing with specific groups of vegetable sources of phytochemicals, either in terms of taxonomy (species of the same family) or in terms of edible parts morphology (*e.g.* leafy and root vegetables). For each species, a short introduction regarding the description of morphology, taxonomy and general information is included, as well as its chemical composition and its main health effects.

Chapter 1 presents the main phytochemicals that have been identified in various roots vegetables consumed throughout the world, including potato, celeriac, turnips, radish, beets, Hamburg parsley, taro, yam, parsnip and salsify. Chapter 2 presents vegetables that belong to the Allium genus. Chapter 3 presents bean, a vegetable of the Fabaceae family, which is one of the main starch and protein sources for most of the world. Chapter 4 demonstrates the chemical composition and health effects of another group of vegetables that all belong to the Cucurbitaceae family. Chapters 5-7 provides a clear insight into a diversified group of vegetables that are consumed for their edible leaves, belonging to Asteraceae and Apiaceae families. Chapter 8 discusses the phytochemicals content, their functionality and breeding tools for the enrichment of cole vegetables (Brassicaceae) in phytochemicals. Chapter 9 presents another important group of fruit vegetable that belongs to the Solanaceae family, namely, tomato, eggplant and pepper. Other important vegetables, such as globe artichoke and okra, are characterized in chapters 10 and 11. Finally chapter 12 deals with a special group of fruit and vegetables, which although they have a regional interest and are less well-known, they present important bioactive properties and health effects.

Spyridon A. Petropoulos
Department of Agriculture, Crop Production and Rural Environment,
University of Thessaly,
Volos, Greece

Isabel C.F.R. Ferreira & Lilian Barros
Centro de Investigação de Montanha (CIMO),
Instituto Politécnico de Bragança,
Bragança, Portugal

List of Contributors

Milica Aćimović	Institute of Field and Vegetable Crops Novi Sad, Serbia
Ryszard Amarowicz	Institute of Animal Reproduction and Food Research of the Polish Academy of Sciences, Poland
Adelar Bracht	Departamento de Bioquímica, Universidade Estadual de Maringá, Maringá, Paraná, Brazil
Angela Cardinali	Institute of Sciences of Food Production, National Research Council, Bari, Italy
Antonios Chrysargyris	Department of Agricultural Sciences, Biotechnology and Food Science, Cyprus University of Technology, Lemesos, Cyprus
Mirjana Cvetković	Institute of Chemistry, Technology and Metallurgy, University of Belgrade, Serbia
Isabella D'Antuono	Institute of Sciences of Food Production, National Research Council, Bari, Italy
Nevena S. Dojčinović	Faculty of Chemistry, University of Belgrade, Serbia
Francesco Di Gioia	Department of Plant Science, Pennsylvania State University, University Park, PA, USA
Massimiliano D'Imperio	Institute of Sciences of Food Production, CNR, Bari, Italy
Rubia Carvalho Gomes Correa	Departamento de Bioquímica, Universidade Estadual de Maringá, Maringá, Paraná, Brazil
Isabel C.F.R. Ferreira	Centro de Investigação de Montanha (CIMO), Instituto Politécnico de Bragança, Bragança, Portugal
Maria Fernanda Francelin	Departamento de Bioquímica, Universidade Estadual de Maringá, Maringá, Paraná, Brazil
Jessica Amanda Andrade Garcia	Departamento de Bioquímica, Universidade Estadual de Maringá, Maringá, Paraná, Brazil
Donato Giannino	Institute of Agricultural Biology and Biotechnology, CNR, Unit of Rome, Italy
Maria Gonnella	Institute of Sciences of Food Production, CNR, Bari, Italy
Marina Z. Kostic	Department of Plant Physiology, Institute for Biological Research "Siniša Stanković", University of Belgrade, Belgrade, Serbia
Vlastimil Kubáň	Department of Food Technology, Faculty of Technology, Tomas Bata University in Zlin, Czech Republic
Raj Kumar	Division of Life Sciences, Plant Molecular Biology and Biotechnology Research Center, Research Institute of Natural Science, Gyeongsang National University, Jinju-52828, Republic of Korea
Vito Linsalata	Institute of Sciences of Food Production, National Research Council, Bari, Italy
Antonio Roberto Giriboni Monteiro	Departamento de Bioquímica, Universidade Estadual de Maringá, Maringá, Paraná, Brazil

Ronald B. Pegg   Department of Food Science and Technology, College of Agricultural and Environmental Sciences, The University of Georgia, USA

Rosane Marina Peralta   Departamento de Bioquímica, Universidade Estadual de Maringá, Maringá, Paraná, Brazil

Rosa Perez-Gregorio   LAQV-REQUIMTE, Departamento de Química e Bioquímica, Faculdade de Ciências da Universidade do Porto, Portugal

Spyridon A. Petropoulos   Department of Agriculture, Crop Production and Rural Environment, University of Thessaly, Volos, Greece

Sofia Plexida   Department of Agriculture, Crop Production and Rural Environment, University of Thessaly, Volos, Greece

Milica M. Rat   Faculty of Sciences, University of Novi Sad, Serbia

Massimiliano Renna   Institute of Sciences of Food Production, CNR, Bari, Italy
Department of Agricultural and Environmental Science, University of Bari Aldo Moro, Bari, Italy

Ana Sofia Rodrigues   Instituto Politécnico de Viana do Castelo, Escola Superior Agrária, Ponte de Lima, Portugal
Centre for Research and Technology of Agro-Environmental and Biological Sciences – CITAB, Vila Real, Portugal

Erin N. Rosskopf   USDA-ARS, U.S. Horticultural Research Laboratory, Fort Pierce, FL, USA

Jesus Simal-Gandara   Nutrition and Bromatology Group, Department of Analytical and Food Chemistry, Faculty of Food Science and Technology, University of Vigo, Spain

Saurabh Singh   Division of Vegetable Science, ICAR-Indian Agricultural Research Institute (IARI), New Delhi, India

Rajender Singh   ICAR-Indian Agricultural Research Institute (IARI), Regional Station, Katrain, Kullu Valley, Himachal Pradesh, India

Marija S. Smiljkovic   Department of Plant Physiology, Institute for Biological Research "Siniša Stanković", University of Belgrade, Belgrade, Serbia

Marina D. Sokovic   Department of Plant Physiology, Institute for Biological Research "Siniša Stanković", University of Belgrade, Belgrade, Serbia

Jovana Stanković   Institute of Chemistry, Technology and Metallurgy, University of Belgrade, Serbia

Dejan S. Stojkovic   Department of Plant Physiology, Institute for Biological Research "Siniša Stanković", University of Belgrade, Belgrade, Serbia

Blanka Svobodová   Department of Food Technology, Faculty of Technology, Tomas Bata University in Zlin, Czech Republic

Vele V. Tešević   Faculty of Chemistry, University of Belgrade, Serbia

Giulio Testone   Institute of Agricultural Biology and Biotechnology, CNR, Unit of Rome, Italy

Prerna Thakur   Department of Vegetable Science, Punjab Agricultural University (PAU), Ludhiana, Punjab, India

Marina Todosijević   Faculty of Chemistry, University of Belgrade, Serbia

Nikolaos Tzortzakis Department of Agricultural Sciences, Biotechnology and Food Science, Cyprus University of Technology, Lemesos, Cyprus

Tatiane Francielli Vieira Departamento de Bioquímica, Universidade Estadual de Maringá, Maringá, Paraná, Brazil

CHAPTER 1

Root Vegetables as a Source of Biologically Active Agents - Lesson from Soil

Dejan S. Stojkovic, Marija S. Smiljkovic, Marina Z. Kostic and **Marina D. Sokovic**[*]

Department of Plant Physiology, Institute for Biological Research "Siniša Stanković", University of Belgrade, Bulevar Despota Stefana 142, 11000 Belgrade, Serbia

Abstract: Natural products and primary and secondary metabolites of plants have many biological functions, many of which are considered as health-beneficial for mankind. This chapter will focus on biologically active ingredients in widely consumed root vegetables, such as potato, celeriac, turnips, radish, beets, Hamburg parsley, taro, yam, parsnip and salsify. A recent update of studies is presented regarding underground parts of the mentioned vegetables – plant underground parts. Chemical constituents responsible for such biological activities, with focus on recent findings for each root vegetable separately are presented.

Keywords: Antidiabetic, Antihypertension, Antimicrobial, Antimutagenic, Antioxidant, Biological activity, Cardioprotective, Chemical constituents, Chemopreventive, Crops, Hepatoprotective, Lectins, Metabolites, Phenolics, Pigments, Polysachrides, Root vegetables, Thnopharmacology, Vitamins.

INTRODUCTION TO BIOLOGICAL ACTIVITY OF NATURAL PRODUCTS: ROOT VEGETABLES

Natural products have historically been an extremely productive source for new medicines in all cultures and continue to deliver a great variety of structural templates for drug discovery and development. Although products derived from natural sources may not necessarily represent active ingredients in their final form, the majority of all drugs in the market have their origin in nature [1, 2]. A significant number of drugs have been derived from plants that were traditionally employed in ethnomedicine or ethnobotany, while others were discovered through random screening of plant extracts for their biological potential and actual application to which our research group was focused in the past decade [3 - 14].

[*] **Corresponding author Marina Sokovic:** University of Belgrade, Department of Plant Physiology, Belgrade, Serbia; Tel: +381 11 2078419; E-mail: mris@ibiss.bg.ac.rs

Spyridon A. Petropoulos, Isabel C.F.R. Ferreira and Lillian Barros (Eds.)

An avenue that may have influenced ethnopharmacology suggests that some traditionally used remedies may have arisen from observations of self-medication by animals [15]. Studies have shown that wild animals often consume plants and other materials for medical rather than nutritional reasons, treating parasitic infections and possible viral and bacterial diseases [16].

Cultivated plants, which have their edible part underground, are called root and tuber crops. Man domesticated various roots and tuber producing species for similar use in different parts of the world and on different elevations, such as yams in Africa and Asia, taro in Asia, cassava, sweet potato, and potato in America in low, medium and high altitudes. Additionally, different parts of the same plant species are used in different regions, such as leaves and petioles, fruit, seeds and roots and tubers which are the topic of interest in the present book chapter. The thickened taproot, the hypocotyl and the epicotyl constitute the edible parts of carrots, beets and some radishes, while early spring radish has its thickened hypocotyl belowground; potato tubers are modified underground stems (stolons), and the marketed parts of onions and garlic are modified thickened leaves (bulbs) [17].

Potato

Potato (*Solanum tuberosum* L.) (Fig. **1**) belonging to the family Solanaceae is the fifth most important crop in the world, it is rich in calories and biologically active phytochemicals (*β*-carotene, polyphenols, ascorbic acid, tocopherol, *α*-lipoic acid, *etc.*) (Table **1**) [18, 19]. The main nutrient in potato is starch, since tubers are the main storage organs of the species [19]. Potatoes could be prepared and used in different ways like baking, boiling, dehydrating, and frying [20].

Potato tubers are proven to have various activities and their consumption could subsequently lead to a healthier population Table **2**, mostly due to numerous chemicals that could be found in the organs of this widely popular crop. Phytochemicals that play an important role in human health as antioxidants are concentrated in potato peel. Their content is higher in potato cultivars with brighter peel colors and frequent consumption of potato increases phenolic content in our nutrition [19]. Phenolic compounds prevent oxidative damage of DNA, reduce gut glucose absorption, suppress adipogenesis, reduce systolic and diastolic blood pressure and prevent proliferation of cancer cells [21 - 24]. It is reported that chlorogenic acid (Fig. **2A**), one of the main phenolics found in potato tubers Table **1**, has strong antioxidant, antidiabetic and antihypertension activity [25, 26]. The lectin StL-20 isolated from potato, showed antimicrobial activity against *Listeria monocytogenes*, *Escherichia coli*, *Salmonella enteritidis*, *Shigella boydii*, *Rhizopus* spp., *Penicillium* spp. and *Aspergillus niger*. Also, lectin has shown antibiofilm activity against *Pseudomonas aeruginosa*, while it reduced biofilm formation by 5-20% in 24h in a dose-dependent manner [27].

Table 1. Chemical constituents in potato tubers.

Root vegetable	Phytochemical group	Individual compounds	References
Solanum tuberosum	lectins	StL-20	[27]
	glycoalkaloids	α-chaconine	[31]
		α -solanine	
		solanidine	
	phenolic acids	caffeic acid	[38]
		chlorogenic acid	
		protocatechuic acid	
		trans-cinnamic acid	
		p-coumaric acid	
		ferulic acid	
		vanillic acid	
		gallic acid	
		syringic acid	
		salicylic acid	
	anthocyanins	3-rutinoside-5-glucoside	[19]
		peonidin-3-rutinoside-5-glycosides	
		petunidin-3-rutinoside-5-glycosides acylated with *p*-coumaric and ferulic acid	
		malvidin	
		delphinidin	
		cyanidin	
		pelargonidin	
	polysaccharides	starch	[19]
	peptides	patatin	[37]

Lectins are reported to induce apoptosis and have potential anticancer activity [28]. Glycoalkaloids have also numerous bioactivities such as antimicrobial, anticancer, anticholesterol and anti-inflammatory [29]. Glycoalkaloids (Table **1**) showed antifungal activity, with α-chaconine being the most active among them [30]. Moreover, α-chaconine (Fig. **2B**) inhibited the growth of *Aspergillus niger*, *Penicillium roqueforti*, and *Fusarium graminearum* [31]. Other glycoalkaloids such as α-solasonine and α-solamargine demonstrated synergistic activity against

Phoma medicaginis and *Rhizoctonia solani* [30]. Ethanolic extract of *S. tuberosum* is active against *Staphylococcus aureus, Streptococcus pyogenes, Klebsiella pneumonia* and *Pseudomonas aeruginosa* with minimal inhibitory concentration (MIC) values in the range of 0.62-10 mg/ml [32]. Potato peel is rich in anthocyanins which play an important role in human health. Many researchers reported antioxidant, anticancer and anti-inflammatory activities of anthocyanins [33 - 36]. Patatin, a peptide present in potato tubers demonstrated antioxidant activity and could also inhibit hydroxyl radical-induced DNA damage *in vitro* [37].

Table 2. Health promoting effects of potato.

Root vegetable	Biological activity	References
Solanum tuberosum	antimicrobial	[27, 30 - 32, 39, 40]
	antibiofilm	[27]
	antioxidant	[39]
	antiproliferative	[39]
	anticancer	[35]
	anti-inflammatory	[36]
	antiobesity	[22]
	antihypertensive	[23]
	antidiabetic	[25]

Fig. (1). Potato tubers (photographed by M. Kostic).

A   B

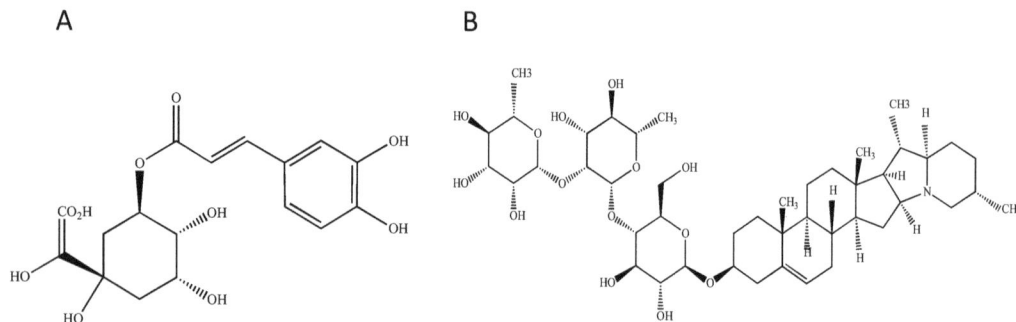

Fig. (2). A) chlorogenic acid B) α-chaconine

Celeriac

Apium graveolens L., commonly known as celery Fig. (**3**), is an edible plant of the family Apiaceae. Literature data indicates that *A. graveolens* has a wide spectrum of biological properties such as antifungal, antioxidant, antihypertensive, antihyperlipidemic, diuretic, and anticancer [41], with roots being less examined than other plant parts.

Researchers identified some phenolic compounds (Fig. **4**) and coumarins from celery root extract which was shown to have antioxidant and anti-inflammatory effects (Table **3**) [42, 43].

Table 3. Chemical constituents of celeriac roots.

Root vegetable	Phytochemical group	Individual compounds	References
Apium graveolens	phenolic compounds	kaempferol glucoside	[43, 45]
		quercetin glucoside	
		genkwanin glucoside	
		apigenin glucoside	
		luteolin glucoside	
		naringenin	
		(*Z*)-3butylidenephthalide, 3-butyl-4,5 dihydrophthalide	
		α-thujene	
	coumarin	6-(3'-methyl-1'oxobutyl)-7-hydroxy coumarin	[42]

Another study indicated that the root extracts of *A. graveolens* significantly decreased CC14-induced acute hepatic injury [44]. The root extracts have also antioxidant effect with EC_{50} value ranging from 2.41-3.14 mg/ml [45]. Besides antioxidant and hepatoprotective effects, no data is available about some other potential benefits suggesting that celeriac roots should be further examined (Table **4**).

Table 4. Health promoting effects of celeriac.

Root vegetable	Biological activity	References
Apium graveolens	antioxidant	[45]
	hepatoprotective	[44]

Fig. (3). Celeriac roots (photographed by M. Kostic).

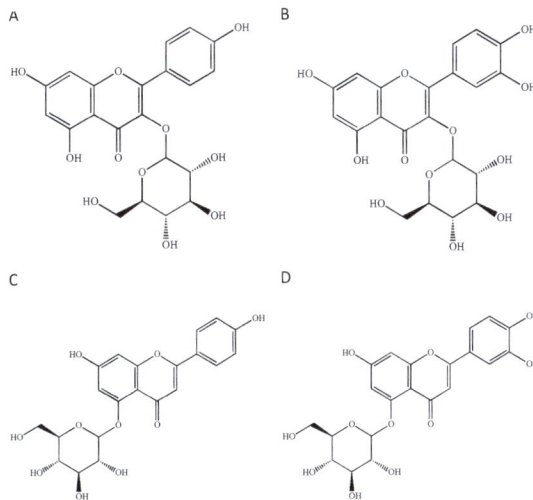

Fig. (4). Phenolic compounds found in celeriac roots A) kaempferol glucoside B) quercetin glucoside C) apigenin glucoside D) luteolin glucoside.

Turnips

Brassicaceae family (turnips, broccoli, Brussels sprouts, cauliflower, and cabbages among others) has been extensively studied due to its nutritional and health benefits. *Brassica rapa,* has many variants: *B. rapa* var. *ruvo* (broccoli raab), *B. rapa* var. *chinensis* (Chinese cabbage), *B. rapa* var. *pekinensis* (turnip greens), *B. rapa* var. *parachinensis* (Chinese flowering cabbage), and *B. rapa* var. *pervidis* (tender greens) among which *Brassica rapa var. rapifera* (turnip) is one of the oldest cultivated vegetables [46]. Turnips are usually consumed as a boiled vegetable, while its root is an ingredient in folk medicine for cold remedy.

From root essential oils 41 compounds were detected by GC and GC/MS analyses, some of these compounds are presented in (Table **5**) [46]. Within terpenes (5.0 – 14.6%), menthol Fig. (**5**) is the most common component (4.9-6.1%) and might be one of the reasons for various biological activities of the species (Table **6**). Essential oils of turnips roots showed antimicrobial effect against *Listeria monocytogenes, Staphylococcus aureus, Salmonella enterica, Escherichia coli, Klebsiella pneumoniae, Pseudomonas aeruginosa, Fusarium culmorum, Aspergillus ochraceus, A. flavus,* and *Candida albicans* with great antifungal activity (MIC values 0.5 to 2 mg/ml) and moderate antibacterial activity (MIC values 2 to 7 mg/ml). Antimicrobial activity can be associated with the presence of menthol, hexahydrofarnesyl aceton, allyl isothiocyanate and 2-phenylethanol [47]. Antioxidant activity of these oils was proven with different tests (DPPH, reducing power, β-carotene bleaching and chelating ability on ferrous ions) [46].

Additionally, ethanolic and aqueous extracts showed antimicrobial activity against *Staphylococcus aureus, Bacillus subtilis, Pseudomonas aeruginosa, Escherichia coli, Candida albicans,* and *Aspergillus niger* with MIC values in the range of 12.5 - 25 mg/ml [48]. The HPLC-DAD and HPLC-UV analysis showed a phenolic and organic acids profile of root aqueous extract [49] with kaempferol 3-*O*-sophoroside-7-*O* glucoside, kaempferol3-*O*-(feruloyl/caffeoyl)-sophoroside-7 glucoside,isorhamnetin 3,7-*O*-diglucoside, isorhamnetin 3-*O*-glucoside, and malic acid as the main components. This extract showed low antioxidant activity with IC_{25}=1.44 mg/ml [49]. The metanolic extract of turnip root showed antioxidant activity and anticancer activity against HT-29 and MCF-7 lines [50].

Taro

Several plants from Araceae family, such as taro (*Colocasia esculenta*), eddoe (*Colocasia antiquorum*), giant taro (*Alocasia macrorrhiza* (L.) Schott), swamp taro (*Cyrtosperma merkusii*), and arrow leaf elephant's ear (*Xanthosoma sagittifolium*), are widely used in subtropics and tropic countries, as energy.

Table 5. Chemical constituents of turnip roots.

Root vegetable	Phytochemical group	Individual compounds	References
Brassica rapa var. *rapifera*	alcohols	(Z)-3-Hexenol	[46]
		(E)-3-Hexenol	
		2-Phenylethanol	
		methoxyvinylphenol	
	phenols	3-*p*-coumaroylquinic acid	[49]
		caffeic acid	
		ferulic acid	
		sinapic acid	
		kaempferol 3-*O*-sophoroside-7-*O*-glucoside	
		kaempferol 3-*O*-sophoroside-7-*O*-sophoroside	
		kaempferol 3-*O*-(feruloyl/caffeoyl)-sophoroside-7-*O*-glucoside	
		kaempferol 3,7-*O*-diglucoside, isorhamnetin 3,7-*O*-diglucoside	
		kaempferol 3-*O*-sophoroside,	
		kaempferol 3-*O*-glucoside	
		isorhamnetin 3-*O*-glucoside	
		1,2-disinapoylgentiobiose	
		1,2'-disinapoyl-2-feruloylgentiobiose	
	norisoprenoids	safranal.	[46]
		β-Cyclocitral	
		α-Ionone	
		geranylacetone	
		β-Ionone	
		hexahydrofarnesyl acetone	
	terpenes	tricyclene	
		α-thujene	
		α-pinene	
		camphene	

(Table 5) cont.....

Root vegetable	Phytochemical group	Individual compounds	References
Brassica rapa var. *rapifera*	terpenes	sabinene	[46]
		β-pinene	
		myrcene	
		α-terpinene	
		limonene	
		m-cymenene	
		isomenthone	
		menthol	
		α-terpineol	
		longifolene	
		phytol	
	organic acids	aconitic acid	[49]
		citric acid	
		ketoglutaric acid	
		malic acid	
		shikimic acid	
		fumaric acid	

sources because rhizomes of these vegetables contain a large amount of starch (85% of total dry matter) [51]. Starch (Fig. **6**) from these plants has a large role in the food industry, and great potential for development of products for industrial uses. Taro can be consumed roasted, baked or boiled. Giant taro is widely distributed in China and other Southeastern Asian countries. There is a record that the giant taro extract is used in folk medicine against appendicitis, chronic bronchitis and atrophic rhinitis [52].

Table 6. Health promoting effects of turnip roots.

Root vegetable	Biological activity	References
Brassica rapa var. *rapifera*	antioxidant	[48, 49]
	cytotoxic effect	[50]

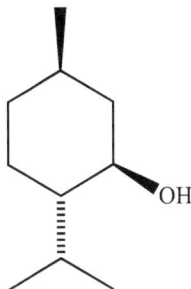

Fig. (5). Menthol, terpene found in turnips root.

Phytochemical investigations of taro roots have been conducted (Table **7**) [52 - 56]. Rahman *et al.* [57] demonstrated that the rhizomes' extract of *A. macrorrhiza* have antihyperglycemic, antioxidant and cytotoxic activity [57]. Indole alkaloids exhibited cytotoxicity against four tested human cancer cell lines (HePG2, Hep-2, HCT-116, MCF-7) with the strongest activity (IC_{50}=10 µM) being observed against Hep-2 larynx cancer cells [54]. Isolated lignanamides and monoindoles (Table **7**) showed no cytotoxicity to RAW 264.7 cells and moderate antiproliferative activity against CNE-1, MGC-803, MCF-7 cancer lines [52]. 2017). Also, piperidine alkaloids isolated from rhizomes of *Alocasia macrorrhiza* (Table **7**) showed cytotoxicity against human cancer cell lines (CNE-1, Detroit 562, Fadu, MGC-803, and MCF-7) [53]. Alocasin showed antifungal activity against *Botrytis cinerea* [58]. Another species, *Cyrtosperma merkusii* is grown in fresh water marshes and swampy areas, its tubers are rich in carotenoids and are known for their high antioxidant activity [59]. Diverse chemical composition might be the cause of a range of activities confirmed for taro roots, from antifungal to antihyperglicemic (Table **8**).

Radish

Radishes (Fig. **7**), an economically important crop belonging to the Brassicaceae family, have root which is edible, widely consumed, especially in salads, and has different compounds important for human health (Table **9**) [60]. Although entire plant is edible, radishes are known for their edible tuberous roots which can vary in shape, size and diameter [61]. Roots can be eaten as raw vegetables or after processing namely pickling, canning or drying [62]. Salted roots are a traditional Japanese food and are consumed in more than 500 000 tones/year, while Daikon (Japanese white radish) served with soy sauce, boiled fish or mushroom [63, 64]. Radishes are widely cultivated due to high-yield, low-labor requirements, short growing season, and pest-resistant nature [65].

Table 7. Chemical constituents of taro roots.

Root vegetable	Phytochemical group	Individual compounds	References
Alocasia macrorrhiza	polysaccharide	starch	[51]
	lignanamides	(±)-(E)-3-(2-(3-hydroxy-5-methoxyphenyl)-3-(hydroxymethyl)-7-methoxy-2,3-dihydrobenzofuran-5-yl)-N-(4-hydroxyphenethyl)acrylamide	[52]
		(±)-(E)-3-(2-(4-hydroxy-3,5-dimethoxyphenyl)-3-(hydroxymethyl)-7-methoxy-2,3-dihydrobenzofuran-5-yl)-N-(4-hydroxyphenethyl)acrylamide	
		(±)-(Z)-3-(2-(3-hydroxy-5-methoxyphenyl)-3-(hydroxymethyl)-7-methoxy-2,3-dihydrobenzofuran-5-yl)-N-(4-hydroxyphenethyl)acrylamide	
		(±)-(Z)-3-(2-(4-hydroxy-3,5-dimethoxyphenyl)-3-(hydroxymethyl)-7-methoxy-2,3-dihydrobenzofuran-5-yl)-N-(4-hydroxyphenethyl)acrylamide	
		(±)-4-(Ethoxy(4-hydroxy-3-methoxyphenyl)methyl)-2-(4-hydroxy-3-methoxyphenyl)-N-(4 hydroxyphenethyl)tetrahydrofuran-3-carboxamide	
	alkaloids	**monoindoles** (1-(2-(5-Hydroxy-1H-indol-3-yl)-2-oxoethyl)-1H-pyrrole-3-carbaldehyde)	[52]
		piperine alkaloids	[53]
		(2S,3R,6R)-2-methyl-6-(1-phenylnonan-4-one-9-yl)piperidin-3-ol	
		(2S,3R,6R)-2-methyl-6-(1-phenylnonan-5-one-9-yl)piperidin-3-ol	
		(2S,3R,6R)-2-methyl-6-(9-phenylnonyl)piperidin-3-ol	
		(2S,3S,6S)-2-methyl-6-(9-phenylnonyl)piperidin-3-ol	
		(2R,3R,4S,6S)-2-methyl-6-(9-phenylnonyl)piperidine-3,4-diol	
		(2R,3R,4R,6R)-2 methyl-6-(9-phenylnonyl)piperidine-3,4-dio	
		indole alkaloid (2-(5-hydroxy-1H-indol-3yl)-2-oxo-acetic acid)	[54]
		alocasin A-E	[54, 56, 58]
		hyrtiosin B	[54]
		hyrtiosulawesin	[54]
	sterols	β-sitosterol	[54]
		β-sitosterol 3-O-β-D-glucoside	
	fatty acids	5-hydroxy-1H-indole-3-carboxylic acid methyl ester	[54]
		α-monopalmitin	
		1-O-β-D-glucopyranosyl-(2S, 3R, 4E, 8Z)-2-[(2(R)-hydroctadecanoyl) amido]-4,-octadecadiene-1,3-diol	
		3-epi-betulinic acid	
		3-epi-ursolic acid	
	ceramide	(2S,3S,4R)-2N-[(2′R)-2′-hydroxy-hexacosanoyl]- tetradecane-1,3,4-triol	[55]
Cyrtosperma merkusii	carotenoids	β-carotene	[59]
		α-carotene	
		β-cryptoxanthin	
		lutein	
		zeaxanthin	

Skin of the taproot can vary in color with red radishes being the most common ones, although there are also pink, white and grey or black skinned varieties [62]. Different varieties include Amethyst (round, purple taproot), Crunchy Royale (round, red taproot), D' Avingnon (cylindrical, red taproot), Miyashige daikon (cylindrical, white taproot), Nero Tondo (round, black taproot), Ping Pong (round, white taproot), Pink Beauty (round, pink taproot), Red Meat (round, red taproot) and these varieties can differ in antioxidant capacity, anthocyanin content, glucosinolate and isothiocyanate content and activation of the antioxidant response element (ARE) [66].

Table 8. Health promoting effects of taro roots.

Root vegetable	Biological activity	References
Alocasia macrorrhiza	antifungal	[58]
	cytotoxic effect	[54]
	antiinflamatory	[52]
	antiproliferative	[53]
	antihyperglycemic	[54]

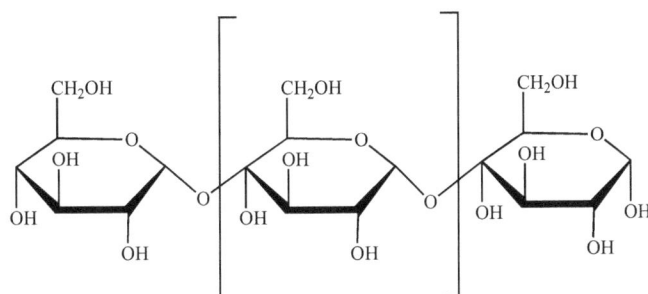

Fig. (6). Starch, compound found abundantly in taro roots.

Different biological activities have been examined for different varieties of radishes (Table 10). Radish has been used in Estonian ethnopharmacology for the relief of tumor symptoms [67]; in India it has been used for health issues like urinary problems and piles [68], while in the Mexican traditional medicine black radish roots are used for the treatment of pigment and cholesterol gallstones, and also for decreasing serum lipid level [69]. It has a potential for probiotic usage due to lactic acid bacterial strains such as *Lactobacillus plantarum* and *Lactobacillus fermentum* which could be isolated from fermented radishes [70].

4-(Methylthio)-3-butenyl isothiocyanate, a compound found widely in radishes, belongs to isothiocyanates Fig. (8), a group of compounds that has been proven to exhibit antimicrobial, antimutagenic and anticarcinogenic activities [64]. N-hexane extract of Daikon has significant antimutagenic effect mainly due to 4-(Methylthio)-3-butenyl isothiocyanate and for the best results it should be eaten not later than 30 min after grating [64, 71]. 4-(Methylthio)-3-butenyl isothiocyanate extracted from Tunisian *R. sativus* showed good chemo-protective effect [72].

Table 9. Chemical constituents of radish roots.

Root vegetable	Phytochemical group	Individual compounds	References
Raphanus sativus var. *niger*	sugars	fructose	[77]
		glucose	
	vitamins	ascorbic acid	
		β-carotene	
		tocopherols	
	phenolics	quercetin	
		kaempferol	
Raphanus sativus, (different varieties)	fatty acids	hexadecanoic acid	[62]
		methyl linolenate	
	isothiocyanates	4-(methylthio)butyl isothiocyanate	
		5-(methylthio)pentyl isothiocyanate	
		4-(methylthio)-3-butenyl isothiocyanate	
		2-phenylethyl isothiocyanate	
		5-(methylthio)-4-pentenenitrile	
	sulfides	dimethyl trisulfide	
Japanese White Radish	isothiocyanates	4-(Methylthio)-3-butenyl Isothiocyanate	[71]
Raphanus sativus red	phenolics	rutin hydrate	[78]
		vanillic acid	
		p-coumaric acid	
		caffeic acid	
		trans-ferulic acid	
		pyrogallol	
		gallic acid	
Raphanus sativus	phenolics	catechin	[79]
		sinapic acid	
Raphanus sativus green	phenolics	quercetin	[80]
		kaempferol	

Root vegetable	Phytochemical group	Individual compounds	References
Raphanus sativus red	phenolics	luteolin	[80]
		kaempferol	
		isorhamnetin	
		luteolin	

Fig. (7). Radish roots (photographed by M. Kostic).

Antibacterial properties that are proven for different parts of radish plants, with root being the most active, are not linked to the total isothiocyanate content, but are positively correlated with levels of individual isothyocyanate classes [73]. Root was found to be the most efficient part of the plant also regarding the inhibition of cell proliferation and induction of apoptosis in human cancer cells, with the hexane extract of root having different classes of isothiocyanates which could be responsible for these beneficial effects [74].

Table 10. Health promoting effects of radish roots.

Root vegetable	Biological activity	References
Raphanus sativus var. *niger*	against flatulence, indigestion and the formation of gallstones	[77]
	stimulation of bile function	
	antiurolithiatic	
	weak hepatoprotective effect	
	beneficial effect in alimentary hyperlipidaemia	
	antioxidant	[65]
	protects from lipid peroxidation	[76]
	protective against acute toxicity	[75]
	treatment of cholesterol gallstones	[69]
	decreasing serum lipids levels	
Raphanus sativus	antibacterial	[73]
	anti-cancer	[74]
Radish juice	antidiabetic	[81]
	hypoglycemic	
Radish different varieties	antioxidant	[79, 66]
Japanese white radish	antimutagenic	[71]
Tunisian radish	chemo-protective	[72]

Glucosinolates are plant secondary metabolites, precursors of isothiocyanates, with glucoraphasatin (4-Methylsulfanyl-3-butenylglucosinolate) accounting for more than 90% of this class in *R. sativus* [71]. Spanish black radish is examined as an attractive source of bioactive glucosinolates [65], and its high glucosinolates level have been found to be beneficial in cases of acute toxicity [75]. Granules from black radish root can protect cell membranes against lipid peroxidation; they also protect membrane changes caused by fat rich diet and have beneficial effect on rat colon mucosa [76].

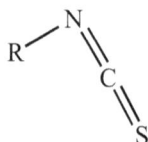

Fig. (8). General structure of isothiocyanates, group of compounds known for their range of biological activities, found abundantly in radish roots.

Beetroot

Beet (*Beta vulgaris*) is part of the Chenopodiaceae family (Fig. **9**). Due to different morphology, cultivated and wild maritime beets are separated in different subspecies. Sea beet (*Beta vulgaris* subsp. *maritima*) is the main member of maritime beets. Sugar beets (*Beta vulgaris* subsp. *saccharifera*), fodder beets (*Beta vulgaris* subsp. *crassa*), leaf beets (*Beta vulgaris* subsp. *cicla*) and garden beets (*Beta vulgaris* subsp. *rubra*) are the members of cultivated beets.

Beta vulgaris subsp. *cicla* (Swiss chard) and *Beta vulgaris* subsp. *rubra* (Red beetroot) are used as food since 1000 B.C. Even Romans ate their leaves, and used the roots in medical purpose [82].

Red beet variety is the cultivated form of *Beta vulgaris* subsp. *vulgaris* (conditiva) and it has twice less sugar than sugar beet. Its taproot has been used in food all over world; it is used as pickles, salad or juice [83].

Red beetroot owes its name to the red color produced by betalain pigment. Various biological activities are confirmed for betalains including antioxidant, anti-cancer, anti-lipidemic and antimicrobial [84].

There is an increasing interest in betalains because of recent trends in the food industry to avoid synthetic colorants, and beet is a valuable source of natural red color. Red color is the result of the mixture of yellow pigments betaxanthins and violet betacyanins (Fig. **10**), both members of betalain group [85].

Beetroot is considered among the 10 best vegetables based on its antioxidant capacity thanks to high phenolic content, but probably also on the synergism between individual phenolic compounds as well as between phenolics and betalains (Table **11**) [86, 87]. Phenolic content varies between different parts of root and decreases in the order of peel, crown, and flesh [88]. These combinations of compounds increase beetroot value as food and attribute to the wide range of health benefits listed in (Table **12**).

Red beet juice is proven to be effective in enhancing athletes' performance mainly due to high nitrates content, so difference found in nitrate levels between different cultivars is important in selecting the best ones for supplements development [83]. Nitrate level of beetroot is also important for its cardioprotective potential since *in vivo* it increases the levels of nitric oxide (NO), which is vasoprotective, it retards angiogenesis and has many other pleiotropic effects [89].

Its beneficial effect can be seen also in obese people and may be related to increased plasma NO concentration where even after a single dose of juice

(140 ml) it attenuated postprandial impairment of flow mediated dilation of brachial artery [90]. In obese patients it also showed beneficial effects on daily systolic blood pressure [91].

Red beetroot juice has beneficial effects in the treatment of obesity [92] and even one shot (70 ml) could improve antioxidant status due to high polyphenol content [93].

Even wastes in food processing like beetroot pomace could be used for its biological activities, which are probably induced by betalaines and phenolic compounds that remain in this by-product (Table **11**) [94]. Its beneficial effects like antioxidant and hepatoprotective have also been confirmed by *in vivo* studies [95]. Beetroot pomace waste was investigated as additive to ginger candies in order to obtain antioxidant rich candy [96], while red beet juice was found to be the most suitable natural colorant for fresh pork sausages [97]. Beetroot can also be used as additive in bread preparation in order to increase its cardioprotective ability [98].

Table 11. Chemical constituents of Beetroot.

Root vegetable	Phytochemical group	Individual compounds	References
Beta vulgaris var. *rubra*	phenolics	4-hydroxybenzoic acid	[86, 87]
		caffeic acid	
		catechin hydrate	
		epicatechin	
		rutin	
		ferulic acid	
		vanillic acid	
		p-hydroxybenzoic acid	
		protocatechuic acid	
	pigments (betalains)	betacyanins	[86, 88, 99]
		betaxanthins	
Beetroot juice		nitrates	[100]
Beta vulgaris varieties	pigments (betalains)	betacyanins	[83]
		betaxanthins	
	phenolics	gallic acid	
		syringic acid	
		caffeic acid	
		ferulic acid	

Table 12. Health promoting effects of Beetroot.

Root vegetable	Biological activity	References
Beetroot juice	enhance exercise performance	[100]
	antioxidant	[93]
	cardioprotective	[89, 101 - 103]
Red beetroot juice and chips	antioxidant	[92]
	antiinflammatory	
Red beetroot juice and bread enriched with red and white beetroot	cardioprotective	[98]
Red beetroot	anti-cancer	[104, 105]
	antioxidant	[87]
	hepatoprotective	[106]
Beetroot pomace extract	antioxidant	[86, 107]
	antimicrobial	[94]
	antiradical	
	cytotoxic properties against Ehrlich carcinoma (EAC)	
	anti-proliferative	[86]
	hepatoprotective	[95]
Beetroot extract	chemopreventive	[108]
	attenuates renal dysfunction and structural damage	[109]

Fig. (9). Beetroot (photographed by M. Kostic).

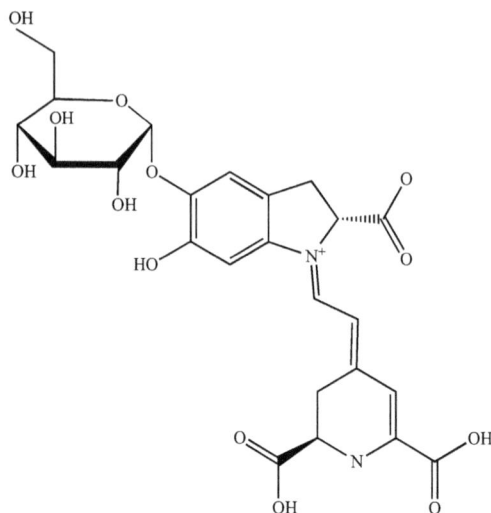

Fig. (10). Betacyanin, pigment found in beetroot, member of betalains responsible for red color.

Parsley

Both leafy and root parsley have been widely used in culinary practice as important spices [110]. Parsley (*Petroselinum crispum* (Mill) Nym) (Fig. **11**) is a biennial plant from the Apiaceae family (Umbelliferae). Parsley was first grown in the Mediterranean region but nowadays is cultivated throughout the world [111]. It has three main types including two types grown for foliage: plain leaf type (ssp. *neapolitanum*, Danert) and the curly leaf type (ssp. *crispum*), and one type grown mainly for its taproots: the turnip-rooted or 'Hamburg' type (ssp. *tuberosum*) [112]. Turnip-rooted parsley has been grown primarily in northern Europe, especially Poland, but now it is becoming popular also in the Mediterranean region [113]. Different components are found in the essential oils from roots of turnip-rooted parsley, with a variety of monoterpenes being dominant (Table **13**). Monoterpenes are known for its diverse biological activities including antimicrobial [114] and antioxidant [115]. Phenylpropene myristicin has hepatoprotective [116] and chemopreventive activity [117]. Apiole (Fig. (**12**) is thought to be a major compound contributing to the antioxidant activity of parsley [118].

Parsnip

Parsnip (*Pastinaca sativa* L.) is part of Apiaceae family, a root vegetable that is frequently used in food preparation and widely consumed partly due to its rich content of fibers (Fig. (**13**) [119]. Parsnips are more distributed in the production of baby food due to their aromatic taste [120]. Different compounds present in parsnip as well as in other apiaceous vegetables are responsible for its chemopreventive activity represented by the inhibition of some carcinogenic activation (Table **14**) [121]. It is used in the traditional medicine of Bulgaria and Italy as cardio-tonic, spasmolytic, hypotensive, coronary dilator, capillarotrophic agent, dietetic, diuretic, and cholagogue [122]. In Bosnia and Herzegovina it is mainly used for stomach malignant diseases [123], which can be explained by its content of polyacetilens falcarinol (Fig. (**14**) and falcarindiol that are found in parsnip roots (Table **15**) and are proven to have preventive effect against the development of colorectal cancer [124].

Table 13. Chemical constituents of parsley roots.

Root vegetable	Phytochemical group	Individual compounds	References
Petroselinum crispum ssp. *tuberosum*	phenylpropenes	apiole	[112, 113]
		myristicin	
	monoterpenes	β-pinene	
		α-phellandrene	
		β-phellandrene	
		β-myrcene	
		p-cymene	
		1,3,8-p-menthatriene	
		terpinolene	
		p-cymenene	
	sesquiterpenoids	β-elemene	

Fig. (11). Parsley roots (photographed by M. Kostic).

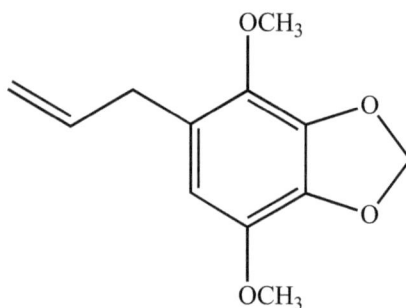

Fig. (12). Apiole, compound found in parsley roots important for its contribution to plant antioxidant activity.

Table 14. Health promoting effects of parsnip roots.

Root vegetable	Biological activity	References
Pastinaca sativa	antioxidant	[126]
	chemopreventive	[121]

Table 15. Chemical constituents of parsnip roots.

Root vegetable	Phytochemical group	Individual compounds	References
Pastinaca sativa	polyacetylene	falcarinol	[125]
		falcarindiol	
		falcarinone	
		falcarinolone	
	furocoumarins	angelicin, isopimpinellin, 5-MOP, 8-MOP, and psoralen	[120]
	phenolics	quercetin 3,7-O diglucoside	[126]
		hydroxycinnamic acid	
		5-O-Caffeoylshikimic acid	
		kaempferol 3-O-rutinoside -7- 0 glucoside	
		kaempferol 3-O-(6"-O-malonylglucoside)-7-Oglucoside	
		quercetin 3-O-rutinoside	
		apigenin 5-O-(6"-O-malonylglucoside)	
		genkwanin 5-O-(6"-O-malonylglucoside)	

Fig. (13). Parsnip roots (photographed by M. Kostic).

Fig. (14). Polyacetilen falcarinol, compound important for parsnip anti-cancer activity.

Yam

This subsection will focus on novel selected studies published in 2017 reporting an update about chemical constituents and biological properties of yam species rhizomes belonging to the genera *Dioscorea*.

Around 613 tuberous climbing plants are described in the genus *Dioscorea* (yam) [127]. According to some authors, only seven to ten species of *Dioscorea* are cultivated on a large scale and only two of them, *D. cayennensis* subsp. *cayennensis* and *D. cayenennsis* subsp. *rotundata* (Poir.) J. Miège, are of primary importance as staple crops for over 100 million people in Western Africa [128 - 130]. It was estimated that approximately around 50 species are eaten as wild-harvested staples or famine food and the genus holds great importance for global food security. *Dioscorea* species have been widely used as traditional medicines in different countries [131].

Four species of *Dioscorea* have been chemically characterized in 2017, namely *D. tokoro, D. bulbifera, D. collettii,* and *D. septemloba* [132 - 135]. The results of chemical analysis are presented in Table **16** Various phytochemical groups of compounds have been found in *Dioscorea* species as presented in Table **16**.

Table 16. Chemical constituents of some *Dioscorea* Spp. Investigated during 2017.

Root vegetable	Phytochemical group	Individual compounds	References
Dioscorea tokoro	saponin	protodioscin	[132]
Dioscorea bulbifera	phytosteroid	diosgenin	[133]
	phytosterol	stigmasterol	
	substituted flavone	3, 7-dimethoxy-5, 3′, 4′-trihydroxyflavone	
	substituted polycyclic aromatic hydrocarbon	6-hydroxy-2, 10, 10-trimethoxy-anthracen-9-one	
		2, 7-dihydroxy-3, 4-dimethoxyphenanthrene	
		3, 7-dihydroxy-2, 4-dimethoxy phenanthrene	
		2, 7-dihydroxy-4-methoxyphenanthrene	
		2, 7-dihydroxy-3, 4-dimethoxy-9, 10-dihydroxy phenanthrene	
		2, 7-dihydroxy-4-methoxy-9, 10-dihydroxy-phenanthrene	
		2, 7-dihydroxy-4-methoxy-9, 10-dihydroxy-phenanthrene	
	dicarboxilic acid	azelaic acid	
	fatty acid ester	pentacosanoic acid 2′, 3 ′-dihydroxypropyl ester	
	aromatic hydrocarbon	3, 5, 4′-trihydroxy-bibenzyl	
	ketone	1, 7-bis-(4-hydroxyphenyl)-4E, 6E-heptadien-3-one	
		1, 7-bis-(4-hydroxyphenyl)-1E, 4E, 6E-heptatrien-3-one	
		6-ethoxy-1 H-pyrimidine-2, 4-dione	
	terpenoid	diosbulbin F	
		diosbulbin B	
		8-epidiosbulbin E acetate	

Root vegetable	Phytochemical group	Individual compounds	References
Dioscorea collettii	ateroid saponins	diosgenin	[134]
		prosapogenin A of dioscin	
		dioscin	
		gracillin	
		yamogenin	
		collettinside III	
	monocyclic phenols	raspberry ketone	
		(-)-rhododendrol	
		E-(4'-hydroxyphenyl)-but-1-en-3-one	
		phloretic acid	
		trans-4-coumaric acid	
		trans-cinnamic acid	
		2,4-dichlorobenzoic acid	
		4-hydroxybenzoic acid	
		protocatechuic acid	
		vanillic acid	
		4-hydroxybenzaldehyde	
		vanillin	
		syringaldehyde	
	flavonoids	formononetin	
		(+)-catechin	
	sterols	β-sitosterol	
		daucosterol	
		stigmasterol	
	cyclodipeptides	cyclo-(L-Pro-L-Leu)	
		cyclo-(L-Leu-L-Ile)	
		cyclo-(L-Leu-L-Leu)	
		cyclo-(L-Phe-L-Val)	
		cyclo-(L-Phe-L-Tyr)	

Root vegetable	Phytochemical group	Individual compounds	References
Dioscorea septemloba	diarylheptanoids	dioscorol A	[135]
		dioscoroside E1	
		dioscoroside E2	
	stilbenes	dioscoroside F1	
		dioscoroside F2	
		1,7-bis(4-hydroxyphenyl)-hepta-4E,6E-dien-3-one	
		1,7-bis(4-hydroxyphenyl)-1,4,6-heptatrien-3-one	
		3,5-dihydroxy-1,7-bis(4-hydroxyphenyl)heptane	
		(3R,5R)-3,5-dihydroxy-1,7-bis(4-hydroxyphenyl)heptane 3-O-β- D-glucopyranoside	
		(3R,5R)-3,5-dihydroxy-1,7-bis(4-hydroxy-3-methoxyphenyl)-heptane 3-O-β-D-glucopyranoside	
		3-O-[α-L-arabinopyranosyl(1→6)-β-D-glucopyranosyl]oct-1-ene-3-ol	

Health promoting effects of *Dioscorea* spp. investigated recently are summarized in Table **17**. Compounds derived from *D. septemloba* expressed biological activities, with clear indication that structures of individual compounds have been related to the expressed effects. It was found that one of three diarylheptanoids was active, while more potent compounds were stilbenes [135]. Methyl protodioscin derived from the rhizomes of *D. collettii* var. *hypoglauca* was explored for the molecular mechanisms by which it induced apoptosis in osteosarcoma cells (MG-63). Cell growth was significantly suppressed when treated with 8 µM of methyl protodioscin (cell viabilities: 22.5 ± 1.9%) [136]. Saponins isolated from *D. collettii*, namely dioscin (Fig. **(15)**, protodioscin, gracillin, and protogracill had an obvious anti-hyperuricemic effect through down-regulation of the URAT1 mRNA and the URAT1 and GLUT9 proteins and up-regulation of the OAT1 and OAT3 proteins [137]. A Glycoprotein (DOT) obtained from *D. opposita* was shown as a potential immunostimulant and DOT exerted its immunomodulatory activity *via* mitogen-activated protein kinases and NF-κB signal pathways [138].

Table 17. Health promoting effects of Yam species investigated recently.

Root vegetable	Biological activity	References
Dioscorea septemloba	increase of glucose consumption	[135]
	triglyceride inhibitory	
Dioscorea collettii **var. *hypoglauca***	apoptosis induction	[136]

Root vegetable	Biological activity	References
Dioscorea collettii	anti-hyperuricemic	[137]
Dioscorea opposita	immunomodulatory	[138]
Dioscorea tokoro	antiproliferative	[132]

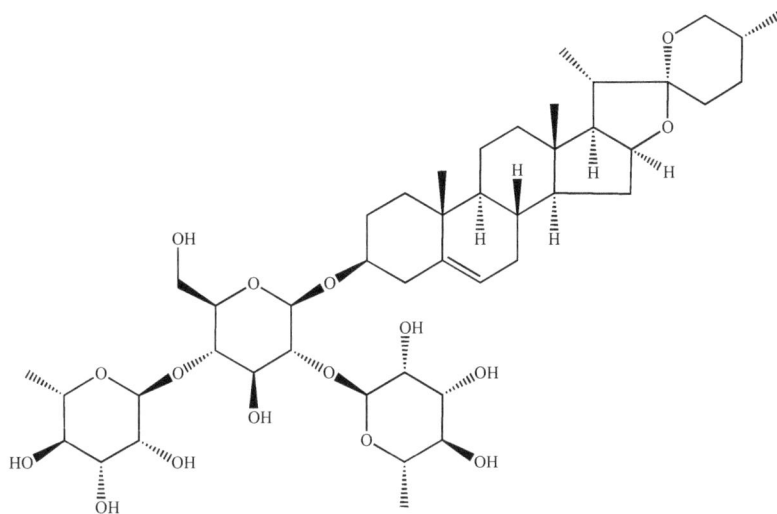

Fig. (15). Dioscin, saponin found in yam toots posesing anti-hyperuricemic activity.

Salsify

Tragopogon porrifolius L. is commonly referred as white salsify, while *Scorzonera hispanica* L. refers to black salsify. Underground parts of both species are consumed as edible or used for their health-beneficial effects. This section will briefly summarize the recent findings of chemical constituents and biological properties of these plants.

T. porrifolius of the Asteraceae family is an edible herb and is commonly known as white salsify, oyster plant, and vegetable oyster. It is an annual or biennial herb of 30–125 cm height with lilac to reddish-purple ligules. All parts of the plant are edible, including its roots, leafy shoots, and open flowers which are consumed both cooked and raw [139]. *T. porrifolius* is widespread throughout the Mediterranean region where it grows wild and it is also cultivated. The nutritional value of this plant has been attributed to its monounsaturated and essential fatty acids, vitamins, polyphenols, and fructooligosaccharides components [140]. White salsify is also considered a medicinal plant as it shows antibilious, diuretic and laxative properties [141, 142].

S. hispanica L. commonly known as black salsify, Spanish salsify or serpent's root is a perennial herbaceous plant belonging to the Asteraceae family [143]. Its natural distribution encompasses Central and Southern Europe, the Caucasus, and Southern Siberia. After removal of its robust black corky skin, fresh underground parts are boiled and eaten together with other vegetables like carrots or served separately with white sauce similar to asparagus. Nowadays underground parts of *S. hispanica* are widely cultivated in Western Europe as a vegetable, particular in Belgium [143]. Underground parts of black salsify were used as a coffee substitute [144], as well as to enhance digestion and perspiration, as a diuretic agent [145], and as a remedy for snakebites; this historic usage explains the vernacular name serpent's root [143].

Since, the whole plant of white salsify is edible, rare studies are conducted focusing on chemical constituents only from roots. Tragoponol, a novel dimeric dihydroisocoumarin was isolated from the roots of *T. porrifolius* and this compound was reported as the first of its kind [146].

Chemical constituents described in the roots of white salsify are presented in the Table **18**. Lipophilic compounds and phenolics were mainly identified.

The cytotoxic activity of the mentioned compounds isolated from black salsify roots was tested as well. It was shown that (-)-syringaresinol Fig. (**16**) was active against myeloma cell lines. (-)-Syringaresinol, puliglutone and 1-oxo- bisabola-(2,10E)-diene-12-carboxylic acid methyl ester were moderately active against the human colon cancer cell line SW480. Unfortunately, (-)- Syringaresinol showed cytotoxicity not only against cancer cell lines but also against peripheral blood mononuclear cells. Thus, puliglutone and 1-oxo- bisabola-(2,10E)-diene-12-carboxylic acid methyl ester could be interesting further investigations as agents or lead compounds to treat colon cancer [147].

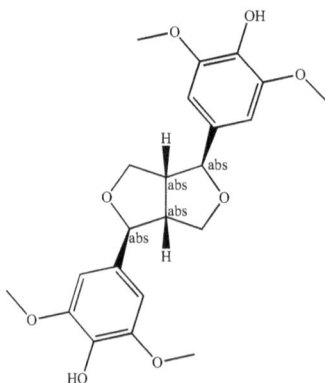

Fig. (16). Syringaresinol, compound showing anti-cancer properties.

Table 18. Chemical constituents in white salsify roots.

Root vegetable	Phytochemical group	Individual compounds	References
Scorzonera hispanica	lipophilic compounds	(-)-syringaresinol	[147]
		1-oxo-bisabola-(2,10E)-diene-12-carboxylic acid	
		1-oxo-bisabola-(2,10E)-diene-12-ol	
		ptilostemonol	
		puliglutone	
		1-oxo-bisabola-(2,10E)-diene-12-carboxylic acid methyl ester	
		9-Hydroxyocta-(10E,12E)-decadienoic acid	
		13-Oxo-(9Z,11E)-octadecadienoic acid	
		9-Oxo-(10E,12Z)-octadecadienoic acid	
		13-Oxo-(9E,11E)-octadecadienoic acid	
		9-Oxo-(10E,12E)-octadecadienoic acid	
		2,9-Epoxycurcumen-12-al isomer	
		2,9-Epoxycurcumen-12-al isomer	
		linoleic acid	
	polyphenols	5-Ocaffeoylquinic acid	
		caffeic acid	
		4-Ocaffeoylquinic acid	
		isoorientin	
		quercetin rhamnohexoside	
		quercetin rhamnohexoside	
		quercetin 3-Ogalactoside	
		quercetin 3-Oglucoside	
		quercetin 3-Oglucuronide	
		quercetin malonylhexoside	
		1,5-ODicaffeoylquinic acid	
		3,5-ODicaffeoylquinic acid	
		4,5-ODicaffeoylquinic acid	

CONCLUSIONS

A wide range of biological activities could be attributed to selected root vegetables: potato, celeriac, turnips, radish, beets, Hamburg parsley, taro, yam, parsnip and salsify described in this chapter. These root vegetables are consumed worldwide and present important root crops. Biological activities of these vegetables could be attributed to their respected extracts and individual compounds identified in each one of them. However, the literature reporting profound studies to confirm *in vivo* effects is still scarce, but the results obtained so far are more than promising. As a general conclusion, it is interesting to underline the necessity for further exploration of biological effects of tuber vegetables that bring a lot of health beneficial effects to mankind, while clinical and *in vivo* studies have to be carried out in order to elucidate the mechanism of action, the main compounds that are responsible for the beneficial effects of these species, as well as the recommended doses in order to achieve these effects.

CONSENT FOR PUBLICATION

Not applicable.

CONFLICT OF INTEREST

The author (editor) declares no conflict of interest, financial or otherwise.

ACKNOWLEDGEMENTS

This work has been supported by the Serbian Ministry of Education, Science and Technological Development for financial support (Grant number 173032).

REFERENCES

[1]   Newman DJ, Cragg GM. Natural products as sources of new drugs over the 30 years from 1981 to 2010. J Nat Prod 2012; 75(3): 311-35.
       [http://dx.doi.org/10.1021/np200906s] [PMID: 22316239]

[2]   Chin YW, Balunas MJ, Chai HB, Kinghorn AD. Drug discovery from natural sources. AAPS J 2006; 8(2): E239-53.
       [http://dx.doi.org/10.1007/BF02854894] [PMID: 16796374]

[3]   Smiljkovic M, Stanisavljevic D, Stojkovic D, *et al.* Apigenin-7-O-glucoside *versus* apigenin: Insight into the modes of anticandidal and cytotoxic actions. EXCLI J 2017; 16: 795-807.
       [PMID: 28827996]

[4]   Stojković D, Barros L, Petrović J, *et al.* Ethnopharmacological uses of *Sempervivum tectorum* L. in southern Serbia: Scientific confirmation for the use against otitis linked bacteria. J Ethnopharmacol 2015; 176: 297-304.
       [http://dx.doi.org/10.1016/j.jep.2015.11.014] [PMID: 26551879]

[5]   Nikolić M, Stojković D, Glamočlija J, *et al.* Could essential oils of green and black pepper be used as food preservatives? J Food Sci Technol 2015; 52(10): 6565-73.
       [http://dx.doi.org/10.1007/s13197-015-1792-5] [PMID: 26396402]

[6]   Ferreira ICFR, Heleno SA, Reis FS, *et al.* Chemical features of *Ganoderma* polysaccharides with antioxidant, antitumor and antimicrobial activities. Phytochemistry 2015; 114: 38-55.
       [http://dx.doi.org/10.1016/j.phytochem.2014.10.011] [PMID: 25457487]

[7]   Reis FS, Ćirić A, Stojković D, *et al.* Effects of different culture conditions on biological potential and metabolites production in three *Penicillium* isolates. Drug Dev Ind Pharm 2015; 41(2): 253-62.

[http://dx.doi.org/10.3109/03639045.2013.858738] [PMID: 24261405]

[8]   Živković J, Stojković D, Petrović J, Zdunić G, Glamočlija J, Soković M. *Rosa canina* L.--new possibilities for an old medicinal herb. Food Funct 2015; 6(12): 3687-92.
[http://dx.doi.org/10.1039/C5FO00820D] [PMID: 26399901]

[9]   Bukvicki D, Stojkovic D, Sokovic M, *et al.* Potential application of *Micromeria dalmatica* essential oil as a protective agent in a food system. Lebensm Wiss Technol 2015; 63(1): 262-7.
[http://dx.doi.org/10.1016/j.lwt.2015.03.053]

[10]   Popović V, Stojković D, Nikolić M, *et al.* Extracts of three *Laserpitium* L. species and their principal components laserpitine and sesquiterpene lactones inhibit microbial growth and biofilm formation by oral *Candida* isolates. Food Funct 2015; 6(4): 1205-11.
[http://dx.doi.org/10.1039/C5FO00066A] [PMID: 25720441]

[11]   Stojković DS, Barros L, Calhelha RC, *et al.* A detailed comparative study between chemical and bioactive properties of *Ganoderma lucidum* from different origins. Int J Food Sci Nutr 2014; 65(1): 42-7.
[http://dx.doi.org/10.3109/09637486.2013.832173] [PMID: 24020451]

[12]   Stojković D, Petrović J, Soković M, Glamočlija J, Kukić-Marković J, Petrović S. In situ antioxidant and antimicrobial activities of naturally occurring caffeic acid, p-coumaric acid and rutin, using food systems. J Sci Food Agric 2013; 93(13): 3205-8.
[http://dx.doi.org/10.1002/jsfa.6156] [PMID: 23553578]

[13]   Stojković D, Reis FS, Barros L, *et al.* Nutrients and non-nutrients composition and bioactivity of wild and cultivated *Coprinus* comatus (O.F.Müll.) Pers. Food Chem Toxicol 2013; 59: 289-96.
[http://dx.doi.org/10.1016/j.fct.2013.06.017] [PMID: 23793036]

[14]   Stojković DS, Zivković J, Soković M, *et al.* Antibacterial activity of *Veronica montana* L. extract and of protocatechuic acid incorporated in a food system. Food Chem Toxicol 2013; 55: 209-13.
[http://dx.doi.org/10.1016/j.fct.2013.01.005] [PMID: 23333716]

[15]   Animal Doctors The Economist 2002.

[16]   Richards S. Natural-born Doctors. Scientist 2012.

[17]   Harlan JR, Ed. Crops and Man. Madison, Wisconsin: American Society of Agronomy-Crop Science Society 1975; p. 279.

[18]   Cray JA, Connor MC, Stevenson A, *et al.* Biocontrol agents promote growth of potato pathogens, depending on environmental conditions. Microb Biotechnol 2016; 9(3): 330-54.
[http://dx.doi.org/10.1111/1751-7915.12349] [PMID: 26880001]

[19]   Visvanathan R, Jayathilake C, Chaminda Jayawardana B, Liyanage R. Health-beneficial properties of potato and compounds of interest. J Sci Food Agric 2016; 96(15): 4850-60.
[http://dx.doi.org/10.1002/jsfa.7848] [PMID: 27301296]

[20]   Camire ME. Potatoes and Human Health. 2nd ed. Adv Potato Chem Technol 2016; pp. 685-704.

[21]   Al-Saikhan MS, Howard LR, Miller JC. Antioxidant activity and total phenolics in different genotypes of potato (*Solanum tuberosum* L.). J Food Sci 1995; 60: 341-3.
[http://dx.doi.org/10.1111/j.1365-2621.1995.tb05668.x]

[22]   Kubow S, Hobson L, Iskandar MM, Sabally K, Donnelly DJ, Agellon LB. Extract of Irish potatoes (*Solanum tuberosum* L.) decreases body weight gain and adiposity and improves glucose control in the mouse model of diet-induced obesity. Mol Nutr Food Res 2014; 58(11): 2235-8.
[http://dx.doi.org/10.1002/mnfr.201400013] [PMID: 25066548]

[23]   McGill CR, Kurilich AC, Davignon J. The role of potatoes and potato components in cardiometabolic health: a review. Ann Med 2013; 45(7): 467-73.
[http://dx.doi.org/10.3109/07853890.2013.813633] [PMID: 23855880]

[24]   Nzaramba MN, Reddivari L, Bamberg JB, Miller JC. Antiproliferative activity and cytotoxicity of

Solanum jamesii tuber extracts on human colon and prostate cancer cells *in vitrol*. J Agric Food Chem 2009; 57(18): 8308-15.
[http://dx.doi.org/10.1021/jf901567k] [PMID: 19711917]

[25]   Bassoli BK, Cassolla P, Borba-Murad GR, *et al.* Chlorogenic acid reduces the plasma glucose peak in the oral glucose tolerance test: effects on hepatic glucose release and glycaemia. Cell Biochem Funct 2008; 26(3): 320-8.
[http://dx.doi.org/10.1002/cbf.1444] [PMID: 17990295]

[26]   Yamaguchi T, Chikama A, Mori K, *et al.* Hydroxyhydroquinone-free coffee: a double-blind, randomized controlled dose-response study of blood pressure. Nutr Metab Cardiovasc Dis 2008; 18(6): 408-14.
[http://dx.doi.org/10.1016/j.numecd.2007.03.004] [PMID: 17951035]

[27]   Hasan I, Ozeki Y, Kabir SR. Purification of a novel chitin-binding lectin with antimicrobial and antibiofilm activities from a bangladeshi cultivar of potato (*Solanum tuberosum*). Indian J Biochem Biophys 2014; 51(2): 142-8.
[PMID: 24980018]

[28]   De Mejía EG, Prisecaru VI. Lectins as bioactive plant proteins: a potential in cancer treatment. Crit Rev Food Sci Nutr 2005; 45(6): 425-45.
[http://dx.doi.org/10.1080/10408390591034445] [PMID: 16183566]

[29]   Milner SE, Brunton NP, Jones PWO, O'Brien NM, Collins SG, Maguire AR. Bioactivities of glycoalkaloids and their aglycones from Solanum species. J Agric Food Chem 2011; 59(8): 3454-84.
[http://dx.doi.org/10.1021/jf200439q] [PMID: 21401040]

[30]   Fewell AM, Roddick JG. Interactive antifungal activity of the glycoalkaloids α-solanine and α-chaconine. Phytochemistry 1993; 33: 323-8.
[http://dx.doi.org/10.1016/0031-9422(93)85511-O]

[31]   Sánchez-Maldonado AF, Schieber A, Gänzle MG. Antifungal activity of secondary plant metabolites from potatoes (*Solanum tuberosum* L.): Glycoalkaloids and phenolic acids show synergistic effects. J Appl Microbiol 2016; 120(4): 955-65.
[http://dx.doi.org/10.1111/jam.13056] [PMID: 26786886]

[32]   Amanpour R, Abbasi-maleki S, Neyriz-naghadehi M, Asadi-samani M. Antibacterial effects of *Solanum tuberosum* peel ethanol extract *in vitrol*. J Herbmed Pharmacol 2015; 4(2): 45-8.

[33]   Brown C, Culley D, Yang CP, Durst R, Wrolstad R. Variation of anthocyanin and carotenoid contents and associated antioxidant values in potato breeding lines. J Am Soc Hortic Sci 2005; 130: 174-80.

[34]   He J, Giusti MM. Anthocyanins: natural colorants with health-promoting properties. Annu Rev Food Sci Technol 2010; 1: 163-87.
[http://dx.doi.org/10.1146/annurev.food.080708.100754] [PMID: 22129334]

[35]   Thompson MD, Thompson HJ, McGinley JN, Neil ES, Rush DK, Holm DG, *et al.* Functional food characteristics of potato cultivars (*Solanum tuberosum* L.): phytochemical composition and inhibition of 1-methyl-1-nitrosourea induced breast cancer in rats. J Food Compos Anal 2009; 22: 571-6.
[http://dx.doi.org/10.1016/j.jfca.2008.09.002]

[36]   Kaspar KL, Park JS, Brown CR, Mathison BD, Navarre DA, Chew BP. Pigmented potato consumption alters oxidative stress and inflammatory damage in men. J Nutr 2011; 141(1): 108-11.
[http://dx.doi.org/10.3945/jn.110.128074] [PMID: 21106930]

[37]   Liu YW, Han CH, Lee MH, Hsu FL, Hou WC. Patatin, the tuber storage protein of potato (*Solanum tuberosum* L.), exhibits antioxidant activity *in vitrol*. J Agric Food Chem 2003; 51(15): 4389-93.
[http://dx.doi.org/10.1021/jf030016j] [PMID: 12848515]

[38]   Reddivari L, Hale A, Miller J. Determination of phenolic content, composition and their contribution to antioxidant activity in specialty potato selections. Am J Potato Res 2007; 84: 275-82.
[http://dx.doi.org/10.1007/BF02986239]

[39]   Ombra MN, Fratianni F, Granese T, Cardinale F, Cozzolino A, Nazzaro F. *in vitrol* antioxidant, antimicrobial and anti-proliferative activities of purple potato extracts (*Solanum tuberosum* cv Vitelotte noire) following simulated gastro-intestinal digestion. Nat Prod Res 2015; 29(11): 1087-91.
[http://dx.doi.org/10.1080/14786419.2014.981183] [PMID: 25420792]

[40]   Rodríguez IF, Sayago JE, Torres S, Zampini IC, Isla MI, Ordóñez RM. Control of citrus pathogens by protein extracts from *Solanum tuberosum* tubers. Eur J Plant Pathol 2015; 141(3): 585-95.
[http://dx.doi.org/10.1007/s10658-014-0566-7]

[41]   Asadi-Samani M, Kafash-Farkhad N, Azimi N, Fasihi A, Alinia-Ahandani E, Rafieian-Kopaei M. Medicinal plants with hepatoprotective activity in Iranian folk medicine. Asian Pac J Trop Biomed 2015; 5(2): 146-57.
[http://dx.doi.org/10.1016/S2221-1691(15)30159-3]

[42]   Popova M, Stoyanova A, Valyovska-Popova N, Bankova V, Peev D. A new coumarin and total phenolic and flavonoids content of Bulgarian celeriac. Izv Him 2014; 46: 88-93.

[43]   Sellami IH. Essential oil and aroma composition of leaves, stalks and roots of celery *(Apium graveolens* var. *dulce)* from Tunisia. J Essent Oil Res 2012; 24(6): 513-21.
[http://dx.doi.org/10.1080/10412905.2012.728093]

[44]   Wu L, Chen Y. Protective effect of celery root against acute liver injury by CCl4. Huaxi Yaoxue Zazhi 2008; 23: 411-2.
[http://dx.doi.org/10.3969/j.issn.1006-0103.2008.04.013]

[45]   Nikolić N, Cvetković D, Todorović Z. A Characterization of content, composition and antioxidant capacity of phenolic compounds in celery roots. Ital J Food Sci 2011; 23(2): 214-9.

[46]   Saka B, Djouahri A, Djerrad Z, *et al.* Chemical Variability and Biological Activities of *Brassica rapa* var. *rapifera* Parts Essential Oils Depending on Geographic Variation and Extraction Technique. Chem Biodivers 2017; 14(6)
[http://dx.doi.org/10.1002/cbdv.201600452] [PMID: 28145061]

[47]   Yu J, Lei J, Yu H, Cai X, Zou G. Chemical composition and antimicrobial activity of the essential oil of *Scutellaria barbata.* Phytochemistry 2004; 65(7): 881-4.
[http://dx.doi.org/10.1016/j.phytochem.2004.02.005] [PMID: 15081288]

[48]   Beltagy AM. Investigation of new antimicrobial and antioxidant activities of *Brassica rapa.* Int J Pharm Pharm Sci 2014; 6(6): 84-8.

[49]   Fernandes F, Valenta P, Sousa C, Pereira JA, Seabra RM, Andrade PB. Chemical and antioxidative assessment of dietary turnip (*Brassica rapa* var. *rapa* L.). Food Chem 2007; 105: 1003-10.
[http://dx.doi.org/10.1016/j.foodchem.2007.04.063]

[50]   Chung IM, Rekha K, Rajakumar G, Thiruvengadam M. Production of glucosinolates, phenolic compounds and associated gene expression profiles of hairy root cultures in turnip (*Brassica rapa* ssp. *rapa*). 3 Biotech 2016; 6(2): 175.
[http://dx.doi.org/10.1007/s13205-016-0492-9] [PMID: 28330247]

[51]   Zhu F. Structure, properties, and applications of aroid starch. Food Hydrocoll 2016; 52: 378-92.
[http://dx.doi.org/10.1016/j.foodhyd.2015.06.023]

[52]   Huang W, Li C, Wang Y, Yi X, He X. Anti-inflammatory lignanamides and monoindoles from *Alocasia macrorrhiza.* Fitoterapia 2017; 117: 126-32.
[http://dx.doi.org/10.1016/j.fitote.2017.01.014] [PMID: 28161134]

[53]   Huang W, Yi X, Feng J, Wang Y, He X. Piperidine alkaloids from *Alocasia macrorrhiza.* Phytochemistry 2017; 143: 81-6.
[http://dx.doi.org/10.1016/j.phytochem.2017.07.012] [PMID: 28780427]

[54]   Elsbaey M, Ahmed KFM, Elsebai MF, Zaghloul A, Amer MMA, Lahloub MI. Cytotoxic constituents of *Alocasia macrorrhiza.* Z Natforsch C J Biosci 2017; 72(1-2): 21-5.

[http://dx.doi.org/10.1515/znc-2015-0157] [PMID: 27497869]

[55]   Tien NQ, Ngoc P, Minh PH, Van Kiem P, Van Minh C, Kim YH. New ceramide from *Alocasia macrorrhiza*. Arch Pharm Res 2004; 27(10): 1020-2.
       [http://dx.doi.org/10.1007/BF02975424] [PMID: 15554257]

[56]   Zhu LH, Chen C, Wang H, Ye WC, Zhou GX. Indole alkaloids from *Alocasia macrorrhiza*. Chem Pharm Bull (Tokyo) 2012; 60(5): 670-3.
       [http://dx.doi.org/10.1248/cpb.60.670] [PMID: 22689406]

[57]   Rahman MM, Hossain MA, Siddique SA, Biplab KP, Uddin MH. Antihyperglycemic, antioxidant and cytotoxic activities of *Alocasia macrorrhiza* (L.) rhizomes extract. Turk J Biol 2012; 36: 574-9.

[58]   Wang HX, Ng TB. Alocasin, an anti-fungal protein from rhizomes of the giant taro *Alocasia macrorrhiza*. Protein Expr Purif 2003; 28(1): 9-14.
       [http://dx.doi.org/10.1016/S1046-5928(02)00604-6] [PMID: 12651101]

[59]   Englberger L, *et al*. Carotenoid and mineral content of Micronesian giant swamp taro (Cyrtosperma) cultivars. J Food Compos Anal 2008; 21: 93-106.
       [http://dx.doi.org/10.1016/j.jfca.2007.09.007]

[60]   Shin T, Ahn M, Kim GO, Park SU. Biological activity of various radish species. Orient Pharm Exp Med 2015; 15: 105-11.
       [http://dx.doi.org/10.1007/s13596-015-0183-9]

[61]   Mitsui Y, Shimomura M, Komatsu K, *et al*. The radish genome and comprehensive gene expression profile of tuberous root formation and development. Sci Rep 2015; 5: 10835.
       [http://dx.doi.org/10.1038/srep10835] [PMID: 26056784]

[62]   Blazević I, Mastelić J. Glucosinolate degradation products and other bound and free volatiles in the leaves and roots of radish (*Raphanus sativus* L.). Food Chem 2009; 113: 96-102.
       [http://dx.doi.org/10.1016/j.foodchem.2008.07.029]

[63]   Gutiérrez RM, Perez RL. *Raphanus sativus* (Radish): their chemistry and biology. Sci World J 2004; 4: 811-37.
       [http://dx.doi.org/10.1100/tsw.2004.131] [PMID: 15452648]

[64]   Nakamura Y, Iwahashi T, Tanaka A, *et al*. 4-(Methylthio)-3-butenyl isothiocyanate, a principal antimutagen in daikon (*Raphanus sativus*; Japanese white radish). J Agric Food Chem 2001; 49(12): 5755-60.
       [http://dx.doi.org/10.1021/jf0108415] [PMID: 11743759]

[65]   Hanlon PR, Webber DM, Barnes DM. Aqueous extract from Spanish black radish (*Raphanus sativus* L. Var. *niger*) induces detoxification enzymes in the HepG2 human hepatoma cell line. J Agric Food Chem 2007; 55(16): 6439-46.
       [http://dx.doi.org/10.1021/jf070530f] [PMID: 17616135]

[66]   Hanlon PR, Barnes DM. Phytochemical composition and biological activity of 8 varieties of radish (*Raphanus sativus* L.) sprouts and mature taproots. J Food Sci 2011; 76(1): C185-92.
       [http://dx.doi.org/10.1111/j.1750-3841.2010.01972.x] [PMID: 21535648]

[67]   Sak K, Jürisoo K, Raal A. Estonian folk traditional experiences on natural anticancer remedies: from past to the future. Pharm Biol 2014; 52(7): 855-66.
       [http://dx.doi.org/10.3109/13880209.2013.871641] [PMID: 24920231]

[68]   Ahmad I, Beg AZ. Antimicrobial and phytochemical studies on 45 Indian medicinal plants against multi-drug resistant human pathogens. J Ethnopharmacol 2001; 74(2): 113-23.
       [http://dx.doi.org/10.1016/S0378-8741(00)00335-4] [PMID: 11167029]

[69]   Castro-Torres IG, Naranjo-Rodríguez EB, Domínguez-Ortíz MA, Gallegos-Estudillo J, Saavedra-Vélez MV. Antilithiasic and hypolipidaemic effects of *Raphanus sativus* L. var. *niger* on mice fed with a lithogenic diet. J Biomed Biotechnol 2012; 2012: 161205.
       [http://dx.doi.org/10.1155/2012/161205] [PMID: 23093836]

[70]   Damodharan K, Palaniyandi SA, Yang SH, Suh JW. *In vitrol* probiotic characterization of Lactobacillus strains from fermented radish and their anti-adherence activity against enteric pathogens. Can J Microbiol 2015; 61(11): 837-50.
[http://dx.doi.org/10.1139/cjm-2015-0311] [PMID: 26382558]

[71]   Kakizaki T, Kitashiba H, Zou Z, *et al.* A 2-Oxoglutarate-Dependent Dioxygenase Mediates the Biosynthesis of Glucoraphasatin in Radish. Plant Physiol 2017; 173(3): 1583-93.
[http://dx.doi.org/10.1104/pp.16.01814] [PMID: 28100450]

[72]   Ben Salah-Abbès J, Abbès S, Ouanes Z, Abdel-Wahhab MA, Bacha H, Oueslati R. Isothiocyanate from the Tunisian radish (*Raphanus sativus*) prevents genotoxicity of Zearalenone *in vivo* and *in vitro*. Mutat Res 2009; 677(1-2): 59-65.
[http://dx.doi.org/10.1016/j.mrgentox.2009.05.017] [PMID: 19501672]

[73]   Beevi SS, Mangamoori LN, Dhand V, Ramakrishna DS. Isothiocyanate profile and selective antibacterial activity of root, stem, and leaf extracts derived from *Raphanus sativus* L. Foodborne Pathog Dis 2009; 6(1): 129-36.
[http://dx.doi.org/10.1089/fpd.2008.0166] [PMID: 19182965]

[74]   Beevi SS, Mangamoori LN, Subathra M, Edula JR. Hexane extract of *Raphanus sativus* L. roots inhibits cell proliferation and induces apoptosis in human cancer cells by modulating genes related to apoptotic pathway. Plant Foods Hum Nutr 2010; 65(3): 200-9.
[http://dx.doi.org/10.1007/s11130-010-0178-0] [PMID: 20652750]

[75]   N'jai AU, Kemp MQ, Metzger BT, *et al.* Spanish black radish (*Raphanus sativus* L. Var. *niger*) diet enhances clearance of DMBA and diminishes toxic effects on bone marrow progenitor cells. Nutr Cancer 2012; 64(7): 1038-48.
[http://dx.doi.org/10.1080/01635581.2012.714831] [PMID: 23061907]

[76]   Sipos P, Hagymási K, Lugasi A, Fehér E, Blázovics A. Effects of black radish root (*Raphanus sativus* L. var *niger*) on the colon mucosa in rats fed a fat rich diet. Phytother Res 2002; 16(7): 677-9.
[http://dx.doi.org/10.1002/ptr.950] [PMID: 12410553]

[77]   Lugasi A, Blázovics A, Hagymási K, Kocsis I, Kéry A. Antioxidant effect of squeezed juice from black radish (*Raphanus sativus* L. var *niger*) in alimentary hyperlipidaemia in rats. Phytother Res 2005; 19(7): 587-91.
[http://dx.doi.org/10.1002/ptr.1655] [PMID: 16161062]

[78]   Goyeneche R, Roura S, Ponce A, *et al.* Chemical characterization and antioxidant capacity of red radish (*Raphanus sativus* L.) leaves and roots. J Funct Foods 2015; 16: 256-64.
[http://dx.doi.org/10.1016/j.jff.2015.04.049]

[79]   Beevi SS, Mangamoori LN, Gowda BB. Polyphenolics profile and antioxidant properties of *Raphanus sativus* L. Nat Prod Res 2012; 26(6): 557-63.
[http://dx.doi.org/10.1080/14786419.2010.521884] [PMID: 21714734]

[80]   Cao J, Chen W, Zhang Y, Zhang Y, Zhao X. Content of Selected Flavonoids in 100 Edible Vegetables and Fruits. Food Sci Technol Res 2010; 16(5): 395-402.
[http://dx.doi.org/10.3136/fstr.16.395]

[81]   Shukla S, Chatterji S, Mehta S, *et al.* Antidiabetic effect of *Raphanus sativus* root juice. Pharm Biol 2011; 49(1): 32-7.
[http://dx.doi.org/10.3109/13880209.2010.493178] [PMID: 20687786]

[82]   Ninfali P, Angelino D. Nutritional and functional potential of *Beta vulgaris cicla* and *rubra*. Fitoterapia 2013; 89: 188-99.
[http://dx.doi.org/10.1016/j.fitote.2013.06.004] [PMID: 23751216]

[83]   Wruss J, Waldenberger G, Huemer S, *et al.* Compositional characteristics of commercial beetroot products and beetroot juice prepared from seven beetroot varieties grown in Upper Austria. J Food Compos Anal 2015; 42: 46-55.

[http://dx.doi.org/10.1016/j.jfca.2015.03.005]

[84]   Gengatharan A, Dykes GA, Choo WS. Betalains: Natural plant pigments with potential application in functional foods. Lebensm Wiss Technol 2015; 64(2): 645-9.
[http://dx.doi.org/10.1016/j.lwt.2015.06.052]

[85]   Nemzer B, Pietrzkowski Z, Spórna A, *et al.* Betalainic and nutritional profiles of pigment-enriched red beet root (*Beta vulgaris* L.) dried extracts. Food Chem 2011; 127(1): 42-53.
[http://dx.doi.org/10.1016/j.foodchem.2010.12.081]

[86]   Vulić J, Čanadanović-Brunet J, Ćetković G, *et al.* Antioxidant and cell growth activities of beet root pomace extracts. J Funct Foods 2012; 4(3): 670-8.
[http://dx.doi.org/10.1016/j.jff.2012.04.008]

[87]   Georgiev VG, Weber J, Kneschke EM, Denev PN, Bley T, Pavlov AI. Antioxidant activity and phenolic content of betalain extracts from intact plants and hairy root cultures of the red beetroot *Beta vulgaris* cv. Detroit dark red. Plant Foods Hum Nutr 2010; 65(2): 105-11.
[http://dx.doi.org/10.1007/s11130-010-0156-6] [PMID: 20195764]

[88]   Kujala TS, Loponen JM, Klika KD, Pihlaja K. Phenolics and betacyanins in red beetroot (*Beta vulgaris*) root: distribution and effect of cold storage on the content of total phenolics and three individual compounds. J Agric Food Chem 2000; 48(11): 5338-42.
[http://dx.doi.org/10.1021/jf000523q] [PMID: 11087483]

[89]   Webb AJ, Patel N, Loukogeorgakis S, *et al.* Acute blood pressure lowering, vasoprotective, and antiplatelet properties of dietary nitrate *via* bioconversion to nitrite. Hypertension 2008; 51(3): 784-90.
[http://dx.doi.org/10.1161/HYPERTENSIONAHA.107.103523] [PMID: 18250365]

[90]   Joris PJ, Mensink RP. Beetroot juice improves in overweight and slightly obese men postprandial endothelial function after consumption of a mixed meal. Atherosclerosis 2013; 231(1): 78-83.
[http://dx.doi.org/10.1016/j.atherosclerosis.2013.09.001] [PMID: 24125415]

[91]   Jajja A, Sutyarjoko A, Lara J, *et al.* Beetroot supplementation lowers daily systolic blood pressure in older, overweight subjects. Nutr Res 2014; 34(10): 868-75.
[http://dx.doi.org/10.1016/j.nutres.2014.09.007] [PMID: 25294299]

[92]   Zielińska-Przyjemska M, Olejnik A, Dobrowolska-Zachwieja A, Grajek W. *In vitro1* effects of beetroot juice and chips on oxidative metabolism and apoptosis in neutrophils from obese individuals. Phytother Res 2009; 23(1): 49-55.
[http://dx.doi.org/10.1002/ptr.2535] [PMID: 18814207]

[93]   Wootton-Beard PC, Ryan L. A beetroot juice shot is a significant and convenient source of bioaccessible antioxidants. J Funct Foods 2011; 3: 329-34.
[http://dx.doi.org/10.1016/j.jff.2011.05.007]

[94]   Vulić JJ, Cebović TN, Canadanović VM, *et al.* Antiradical, antimicrobial and cytotoxic activities of commercial beetroot pomace Food Funct 2013; 4(5): 713-21.

[95]   Vulić JJ, Ćebović TN, Čanadanović-Brunet JM, *et al. in vivo* and *in vitro* antioxidant effects of beetroot pomace extracts. J Funct Foods 2014; 6: 168-75.
[http://dx.doi.org/10.1016/j.jff.2013.10.003]

[96]   Kumar V, Kushwaha R, Goyal A, Tanwar B, Kaur J. Process optimization for the preparation of antioxidant rich ginger candy using beetroot pomace extract. Food Chem 2018; 245: 168-77.
[http://dx.doi.org/10.1016/j.foodchem.2017.10.089] [PMID: 29287358]

[97]   Martinez L, Cilla I, Beltran JA, Roncales P. Comparative effect of red yeast rice (*Monascus purpureus*), red beet root (*Beta vulgaris*) and betanin (E-162) on colour and consumer acceptability of fresh pork sausages packaged in a modified atmosphere. J Sci Food Agric 2006; 86(4): 500-8.
[http://dx.doi.org/10.1002/jsfa.2389]

[98]   Hobbs DA, Kaffa N, George TW, Methven L, Lovegrove JA. Blood pressure-lowering effects of beetroot juice and novel beetroot-enriched bread products in normotensive male subjects. Br J Nutr

2012; 108(11): 2066-74.
[http://dx.doi.org/10.1017/S0007114512000190] [PMID: 22414688]

[99]   Nistor OV, Seremet Ceclu L, Andronoiu DG, Rudi L, Botez E. Influence of different drying methods on the physicochemical properties of red beetroot (*Beta vulgaris* L. var. Cylindra). Food Chem 2017; 236: 59-67.
[http://dx.doi.org/10.1016/j.foodchem.2017.04.129] [PMID: 28624090]

[100] Ormsbee MJ, Bach CW, Baur DA. Pre-exercise nutrition: the role of macronutrients, modified starches and supplements on metabolism and endurance performance. Nutrients 2014; 6(5): 1782-808.
[http://dx.doi.org/10.3390/nu6051782] [PMID: 24787031]

[101] Das S, Filippone SM, Williams DS, Das A, Kukreja RC. Beet root juice protects against doxorubicin toxicity in cardiomyocytes while enhancing apoptosis in breast cancer cells. Mol Cell Biochem 2016; 421(1-2): 89-101.
[http://dx.doi.org/10.1007/s11010-016-2789-8] [PMID: 27565811]

[102] Coles LT, Clifton PM. Effect of beetroot juice on lowering blood pressure in free-living, disease-free adults: a randomized, placebo-controlled trial. Nutr J 2012; 11: 106.
[http://dx.doi.org/10.1186/1475-2891-11-106] [PMID: 23231777]

[103] Clifford T, Howatson G, West DJ, Stevenson EJ. The potential benefits of red beetroot supplementation in health and disease. Nutrients 2015; 7(4): 2801-22.
[http://dx.doi.org/10.3390/nu7042801] [PMID: 25875121]

[104] Kapadia GJ, Azuine MA, Rao GS, Arai T, Iida A, Tokuda H. Cytotoxic effect of the red beetroot (*Beta vulgaris* L.) extract compared to doxorubicin (Adriamycin) in the human prostate (PC-3) and breast (MCF-7) cancer cell lines. Anticancer Agents Med Chem 2011; 11(3): 280-4.
[http://dx.doi.org/10.2174/187152011795347504] [PMID: 21434853]

[105] Kapadia GJ, Tokuda H, Konoshima T, Nishino H. Chemoprevention of lung and skin cancer by *Beta vulgaris* (beet) root extract. Cancer Lett 1996; 100(1-2): 211-4.
[http://dx.doi.org/10.1016/0304-3835(95)04087-0] [PMID: 8620443]

[106] Váli L, Stefanovits-Bányai E, Szentmihályi K, *et al.* Liver-protecting effects of table beet (*Beta vulgaris* var. *rubra*) during ischemia-reperfusion. Nutrition 2007; 23(2): 172-8.
[http://dx.doi.org/10.1016/j.nut.2006.11.004] [PMID: 17234508]

[107] Čanadanović-Brunet JM, Savatović SS, Ćetković GS, *et al.* Antioxidant and antimicrobial activities of beet root pomace extracts. Czech J Food Sci 2011; 29: 575-85.
[http://dx.doi.org/10.17221/210/2010-CJFS]

[108] Kapadia GJ, Azuine MA, Sridhar R, *et al.* Chemoprevention of DMBA-induced UV-B promoted, NOR-1-induced TPA promoted skin carcinogenesis, and DEN-induced phenobarbital promoted liver tumors in mice by extract of beetroot. Pharmacol Res 2003; 47(2): 141-8.
[http://dx.doi.org/10.1016/S1043-6618(02)00285-2] [PMID: 12543062]

[109] El Gamal AA, AlSaid MS, Raish M, *et al.* Beetroot (*Beta vulgaris* L.) extract ameliorates gentamicin-induced nephrotoxicity associated oxidative stress, inflammation, and apoptosis in rodent model. Mediators Inflamm 2014; 2014: 983952.
[http://dx.doi.org/10.1155/2014/983952] [PMID: 25400335]

[110] Kaefer CM, Milner JA. The role of herbs and spices in cancer prevention. J Nutr Biochem 2008; 19(6): 347-61.
[http://dx.doi.org/10.1016/j.jnutbio.2007.11.003] [PMID: 18499033]

[111] Farzaei MH, Abbasabadi Z, Ardekani MR, Rahimi R, Farzaei F. Parsley: a review of ethnopharmacology, phytochemistry and biological activities. J Tradit Chin Med 2013; 33(6): 815-26.
[http://dx.doi.org/10.1016/S0254-6272(14)60018-2] [PMID: 24660617]

[112] Petropoulos SA, Daferera D, Akoumianakis CA, Passam HC, Polissiou MG. The effect of sowing date and growth stage on the essential oil composition of three types of parsley (*Petroselinum crispum*). J

Sci Food Agric 2004; 84: 1606-10.
[http://dx.doi.org/10.1002/jsfa.1846]

[113] Petropoulos S, Daferera D, Polissiou M, Passam H. The effect of salinity on the growth, yield and essential oils of turnip-rooted and leaf parsley cultivated within the Mediterranean region. J Sci Food Agric 2009; 89(9): 1534-42.
[http://dx.doi.org/10.1002/jsfa.3620]

[114] Marchese A, Arciola CR, Barbieri R, *et al.* Update on Monoterpenes as Antimicrobial Agents: A Particular Focus on p-Cymene. Materials (Basel) 2017; 10(8): 947.
[http://dx.doi.org/10.3390/ma10080947] [PMID: 28809799]

[115] de Oliveira TM, de Carvalho RB, da Costa IH, *et al.* Evaluation of p-cymene, a natural antioxidant. Pharm Biol 2015; 53(3): 423-8.
[http://dx.doi.org/10.3109/13880209.2014.923003] [PMID: 25471840]

[116] Morita T, Jinno K, Kawagishi H, *et al.* Hepatoprotective effect of myristicin from nutmeg (*Myristica fragrans*) on lipopolysaccharide/d-galactosamine-induced liver injury. J Agric Food Chem 2003; 51(6): 1560-5.
[http://dx.doi.org/10.1021/jf020946n] [PMID: 12617584]

[117] Zheng GQ, Kenney PM, Lam LKT. Myristicin: a potential cancer chemopreventive agent from parsley leaf oil. J Agric Food Chem 1992; 40(1): 107-10.
[http://dx.doi.org/10.1021/jf00013a020]

[118] Zhang H, Chen F, Wang X, Yao HY. Evaluation of antioxidant activity of parsley (*Petroselinum crispum*) essential oil and identification of its antioxidant constituents. Food Res Int 2006; 39(8): 833-9.
[http://dx.doi.org/10.1016/j.foodres.2006.03.007]

[119] Castro A, Bergenståhl B, Tornberg E. Parsnip (*Pastinaca sativa* L.): Dietary fibre composition and physicochemical characterization of its homogenized suspensions. Food Res Int 2012; 48(2): 598-608.
[http://dx.doi.org/10.1016/j.foodres.2012.05.023]

[120] Ostertag E, Becker T, Ammon J, Bauer-Aymanns H, Schrenk D. Effects of storage conditions on furocoumarin levels in intact, chopped, or homogenized parsnips. J Agric Food Chem 2002; 50(9): 2565-70.
[http://dx.doi.org/10.1021/jf011426f] [PMID: 11958623]

[121] Peterson S, Lampe JW, Bammler TK, Gross-Steinmeyer K, Eaton DL. Apiaceous vegetable constituents inhibit human cytochrome P-450 1A2 (hCYP1A2) activity and hCYP1A2-mediated mutagenicity of aflatoxin B1. Food Chem Toxicol 2006; 44(9): 1474-84.
[http://dx.doi.org/10.1016/j.fct.2006.04.010] [PMID: 16762476]

[122] Leporatti ML, Ivancheva S. Preliminary comparative analysis of medicinal plants used in the traditional medicine of Bulgaria and Italy. J Ethnopharmacol 2003; 87(2-3): 123-42.
[http://dx.doi.org/10.1016/S0378-8741(03)00047-3] [PMID: 12860298]

[123] Redzić SS. The ecological aspect of ethnobotany and ethnopharmacology of population in Bosnia and Herzegovina. Coll Antropol 2007; 31(3): 869-90.
[PMID: 18041402]

[124] Kobaek-Larsen M, El-Houri RB, Christensen LP, Al-Najami I, Fretté X, Baatrup G. Dietary polyacetylenes, falcarinol and falcarindiol, isolated from carrots prevents the formation of neoplastic lesions in the colon of azoxymethane-induced rats. Food Funct 2017; 8(3): 964-74.
[http://dx.doi.org/10.1039/C7FO00110J] [PMID: 28197615]

[125] Roman M, Baranski R, Baranska M. Nondestructive Raman analysis of polyacetylenes in apiaceae vegetables. J Agric Food Chem 2011; 59(14): 7647-53.
[http://dx.doi.org/10.1021/jf202366w] [PMID: 21682272]

[126] Nikolić NC, Lazić MM, Karabegović IT, Stojanović GS, Todorović ZB. A characterization of content,

composition and scavenging capacity of phenolic compounds in parsnip roots of various weight. Nat Prod Commun 2014; 9(6): 811-4.
[PMID: 25115085]

[127]   Govaerts R, Wilkin P, Saunders RMK. World Checklist of Dioscoreales: Yams and Their Allies. (Kew Publishing, 2007) Available at: http://apps.kew.org/wcsp/home.do(Accessed: 23rd Sept. 2013)

[128]   Lebot V. Yams.Tropical Root and Tuber Crops: Cassava, Sweet Potato, Yams and Aroids. Wallingford: CABI 2008; pp. 181-275.

[129]   Asiedu R, Sartie A. Crops that feed the world 1. Yams Food Secur 2010; 2: 305-15.
[http://dx.doi.org/10.1007/s12571-010-0085-0]

[130]   Mignouna HD, Abang MM, Asiedu R. Harnessing modern biotechnology for tropical tuber crop improvement: yam (*Dioscorea* spp.) molecular breeding. Afr J Biotechnol 2003; 2: 478-85.
[http://dx.doi.org/10.5897/AJB2003.000-1097]

[131]   Shah NC. My experiences with the herbal plants & drugs as I knew Part XVI: Dioscorea & Costus. Herb Tech Ind 2010; pp. 21-30.

[132]   Oyama M, Tokiwano T, Kawaii S, *et al.* Protodioscin, Isolated from the Rhizome of *Dioscorea tokoro* Collected in Northern Japan is the Major Antiproliferative Compound to HL-60 Leukemic Cells. Curr Bioact Compd 2017; 13(2): 170-4.
[http://dx.doi.org/10.2174/1573407213666170113123428] [PMID: 28579930]

[133]   Liu JS, Gao WN, Zheng J, Wang GK, Yang QS. Chemical constituents from fresh tubers of *Dioscorea bulbifera.*. Zhongguo Zhongyao Zazhi 2017; 42(3): 510-6.
[PMID: 28952257]

[134]   Jing SS, Wang Y, Li XJ, *et al.* Phytochemical and chemotaxonomic studies on *Dioscorea collettii*. Biochem Syst Ecol 2017; 71: 10-5.
[http://dx.doi.org/10.1016/j.bse.2017.01.010]

[135]   Zhang Y, Ruan J, Li J, *et al.* Bioactive diarylheptanoids and stilbenes from the rhizomes of *Dioscorea septemloba* Thunb. Fitoterapia 2017; 117: 28-33.
[http://dx.doi.org/10.1016/j.fitote.2017.01.004] [PMID: 28065697]

[136]   Tseng SC, Shen TS, Wu CC, *et al.* Methyl Protodioscin Induces Apoptosis in Human Osteosarcoma Cells by Caspase-Dependent and MAPK Signaling Pathways. J Agric Food Chem 2017; 65(13): 2670-6.
[http://dx.doi.org/10.1021/acs.jafc.6b04800] [PMID: 28301149]

[137]   Zhu L, Dong Y, Na S, Han R, Wei C, Chen G. Saponins extracted from *Dioscorea collettii* rhizomes regulate the expression of urate transporters in chronic hyperuricemia rats. Biomed Pharmacother 2017; 93: 88-94.
[http://dx.doi.org/10.1016/j.biopha.2017.06.022] [PMID: 28624426]

[138]   Niu X, He Z, Li W, *et al.* Immunomodulatory Activity of the Glycoprotein Isolated from the Chinese Yam (*Dioscorea opposita* Thunb). Phytother Res 2017; 31(10): 1557-63.
[http://dx.doi.org/10.1002/ptr.5896] [PMID: 28840617]

[139]   Gupta K, Talwar G, Jain V, Dhawan K, Jain S. Salad crops root, bulb, and tuber crops Encyclopaedia of Food Science 2000; 5060-73.

[140]   Formisano C, Rigano D, Senatore F, Bruno M, Rosselli S. Volatile constituents of the aerial parts of white salsify (*Tragopogon porrifolius* L., Asteraceae). Nat Prod Res 2010; 24(7): 663-8.
[http://dx.doi.org/10.1080/14786410903172106] [PMID: 20401798]

[141]   Formisano C, Rigano D, Senatore F, Bruno M, Rosselli S. Volatile constituents of the aerial parts of white salsify (*Tragopogon porrifolius* L., Asteraceae). Nat Prod Res 2010; 24(7): 663-8.
[http://dx.doi.org/10.1080/14786410903172106] [PMID: 20401798]

[142]   Spina M, Cuccioloni M, Sparapani L, *et al.* Comparative evaluation of flavonoid content in assessing

quality of wild and cultivated vegetables for human consumption. J Sci Food Agric 2008; 88(2): 294-304.
[http://dx.doi.org/10.1002/jsfa.3089]

[143] Strzelecka H, Kowalski J, Eds. Encyklopedia zielarstwa i ziołolecznictwa. Warsaw: Wydawnictwo Naukowe PWN 2000.

[144] Mabberley DJ, Ed. Mabberley's plant-book. Cambridge: Cambridge University Press 2008.

[145] Blaschek W, Ebel S, Hackenthal E, Eds. Hagers Enzyklopädie der Arzneistoffe und Drogen (Vol. 14) Stuttgart: Wissenschaftliche Verlagsgesellschaft mbH Stuttgart. 2007.

[146] Zidorn C, Petersen BO, Sareedenchai V, Ellmerer EP, Duus JØ. Tragoponol, a dimeric dihydroisocoumarin from *Tragopogon porrifolius* L. Tetrahedron Lett 2010; 51(10): 1390-3.
[http://dx.doi.org/10.1016/j.tetlet.2010.01.016]

[147] Granica S, Lohwasser U, Jöhrer K, Zidorn C. Qualitative and quantitative analyses of secondary metabolites in aerial and subaerial of *Scorzonera hispanica* L. (black salsify). Food Chem 2015; 173: 321-31.
[http://dx.doi.org/10.1016/j.foodchem.2014.10.006] [PMID: 25466029]

CHAPTER 2

Effects of Pre- and Post-Harvest, Technological and Cooking Treatments on Phenolic Compounds of the Most Important Cultivated Vegetables of the Genus *Allium*

Rosa Perez-Gregorio[a]**, Ana Sofia Rodrigues**[b,c] **and Jesus Simal-Gandara**[*, d]

[a] *LAQV-REQUIMTE, Departamento de Química e Bioquímica, Faculdade de Ciências da Universidade do Porto, Rua do Campo Alegre 687, Porto, Portugal*

[b] *Instituto Politécnico de Viana do Castelo, Escola Superior Agrária, Ponte de Lima, Portugal*

[c] *Centre for Research and Technology of Agro-Environmental and Biological Sciences – CITAB, Vila Real, Portugal*

[d] *Nutrition and Bromatology Group, Department of Analytical and Food Chemistry, Faculty of Food Science and Technology, University of Vigo – Ourense Campus, E-32004Ourense, Spain*

Abstract: The genus *Allium* is one of the richest sources of phenols among vegetable foods and is highly ranked for its contribution of phenolic compounds to human diet. We thoroughly studied the chemical composition and quality of phenolic compounds in various *Allium* species with special emphasis on their bioactive properties. Pre-harvest (genotype and cultivation practices) and post-harvest conditions (storage and processing) had strong effects on the chemical composition and bioactive potency of the phenol fraction in *Allium* vegetables. In conclusion, genetic variability among *Allium* populations and ecotypes should further be examined to facilitate the selection of elite germplasms with increased contents in bioactive compounds in order to improve the preharvest quality of Allium and Allium-based foodstuffs. Proper cultivation practices and postharvest treatments have to be implemented in order to retain the preferred quality features and phenolic compounds in particular.

Keywords: *Allium*, Antioxidants, Cooking Treatments, Industrial Treatments, Phenolic Compounds, Post-Harvest Effects, Pre-Harvest Effects.

[*] **Corresponding author Jesus Simal-Gandara:**University of Vigo – Ourense Campus, Nutrition and Bromatology Group, Department of Analytical and Food Chemistry, Faculty of Food Science and Technology; Tel: +34988 387 000; E-mail: jsimal@uvigo.es

Spyridon A. Petropoulos, Isabel C.F.R. Ferreira and Lillian Barros (Eds.)

PHENOLIC ANTIOXIDANTS IN THE GENUS *ALLIUM*

A plethora of scientific data suggests that plant species of the genus *Allium* (*Alliaceae*) have various therapeutic effects. For example, garlic (*Allium sativum* L.), onion (*Allium cepa* L.) and the bioactive compounds they contain have since ancient times been used to prevent some diseases [1 - 7]. There is a growing trend in the food science to seeking evidence in support of food functionality with studies on the chemical composition and biological properties of vegetable foods. Such studies have focused on the influence of lifestyle and dietary habits on the prevalence of cancer, type 2 diabetes, metabolic disorders, and cardiovascular and/or neurological diseases, among others [8 - 10]. Vegetables of the genus *Allium* such as garlic (*A. sativum*), onion (*A. cepa*) and leek (*A. porrum*) are widely consumed not only for their health-promoting benefits, but also for their characteristic flavour. According to FAOSTAT [11], *per capita* onion annual consumption in 2013 amounted to 10.8 kg in the world and 8.97 kg in Europe. In fact, onions are among the most consumed vegetables and garlic has been ranked 17th based on *per capita* consumption worldwide [12]. Consumption of allium species has increased steadily over the past decade [11], probably in response to campaigns encouraging the intake of more fruit and vegetables containing healthy compounds. Plant-derived compounds that render health-enhancing effects upon consumption are known as "phytochemicals". Phytochemicals from Allium, and particularly organosulfur and phenolic compounds, have gained increasing popularity thanks to their protective properties against various chronic diseases.

Garlic has been used as a medicinal food in traditional remedies for some types of cancer (particularly those of the gastrointestinal tract) [13 - 16]. In addition, garlic consumption has been associated to a reduced risk of cardiovascular disease (CVD). Existing pre-clinical and clinical evidence suggests that garlic lowers blood pressure and cholesterol levels, and inhibits platelet aggregation [17, 18]. In addition, garlic has been ascribed with antimicrobial, antioxidant and anti-inflammatory properties. Indeed, garlic has been deemed one of the most effective vegetables for promoting good health and longevity not only within the genus *Allium*, but also in the whole plant kingdom [19, 20]. Crude garlic bulbs contain mainly organosulphur compounds (γ-glutamylcystein peptides, alliin and its degradation products formed upon disruption of cell membranes) in addition to flavonoids (nobiletin, tangeretin, rutin) and steroids that may significantly contribute to pharmacological activity through pharmacokinetic modulation of transporter-enzyme [21].

Studies on *Allium* species have focused not only on garlic, but also on other well-known species such as onion and leek. Leek (*Allium ampeloprasum* var. *porrum*) is predominantly a European crop with significant cultivation in Turkey

(9000 ha), France (5800 ha), Belgium (4800 ha) and Poland (4400 ha). Although worldwide, Indonesia is the largest producer of leek as stated by FAO. It is grown for its cylindrical pseudo stem, which is blanched white from growing underground and is made up of long leaf bases. The white shaft is used in many culinary preparations, whereas the green leaves are considered inferior and are, therefore, usually only used in soups or discarded during harvesting and processing of the fresh produce for the market. With regard to its health aspects, epidemiologic studies elucidated the reduction of the risk of prostate, colorectal, stomach and breast cancer upon the consumption of leek. These health benefits are linked to a range of phytochemicals, including 4 important chemical groups, *i.e.* (1) S-alk(en)yl-L-cysteine sulfoxides, (2) polyphenols, (3) vitamins and (4) fructans [22].

As stated above, the broad spectrum of health benefits associated to *Allium* has been ascribed mainly to the presence of phenolic and organosulfur compounds. In fact, phenolic compounds are pivotal to maintaining human physiological functions and preventing cardiovascular disease and cancer [23 - 26]. These compounds are phytochemicals resulting from secondary plant metabolism and acting as physical and chemical barriers in plant defence mechanisms. Moreover, phenols are highly abundant and play a central role as monomers in lignin synthesis and as chemical agents. Phenolic compounds have received much attention by virtue of their powerful antioxidant, antimicrobial, anti-proliferative and antiviral activities, and also of their ability to protect neurons against oxidative stress, stimulate vasodilatation, reduce vascularization and improve insulin secretion [27]. In addition, they are largely responsible for the sensory properties (particularly colour and taste) of a number of plant-derived foods and beverages [28, 29].

Flavonoids constitute the most prevalent family of phenolic compounds; in fact, they account for more than 9000 compounds identified in plants and the list continues to grow [30]. They share a common backbone consisting of two phenyl rings connected by a three-carbon bridge that is usually part of a six-membered oxygen heterocycle. Based on the degree of oxidation and saturation in the heterocyclic C-ring, the flavonoids may be divided into 8 groups (flavan, flavanone, flavones, flavonol, dihydroflavonol, flavan-3-ol, flavan-4-ol and flavan-3,4-diol) [31]. Besides other classifications have been proposed, most accepted nomenclature classified the flavonoids into flavonols, flavanonols, flavones, flavan-3-ols, flavanones, isoflavones and anthocyanidins, depending on the degree of unsaturation and oxidation of the three-carbon bridge (Table **1**) [32]. The hydroxyl functional groups found on all three rings are potential sites for links to carbohydrates, and if bound to one or more sugar molecules, they are known as flavonoid glycosides, whereas those that are not bound to a sugar molecule are called aglycones. The structural complexity of flavonoids is further increased with the linking of acetyl and malonyl groups to the sugar conjugates [33, 34]. Flavonols are a major group of flavonoids, which occur mainly in the form of glycosides in plants. In vegetables, quercetin glycosides predominate, but glycosides of kaempferol, luteolin and apigenin are also present. Each flavonoid class comprises a wide variety of derivatives with a variable number of different substituents on the flavonoid nucleus. Most flavonoids occur as glycosides and other conjugates, which makes them highly complex and diverse.

A study of the flavonoids myricetin, quercetin, kaempferol, luteolin and apigenin in 62 edible tropical plants revealed the highest total flavonoid content (TFC) (expressed in dry weight) in onion leaves (*Allium fistulosum*) with 2720.5 mg/kg (1497.5 mg/kg quercetin, 391.0 mg/kg luteolin and 832.0 mg/kg kaempferol), followed by Semambu leaves (*Calamus scipronum*) (2041.0 mg/kg), bird chili (*Capsicum frutescens*) (1663.0 mg/kg), black tea (*Camellia chinensis*) (1491.0 mg/kg), papaya shoots (*Carica papaya*) (1264.0 mg/kg) and guava (*Psidium guajava*) (1128.5 mg/kg). The major flavonoid in the extracts was quercetin, followed by myricetin and kaempferol. All *Allium* vegetables studied, onion leaves, Chinese chives leaves (*Allium odorum*) and garlic (*Allium sativum*), exhibited high flavonoid levels. For example, garlic contained 639.0 mg/kg myricetin, 47.0 mg/kg quercetin and 217.0 mg/kg apigenin [35].

Table 1. Structure of the main *Allium* flavonoids.

FLAVONOIDS				
GROUP	**STRUCTURE**	**R**		**COMPOUND**
Flavonols		**R₁**	**R₂**	
		H	H	Kaempferol
		OH	H	Quercetin
		OH	OH	Myrecitin
		OCH₃	H	Isorhamnetin
Flavononols and Flavones		H		Engeletin
		OH		Astilbin

(Table 1) cont.....

Antocyanidins*			R_1	R_2	
			OH	H	Cyanidin
			OCH_3	H	Peonidin
			OH	OH	Delfinidin
			OCH_3	OH	Petunidin
			OCH_3	OCH_3	Malvidin
Flavanols	*Tannins*		H		(+)-Catequin
			OH		(+)-Galocatequin
			H		(-)-Catequin
			OH		(-)-Galocatequin
			H		(-)-Epicatequin
			OH		(-)-Epigalocatequin
			H		(+)-Epicatequin
			OH		(+)-Epigalocatequin
	Procyanidins		R_1	R_2	R_3	
			H	H OH O-galate	H OH	Procyanidins
			OH	H OH O-galate	H OH	Prodelfinidins

*Anthocyanidins appears glycosylated (anthocyanins)

The biological activity of flavonoids was initially ascribed to their antioxidant activity against free radical-related damage. In fact, flavonoids are effective scavengers for reactive oxygen, nitrogen and chlorine species [36, 37]. In addition, they can chelate metal ions and consequently diminish their pro-oxidant activity [38]. Overall, flavonoids mitigate oxidative stress by preserving hemodynamic and biochemical function.

In any case, apparent evidence on dietary flavonoids should be interpreted with caution since it has been established mostly from *in vitro* studies of their unmodified forms, whereas their metabolites occur *in vitro*. The antioxidant activity of flavonoids can be diminished by methylation, sulphation and glucuronidation reactions [39]. Studies on the potential antioxidant effects of dietary flavonoids *in vitro* have produced confusing and conflicting data. Thus, a negative correlation between the intake of dietary flavonoids and the incidence of several diseases has been found in a number of epidemiological studies. Although the vast majority of scientific knowledge on flavonoids has been obtained from intact forms, some studies have focused on the effect of their metabolites [40]. However, confusing data have also arisen from *in vitro* studies only. Thus, some cross-sectional and cohort studies have found no difference between a flavonoid-rich and a flavonoid-poor diet. These conflicting data may have resulted at least partly from the biomarkers used for lipid and protein peroxidation being far from ideal. In addition, isolating the effect of flavonoids from those of other diet components is very difficult. Therefore, it remains a gap in research on the preventive or protective role of flavonoids against some diseases that should be filled with new studies to better elucidate their actual effects on human health [41].

There also remains a need to find effective procedures for maintaining phenolic compounds extracted from plants at high levels to further investigate their consumption and bioavailability. The phenolic composition of plants is directly related to intrinsic and extrinsic factors such as crop environment (weather conditions and agricultural practices), post-harvest practices (storage conditions) or even industrial and cooking treatments. This chapter deals with the phenolic profile of the genus *Allium*, focusing on the most important cultivated species of the genera, namely garlic, onion and leek, as well as with the effects of pre- and post-harvest, technological and cooking treatments on their antioxidant properties.

ALLIUM CHEMISTRY: PHENOLIC COMPOUNDS AND ANTIOXIDANT ACTIVITY DETERMINATION

Allium is a genus of monocotyledonous flowering plants that encompass hundreds of species including onion, garlic, scallion, shallot, leek and chives. The Allium vegetables most widely consumed in Europe are onion, garlic and leek. These vegetables contain a number of bioactive compounds (particularly polyphenols); also, they are rich sources of such compounds for human diet and provide consumers with substantial health benefits. The growing demand for healthy foods such as *Allium* vegetables has boosted research into their chemical properties with a view to identifying the specific compounds responsible for their health benefits. This section discusses the influence of management procedures on bioactive polyphenols in onion, garlic and leek.

Structurally, dietary polyphenols range from simple molecules of monomers and oligomers to complex polymers. Polyphenols are typically classified as flavonoids or non-flavonoids. Non-flavonoids are either phenolic acids or stilbenes, while flavonoids encompass a much wider variety of compounds including flavonols, flavan-3-ols, flavones, isoflavones, flavanones, chalcones and anthocyanins. Phenols in onions are mainly flavonols (particularly quercetin derivatives), and so are those in garlic —except for the flavonoid group; by contrast, leek contains both flavonoids and non-flavonoids in roughly the same proportion [42].

The importance of the phenolic composition of *Allium* vegetables has promoted the development of robust, routinely manageable, sensitive methods for its determination. Phenolic compounds are usually extracted with organic solvents such as acetone, methanol or acetonitrile mixed in different proportions with water. Anthocyanins, however, often require using formic or acetic acid. Although phenols differ in their optimal extraction and analysis conditions, they are usually extracted and determined with a common method. In fact, most available methods for analysing phenolic compounds from *Allium* vegetables involve the spectrophotometric determination of total phenols, total flavonols, total anthocyanins, and so forth, and very few protocols use chromatography for separation and analysis of each individual phenol. Reversed-phase HPLC methods in combination with UV–Vis and ESI–MS detection have emerged as an advantageous choice for determining a variety of phytochemicals including phenolic compounds from various biological matrices. Total composition analyses provide useful information about the general behaviour and/or activity of phenolic compounds as antioxidants. However, specific compounds may have a special influence on the overall biological effects of the phenols. Developing fast, sensitive methods for assessing the individual contribution of phenolic compounds is therefore imperative.

The identity of phytochemicals is usually confirmed by using appropriate standards or LC–MS; however, the lack of standards for many phenolic compounds has rendered their identification difficult. Acid and basic hydrolysis provide alternative identification tools because most flavonoids in *Allium* species are glycosylated or acylated derivatives. By virtue of their structure, polyphenols are effective radical scavengers and hence potentially useful against oxidative stress. Antioxidant activity, however, should not be assessed in terms of a single antioxidant test, but rather be determined with several available *in vitro* tests for this purpose. Natural polyphenols neutralize reactive oxygen species by electron or hydrogen atom transfer. Common *in vitro* tests for assessing antioxidant activity in simple and complex phenolic extracts include the hydrogen peroxide scavenging assay, nitric oxide scavenging assay, peroxynitrite radical scavenging test, Trolox equivalent antioxidant capacity (TEAC) method/ABTS radical cation decolourization assay, total radical-trapping antioxidant parameter (TRAP) method, ferric reducing-antioxidant power (FRAP) assay, superoxide radical scavenging activity (SOD) test, hydroxyl radical scavenging activity assay, hydroxyl radical averting capacity (HORAC) method, oxygen radical absorbance capacity (ORAC) method, reducing power (RP) method, phosphomolybdenum method, ferric thiocyanate (FTC) method, thiobarbituric acid (TBA) method, DMPD (*N,N*-dimethyl-*p*-phenylene diamine dihydrochloride) method, xanthine oxidase method, cupric ion reducing antioxidant capacity (CUPRAC) method, β-carotene linoleic acid method/conjugated diene assay, and metal chelating activity or antiradical activity test for the free radical 2,2-diphenyl-1-picrylhydrazyl (DPPH*). Each method measures a specific property that may differ under various biological conditions and should thus be confirmed with an *in vitro* antioxidant test. To date, major studies have focused on *in vitro* antioxidant activity as measured with one or at most two different *in vitro* methods, so further research to compare their results with those obtained *in vitro* is required to accurately assess the antioxidant capacity of phenolic compounds.

Because the vast majority of studies in this field have been conducted on complex extracts, synergies and antagonisms require careful evaluation. Thus, a recent method involving HPLC coupled to a post-column reaction system involving ABTS and bleaching was used to assess the chemical activity of radical scavenging compounds in various vegetables. The system afforded structural analysis of the target compounds. This widely applicable method may be a useful tool for the rapid screening and quantification of specific phytochemicals in food extracts and biological fluids, and hence for facilitating future biochemical and physiological research.

GENETIC FACTORS INFLUENCING THE DISTRIBUTION OF PHENOLIC COMPOUNDS IN *ALLIUM*

Genetic, pre-harvest (genotype, harvest season, weather, cultivation practices) and post-harvest factors (curing methods, storage and domestic preparation practices such as peeling and cooking) have clear-cut effects on total phenolic and flavonoid content [43 - 46].

Some studies revealed that onions had higher radical scavenging activity than garlic [47, 48] Antioxidant activity in onion, garlic and leek can be samples classified as follows: onion > leek > garlic and TPC as onion > garlic > leek [48]; also, red onion proved being more active than yellow onion and onion skin extracts possessed the highest activity [49]. However, other studies [50], antioxidant activity of garlic is about six fold higher that of onion. These differences are probably at least partially due to the different methods of extraction and/or quantification used.

The existence of a variety of ecotypes comprising different cytotypes which have been cultivated for many decades in some areas may raise quality issues in terms of uniformity and contents in bioactive compounds [51]. Selecting an appropriate cultivar may be a useful way of increasing the contents in total phenolics and ferulic acid irrespective of the cropping conditions [51]. Thus, Chen *et al.* [52] examined 43 garlic cultivars for phenolic composition (total phenolics and flavonoids) and antioxidant activity with various assays and found large differences among cultivars. In addition, they identified three segregated groups of cultivars differing markedly in composition and antioxidant activity. Fanaei *et al.* [53] also reported significant differences in pungency and in suitability for long storage and cooking among garlic genotypes.

Significant differences in phenolic composition were found among the Italian garlic varieties Bianco, Torella, Salomone, Schiacciato and Ufita, which contained various phenolic acids (*e.g.*, caffeic, ferulic, *p*-coumaric, gallic, chlorogenic), flavan-3-ols 16 (*e.g.*, epicatechin), flavones (*e.g.,* apigenin and luteolin), flavanones (*e.g.*, naringenin) and flavonols (*e.g.*, quercetin and its derivatives rutin and hyperoside). A few compounds including gallic acid, chlorogenic acid were present in all endemic varieties [54].

Garlic bulb extracts from more than 40 cultivars were evaluated for total phenolic content and antioxidant activities in order to assess the extent of genotypic variation. The coefficient of variation in Total Phenolic Content (TPC) of bulbs and bolts, 3.06%, suggested that TPC differed considerably among cultivars. Thus, TPC in bulbs ranged from 17.16 to 42.53 mg GAE/g. Also, the bulbs of cultivar '74-x' had the highest TPC, followed by the 'Hanzhong purple' cultivar; on the other hand, cultivar 'Gailiang' had the lowest TPC. These results suggest that '74-x' may be a better source of phenolic compounds than other garlic cultivars. Total phenolic acid content (TPAC) in bulbs also differed considerably (by a factor of more than 4) among cultivars, with a coefficient of variation of 2.70%. TPAC ranged from 0.15 mg/g rutin DW for cultivar 'No. 105 from Korea' to 0.60 mg rutin DW/g for '74-x'. The latter cultivar also had the highest TPAC in bulbs, followed by 'Hanzhong purple'; on the other hand, cultivars 'No. 105 from Korea' and 'Gailiang' had the lowest TPAC. This result is consistent with that for TPC. The variability of TPC and TPAC among cultivars can be ascribed to various characteristics such as clove size, which indirectly affects the final concentration of phenolic compounds. Consistent with these results, previous reports showed different garlic cultivars to exhibit also different yields, allicin contents and polyphenol contents; by contrast, garlic variety and origin had little influence on the agricultural traits of this crop.

TPC in garlic bolts of 32 cultivars was found to range from 10.17 mg GAE/g for cultivar 'Xiangfan ershui early' to 22.66 mg GAE/g for 'Hanzhong purple'; these TPC levels are much higher than those in leaves of *Allium roseum* L., where the TPC was 1.23 mg GAE/g aged extract. TPC was higher in bulbs than in bolts. TPAC in bolts, expressed as mg rutin DW/g, ranged from 0.06 in 'Baihe early' to 0.67 in 'Sicuan red'; also, 'Hanzhong purple' exhibited the highest TPC and increased TPAC levels. The coefficients of variation for TPC and TPAC were 3.47% and 5.85%, respectively [52].

A study intended to assess varietal differences in phenolic and flavonoid contents, and in antioxidant and anti-proliferative activities of onions was conducted on shallots and ten onion varieties usually available in the United States, namely: Western Yellow, Northern Red, New York Bold, Western White, Peruvian Sweet, Empire Sweet, Mexico, Texas 1015, Imperial Valley Sweet and Vidalia. TPC were strongly correlated with flavonoid contents and total antioxidant activity. Specifically, shallots had the highest TPC values (114.7 mg/100 g sample) among the onion varieties studied, with a 6-fold difference from the variety with the lowest content (Vidalia, $p < 0.05$). Western Yellow exhibited the highest TFC (69.2 mg/100 g onion), with an 11-fold difference relative to the variety with the lowest content: Western White ($p < 0.05$) [55].

Nuutila *et al.* [47] found the quercetin content of onion (*A. cepa*) to range from 7 mg to 83 mg/kg, depending on the variety and particular plant part —quercetin levels were higher in the outer layers. A distinct gradient in TFC was found between the outer, central and inner edible scales and along the longitudinal axis of the bulb and differences in flavonol levels between small- and large-sized onions [58]. Sultana and Anwar [56] found a quercetin content of 104.8 mg/kg in onion, which was low relative to red and yellow spring onion (113 and 1274 mg/kg, respectively) but high as compared to the values reported by Nuutila *et al.* [47] for giant onion (85 mg/kg).

Kwak *et al.* [57] quantify the contents of individual quercetin glycosides in red, yellow and chartreuse onion. Contents of total quercetin glycosides varied extensively among three varieties (ranged from 16.10 to 103.93 mg/g DW). Quercetin was the predominant compound that accounted mean 32.21 mg/g DW in red onion (43.6% of the total) and 127.92 mg/g DW in chartreuse onion (78.3% of the total).

TFC differs markedly with the particular plant phenophasis Table (**2**); thus, it is especially high in immature garlic plants. The decrease in total phenolic and flavonoid contents is most probably the result of increased levels of sulphur compounds and terpenoids present in essential oil from mature garlic bulbs [6].

Total polyphenol content (TPC) and antioxidant capacity differed among anatomical parts of bear garlic (Allium ursinum) harvested at different stages of ripeness. Thus, TPC ranged from 6.5 mg/100 g FW in bulbs to 42.5 mg/100 g FW in leaves. The highest DPPH, ABTS and FRAP antioxidant capacity was found in leaves (mean 202.3, 7971.5 and 856.8 µmol Trolox/100 g FW, respectively), and the lowest in bulbs (mean 105.3, 3228.5 and 196.5 µmol Trolox/100 g FW) [58].

The high TOC values of raw garlic can be ascribed not only to the presence of phenolic compounds, but also to the ability of allicin to react with the Folin–Ciocalteu reagent as confirmed by tests with the pure compounds, where TPC in raw samples was overestimated [59]. A study of 103 garlic cultivars collected from various regions throughout the world revealed large differences among geographic collection sites in the contents in *S*-allyl-cysteine sulphoxide and total phenolics that were ascribed to adaptation of the species under different environmental conditions during the domestication and expansion process. The differences among studies may have arisen from differences in the experimental conditions (*e.g.*, sample preparation, extraction method), and from natural qualitative and quantitative variability in the raw material.

Choosing an appropriate genotype among the many available for a variety of soils and climatic conditions is pivotal for ensuring high quality without compromising yield.

PRE-HARVEST EFFECTS ON PHENOLIC COMPOUNDS

The chemical composition of *Allium* vegetables is influenced by various factors involved in the production process, which could therefore be used to enhance the quality and bioactive properties of the product.

Cultivation practices (particularly irrigation regime and fertilization schedule) are also important since they not only contribute to fulfilling water and nutrient requirements —and hence avoiding nutrient deficiencies or water stress— but also can have beneficial effects on the chemical composition and quality of the end-product [60].

Beato *et al.* [12] studied the effect of growing conditions on ten garlic genotypes cultivated at four different locations and found that, although the location had no effect on total phenolics or ferulic acid, it considerably influenced the contents in caffeic, *p*-coumaric, *p*-hydroxybenzoic and vanillic acids. Despite its impact on the quality of most vegetables, no study on the effect of irrigation on the chemical composition and quality of *Allium* species appears to have been reported to date. Csiszár *et al.* [61] observed a mild water deficiency (specifically, a decrease in soil water content by 40%) during the growing season; thus, water holding for

1 week at the 3–5 leaf growth stage caused significant changes in antioxidants and in the activities of antioxidant enzymes.

Application of selenium and humic acids to garlic plants boosts their antioxidant activity but reduces their flavonoid contents [62]. Based on data from a 2008 study, selenium was assumed to influence the antioxidant potential of 'Ziemiai' garlic; thus, one selenium treatment decreased the content in ascorbic acid content and all reduced the total phenolic content. In a 2009 study, however, selenium increased phenol levels and a positive correlation between Se dose and antioxidant capacity was observed ($n = 32$; $r = 0.453$; $P < 0.01$). No contribution of total phenolics, ascorbic acid or pungency to the antioxidant potential of the garlic was apparent, however. Possibly, such a potent antioxidant as Se overshadowed the ability of other bioactive compounds to scavenge free radicals (*e.g.*, ABTS). Based on the results, leaf fertilization with a Na_2SeO_4 solution containing 10 or 50 μg Se/mL can increase bulb size and antioxidant capacity in garlic. Selenium was also found to replace sulphur in garlic plant metabolism and to decrease the contents of the most essential macronutrients in garlic bulbs. These results should be considered if high Se doses are to be used or Se is to be applied as fertilizer to soil [63]. A study of the effect of salinity on the phenolic contents of five garlic cultivars differing in salinity tolerance revealed that the initial contents of phenolic compounds and sulphur were comparatively low in salinity-tolerant cultivars but increased with increasing salinity level, whereas the salt-sensitive cultivars exhibited the opposite behaviour [64]. In another study, large amounts of nitrogen in the form of ammonium sulphate and urea increased garlic pungency as expressed by a high pyruvic content [65]. Studies with onion bulbs demonstrate a positive correlation between N, S and Fe fertilization and flavonoids content [66, 67]. Mogren *et al.* [68, 69] concluded that nitrogen fertiliser do not affect the flavonoid content or composition in onion neither at start of storage nor during 5 months of cold storage, which means that it may be possible to grow onions with limited nitrogen leakage without reduced yield or polyphenol concentration. Soil inoculation with arbuscular mycorrhizal inoculum and NH_4^+:NO_3^- ratios as nitrogen source variations in the root environment affect the content of health-related organosulfur compounds, total phenolic compounds, and flavonol glycoside concentrations in onions. Antioxidant activity, pyruvic acid, an indicator for organosulfur compounds, was significantly increased at dominant NO_3^- supply and also total phenolic concentration in the onion bulb appeared to be higher at dominant NO_3^- supply and colonization, particularly quercetin glucosides [70]. Antioxidant activity (by DPPH and FRAP), total flavonol content, and levels of quercetin glucosides were higher in onion under organic production compared to conventional management. Differences were primarily due to soil management practices used rather than pesticide/ herbicide application [71].

TPC in the white shaft and green leaves of the 30 leek cultivars varied from 5 to 14 mg GAE/g DW and from 5 to 15 mg GAE/g DW, respectively. In general, the white shaft of the summer, autumn and winter leek types had a mean TP content of 7.8, 7.3 and 8.9 mg GAE/g DW, respectively and did not show significant differences. Furthermore, the green leaves of the summer, autumn and winter types contained mean TPC levels of 8.9, 8.1 and 9.9 mg GAE/g DW, respectively. The green leaves of the winter types exhibited a significantly higher phenolic content compared to the autumn types, while it was not significantly higher compared to content of the summer cultivars. Only Q34'G, K3G and K could be quantified in the white shaft of some cultivars. In the green leaves of the leek samples, *i.e.* Q34'G, K3G, I3G, Q3G, Q, K and I. A lot of variation in polyphenol content was observed between the green leaves of the 30 leek cultivars. Cultivars Uyterhoeven, Pretan and Fahrenheit F1 gave the highest oxygen radical absorbance capacity (ORAC), 2,2-diphenyl-1-picrylhydrazyl (DPPH) free radical scavenging activity and ferric reducing antioxidant potential (FRAP) value, respectively, while cultivars Toledo and Breugel F1 had the highest polyphenol levels. Fahrenheit F1 contained the highest AA levels, while Apollo F1 and Artico were especially rich in ACSOs. Zeus F1 was the cultivar with the highest fructan content [22].

Relating to harvest time, Mogren *et al.* [68, 69] reported an annual variation in quercetin content of onion at lifting and observed that late lifting time resulted in higher quercetin content.

Although farming systems (intensive, conventional, organic) affect not only plant growth and yield, but also end-product quality in various crops, they seemingly have no appreciable effect on chemical composition or quality-related features such as the allicin content of garlic [72].

POST-HARVEST EFFECTS ON PHENOLIC COMPOUNDS

Ensuring proper storage conditions is crucial for garlic and onion bulbs, and for their by-products, in order to retain their high quality. The natural sensitivity of the most abundant bioactive components in garlic has been ascribed to their high susceptibility to thermal degradation, which diminishes their potency. Industrial advances have allowed the shelf life of many foodstuffs to be improved. Thus, modified and controlled atmospheres, and irradiation, are useful to maintain the stability, microbiological, physical–chemical and sensory characteristics of garlic during the post-harvest period.

Curing is a widespread traditional practice involving heating prior to further processing. The procedure has favourable effects on fruit quality and reduces storage losses without altering the acidity or colour index of foods. Curing *Allium* vegetables can also have promissory benefits such as extending their shelf life and improving their stability during storage. The flavour of garlic is improved by curing but too high temperatures can cause unpleasant changes in its sensory properties [60, 73]. Field curing resulted in increases in quercetin content compared to levels at lifting, especially important for all white bulbs (33–40% increase). Flavonol and anthocyanin levels in onions cured in the dark were similar to those obtained in bulbs cured in the light. Further knowledge in this respect could be acquired by exploring curing under variable conditions of temperature, exposure time and relative humidity [74].

The TPC and TFC of garlic scales from different varieties stored in a controlled atmosphere (O_2 = 2–5%, CO_2 = 3–6%) at 0 ± 0.5 °C and RH = 85–95% for 140–224 days decreased with time —by exception, allicin exhibited the opposite response [75].

Storage time has a crucial impact on the bioactive properties of garlic. As noted earlier, the antioxidant capacity of garlic cloves stored at 20 ± 2 °C peaks after 8 weeks, and those of organosulfur compounds and polyphenols after 6–8 weeks of storage, with a considerable subsequent decrease [76]. Allicin content decreases with increasing storage time, whereas antioxidant activity and total phenolic content change in the opposite direction [77]. Processing treatments should be adapted to the intended end use of the garlic raw material; thus, storage for several weeks prior to processing is one of the most widely recommended treatments for garlic-based dietary supplements.

Storage temperature can also affect the chemical composition and final bioactivity of garlic. Storing garlic cloves at low temperatures (5 °C), a technique which is known as "conditioning", has been reported to affect the expression of the 1-SST gene, which is associated to fructan metabolism, and hence to carbohydrate and total soluble solid contents [78]. However, the antioxidant potential of garlic decreases with increasing temperature. Changes in the contents in chlorophyll, carbohydrates, amylase and invertase enzymes of garlic at low temperatures were assessed, and qualitative and quantitative changes in sugars that boosted sprouting incidence; also, chilling significantly increased the contents in chlorophyll, carotenoids, amylase and invertase, which peaked after 30 days of treatment [79]. Leek could be stored for 13 days under refrigerated conditions without a negative impact on the antioxidant properties [22].

Physical–chemical conditions such as pH, which reflects the acidity or alkalinity of many products, have substantial effects on the final quality of harvested products, their chemical composition and, consequently, their biological potential. Some authors have shown that pH can affect or even control the formation of volatile compounds (especially thiosulphinates) [80] and hence their release upon rupture of cloves [81]. The potential of modified/controlled conditions for maintaining or improving the quality of garlic and its by-products has been thoroughly examined. Thus, a study of the effects of different storage conditions (O_2 and CO_2 levels, mainly) on the final quality, sprout growth, decay and discoloration of garlic bulbs suggested that atmospheres containing CO_2 have a beneficial effect relative to those containing O_2 alone [82]. In addition, with more than 15% CO_2, bulbs showed signs of injury after 4–6 months of storage. Thus, an atmosphere containing 5–15% CO_2 and/or little O_2 (1–3%) is the most appropriate for delaying discoloration and decay of fresh peeled garlic cloves at 5 or 10 °C beyond 3 weeks of storage [82].

EFFECTS OF INDUSTRIAL TREATMENTS ON PHENOLIC COMPOUNDS

Agro-industrial processing techniques are widely used to improve the efficacy and bioavailability of many foods, and to alleviate some unpleasant features when present. For example, some civilizations used to soak garlic bulbs in various solvents such as alcohol, wine, milk or vinegar to extract bioactive ingredients [83]. In addition, a number of recent studies have shown biochemical changes and inter-conversions to occur during processing. In fact, some authors have found considerable differences in final bioactive potential depending on the type of preparation [65, 84].

Onion, garlic or leek can be subjected to different industrial treatments affecting their phenolic composition before reaching consumers. Vegetable products used as condiments or spices for cooking, or for medicinal purposes, should be stored in an appropriate, safe manner [85 - 87].

Garlic processing usually includes blanching, by which peeled cloves are exposed to high temperatures by using hot water, steam, microwaves, radio frequency or infrared irradiation [88]. Kinalski and Noreña [89] found blanching to considerably decrease thiosulphinate content and antioxidant activity, and their decrease rate increased with increasing time and temperature. Garlic can be consumed in the form of extracts from raw or dried powder garlic cloves. Lemar *et al.* [90] found fresh garlic extract preparations to possess a higher anti-Candida potential than dried garlic powder extracts and that freeze-dried garlic had decreased contents in bioactive compounds (particularly phenols). Also, other authors have reported that fresh garlic extracts are very effective against microbial infections [91], and so are aqueous extracts [92]. Other processing methods routinely used by the biotechnological and food industries (*e.g.*, boiling, frying, microwaving) apparently have no substantial effect on the content in bioactive compounds (anthocyanins, ascorbic acid, flavonoids, flavanols, polyphenols and tannins) or antioxidant activity of garlic [93]. The wide variety of garlic supplements available fall largely into one of the following four general categories: dehydrated garlic powder, garlic oil, garlic oil macerate or aged garlic extract [83]. Manufacturers must ensure that garlic supplements are safe, stable and efficient, and supply pertinent information and certification for all garlic by-products.

Irradiation techniques have been also assessed for their ability to extend garlic shelf life. Some studies have confirmed that they inhibit garlic sprouting and mitosis [94]. Thus, low gamma ray doses had no effects, whereas a 10 Gy dose considerably reduced garlic sprouting and stopped mitosis [94]. The effects of gamma rays on garlic bulbs have also been examined [95]. A dose of 60 Gy applied for 8 months considerably diminished lipid and fatty acid contents, and concomitantly reduced sprouting of garlic bulbs. Lipids and fatty acids are thus seemingly involved in the biosynthetic process behind sprout growth, so the long-term effects of irradiation can be interpreted as delaying or slowing down sprouting [95]. However, irradiation has been used to prevent microbial contamination during storage, and also to replace chemical fungicides during the post-harvest period [96].

Fresh-cut fruit and vegetable products have hardly increased their presence on the marketplace owing to the increasing demands of consumers. The fresh-cut food industry is widely expected to experience unprecedented growth in the coming years. However, processors of fresh-cut fruit products face a number of challenges rarely encountered during fresh-cut vegetable processing that require a higher level of technical and operational sophistication. Physical changes resulting from minimally processed food production could induce physiological —and hence compositional— changes potentially affecting the final quality of foodstuffs. The effect of minimal processing on the flavonoid content of onion in each step of the food processing has been examined. Wounding stress increases the phenolic content and antioxidant activity of vegetables [97 - 99]. According to Cantos *et al.* [100], the activity of the three most important enzymes involved in phenol metabolism (*viz.*, polyphenol oxidase, peroxidase and phenylalanine ammonia-lyase) remains unaltered after wounding. Reyes *et al.* [101] further confirmed that the effect of this source of stress depends on the particular vegetable. Thus, wounding increases the phenolic content and antioxidant activity of sliced onions [87].

Because of the way onion flavonoids are distributed in bulb tissues, the effect of wounding is also influenced by the cutting technology used. Thus, the outer layers typically have the highest flavonoid levels [87, 102, 103]. The greatest losses occur during pre-processing steps such as peeling or trimming. Because only the edible part of onion is consumed by humans, the brown outer leafs are not considered to this end. As noted earlier, flavonoid distribution in edible onion bulbs is not uniform. As a result, the initial flavonoid content of onion, and its changes, depend on the cutting technique used. Overall, chopping [87] or slicing onion [104] generally increases flavonoid content. However, onions can also be diced or cut into half-rings, rings or julienne strips. Recent studies have examined the effect of the cutting method on flavonoid contents [105] and found slicing to increase anthocyanin levels relative to dicing. However, the effect of cutting on the flavonoid content of onion is not unambiguous. The influence of temperature, the presence or absence of light, and the storage time have usually been studied in parallel with that of cutting. Some authors believe that only the storage time is influential [105], whereas others have found temperature changes to also result in differences [87]. Further research is therefore required to confirm differences in flavonoid evolution and the mechanisms behind the changes in different tissues.

A number of sanitizing technologies for disinfecting fresh-cut food prior to packaging have recently emerged in food science. The USDA and FDA definitions for "fresh" and "minimally processed" fruits and vegetables imply that fresh-cut (pre-cut) products have been freshly cut, washed, packaged and stored refrigerated. Fresh-cut products are raw and, even processed (*viz.*, physically altered from their original forms), they remain in a fresh state, ready to eat or cook, without freezing, thermal processing, or treatments involving additives or preservatives [106]. Because fresh-cut products are not subjected to any thermal processing, they require some sanitization in order to preserve the hygienic quality of the raw food. Washing is one of the most important processing operations in this field and usually involves physical and chemical treatments to completely remove or at least reduce the populations of pathogenic and spoilage-inducing microorganisms. However, according to Perez-Gregorio *et al.* [86], the main source of flavonol losses from fresh-cut onion slices is the solubility of flavonols in water; thus, immersing slices in water at 4 to 50 °C causes them to lose 17 to 23% of their flavonoid content. Although sodium hypochlorite cannot be used to sanitize fresh-cut vegetables in some European countries, it remains the most widely used sanitizer for this purpose by virtue of its low cost, ease of use and broad spectrum of activity [86]. Chlorine can oxidize organic matter in foods or in water, its reaction with water yielding by-products such haloforms and haloacetic acids that are potentially carcinogenic and mutagenic [107]. Organochlorine compounds such as sodium dichloroisocyanurate, potassium dichloroisocyanurate, dichloroisocyanuric acid and trichloroisocyanuric acid have gained increasing interest as alternative sanitizing agents in recent years [108]; however, they exhibit lower antimicrobial efficacy on onions than does hydrogen peroxide [109]. In any case, chlorine, organic chlorine and hydrogen peroxide considerably diminish the flavonoid content of onion [86].

Alternative treatments based on nisin and citric acid in combination have be used to sanitize fresh-cut onion. Both substances are generally recognized as safe (GRAS) for use as food ingredients [110], which is an advantage with a view to removing microbes from fresh-cut onion. In fact, Cheng *et al.* [104] found washing fresh-cut onions with nisin and citric acid to increase their total phenolic content and antioxidant activity. Therefore, this combination may provide a safe preservative for this type of product with the added advantage of an increased phenolic content. Post-harvest diseases can also be controlled with other treatments such as UV-C irradiation. In fact, UV light over the wavelength range of 250–260 nm is lethal to most microorganisms (bacteria, viruses, protozoa, mycelial fungi, yeasts, algae); also, it raises the flavonoid content of onions [86]. Ozone has also been used as a sanitizer but never to date has been assessed for its effect on onion flavonoid content [111].

As noted earlier, fresh-cut technology can promote some physiological changes that may induce microbial spoilage. Also, it can cause colour changes, softening, surface dehydration, water loss, translucency, off-flavour and off-odour development, all of which detract from the quality of the end-product. Currently, innovative modified atmospheres and edible coatings are especially favoured over other methods for maintaining freshness and safety in fresh-cut fruits and vegetables. The effectiveness of these techniques has been confirmed by several studies. However, the effects of treatment and storage conditions on fresh-cut fruits remain largely unexplored. For example, little is known about the potential effects of packaging procedures (*e.g.*, 'ready-to-eat' packaging). Flavonoid stability during fresh-cut onion storage in perforated films [112], and polyethylene or polyethylene terephthalate cups [87], was examined and onions found to undergo changes in flavonoid content as a result. The changes were governed by storage conditions such as temperature, time and the presence or absence of light. Thus, anthocyanin levels were greater under light than in the dark [87]. Also, stability differed markedly among individual flavonoids, malonated anthocyanins being much more stable than the corresponding non-acylated pigments and arabinosides proving less stable than the corresponding glucosides [112].

A gap in knowledge remains as to how the packaging material can influence changes in onion flavonoids during storage. A need also exists for a better understanding of the influence of the package atmosphere and for package selection criteria with a view to preserving flavonoid levels in onions.

EFFECTS OF COOKING TREATMENTS ON PHENOLIC COMPOUNDS

As stated above, the genus *Allium* has been ascribed with medicinal properties since ancient times. In addition, *Allium* vegetables are widely appreciated for their cooking uses, whether fresh, cooked or dehydrated.

Domestic cooking practices may significantly decrease the contents in bioactive compounds and proteins of garlic, and hence its antioxidant activity, especially after heating at 100 °C for more than 20 min [93]. A recent study examined the effects of various pre-cooking (chopping, crushing and no treatment) and cooking treatments (rolling-boiling, simmering, stir-frying and no cooking) on major organosulphur compounds (OSCs) in garlic cloves [113]. The results revealed significant differences among treatments that were largely due to differences in extent of processing and in exposure to high temperatures. A substantial decrease in bioactive compounds and antioxidant activity was found after boiling or frying of garlic [114]. The differences, however, may have resulted from long exposure to high temperatures relative to other studies [113]. Moreover, crushing the garlic cloves prior to cooking (over-heating, boiling or microwaving) seemingly alleviated losses of antiplatelet activity and the reduction of thiosulphinate content [115, 116].

The chemo-preventive effects of garlic extracts, including its individual constituents (organosulphur compounds, mainly), have been increasingly investigated. Thus, a study examined the antioxidant and antigenotoxic effects of various garlic extracts and revealed increased contents of total phenolic compounds in aged-garlic extracts (AGE) relative to raw and heated garlic extracts (RGE and HGE, respectively) [117]. In addition, despite the decrease in total phenolic content and antioxidant activity upon heating, the extracts retained their antioxidant and health-protective properties irrespective of processing method. Antioxidant activity and phenolic contents were also substantially increased by short-time fermentation. Organosulphur compounds are among the most common and widely studied garlic constituents [118]; thus, apart from their well-known bioactive properties, most of these odorous compounds are largely unstable and easily decomposed. As a result, even mild processing of garlic cloves and onions [74] can cause some OGSs to be transformed or even suppressed.

Aged garlic was found to exhibit more potent antiglycation and antioxidant properties than fresh garlic extract *in vitro* in a cell-free system. In addition, the TPC, total flavonoids and flavonols were significantly higher in aged garlic than in fresh garlic extract (Table **2**). Aged garlic is produced by soaking sliced raw garlic in 15–20% aqueous ethanol at room temperature for up to 20 months. The extract is then filtered and concentrated under reduced pressure at a low temperature. The ageing procedure converts odorous and harsh irritating compounds in garlic into odourless, non-irritating, safe sulphur compounds. The process causes the loss of allicin and increases the activity of other water-soluble organosulphur compounds such as *S*-allyl cysteine (SAC), *S*-allyl mercaptocysteine (SAMC) and allixin, and of selenium, all of which have antioxidant properties. Aged garlic extract additionally contains *N*-fructosyl arginine, which is not present in raw or heat-treated garlic preparations [119].

TPC and TFC of garlic subjected to different thermal procedures were found to exceed those of fresh garlic. Hydroxycinnamic acid derivatives were the major phenolic acids in garlic in different processing steps. In addition, flavanols were the prevailing flavonoid class, followed by flavanones and flavones. The contents in free polyphenols and flavonoids of heated garlic were significantly higher than those of fresh and steamed garlic. Heating increased phenol contents through cleavage of bound (esterified and glycosylated) forms, which increased the amounts of free forms. One other plausible reason for the increased phenol levels in heated garlic was diminished or inhibited enzymatic oxidation involving the antioxidant compounds in the raw plant material. An increase in TPC may arise from increased levels of complex polyphenols from a late stage of the browning reaction [120].

Table 2. Flavonoids, Flavonols and Total Phenolics in Various *Allium* Sources

Source	Total phenolics	Total flavonoids	Total flavonols	Ref.
Chinese chives (*Allium odorum*)		160 mg/kg DW		[35]
Chives cv. Pražská (*Allium schoenoprasum*)	1591 mg/kg			[139]
Bear garlic (*Allium ursinum*) Leaves Bulbs	425 mg/kg FW 65 mg/kg FW			[58]
Garlic (*Allium sativum*)		957 mg/kg DW		[35]
Garlic extract fresh	56000 mg/kg	47±6 mg/g	43±3.3 mg/g	[119]
Garlic extract aged	1290008 mg/kg	101±1.8 mg/g	94±0.9 mg/g	[119]
Garlic fresh Black: 90 °C/100% RH/34h Black: 60 °C/60% RH/6 h Black: 75 °C/70% RH/48 h Black: 70 °C/60% RH/60 h Black: 65 °C/50% RH/192h	105.73 mg GAE/kg DW 412.36 509.87 696.66 919.88 982.14	595.38 mg GAE/kg DW 646.99 673.82 741.95 834.16 869.94		[120]
Garlic bulbs fresh	493μg GAE/kg FW			[122]
Garlic bulbs heated 100 °C 20 min	451 μg GAE/kg FW			[122]
Garlic bulbs heated 100 °C 40 min	34 μg GAE/kg FW			[122]
Garlic bulbs heated 100 °C 60 min	324μg GAE/kg FW			[122]

(Table 2) cont.....

Source	Total phenolics	Total flavonoids	Total flavonols	Ref.
Garlic Red cv. Rubí -raw -slowly simmered -rolled -stir-fried -chopped garlic stir-fried in synthetic solution	mg GAE/kg DW 112.1 16.9 40.2 24.3 93.7			[59]
Garlic immature plants	980 GAE mg/kg DW	6.99 ± 0.01 µg QE/g		[138]
Garlic bulbs ground and air-dried (as Aged Garlic Extract)	180 GAE mg/kg DW	5.78 ± 0.09 µg QE/g		[138]
Garlic bulbs fresh	50 GAE mg/kg DW	4.16 ± 0.03 µg QE/g		[138]
Garlic cv. Mojmír	1051 mg/kg			[137]
Ramson (Wild garlic)	871 mg/kg			[137]
Garlic	812 mg/ kg	124 mg/ kg	16.9 mg/ kg	[139]
Garlic Tunisian	436 mg GAE/kg	132 mg QE/kg		[140]
Garlic	2.0 g GAE/kg DW			[141]
Leek (*Allium porrum*) cv. Atal	416 mg/ kg	101 mg/ kg	10.1 mg/ kg	[139]
Leek cv. Rossa di Trento	882 mg/ kg	280 mg/ kg	5.3 mg/ kg	[139]
Leek cv. Romana	547 mg/ kg	387 mg/ kg	9.8 mg/ kg	[139]
Leek cv. 'Inegol-92' - fresh - dried	116.43 mg RE/100 g DW 26.33 mg RE/100 g DW			[135]
Leek cv.Varna - Leaf - Stem	45.39 mg GAE/g DW 69.46 mg GAE/g DW	10.24 mg CE/g 33.53 mg CE/g		[136]
Onion (*Allium cepa*) cv. Western yellow		692 mg/kg		[55]
Onion bulbs (*Allium cepa*)			104.8 mg/kg DW	[56]
Onion red cv. Red mate	1313 mg/kg			[137]
Onion yellow cv. Sherpa	935.2 mg/kg			[137]
Onion white (White solid)	444.3 mg/kg			[137]
Onion white cv. Bianca di maggio	236 mg/ kg	64 mg/ kg	2.8 mg/ kg	[139]
Onion red cv. Rossa di tropea	428 mg/ kg	36 mg/ kg	2.1 mg/ kg	[139]
Onion yellow cv. Grano de Oro	4388,8 mg/ kg DW			[132]
Onion	350 mg/ kg FW			[137]

(Table 2) cont.....

Source	Total phenolics	Total flavonoids	Total flavonols	Ref.
Onion red (two cv.)	2536-3108 mg GAE/kg FW			[139]
Onion white	2167 mg GAE/ kg FW			[139]
Onion red			580.9 mg/ kg FW	[140]
Onion yellow			285.5-516.4 mg/ kg FW	[140]
Shallots (*Allium cepa* var. *aggregatum*)	1147 mg GAE/kg FW	344 mg CE/kg FW		[55]
Onion leaves (*Allium fistulosum*)		2720.5 mg/kg DW		[35]
Spring onion (*Allium fistulosum*)	4.2 g GAE/kg DW			[141]

Catechin Equivalents (CE); Gallic Acid Equivalents (GAE), Fresh Weight (FW); Dry Weight (DW); Quercetin Equivalents (QE); Rutin Equivalent (RE).

Total flavonol content decreased by 18.8% in the boiled onion compared to the raw, but the relative decrease between cooked and raw was not the same for the mono- and diglucosides. Compared to the raw sample, the diglucoside content in the boiled onion decreased by 14.8% and the monoglucoside content decreased by 20.1%, possibly indicating that the monoglucoside has a greater tendency to leach into the cook water than the diglucoside, or that under boiling conditions that the monoglucoside is less stable [121].

In order to preserve its major properties, garlic must be added to cooking dishes not earlier than 20 min before the end of the cooking process. The decrease in the content of total polyphenols and in related total antioxidant (FRAP and DPPH) activities only becomes significant after cooking at 100 °C for 40 and 60 min, respectively [122]. Black garlic is obtained by fermenting fresh garlic at a high temperature (60–90 °C) under controlled, high humidity (80–90%) conditions; unlike fresh garlic, black garlic releases no strong offensive flavours thanks to its low allicin content. The increased bioactivity of black garlic has been ascribed to changes in physical–chemical properties [123] such as increased polyphenol and flavonoid contents by a factor of 4.19 and 4.77, respectively [124].

The impact of common domestic and technological treatments on flavonoid composition in onions has also been studied [125 - 131]. Onion tissue undergoes major chemical and biochemical reactions by effect of technological and cooking treatments. Such reactions may have an impact on flavonoid structure, and alter the bioavailability and activity of flavonoids as a result [132]. In general, cooking onions decreases their total flavonol content to an extent dependent on the particular treatment (frying, boiling, roasting) and on its duration. Overall, mild conditions have no effect on flavonol content, but strong treatments can cause flavonol losses by 16 to 30% [131]. Boiling onions results in increased losses of quercetin glycosides, which can leach to the boiling water by up to 53% in specially strong treatments [131]. Also, quercetin degradation is more marked with diglucosides than it is with monoglucosylated quercetin derivatives, whereas anthocyanins undergo greater losses by exposure to cooking temperatures [131].

In leek, steaming of the green leaves resulted in an increase of antioxidant capacity, while boiling had a negative effect on antioxidant capacity and in TPC in the white shaft and green leaves. In general, steaming seems to be responsible for a better retention of the bioactive compounds present in leek and the losses in processed leek upon boiling are generally not attributed to a chemical breakdown of flavonoid conjugates or formation of new compounds, but rather to the leaching of phenolic compounds into the cooking water [22].

Fermentation of leek resulted in a higher antioxidant capacity and TPC especially for the green leaves. After 21 days of fermentation, new polyphenolic compounds were found such as hydroferulic acid, quercetin 3-O-rutinoside, quercetin 3- O-arabinoside, naringenin and dihydroquercetin, while sinapinic acid disappeared. The contents of ferulic acid, kaempferol 3-O-glucoside, luteolin and naringenin increased significantly after a leek fermentation process of 21 days compared with the initial concentration, while the caffeic acid content decreased. The qualitative changes in polyphenols during fermentation could indicate that lactic acid bacteria are capable of producing β-glucosidase, which catalyses the cleavage of sugar linkages during fermentation [22].

A trend currently exists to produce new pre-processed vegetables in addition to ready-to-use vegetables. Many people cannot eat fresh vegetables every day and use frozen vegetables instead. Frozen storage increases the shelf life of vegetables and reduces wastage of unused products. However, frozen vegetables may have a lower nutritional value than their respective commodities. Although fairly little is known about the potential effect of this technology on onion flavonoids, some authors have concluded that freezing raises flavonoid levels in onion [131, 133].

Although industrial freezing and drying processes facilitate long-term storage, the health-promoting capacity and nutritional characteristics of some plant products depend on the particular processing method used. Thus, dried onions can be marketed as powder for cooking purposes [134]. Consumers in search of healthy, fresh-like, convenient foods have driven developments in drying technology by increasingly demanding processed foods that will better retain their original characteristics. Industrially, this requires minimizing the adverse effects of processing. With food drying, this involves preventing the loss of volatiles and flavours, changes in colour and texture, and a decrease in nutritional value. A study examined the effect of dehydration on onion quality [135]. Dried foods can be produced by using convective dryers. This process, however, detracts from colour, flavour (taste and aroma) and texture; also, it often makes rehydration difficult. Freeze-drying, which is a far more expensive procedure than convective drying, is used mainly to obtain small amounts of high-value arable and horticultural produce; however, it is known to provide the highest-quality dried foodstuffs. This is largely because the procedure does not damage food structure as severely as other preservation procedures. Irrespective of drying procedure, dried food has residual enzyme and microbial activities, which are essential to extend its shelf life. In addition, the fact that dehydration minimizes enzyme activity may have an impact on quality-related factors such as antioxidant activity and flavonoid content. In fact, the flavonoid content of onion has been found to be raised by freeze-drying [136]. In leek, analysis of individual polyphenols revealed that, air-dried samples contained higher quantities of polyphenols than freeze-dried leek, demonstrating the thermostability of polyphenols. Freeze-dried leek on his turn exhibited higher levels of polyphenols compared to refractance window dried samples. Although air-drying was the best drying technique in retaining the antioxidant capacity and polyphenols, resulted in high losses of the ACSOs compared to freeze-drying [22].

The food industry has increasingly sought new cooking ingredients with healthy properties in recent times. Among these ingredients, spices are widely appreciated for their flavouring and colouring potential. Spices may contain phenolic compounds and contribute to the intake of natural antioxidants. Therefore, incorporating purified extracts of bioactive compounds into foods may be effective to increase consumption of these substances and allow the population to benefit from their presumed positive effects. For example, onion can be used as freeze-dried powder to improve the antioxidant capacity of foods and their onion flavour. Further research is needed, however, to improve existing knowledge about how flavonoids in onions are affected by domestic and industrial treatments. Scientific evidence on flavonoid content could be exploited by developing industrial processes allowing the production of foods with high quality and benefit.

CONCLUSIONS ON THE MANAGEMENT OF PHENOLIC RICHNESS IN *ALLIUM*

The quality of *Allium* vegetables in terms of chemical composition and bioactive compounds is strongly dependent on the particular pre- and post-harvest treatments they receive. Their quality can be maximized by choosing appropriate cultivation practices, genotypes and growing conditions, and by optimizing the processing chain. In addition, further exploring genetic variability among *Allium* populations and ecotypes may facilitate selection of germplasms with increased contents in bioactive compounds with a view to improving the quality of garlic and garlic-related foodstuffs. This is especially important in temperate regions, where asexual propagation and genetic preservation are usually easier.

CONSENT FOR PUBLICATION

Not applicable.

CONFLICT OF INTEREST

The author (editor) declares no conflict of interest, financial or otherwise.

ACKNOWLEDGEMENTS

This work was funded by FEDER under the programme Interreg V Spain-Portugal (POCTEP) 2014-2020 (ref. 0377-IBERPHENOL-6-E).

REFERENCES

[1]   Lau BHS. Suppression of LDL oxidation by garlic compounds is a possible mechanism of cardiovascular health benefit. J Nutr 2006; 136(3) (Suppl.): 765S-8S.
[http://dx.doi.org/10.1093/jn/136.3.765S] [PMID: 16484559]

[2]   Neri-Numa IA, Soriano Sancho RA, Pereira APA, Pastore GM. Small Brazilian wild fruits: Nutrients, bioactive compounds, health-promotion properties and commercial interest. Food Res Int 2018; 103: 345-60.
[http://dx.doi.org/10.1016/j.foodres.2017.10.053] [PMID: 29389624]

[3]   Perrone D, Fuggetta MP, Ardito F, *et al.* Resveratrol (3,5,4′-trihydroxystilbene) and its properties in oral diseases. Exp Ther Med 2017; 14(1): 3-9. [Review].
[http://dx.doi.org/10.3892/etm.2017.4472] [PMID: 28672886]

[4]   Zhu F, Du B, Xu B. Anti-inflammatory effects of phytochemicals from fruits, vegetables, and food legumes: A review. Crit Rev Food Sci Nutr 2017; : 1-11.
[PMID: 28605204]

[5]   Yeh YY, Yeh SM. Homocysteine-lowering action is another potential cardiovascular protective factor of aged garlic extract. J Nutr 2006; 136(3) (Suppl.): 745S-9S.
[http://dx.doi.org/10.1093/jn/136.3.745S] [PMID: 16484555]

[6]   Lanzotti V. The analysis of onion and garlic. J Chromatogr A 2006; 1112(1-2): 3-22.
[http://dx.doi.org/10.1016/j.chroma.2005.12.016] [PMID: 16388813]

[7]   Ayrle H, Mevissen M, Kaske M, Nathues H, Grützner N, Melzig M, *et al.* Medicinal plants for gastrointestinal and respiratory diseases in calves and piglets. Tierarztl Umsch 2017; 72(10): 368-77.

[8]   Tapsell LC, Hemphill I, Cobiac L, *et al.* Health benefits of herbs and spices: the past, the present, the future. Med J Aust 2006; 185(4) (Suppl.): S4-S24.
[PMID: 17022438]

[9]   Wilson DW, Nash P, Buttar HS, *et al.* The role of food antioxidants, benefits of functional foods, and influence of feeding habits on the health of the older person: An overview. Antioxidants 2017; 6(4): E81.
[http://dx.doi.org/10.3390/antiox6040081] [PMID: 29143759]

[10]   Mora N, Golden SH. Understanding Cultural Influences on Dietary Habits in Asian, Middle Eastern, and Latino Patients with Type 2 Diabetes: A Review of Current Literature and Future Directions. Curr Diab Rep 2017; 17(12): 126.
[http://dx.doi.org/10.1007/s11892-017-0952-6] [PMID: 29063419]

[11]   2017. Available from http://www.fao.org/faostat/en/#compare

[12]   Beato VM, Orgaz F, Mansilla F, Montaño A. Changes in phenolic compounds in garlic (*Allium sativum* L.) owing to the cultivar and location of growth. Plant Foods Hum Nutr 2011; 66(3): 218-23.
[http://dx.doi.org/10.1007/s11130-011-0236-2] [PMID: 21667145]

[13]   Rivlin RS. Historical perspective on the use of garlic. J Nutr 2001; 131(3s) (Suppl.): 951S-4S.
[http://dx.doi.org/10.1093/jn/131.3.951S] [PMID: 11238795]

[14]   Guercio V, Galeone C, Turati F, La Vecchia C. Gastric cancer and allium vegetable intake: a critical review of the experimental and epidemiologic evidence. Nutr Cancer 2014; 66(5): 757-73.
[http://dx.doi.org/10.1080/01635581.2014.904911] [PMID: 24820444]

[15]   Nicastro HL, Ross SA, Milner JA. Garlic and onions: their cancer prevention properties. Cancer Prev Res (Phila) 2015; 8(3): 181-9.
[http://dx.doi.org/10.1158/1940-6207.CAPR-14-0172] [PMID: 25586902]

[16]   Le Bon AM. Allium vegetables and cancer prevention. Phytotherapie 2016; 14(3): 159-64.
[http://dx.doi.org/10.1007/s10298-016-1041-8]

[17]   Beretta HV, Bannoud F, Insani M, *et al.* Relationships between bioactive compound content and the antiplatelet and antioxidant activities of six Allium vegetable species. Food Technol Biotechnol 2017; 55(2): 266-75.
[http://dx.doi.org/10.17113/ftb.55.02.17.4722] [PMID: 28867958]

[18]   Mota AH. A review of medicinal plants used in therapy of cardiovascular diseases. Int J Pharmacogn Phytochem Res 2016; 8(4): 572-91.

[19]   Shi Z, Zhang T, Byles J, Martin S, Avery JC, Taylor AW. Food habits, lifestyle factors and mortality among oldest old Chinese: The Chinese longitudinal healthy longevity survey (CLHLS). Nutrients 2015; 7(9): 7562-79.
[http://dx.doi.org/10.3390/nu7095353] [PMID: 26371039]

[20]   Huang CH, Hsu FY, Wu YH, *et al.* Analysis of lifespan-promoting effect of garlic extract by an integrated metabolo-proteomics approach. J Nutr Biochem 2015; 26(8): 808-17.
[http://dx.doi.org/10.1016/j.jnutbio.2015.02.010] [PMID: 25940980]

[21]   Berginc K, Milisav I, Kristl A. Garlic flavonoids and organosulfur compounds: impact on the hepatic pharmacokinetics of saquinavir and darunavir. Drug Metab Pharmacokinet 2010; 25(6): 521-30.
[http://dx.doi.org/10.2133/dmpk.DMPK-10-RG-053] [PMID: 20930421]

[22]   Bernaert N. Bioactive compounds in leek (Allium ampeloprasum var porrum): analysis as a function of the genetic diversity, harvest time and processing techniques. Ghent University 2013.

[23]   Yoo SR, Jeong SJ, Lee NR, Shin HK, Seo CS. Simultaneous determination and anti-inflammatory effects of four phenolic compounds in Dendrobii Herba. Nat Prod Res 2017; 31(24): 2923-6.
[http://dx.doi.org/10.1080/14786419.2017.1300798] [PMID: 28281361]

[24]   Kohoude MJ, Gbaguidi F, Agbani P, Ayedoun MA, Cazaux S, Bouajila J. Chemical composition and biological activities of extracts and essential oil of Boswellia dalzielii leaves. Pharm Biol 2017; 55(1): 33-42.
[http://dx.doi.org/10.1080/13880209.2016.1226356] [PMID: 27650786]

[25]   Fernandes I, Pérez-Gregorio R, Soares S, Mateus N, de Freitas V, Santos-Buelga C, *et al.* Wine flavonoids in health and disease prevention. Molecules 2017; 22(2): E292.
[http://dx.doi.org/10.3390/molecules22020292] [PMID: 28216567]

[26]   Dziri S, Hassen I, Fatnassi S, Mrabet Y, Casabianca H, Hanchi B, *et al.* Phenolic constituents,

antioxidant and antimicrobial activities of rosy garlic (Allium roseum var. odoratissimum). J Funct Foods 2012; 4(2): 423-32.
[http://dx.doi.org/10.1016/j.jff.2012.01.010]

[27]   Del Rio D, Costa LG, Lean MEJ, Crozier A. Polyphenols and health: what compounds are involved? Nutr Metab Cardiovasc Dis 2010; 20(1): 1-6.
[http://dx.doi.org/10.1016/j.numecd.2009.05.015] [PMID: 19713090]

[28]   Caponio F, Gomes T, Pasqualone A. Phenolic compounds in virgin olive oils: influence of the degree of olive ripeness on organoleptic characteristics and shelf-life. Eur Food Res Technol 2001; 212(3): 329-33.
[http://dx.doi.org/10.1007/s002170000268]

[29]   Es-Safi N-E, Cheynier V, Moutounet M. Implication of phenolic reactions in food organoleptic properties. J Food Compos Anal 2003; 16(5): 535-53.
[http://dx.doi.org/10.1016/S0889-1575(03)00019-X]

[30]   Williams CA, Grayer RJ. Anthocyanins and other flavonoids. Nat Prod Rep 2004; 21(4): 539-73.
[http://dx.doi.org/10.1039/b311404j] [PMID: 15282635]

[31]   Grotewold E. The science of flavonoids New York: SPringer 2006; pp. 1-273.
[http://dx.doi.org/10.1007/978-0-387-28822-2]

[32]   Santos EL, Maia BHLNS, Ferriani AP, Teixeira SD. Flavonoids: Classification, Biosynthesis and Chemical Ecology, Flavonoids.From Biosynthesis to Human Health. London: InTech 2017; pp. 3-16.
[http://dx.doi.org/10.5772/67861]

[33]   Beecher GR. Overview of dietary flavonoids: nomenclature, occurrence and intake. J Nutr 2003; 133(10): 3248S-54S.
[http://dx.doi.org/10.1093/jn/133.10.3248S] [PMID: 14519822]

[34]   Erdman JW Jr, Balentine D, Arab L, *et al.* Flavonoids and heart health: proceedings of the ILSI North America Flavonoids Workshop, May 31-June 1, 2005, Washington, DC. J Nutr 2007; 137(3) (Suppl. 1): 718S-37S.
[http://dx.doi.org/10.1093/jn/137.3.718S] [PMID: 17311968]

[35]   Miean KH, Mohamed S. Flavonoid (myricetin, quercetin, kaempferol, luteolin, and apigenin) content of edible tropical plants. J Agric Food Chem 2001; 49(6): 3106-12.
[http://dx.doi.org/10.1021/jf000892m] [PMID: 11410016]

[36]   Guo YJ, Sun LQ, Yu BY, Qi J. An integrated antioxidant activity fingerprint for commercial teas based on their capacities to scavenge reactive oxygen species. Food Chem 2017; 237: 645-53.
[http://dx.doi.org/10.1016/j.foodchem.2017.05.024] [PMID: 28764047]

[37]   Assadpour S, Nabavi SM, Nabavi SF, Dehpour AA, Ebrahimzadeh MA. *in vitro* antioxidant and antihemolytic effects of the essential oil and methanolic extract of Allium rotundum L. Eur Rev Med Pharmacol Sci 2016; 20(24): 5210-5.
[PMID: 28051246]

[38]   Vanitha T, Sumathy H, Sangeetha J, Devaki B, Vijayalakshmi K. Phytochemical analysis of Allium ascalonicum. Biomedicine 2009; 29(1): 22-5.

[39]   Fernandes I, Marques F, de Freitas V, Mateus N. Antioxidant and antiproliferative properties of methylated metabolites of anthocyanins. Food Chem 2013; 141(3): 2923-33.
[http://dx.doi.org/10.1016/j.foodchem.2013.05.033] [PMID: 23871042]

[40]   Cruz L, Fernandes I, Évora A, De Freitas V, Mateus N. Synthesis of the Main Red Wine Anthocyanin Metabolite: Malvidin-3-O-β-Glucuronide. Synlett 2017; 28(5): 593-6.
[http://dx.doi.org/10.1055/s-0036-1588673]

[41]   Pérez-Gregorio R, Simal-Gándara J. A critical review of bioactive food components, and of their functional mechanisms, biological effects and health outcomes. Curr Pharm Des 2017; 23(19): 2731-41.

[http://dx.doi.org/10.2174/1381612823666170317122913] [PMID: 28317483]

[42]   Dragović-Uzelac V, Bursać Kovačević D, Levaj B, Pedisić S, Mezak M, Tomljenović A. Polyphenols and antioxidant capacity in fruits and vegetables common in the Croatian diet. ACS Agric Conspec Sci 2009; 74(3): 175-9.

[43]   Pérez-Gregorio RM, García-Falcón MS, Simal-Gándara J, Rodrigues AS, Almeida DP. Identification and quantification of flavonoids in traditional cultivars of red and white onions at harvest. J Food Compos Anal 2010; 23(6): 592-8.
[http://dx.doi.org/10.1016/j.jfca.2009.08.013]

[44]   Rodrigues AS, Pérez-Gregorio MR, García-Falcón MS, Simal-Gándara J, Almeida DPF. Effect of meteorological conditions on antioxidant flavonoids in Portuguese cultivars of white and red onions. Food Chem 2011; 124(1): 303-8.
[http://dx.doi.org/10.1016/j.foodchem.2010.06.037]

[45]   Rodrigues AS, García-Falcón S, Simal-Gándara J, Almeida D. Effect of Meteorological Conditions on Flavonoids in Portuguese Landrace Varieties of Onion. Symposium Edible Alliaceae. 7.

[46]   Rodrigues ASP-G. Determination of flavonoids in different onion varieties Actas do 7° encontro de quActas do 7° encontro de química dos alimentosmica dos alimentos 2002; 3

[47]   Nuutila AM, Puupponen-Pimiä R, Aarni M, Oksman-Caldentey K-M. Comparison of antioxidant activities of onion and garlic extracts by inhibition of lipid peroxidation and radical scavenging activity. Food Chem 2003; 81(4): 485-93.
[http://dx.doi.org/10.1016/S0308-8146(02)00476-4]

[48]   Kavalcová P, Bystrická J, Tomáš J, Karovičová J, Kuchtová V. Evaluation and comparison of the content of total polyphenols and antioxidant activity in onion, garlic and leek. Potravinarstvo Slovak J Food Sci 2014; 8(1): 272-6.

[49]   Nuutila AM, Kammiovirta K, Oksman-Caldentey KM. Comparison of methods for the hydrolysis of flavonoids and phenolic acids from onion and spinach for HPLC analysis. Food Chem 2002; 76(4): 519-25.
[http://dx.doi.org/10.1016/S0308-8146(01)00305-3]

[50]   Miller HE, Rigelhof F, Marquart L, Prakash A, Kanter M. Antioxidant content of whole grain breakfast cereals, fruits and vegetables J Am Coll Nutr 2000; 19 (sup3): 312S-9S.

[51]   Figliuolo G, Candido V, Logozzo G, Miccolis V, Zeuli PS. Genetic evaluation of cultivated garlic germplasm (*Allium sativum* L. and A. ampeloprasum L.). Euphytica 2001; 121(3): 325-34.
[http://dx.doi.org/10.1023/A:1012069532157]

[52]   Chen S, Shen X, Cheng S, *et al.* Evaluation of garlic cultivars for polyphenolic content and antioxidant properties. PLoS One 2013; 8(11): e79730.
[http://dx.doi.org/10.1371/journal.pone.0079730] [PMID: 24232741]

[53]   Fanaei H, Narouirad M, Farzanjo M, Ghasemi M. Evaluation of Yield and Some Agronomical Traits in Garlic Genotypes (*Allium sativum* L). Annu Res Rev Biol 2014; 4(22): 3386-91.
[http://dx.doi.org/10.9734/ARRB/2014/9090]

[54]   Fratianni F, Ombra MN, Cozzolino A, Riccardi R, Spigno P, Tremonte P, *et al.* Phenolic constituents, antioxidant, antimicrobial and anti-proliferative activities of different endemic Italian varieties of garlic (*Allium sativum* L.). J Funct Foods 2016; 21: 240-8.
[http://dx.doi.org/10.1016/j.jff.2015.12.019]

[55]   Yang J, Meyers KJ, van der Heide J, Liu RH. Varietal differences in phenolic content and antioxidant and antiproliferative activities of onions. J Agric Food Chem 2004; 52(22): 6787-93.
[http://dx.doi.org/10.1021/jf0307144] [PMID: 15506817]

[56]   Sultana B, Anwar F. Flavonols (kaempferol, quercetin, myricetin) contents of selected fruits, vegetables and medicinal plants. Food Chem 2008; 108(3): 879-84.
[http://dx.doi.org/10.1016/j.foodchem.2007.11.053] [PMID: 26065748]

[57] Kwak J-H, Seo JM, Kim N-H, *et al.* Variation of quercetin glycoside derivatives in three onion (*Allium cepa* L.) varieties. Saudi J Biol Sci 2017; 24(6): 1387-91.
[http://dx.doi.org/10.1016/j.sjbs.2016.05.014] [PMID: 28855836]

[58] Lachowicz S, Kolniak-Ostek J, Oszmiański J, Wiśniewski R. Comparison of phenolic content and antioxidant capacity of bear garlic (Allium ursinum L.) in different maturity stages. J Food Process Preserv 2017; 41(1): 1-10.
[http://dx.doi.org/10.1111/jfpp.12921]

[59] Locatelli DA, Nazareno MA, Fusari CM, Camargo AB. Cooked garlic and antioxidant activity: Correlation with organosulfur compound composition. Food Chem 2017; 220: 219-24.
[http://dx.doi.org/10.1016/j.foodchem.2016.10.001] [PMID: 27855892]

[60] Martins N, Petropoulos S, Ferreira IC. Chemical composition and bioactive compounds of garlic (*Allium sativum* L.) as affected by pre- and post-harvest conditions: A review. Food Chem 2016; 211: 41-50.
[http://dx.doi.org/10.1016/j.foodchem.2016.05.029] [PMID: 27283605]

[61] Csiszár J, Lantos E, Tari I, Madosa E, Wodala B, Vashegyi Á, *et al.* Antioxidant enzyme activities in Allium species and their cultivars under water stress. Plant Soil Environ 2007; 53(12): 517-23.
[http://dx.doi.org/10.17221/2192-PSE]

[62] Ghasemi K, Bolandnazar S, Tabatabaei S, Pirdashti H, Arzanlou M, Ebrahimzadeh M, *et al.* Antioxidant properties of garlic as affected by selenium and humic acid treatments. New Zeal J Crop Hort 2015; 43(3): 173-81.
[http://dx.doi.org/10.1080/01140671.2014.991743]

[63] Põldma P, Tõnutare T, Viitak A, Luik A, Moor U. Effect of selenium treatment on mineral nutrition, bulb size, and antioxidant properties of garlic (*Allium sativum* L.). J Agric Food Chem 2011; 59(10): 5498-503.
[http://dx.doi.org/10.1021/jf200226p] [PMID: 21495721]

[64] Siddiqui S, Gupta K, Yadav A, Mangal J. Soil salinity effect on soluble saccharides, phenol, fatty acid and mineral contents, and respiration rate of garlic cultivars. Biol Plant 1996; 38(4): 611-5.
[http://dx.doi.org/10.1007/BF02890618]

[65] Rodrigues AS, Pérez-Gregorio MR, García-Falcón MS, Simal-Gándara J, Almeida DPF. Effect of post-harvest practices on flavonoid content of red and white onion cultivars. Food Control 2010; 21(6): 878-84.
[http://dx.doi.org/10.1016/j.foodcont.2009.12.003]

[66] Golisová A, Slamka P, Kóňa J. Content of flavonoids in onion (*Allium cepa* L.) under various fertilization. Acta Hortic Reg 2008; 11(2): 54-6.

[67] Golisová A, Slamka P, Ložek O, Hanáčková E. Content of phenols in onion (*Allium cepa* L.) fertilized with nitrogen, sulphur and iron. Agrochémia 2009; 13(3): 14-8.

[68] Mogren LM, Olsson ME, Gertsson UE. Quercetin content in stored onions (*Allium cepa* L.): effects of storage conditions, cultivar, lifting time and nitrogen fertiliser level. J Sci Food Agric 2007; 87(8): 1595-602.
[http://dx.doi.org/10.1002/jsfa.2904]

[69] Mogren LM, Olsson ME, Gertsson UE. Effects of cultivar, lifting time and nitrogen fertiliser level on quercetin content in onion (*Allium cepa* L.) at lifting. J Sci Food Agric 2007; 87(3): 470-6.
[http://dx.doi.org/10.1002/jsfa.2735]

[70] Perner H, Rohn S, Driemel G, *et al.* Effect of nitrogen species supply and mycorrhizal colonization on organosulfur and phenolic compounds in onions. J Agric Food Chem 2008; 56(10): 3538-45.
[http://dx.doi.org/10.1021/jf073337u] [PMID: 18457399]

[71] Ren F, Reilly K, Kerry JP, Gaffney M, Hossain M, Rai DK. Higher Antioxidant Activity, Total Flavonols, and Specific Quercetin Glucosides in Two Different Onion (*Allium cepa* L.) Varieties

Grown under Organic Production: Results from a 6-Year Field Study. J Agric Food Chem 2017; 65(25): 5122-32.
[http://dx.doi.org/10.1021/acs.jafc.7b01352] [PMID: 28612608]

[72]   Mirzaei R, Liaghati H, Damghani AM. Evaluating yield quality and quantity of garlic as affected by different farming systems and garlic clones. Pak J Biol Sci 2007; 10(13): 2219-24.
[http://dx.doi.org/10.3923/pjbs.2007.2219.2224] [PMID: 19070185]

[73]   Tiwari U, Cummins E. Factors influencing levels of phytochemicals in selected fruit and vegetables during pre-and post-harvest food processing operations. Food Res Int 2013; 50(2): 497-506.
[http://dx.doi.org/10.1016/j.foodres.2011.09.007]

[74]   Rodrigues AS, Pérez-Gregorio MR, García-Falcón MS, Simal-Gándara J. Effect of curing and cooking on flavonols and anthocyanins in traditional varieties of onion bulbs. Food Res Int 2009; 42(9): 1331-6.
[http://dx.doi.org/10.1016/j.foodres.2009.04.005]

[75]   Naheed Z, Cheng Z, Wu C, Wen Y, Ding H. Total polyphenols, total flavonoids, allicin and antioxidant capacities in garlic scape cultivars during controlled atmosphere storage. Postharvest Biol Technol 2017; 131: 39-45.
[http://dx.doi.org/10.1016/j.postharvbio.2017.05.002]

[76]   Fei ML, Tong L, Wei L, De Yang L. Changes in antioxidant capacity, levels of soluble sugar, total polyphenol, organosulfur compound and constituents in garlic clove during storage. Ind Crops Prod 2015; 69: 137-42.
[http://dx.doi.org/10.1016/j.indcrop.2015.02.021]

[77]   Siddiqui MW, Chakraborty I, Ayala-Zavala J, Dhua R. Advances in minimal processing of fruits and vegetables: a review 2011; 70(10): 823-34.

[78]   Benkeblia N, Shiomi N. Hydrolysis kinetic parameters of DP 6, 7, 8, and 9-12 fructooligosaccharides (FOS) of onion bulb tissues. effect of temperature and storage time. J Agric Food Chem 2006; 54(7): 2587-92.
[http://dx.doi.org/10.1021/jf052848i] [PMID: 16569048]

[79]   Atashi S, Akbarpour V, Mashayekhi K, Mousavizadeh SJ. Garlic physiological characteristics from harvest to sprouting in response to low temperature. J Stored Prod Postharvest Res 2011; 2(15): 285-91.

[80]   Khanum F, Anilakumar KR, Viswanathan KR. Anticarcinogenic properties of garlic: a review. Crit Rev Food Sci Nutr 2004; 44(6): 479-88.
[http://dx.doi.org/10.1080/10408690490886700] [PMID: 15615431]

[81]   Rahman K. Effects of garlic on platelet biochemistry and physiology. Mol Nutr Food Res 2007; 51(11): 1335-44.
[http://dx.doi.org/10.1002/mnfr.200700058] [PMID: 17966136]

[82]   Cantwell M, Hong G, Kang J, Nie X. Controlled atmospheres retard sprout growth, affect compositional changes, and maintain visual quality attributes of garlic. Acta Hortic 2003; (600): 791-4.
[http://dx.doi.org/10.17660/ActaHortic.2003.600.122]

[83]   Amagase H. Clarifying the real bioactive constituents of garlic. J Nutr 2006; 136(3) (Suppl.): 716S-25S.
[http://dx.doi.org/10.1093/jn/136.3.716S] [PMID: 16484550]

[84]   Rodrigues AS, Almeida DPF, García-Falcón MS, Simal-Gándara J, Pérez-Gregorio MR. Postharvest storage systems affect phytochemical content and quality of traditional Portuguese onion cultivars. Acta Hortic 2012; (934): 1327-34.
[http://dx.doi.org/10.17660/ActaHortic.2012.934.180]

[85]   Pérez-Gregorio MR, Regueiro J, González-Barreiro C, Rial-Otero R, Simal-Gándara J. Changes in

antioxidant flavonoids during freeze-drying of red onions and subsequent storage. Food Control 2011; 22(7): 1108-13.
[http://dx.doi.org/10.1016/j.foodcont.2011.01.006]

[86] Pérez-Gregorio MR, González-Barreiro C, Rial-Otero R, Simal-Gándara J. Comparison of sanitizing technologies on the quality appearance and antioxidant levels in onion slices. Food Control 2011; 22(12): 2052-8.
[http://dx.doi.org/10.1016/j.foodcont.2011.05.028]

[87] Pérez-Gregorio MR, García-Falcón MS, Simal-Gándara J. Flavonoids changes in fresh-cut onions during storage in different packaging systems. Food Chem 2011; 124(2): 652-8.
[http://dx.doi.org/10.1016/j.foodchem.2010.06.090]

[88] Szymanek M. Effects of blanching on some physical properties and processing recovery of sweet corn cobs. Food Bioprocess Technol 2011; 4(7): 1164-71.
[http://dx.doi.org/10.1007/s11947-009-0246-3]

[89] Kinalski T, Noreña CPZ. Effect of blanching treatments on antioxidant activity and thiosulfinate degradation of garlic (*Allium sativum* L.). Food Bioprocess Technol 2014; 7(7): 2152-7.
[http://dx.doi.org/10.1007/s11947-014-1282-1]

[90] Lemar KM, Turner MP, Lloyd D. Garlic (*Allium sativum*) as an anti-Candida agent: a comparison of the efficacy of fresh garlic and freeze-dried extracts. J Appl Microbiol 2002; 93(3): 398-405.
[http://dx.doi.org/10.1046/j.1365-2672.2002.01707.x] [PMID: 12174037]

[91] Chudzik B, Malm A, Rajtar B, Kołodziej S, Polz-Dacewicz MA. The fresh extracts of Allium species as potential *in vitro* agents against planktonic and adherent cells of Candida spp. Ann Univ Mariae Curie Sklodowska Med 2010; 23(1): 73-8.

[92] Belguith H, Kthiri F, Chati A, Sofah AA, Hamida JB, Ladoulsi A. Inhibitory effect of aqueous garlic extract (*Allium sativum*) on some isolated Salmonella serovars. Afr J Microbiol Res 2010; 4(5): 328-38.

[93] Gorinstein S, Jastrzebski Z, Leontowicz H, Leontowicz M, Namiesnik J, Najman K, *et al.* Comparative control of the bioactivity of some frequently consumed vegetables subjected to different processing conditions. Food Control 2009; 20(4): 407-13.
[http://dx.doi.org/10.1016/j.foodcont.2008.07.008]

[94] Pellegrini C, Cróci C, Orioli G. Morphological changes induced by different doses of gamma irradiation in garlic sprouts. Radiat Phys Chem 2000; 57(3): 315-8.
[http://dx.doi.org/10.1016/S0969-806X(99)00397-7]

[95] Pérez MB, Aveldano MI, Croci CA. Growth inhibition by gamma rays affects lipids and fatty acids in garlic sprouts during storage. Postharvest Biol Technol 2007; 44(2): 122-30.
[http://dx.doi.org/10.1016/j.postharvbio.2006.08.018]

[96] Thomas P. Control of post-harvest loss of grain, fruits and vegetable by radiation processing. Irradiat Food Saf Qual 2001; 93: 24-6.

[97] Reyes LF, Cisneros-Zevallos L. Wounding stress increases the phenolic content and antioxidant capacity of purple-flesh potatoes (Solanum tuberosum L.). J Agric Food Chem 2003; 51(18): 5296-300.
[http://dx.doi.org/10.1021/jf034213u] [PMID: 12926873]

[98] Ke D, Saltveit ME. Wound-induced ethylene production, phenolic metabolism and susceptibility to russet spotting in iceberg lettuce. Physiol Plant 1989; 76(3): 412-8.
[http://dx.doi.org/10.1111/j.1399-3054.1989.tb06212.x]

[99] Cisneros-Zevallos L. The Use of Controlled Postharvest Abiotic Stresses as a Tool for Enhancing the Nutraceutical Content and Adding-Value of Fresh Fruits and Vegetables. J Food Sci 2003; 68(5): 1560-5.
[http://dx.doi.org/10.1111/j.1365-2621.2003.tb12291.x]

[100] Cantos E, Espín JC, Tomás-Barberán FA. Effect of wounding on phenolic enzymes in six minimally processed lettuce cultivars upon storage. J Agric Food Chem 2001; 49(1): 322-30.
[http://dx.doi.org/10.1021/jf000644q] [PMID: 11170594]

[101] Reyes LF, Villarreal JE, Cisneros-Zevallos L. The increase in antioxidant capacity after wounding depends on the type of fruit or vegetable tissue. Food Chem 2007; 101(3): 1254-62.
[http://dx.doi.org/10.1016/j.foodchem.2006.03.032]

[102] Feng X, Liu W. Variation of quercetin content in different tissues of welsh onion (*Allium fistulosum* L.). Afr J Agric Res 2011; 6(26): 5675-9.

[103] Qureshi AA, Lawande KE, Mani S, Patil VB. Colour and tissue differences in distribution of quercetin in Indian onions (*Allium cepa*). Indian J Agric Sci 2012; 82(7): 629-31.

[104] Chen C, Hu W, Zhang R, Jiang A, Zou Y. Levels of phenolic compounds, antioxidant capacity, and microbial counts of fresh-cut onions after treatment with a combination of nisin and citric acid. Hortic Environ Biotechnol 2016; 57(3): 266-73.
[http://dx.doi.org/10.1007/s13580-016-0032-x]

[105] Berno ND, Tezotto-Uliana JV, dos Santos Dias CT, Kluge RA. Storage temperature and type of cut affect the biochemical and physiological characteristics of fresh-cut purple onions. Postharvest Biol Technol 2014; 93: 91-6.
[http://dx.doi.org/10.1016/j.postharvbio.2014.02.012]

[106] AMS [U.S. Department of Agriculture AMS. Quality through verification program for the fresh-cut produce industry. Fed Regist 1998; 63: 47220-4.

[107] Ölmez H, Kretzschmar U. Potential alternative disinfection methods for organic fresh-cut industry for minimizing water consumption and environmental impact. Lebensm Wiss Technol 2009; 42(3): 686-93.
[http://dx.doi.org/10.1016/j.lwt.2008.08.001]

[108] Zhang J, Yang H. Effects of potential organic compatible sanitisers on organic and conventional fresh-cut lettuce (Lactuca sativa Var. Crispa L). Food Control 2017; 72: 20-6.
[http://dx.doi.org/10.1016/j.foodcont.2016.07.030]

[109] Beerli KMC, Boas V, de Barros EV, Piccoli RH. Effect of sanitizers on the microbial, physical and physical-chemical characteristics of fresh-cut onions (*Allium cepa* L). Cienc Agrotec 2004; 28(1): 107-12.
[http://dx.doi.org/10.1590/S1413-70542004000100014]

[110] Stevens KA, Sheldon BW, Klapes NA, Klaenhammer TR. Nisin treatment for inactivation of Salmonella species and other gram-negative bacteria. Appl Environ Microbiol 1991; 57(12): 3613-5.
[PMID: 1785933]

[111] Carletti L, Botondi R, Moscetti R, Stella E, Monarca D, Cecchini M, *et al.* Use of ozone in sanitation and storage of fresh fruits and vegetables. J Food Agric Environ 2013; 11(3-4): 585-9.

[112] Ferreres F, Gil MI, Tomas-Barberan FA. Anthocyanins and flavonoids from shredded red onion and changes during storage in perforated films. Food Res Int 1996; 29(3): 389-95.
[http://dx.doi.org/10.1016/0963-9969(96)00002-6]

[113] Locatelli D, Altamirano J, González R, Camargo A. Home-cooked garlic remains a healthy food. J Funct Foods 2015; 16: 1-8.
[http://dx.doi.org/10.1016/j.jff.2015.04.012]

[114] Queiroz YS, Antunes PB, Vicente SJ, Sampaio GR, Shibao J, Bastos DH, *et al.* Bioactive compounds, *in vitro* antioxidant capacity and Maillard reaction products of raw, boiled and fried garlic (*Allium sativum* L.). Int J Food Sci Technol 2014; 49(5): 1308-14.
[http://dx.doi.org/10.1111/ijfs.12428]

[115] Cavagnaro PF, Camargo A, Galmarini CR, Simon PW. Effect of cooking on garlic (*Allium sativum* L.)

antiplatelet activity and thiosulfinates content. J Agric Food Chem 2007; 55(4): 1280-8.
[http://dx.doi.org/10.1021/jf062587s] [PMID: 17256959]

[116]   Song K, Milner JA. The influence of heating on the anticancer properties of garlic. J Nutr 2001;
131(3s): 1054S-7S.
[http://dx.doi.org/10.1093/jn/131.3.1054S] [PMID: 11238815]

[117]   Park J-H, Park YK, Park E. Antioxidative and antigenotoxic effects of garlic (*Allium sativum* L.)
prepared by different processing methods. Plant Foods Hum Nutr 2009; 64(4): 244-9.
[http://dx.doi.org/10.1007/s11130-009-0132-1] [PMID: 19711184]

[118]   Higuchi O, Tateshita K, Nishimura H. Antioxidative activity of sulfur-containing compounds in
Allium species for human low-density lipoprotein (LDL) oxidation *in vitro*. J Agric Food Chem 2003;
51(24): 7208-14.
[http://dx.doi.org/10.1021/jf034294u] [PMID: 14611195]

[119]   Elosta A, Slevin M, Rahman K, Ahmed N. Aged garlic has more potent antiglycation and antioxidant
properties compared to fresh garlic extract *in vitro*. Sci Rep 2017; 7: 39613.
[http://dx.doi.org/10.1038/srep39613] [PMID: 28051097]

[120]   Kim J-S, Kang O-J, Gweon O-C. Comparison of phenolic acids and flavonoids in black garlic at
different thermal processing steps. J Funct Foods 2013; 5(1): 80-6.
[http://dx.doi.org/10.1016/j.jff.2012.08.006]

[121]   Lombard K, Peffley E, Geoffriau E, Thompson L, Herring A. Quercetin in onion (*Allium cepa* L.) after
heat-treatment simulating home preparation. J Food Compos Anal 2005; 18(6): 571-81.
[http://dx.doi.org/10.1016/j.jfca.2004.03.027]

[122]   Jastrzebski Z, Leontowicz H, Leontowicz M, *et al.* The bioactivity of processed garlic (*Allium sativum*
L.) as shown *in vitro* and *in vivo* studies on rats. Food Chem Toxicol 2007; 45(9): 1626-33.
[http://dx.doi.org/10.1016/j.fct.2007.02.028] [PMID: 17408832]

[123]   Kimura S, Tung Y-C, Pan M-H, Su N-W, Lai Y-J, Cheng K-C. Black garlic: A critical review of its
production, bioactivity, and application. Yao Wu Shi Pin Fen Xi 2017; 25(1): 62-70.
[http://dx.doi.org/10.1016/j.jfda.2016.11.003] [PMID: 28911544]

[124]   Choi IS, Cha HS, Lee YS. Physicochemical and antioxidant properties of black garlic. Molecules
2014; 19(10): 16811-23.
[http://dx.doi.org/10.3390/molecules191016811] [PMID: 25335109]

[125]   Ozyurt D, Goc B, Demirata B, Apak R. Effect of Oven and Microwave Heating on the Total
Antioxidant Capacity of Dietary Onions Grown in Turkey. Int J Food Prop 2013; 16(3): 536-48.
[http://dx.doi.org/10.1080/10942912.2011.555900]

[126]   Ewald C, Fjelkner-Modig S, Johansson K, Sjöholm I, Åkesson B. Effect of processing on major
flavonoids in processed onions, green beans, and peas. Food Chem 1999; 64(2): 231-5.
[http://dx.doi.org/10.1016/S0308-8146(98)00136-8]

[127]   Harris S, Brunton N, Tiwari U, Cummins E. Human exposure modelling of quercetin in onions
(*Allium cepa* L.) following thermal processing. Food Chem 2015; 187: 135-9.
[http://dx.doi.org/10.1016/j.foodchem.2015.04.035] [PMID: 25977008]

[128]   Islek M, Nilufer-Erdil D, Knuthsen P. Changes in Flavonoids of Sliced and Fried Yellow Onions
(*Allium cepa* L. var. zittauer) During Storage at Different Atmospheric, Temperature and Light
Conditions. J Food Process Preserv 2015; 39(4): 357-68.
[http://dx.doi.org/10.1111/jfpp.12240]

[129]   Juániz I, Ludwig IA, Huarte E, *et al.* Influence of heat treatment on antioxidant capacity and
(poly)phenolic compounds of selected vegetables. Food Chem 2016; 197(Pt A): 466-73.
[http://dx.doi.org/10.1016/j.foodchem.2015.10.139] [PMID: 26616976]

[130]   Makris DP, Rossiter JT. Domestic processing of onion bulbs (*Allium cepa*) and asparagus spears
(Asparagus officinalis): effect on flavonol content and antioxidant status. J Agric Food Chem 2001;

49(7): 3216-22.
[http://dx.doi.org/10.1021/jf001497z] [PMID: 11453754]

[131] Rodrigues A, Pérez-Gregorio M, García-Falcón M, Simal-Gándara J. Effect of curing and cooking on flavonols and anthocyanins in traditional varieties of onion bulbs. Food Res Int 2009; 42(9): 1331-6.
[http://dx.doi.org/10.1016/j.foodres.2009.04.005]

[132] Ross JA, Kasum CM. Dietary flavonoids: bioavailability, metabolic effects, and safety. Annu Rev Nutr 2002; 22(1): 19-34.
[http://dx.doi.org/10.1146/annurev.nutr.22.111401.144957] [PMID: 12055336]

[133] Pinho C, Soares MT, Almeida IF, Aguiar AARM, Mansilha C, Ferreira IMPLVO. Impact of freezing on flavonoids/radical-scavenging activity of two onion varieties. Czech J Food Sci 2015; 33(4): 340-5.
[http://dx.doi.org/10.17221/704/2014-CJFS]

[134] Alezandro MR, Lui MCY, Lajolo FM, Genovese MI. Commercial spices and industrial ingredients: Evaluation of antioxidant capacity and flavonoids content for functional foods development. Cienc Tecnol Alime 2011; 31(2): 527-33.
[http://dx.doi.org/10.1590/S0101-20612011000200038]

[135] Sahoo NR, Bal LM, Pal US, Sahoo D. Impact of pretreatment and drying methods on quality attributes of onion shreds. Food Technol Biotechnol 2015; 53(1): 57-65.
[http://dx.doi.org/10.17113/ftb.53.01.15.3598] [PMID: 27904332]

[136] Pérez-Gregorio M, Regueiro J, González-Barreiro C, Rial-Otero R, Simal-Gándara J. Changes in antioxidant flavonoids during freeze-drying of red onions and subsequent storage. Food Control 2011; 22(7): 1108-13.
[http://dx.doi.org/10.1016/j.foodcont.2011.01.006]

[137] Lenkova M, Bystrická J, Tomáš T, Hrstkova M. Evaluation and comparison of the content of total polyphenols and antioxidant activity of selected species of the genus Allium. J Cent Eur Agric 2016; 17(4)

[138] Bozin B, Mimica-Dukic N, Samojlik I, Goran A, Igic R. Phenolics as antioxidants in garlic (*Allium sativum* L., Alliaceae). Food Chem 2008; 111(4): 925-9.
[http://dx.doi.org/10.1016/j.foodchem.2008.04.071] [PMID: 26050009]

[139] Ninfali P, Mea G, Giorgini S, Rocchi M, Bacchiocca M. Antioxidant capacity of vegetables, spices and dressings relevant to nutrition. Br J Nutr 2005; 93(2): 257-66.
[http://dx.doi.org/10.1079/BJN20041327] [PMID: 15788119]

[140] Chekki RZ, Snoussi A, Hamrouni I, Bouzouita N. Chemical composition, antibacterial and antioxidant activities of Tunisian garlic (*Allium sativum*) essential oil and ethanol extract. Mediterr J Chem 2014; 3(4): 947-56.
[http://dx.doi.org/10.13171/mjc.3.4.2014.09.07.11]

[141] Cai Y, Luo Q, Sun M, Corke H. Antioxidant activity and phenolic compounds of 112 traditional Chinese medicinal plants associated with anticancer. Life Sci 2004; 74(17): 2157-84.
[http://dx.doi.org/10.1016/j.lfs.2003.09.047] [PMID: 14969719]

CHAPTER 3

Beans (*Phaseolus vulgaris* L.) as a Source of Natural Antioxidants

Ryszard Amarowicz[1,*] and Ronald B. Pegg[2]

[1] *Institute of Animal Reproduction and Food Research of the Polish Academy of Sciences, Tuwima St. 10, 10-748 Olsztyn, Poland*

[2] *Department of Food Science and Technology, College of Agricultural and Environmental Sciences, The University of Georgia, 100 Cedar Street, Athens, GA30602-2610, USA*

Abstract: Beans are a staple food in many Latin American and African countries, as well as an important foodstuff in vegetarian diets. This chapter provides an update on the most recent scientific literature pertaining to the phenolics of different types of edible beans, their free radical-scavenging and antioxidant capacities, and how processing and germination affect the endogenous phenolics of beans. Indicated as well are the findings reported from a myriad of *in vitro* antioxidant assays one can perform to characterize the antioxidant potential of an extract prepared from the different bean types. It is noteworthy that variability in antioxidant determinations of the phenolics from bean crude extracts can exist, as the results are greatly impacted by the extraction methodologies employed and the details of the colorimetric or HPLC assays performed.

Keywords: Antioxidant Activity, Beans, Flavonoids, Germination, Legumes, *Phaseolus Vulgaris*, Phenolic Compounds, Phenolic Acids, Tannins, Technological Process.

INTRODUCTION

The seeds of common bean (*Phaseolus vulgaris* L.) are an important source of nutrients such as protein, starch, dietary fiber, and minerals [1 - 3]. They are widely consumed throughout the world and are recognized as a staple foodstuff in many Latin American and African countries [1, 4]. Beans contain a number of bioactive compounds including phenolics, which possess marked antioxidant properties, and are very important from nutritional and technological points of view [5].

* **Corresponding author Ryszard Amarowicz:** Institute of Animal Reproduction and Food research, Polish Academy of Sciences, Olsztyn, Poland; Tel: +48 89 523 46 27; E-mail: r.amarowicz@pan.olsztyn.pl

Spyridon A. Petropoulos, Isabel C.F.R. Ferreira and Lillian Barros (Eds.)

Research findings suggest that oxidative stress is closely associated with a diverse assortment of diseases like cancer and cardiovascular disease. Consumption of beans can potentially offer beneficial effects toward human health [6, 7]. Technological processing and seed germination can decidedly impact the level of phenolic compounds in leguminous seeds. An important point for consideration also must be the high content of phenolic compounds present in bean seed coats.

This book chapter reviews the chemical profile and level of phenolic acids, flavonoids, and tannins found in various types of bean seeds. The antioxidant activity and free radical-scavenging capacity of these beans or their solvent extracts are described. Finally, the influence of technological processing and germination on the content of phenolic compounds endogenous to beans and their antioxidative capacities are discussed.

Content of Total Phenolics, Flavonoids, and Tannins in Bean Seeds

The total phenolics content (TPC), total flavonoids content (TFC) and tannins content in bean seeds inform us about the potential antioxidant capacity of the seeds or antioxidant activity of extracts derived therefrom. The determination of phenolic compounds in bean seeds includes a key extraction step, followed by a colorimetric reaction. For the TPC determination, the extracted phenolic constituents react under alkaline conditions with Folin-Ciocalteu's phenol reagent [8]. The absorbance reading of the chromogen formed, likely a Keggin structure, are reported as the quantity of equivalents of standard compounds (*i.e.*, typically Gallic acid or (+)-catechin) per mass unit of raw material or extract. For the TFC determination, flavonoid-aluminium chloride ($AlCl_3$) complexation is applied [9].

Condensed tannins (*i.e.*, proanthocyanidins or PACs) are flavan-3-ol-based biopolymers that, at high temperature in alcohol solutions of strong mineral acids, release anthocyanidins and catechins as terminal end groups. The most common method employed for condensed tannins analysis in beans is the vanillin/HCl method [10]. The basic chemical structure of a PAC is depicted in (Fig. **1**).

The types of solvent used for extraction of phenolic compounds from beans can vary markedly and have included water [11]; methanol [12]; 0.16 M HCl in 80% methanol [13]; methanol + 0.05% HCl [14]; 70% methanol [15]; 80% methanol [16 - 18]; 60% ethanol [19]; 70% ethanol [20]; and 70% acetone [21]. Furthermore, Veggi *et al.* [22] obtained phenolic compounds from bean seeds employing supercritical and subcritical extraction using pure CO_2, CO_2 with an ethanol modifier, and CO_2 with water. For the extraction of phenolics from red kidney beans and black beans, Xu and Chang [23] tried 50% acetone; 80% acetone; acidic (+0.5% acetic acid) 70% acetone; 70% methanol; 70% ethanol; and 100% ethanol. From their study, the highest contents of total phenolics, flavonoids, and condensed tannins were obtained when acidic 70% acetone was the solvent system employed.

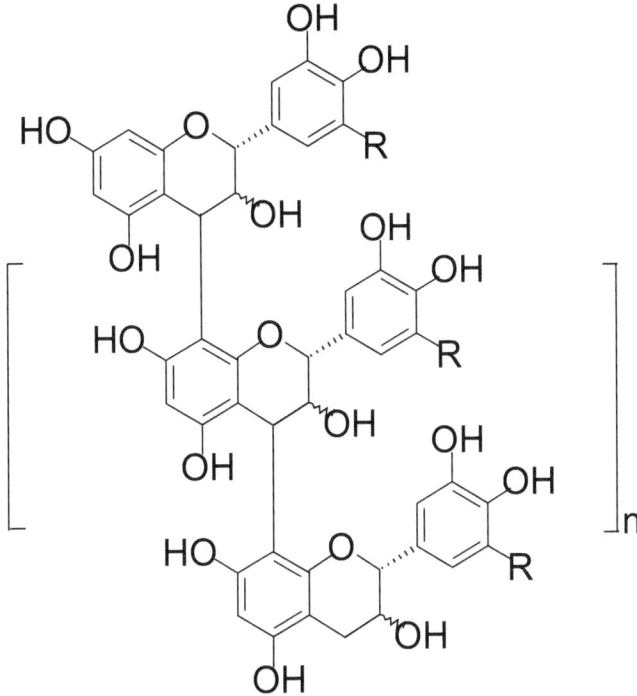

Fig. (1). Chemical structure of condensed tannins (proanthocyanidins/PACs), where represents residues typically from 2 to 14.

The TPC in beans and extracts derived therefrom are reported in (Table **1**). It is clear when examining the data that wide variations exist in the TPC, depending on the variety of bean investigated. Unquestionably the choice of solvent system for extraction of the phenolics also influenced the observed variations in the TPC of leguminous seeds as well as how the seeds had been processed prior to extraction.

Table 1. Content of total phenolics found in selected bean seeds and their extracts.

Material	Unit	Content or Range	Reference
Black turtle bean extract and its fractions	mg GAE/g extract mg GAE/g fraction	60.03 ± 0.28 15.28 - 599.22	[27]
Extracts of 12 common beans from Italy	µg GAE/g extract	135 - 1250	[27]
23 common beans from India	mg GAE/ 100 g seeds	14.54 - 62.66	[4]

(Table 1) cont.....

Material	Unit	Content or Range	Reference
Black bean Kidney bean Pearl bean Red bean Red kidney bean	mg GAE/g seeds DW	1.540 ± 0.007 1.191 ± 0.005 1.190 ± 0.013 0.690 ± 0.007 0.942 ± 0.007	[29]
Red kidney bean	mg GAE/g seeds	2.38 ± 0.12	[30]
Purple variety Green variety Yellow variety	mg GAE/100 g seeds FW	23.2 ± 0.12 8.30 ± 0.11 7.81 ± 0.15	[31]
Coats of two black bean cultivars	mg GAE/g coat	15.83 ± 0.21 5.63 ± 0.18	[32]
10 varieties of white bean from Turkey	mg GAE/g extract mg GAE/g seeds	$2.79 - 4.91$ $0.33 - 0.63$	[33]
26 varieties of bean from Mexico	mg GAE/g seeds DW	$1.3 - 5.4$	[34]
Two cranberry bean varieties	µg/g seeds DW (results of HPLC analysis)	196.2 ± 14.21 1211.8 ± 75.03	[35]
Kidney bean	g/kg seeds DW	3.92 ± 0.12	[36]
Dry common beans from Italy (12 samples)	mg GAE/g seeds DW	$1.17 - 4.40$	[20]
Bean hulls Dehulled beans	mg CE/g extract mg CE/g seeds mg CE/g extract mg CE/g seeds	56.31 2.55 2.08 1.32	[37]
Black bean Navy bean Pinto bean Red kidney bean Red bean Small red bean	mg GAE/100 g FW	8.50 2.33 10.23 12.47 11.85 2.44	[24]
Black bean, whole seed Black bean, cotyledon Black bean, seed coat	mg GAE/100 g FW	213 ± 20 43 ± 2 2971 ± 82	[25]
Soaked and autoclaved seeds	g TAE/kg seeds DM	1.49 ± 0.01	[38]
Black bean (cooked)	mg GAE/g seeds DM	1.2 ± 0.05	[39]
9 common beans from Ital	mg GAE/g seeds DM	$9.8 - 32.0$	[40]
Two common beans from Mexico	mg GAE/g seeds DM	57.1; 51.6 33.4; 31.7	[41]

(Table 1) cont.....

Material	Unit	Content or Range	Reference
Black turtle Eclipse Black turtle T-39 Navy bean Pinto bean Red kidney Pink bean Small red	mg GAE/g seeds	6.98 ± 0.48 3.37 ± 0.15 0.57 ± 0.05 3.76 ± 0.06 4.05 ± 0.05 3.77 ± 0.19 5.76 ± 0.38	[42]
Small red bean Red kidney bean Black kidney bean Navy bean	mg GAE/g extract	45.7 ± 1.8 27.1 ± 3.0 32.9 ± 0.1 11.6 ± 0.01	[19]
7 Common beans from Mexico Extract (70% acetone) Extract (50% methanol)	mg CE/g extract	$18.88 - 25.35$ $11.23 - 16.94$	[43]
Beans from Brazil	mg GAE/ g extract	164 ± 26; 93 ± 12; 338 ± 3	[22]
Red bean Brown bean Black bean	mg CE/g extract	93.6 91.4 44.0	[44]
White bean extract fractions	mg CE/g fraction	$1.11 - 6.69$	[17]
White bean	mg CE/g extract	1.01	[45]
Seed coat of Brazilian and Peruvian bean cultivars	mg CE/ g seed coat	$0.46 - 86$	[15]

Abbreviations: CE, (+)-catechin equivalents; QU, quercetin equivalents; RE, rutin equivalents; equivalents; DW, dry weight.

The highest quantities of TPC were reported by Mojica *et al.* [24] for red kidney bean, red bean, and pinto bean. Noteworthy is that the content of phenolics in seed coats is significantly greater than that available in the endosperm/cotyledon [25, 26].

The TFC in beans also varied across a broad range, as evident in (Table **2**). Column chromatography of crude extracts helps to obtain individual fractions rich in flavonoids [27]. A high content of flavonoids was determined in Brazilian bean [12] and in its seed coats [32]. An important class of flavonoid present in beans with colour is anthocyanins. The anthocyanins contents of bean extracts are summarized in (Table **3**). According to HPLC data, the main anthocyanins existing in beans are cyanidin-3-*O*-glucoside, pelargonidin-3-*O*-glucoside, and pelargonidin-3-*O*-malonylglucoside [16]. In beans, the anthocyanins are located mainly in seed coats [15, 24].

Table 2. Content of total flavonoids in bean seeds and their extracts.

Material	Unit	Content	Reference
Black turtle bean and its fractions	mg CE/g extract mg CE/g fraction	70.21 ± 0.53 15.31 - 295.31	[27]
Extracts of 12 common bean from Italy	µg QE/g extract	51.5 - 925.7	[21]
Coats of two black bean cultivars	mg RE/g coat	1.09 ± 0.15 1.53 ± 0.52	[32]
Dry common beans from Italy (12 samples)	mg CE/g seeds DW	0.22 - 1.43	[20]
Coats of 15 bean cultivars from Mexico and Brazil	mg RUE/g coat	0.083 - 0.694	[24]
Two common beans from Mexico	mg RUE/g seeds DM	7.43; 7.00 4.34; 4.19	[41]
Black turtle Eclipse Black turtle T-39 Navy bean Pinto bean Red kidney Pink bean Small red	mg CE/ g seeds	3.30 ± 0.11 2.51 ± 0.12 0.92 ± 0.02 2.99 ± 0.12 3.39 ± 0.09 3.65 ± 0.13 4.24 ± 0.10	[42]
Bean from Brazil	mg CE/g extract	143 ± 6; 138 ± 0.6; 109 ± 10	[22]

Abbreviations: CE, (+)-catechin equivalents; QU, quercetin equivalents; RE, rutin equivalents; equivalents; DW, dry weight.

Table 3. Content of anthocyanins in bean seeds and their extracts.

Material/compound	Unit	Content	Reference
Extracts of 12 common bean from Italy	µg C3G/g extract	0 - 62.3	[21]
Coats of two black bean cultivars	mg/g extract (results of HPLC analysis)	1.934 1.148	[32]
26 varieties of bean from Mexico	mg C3G/g seeds DW	1.94 - 3.47	[34]
Coats of 15 bean cultivars from Mexico and Brazil	mg/100 g coat (results of HPLC analysis)	1.0 – 250.0	[24]
Black bean from Mexico	mg/100 g bean flour (result of HPLC analysis)	0.41 ± 0.02	[46]
Black bean	mg/100 g seeds FW	213 ± 2	[47]
Kidney bean: Cyanidyn 3-*O*-glucoside Pelargonidyn 3-*O*-glucoside Pelargonidin 3-*O*-malonylglucoside Malvidin derivative	µg/g seeds	0.31 ±0.00 1.90 ±0.02 0.02 ±0.00 0.06 ±0.00	[16]

(Table 3) cont.....

Material/compound	Unit	Content	Reference
Seed coat of Brazilian and Peruvian bean cultivars	mg/100 seed coats	1.79 - 558	[15]
Black bean from Mexico	mg/100 seeds mg/ seed coat	38.84 - 70.59 10.05 - 17.28	[48]

Abbreviations: C3G, cyanidin-3-*O*-glucoside equivalents; CE, (+)-catechin equivalents; DW, dry weight; FW, fresh weight.

The condensed tannins contents (CTC) in selected leguminous seeds or their extracts are reported in (Table **4**). According to Aparicio-Fernandez *et al.* [51], the main PACs of jampa beans are in form of dimers and trimers and comprise (+)-catechin/(–)-epicatechin, gallocatechin/epigallocatechin, and afzelechin/epiafzelechin. PACs from red kidney bean, pinto bean, and small red bean are all of the B-type (C4–C8 bond and C4–C6 linkage) with varying degrees of polymerization [52]. Using HPLC-ESI-MS, Amarowicz *et al.* [53] identified in a red bean extract three prodelphinidin dimers, three procyanidin dimers, a procyanidin trimer, and a digallate procyanidin dimer.

According to Xu and Chang [23], a black bean extract is very rich in tannin constituents. Madhujith *et al.* [44] reported the content of tannins in common bean hulls to be several times greater than that of the whole bean extract. The high-molecular-weight fraction, isolated by Sephadex LH-20 column chromatography from a red bean extract, possessed a greater content of PACs than those separated in the same manner from green lentil, red lentil, vetch, adzuki bean, faba bean, field pea, and broad bean [54]. The presence of tannins (86.6 mg/g) was determined in the total indigestible fraction of cooked black bean [25].

Table 4. Content of tannins in bean seeds and their extracts.

Material	Unit	Content	Reference
Black turtle bean extract and its fractions	mg CE/g extract mg CE/g fraction	40.69 ± 0.75 571.21 - 906.12	[27]
26 varieties of bean from Mexico	mg CE/g seeds DM	5.9 - 21.5	[34]
Kidney bean	g CE/ kg seeds DM	3.59 ± 0.05	[36]
Beans cultivated in Mexico (5 cultivars)	mg CE/g seeds	16.8 - 38.1	[49]
Bean from Mexico	mg CE/100 g seeds	2619	[50]
Coats of 15 bean cultivars from Mexico and Brazil	mg CE/g coat	60.6 - 369.3	[24]
Black bean (cooked)	mg CE/g seeds DM	54.6 ± 0.3	[25]

Abbreviations: CE, (+)-catechin equivalents; DM, dry matter.

Composition of Phenolic Acids and Flavonoids in Beans

The dominant phenolic compounds endogenous to bean seeds are phenolic acids and flavonoids. Figs. (2) and (3) depict the chemical structures of the main phenolic acids and flavonoids typically found in beans. Moreover, bean seeds with colour coats are also rich in anthocyanins, another type of flavonoid. The structures of three anthocyanin aglycones (*i.e.,* anthocyanidins) typically found in bean coats are depicted in (Fig. 4). Finally, the content of phenolic acids and flavonoids in bean seeds and their extracts are listed in (Tables 5 and 6).

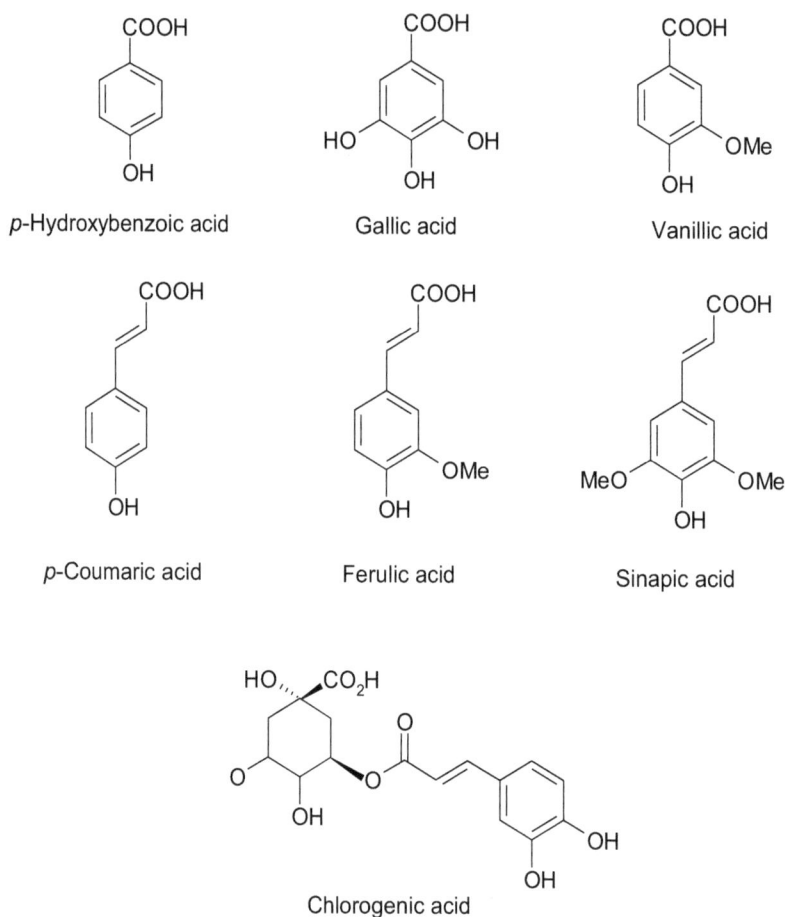

Fig. (2). Chemical structure of phenolic acids typically found in beans.

Based on the study of Sosulski and Dabrowski [55], flour of navy bean contained phenolic acids in the form of soluble esters. From these, alkaline hydrolysis liberated ferulic, *p*-coumaric, and syringic acids. The content of these phenolic acids in navy bean was greater than those found in mung bean, field bean, lentil, faba bean, and pigeon pea. Madhujith *et al.* [56] reported vanillic, caffeic, *p*-coumaric, ferulic, and sinapic acids as the main phenolic acids identified in bean hull extracts. The extract of red bean was characterized by the presence of *p*-coumaric acid, a *p*-coumaric acid derivative, *p*-coumaroyl malic acid, ferulic, acid, protocatechuic acid, *p*-hydroxybenzoic acid, vanillic acid, and sinapic acid. Among these, the dominant phenolic acids were ferulic and *p*-hydroxybenzoic acids [53]. RP-HPLC analysis of the extracts of green- and yellow-podded bean varieties showed the presence of four phenolic acids: they were identified as caffeic acid, *p*-coumaric acid ferulic acid, and sinapic acid [57]. The existence of gallic acid in beans has been confirmed by Huber *et al.* [58], Ramírez-Jiménez *et al.* [12], and Tan *et al.* [27]. Korus *et al.* [13] reported the presence of chlorogenic acid in three Polish bean cultivars. Duénas *et al.* [16] found sinapoyl aldaric acid to be the dominant phenolic acid in kidney beans.

Non-anthocyanin/anthocyanidin flavonoids endogenous to bean seeds are flavonols and flavan-3-ols (catechin/epicatechin). A majority of these, however, are present as glycosides in the seeds. Díaz-Batalla *et al.* [59] also detected the presence of isoflavones in germinated beans. The main flavan-3-ol typical for beans is (+)-catechin [12, 27, 35] and catechin-*O*-hexoside [16, 53, 58]. Amarowicz *et al.* [53] reported a high level of quercetin arabinoglucoside and rutinoside. Furthermore, an extract of white bean was characterized by a high content of kaempferol-3-*O*-glucoside and kaempferol-3-*O*-rutinoside [58].

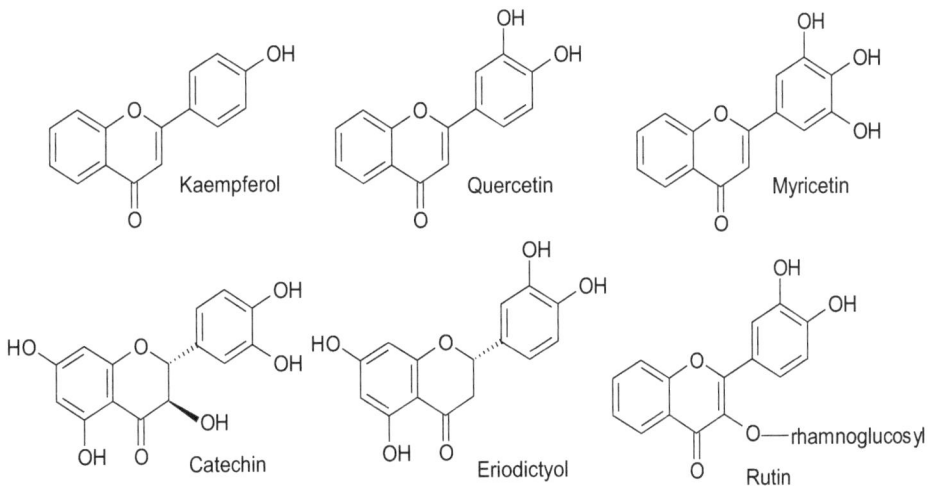

Fig. (3). Chemical structure of flavonoids typical found in beans.

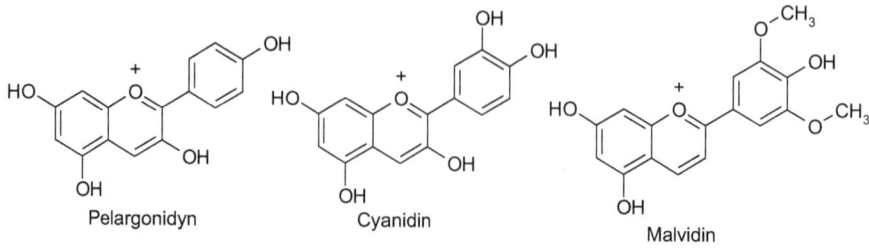

Fig. (4). Chemical structure of anthocyanidins typical found in beans.

Table 5. Content of phenolic acids in bean seeds and their extracts.

Material	Phenolic acid	Unit	Content	Reference
Fractions of black turtle bean extract	Gallic Ferulic Sinapic Syringic	µg/g fraction	2644 ± 13 3021 ±4 641 ± 45 788 ± 11	[27]
Cranberry bean	*p*-Coumaric Ferulic	mg/g extract	0.56 ± 0.01 3.10 ± 0.09	[35]
Dry common beans from Italy (12 samples)	*p*-Coumaric	mg/g seeds DM	0.022 - 0.084	[20]
Four wild bean cultivars from Mexico 10 bean cultivars from Mexico	Sum of phenolic acids determined by HPLC	µg/g seeds DM	12.95 - 22.66	[59]
4 bean cultivars from Mexico	*p*-Hydroxybenzoic Vanillic *p*-Coumaric Ferulic *p*-Hydroxybenzoic Vanillic *p*-Coumaric Ferulic	mg/100g bean flour µg/g seeds DM	5.7 - 13.8 5.2 - 14.11 4.8 - 7.1 20.4 - 36.0 8.3 - 11.3 7.4 - 16.6 3.2 - 5.6 24.6 - 28.4	[46]
Bean Negro 8025 flour Bean Madero Flour	Gallic Chlorogenic Caffeic *p*-Coumaric Ferulic Gallic Chlorogenic Caffeic Ferulic *p*-Coumaric	µg/g flour	146.9; 105.8; 184.6 150.8; 94.2; 170.6 11.55; 8.41; 5.92 -; 3.79;1.94 4.16; -;- 113.3; 65.6; 115.1 137.3; 119.7; 16.57 7.71; 5.91; 16.66 -; -; - 1.37; -; -	[12]

(Table 5) cont.....

Material	Phenolic acid	Unit	Content	Reference
White bean	Vanillic Gallic Chlorogenic Sinapic	µg/g extract	377 2602 3782 146	[58]
Beans of three Polish cultivars	Chlorogenic Caffeic Ferulic p-Coumaric	mg/100 g seeds	17.31; 19.20; 17.87 0.49; 0.73; 0.37 4.41; 5.51; 6.44 0.89; 4.02; 3.95	[13]
Kidney bean (water extract)	Feruloyl hexoside acid Feruloyl aldaric acid p-Hydroxybenzoic p-Coumaric Ferulic Feruolyl quinic acid	µg/g seeds	1.26 ± 0.03 0.79 ± 0.04 3.48 ± 0.10 4.19 ± 0.32 10.56 ± 2.46 1.36 ± 0.42	[18]
Kidney bean	Feruloyl hexoside acid p-Coumaroyl aldaric acid Sinaopyl aldaric acid Ferulic	µg/g seeds	4.03 ± 0.43 6.43 ± 0.61 39.14 ± 5.54 2.71 ± 0.06	[16]
Red bean	p-Coumaric acid p-Coumaric acid derivative p-Coumaroyl malic acid Ferulic Protocatechuic p-Hydroxybenzoic Vanillic Sinapic	µg/g extract	6.19 ± 0.56 9.58 ± 0.92 8.82 ± 0.74 75.98 ± 7.01 45.73 ± 5.23 84.75 ± 6.31 24.21 ± 1.64 0.25 ± 0.03	[53]

Abbreviations: DM, dry matter.

Table 6. Content of flavonoids in bean seeds and their extracts.

Material	Flavonoid	Unit	Content	Reference
Fractions of black turtle bean extract	Myricetin Catechin Epicatechin Quercetin-3-*O*-glucoside Kaempferol-3-*O*-rutinoside Kaempferol-3-*O*-glucoside	µg/g fraction	2.79 ± 0.002 107.4 ± 0.75 93.16 ±0.15 32.78 ±0.04 5.20 ±0.01 19.62 ± 0.04	[27]
Cranberry bean	Catechin glucoside Catechin Epicatechin	mg/g extract	10.46 ± 0.59 44.38 ± 0.07 17.61 ± 0.30	[35]
Dry common beans from Italy (12 samples)	Sum of flavonoids determined by HPLC	mg/g seeds DM	traces-1.274	[20]

(Table 6) cont.....

Material	Flavonoid	Unit	Content	Reference
Four bean cultivars from Mexico	Sum of flavan-3-ols determined by HPLC Sum of flavonols determined by HPLC Sum of flavones determined by HPLC	mg/100g bean flour	0.59 - 27.91 3.26 - 14.11 0.17 - 1.11	[46]
10 bean cultivars from Mexico 4 bean wild cultivars from Mexico	Quercetin Kaempferol Quercetin Kaempferol	µg/g seeds DM	6.9 - 23.5 13.8 - 209.4 10.4 - 17.9 16.1 - 37.0	[59]
Bean Negro 8025 Flour Bean Madero flour	Catechin Rutin Quercetin Catechin Rutin Quercetin		44.3; 246; 71.9 -; 11.03; 5.75 445; 97.5; 471 78.0; 73.1; 196 8.11; 11.3; 7.61 21.1; 509; 570	[12]
White bean	Kaempferol Kaempferol-3-glucoside Kaempferol-3-rutinoside Catechin Quercetin-3-glucoside	µg/g extract	0.008 9.20 72.27 298.39 14.58	[58]
Beans of three Polish cultivars	Myricetin Quercetin Kaempferol	mg/100 g seeds	0; 1.37; 0 16.98; 3.24; 0.84 50.20; 3.24; 0.91	[13]
Kidney bean (water extract)	Hesperetin glucuronide-hexose Quercetin-3-*O*-glucoside Eriodictyol *O*-hexoside Kaemoferol-3-*O*-rutinoside Quercetin-*O*-hexoside Isorhamnetin-*O*-hexoside Catechin Catechin *O*-hexoside	µg/g extract	3.50 ± 0.05 1.27 ± 0.07 1.35 ± 0.14 0.63.± 0.01 0.43 ±0.05 0.03 ±0.02 7.88 ± 1.59 2.76 ± 0.15	[18]
Kidney bean (water extract)	Flavanones Flavonols Catechins	µg/g extract	6.11 ± 0.41 2.37 ± 0.15 10.64 ± 1.74	[18]
Kidney bean	Catechin *O*-hexoside Catechin *O*-acethylhexoside Catechin Eriodictyol *O*-hexoside	µg/g seeds	55.11 ± 5.20 16.51 ± 0.70 32.15 ± 3.94 1.98 ± 0.09	[16]
Seed coat of Brazilian and Peruvian bean cultivars	Quercetin Kaempferol	mg/100 seed coats	0.50 - 69 0 - 750	[15]

(Table 6) cont.....

Material	Flavonoid	Unit	Content	Reference
Red bean	Dihidroquercetin derivative	μg/g extract	0.49 ± 0.02	[53]
	Dihidroquercetin derivative		0.56 ± 0.05	
	Quercetin arabinglucoside		510.1 ± 25.4	
	Quercetin rutinoside		190.5 ± 9.2	
	Quercetin rhamnoside		42.2± 2.6	
	Kaempferol rutinoside		84.7 ± 5.1	
	Kaempferol hexose		66.9 ± 7.2	
	Kaempferol hexose		64.5 ± 5.9	
	Quercetin		86.9 ± 6.6	
	Catechin glucoside		57.4 ±1.2	
	Catechin		10.3 ± 0.9	
	Epicatechin		74.3 ± 6.2	

Antioxidant Activity of Bean Seeds or Their Extracts

Antioxidant and antiradical activities of leguminous seed extracts have been investigated using a variety of methods including storage studies of oils [44, 60];liposomes [61, 62]; a -carotene-linoleate model system [13, 33, 37, 44, 53, 63 - 65]; enhanced chemiluminescence [45]; the DPPH radical-scavenging assay [12, 15, 20, 28, 33, 37, 42, 53, 58, 63 - 66]; the ABTS$^{+•}$ assay [4, 12, 58, 33, 56, 65]; the reducing power assay [19]; low-density lipoprotein cholesterol oxidation [67, 68]; Fe^{2+}-chelating capacity assay [69]; the FRAP assay [33, 42, 53, 65, 70]; and the hydrophilic-ORAC$_{FL}$ assay [42, 71]. Some findings from selected methodologies are reported in (Table **7**).

Based on the research of Madhujith *et al.* [56], coloured beans possessed superior antioxidant activity compared to white counterpart samples. Evaluation of the antioxidant activity in an emulsion system revealed that red, brown, and black whole bean extracts were capable of inhibiting the oxidation of β-carotene (*ca.* 33-52%), as compared to that of the control. In a corn oil model system, red, brown, and black bean extracts inhibited the formation of conjugated dienes (20-28%), 2-thiobarbituric acid reactive substances (44-52%), and hexanal formation (68-84%), when used at a 100-ppm level as (+)-catechin equivalents.

The hydrogen peroxide scavenging capacity of different beans (*i.e.,* white kidney, red pinto, Swedish brown, black kidney) ranged from 58 to 67% at 50 ppm, and 65 to 76% at 100 ppm. All extracts employed retarded human low-density lipoprotein cholesterol oxidation by 61.4 to 99.9% at 2 to 50 ppm levels of (+)-catechin equivalents [67]. Based on experiments by Xu *et al.* [68] with low-density lipoprotein oxidation, the extracts of black bean, red kidney bean, and pinto bean exhibited antioxidant capacities of 29.43, 22.55, and 14.98 μmol Trolox equivalents/g seeds D.M., respectively.

According to Beninger and Hosfield [62] (2003), pure flavonoids such as anthocyanidins, quercetin glycosides, and condensed tannins, which are all present in the seed coats of common beans, have significant antioxidant activity relative to butylated hydroxytoluene (BHT). In fact, in a storage study (26 and 37 °C, 9 months) utilizing soy and sunflower oils, navy bean hull extracts proved to offer superior antioxidant activity than a mixture of butylated hydroxyanisole and BHT, when used at similar concentrations [60].

The anthocyanin pigments isolated from *Phaseolus vulgaris* seed coats (*i.e.*, pelargonidin 3-*O*-β-D-glucoside, cyanidin 3-*O*-β-D-glucoside, and delphinidin 3-*O*-β-D-glucoside) and their corresponding aglycones (*i.e.*, pelargonidin chloride, cyanidin chloride, and delphinidin chloride) exhibited strong antioxidant activities in a liposomal system, and reduced the formation of malondialdehyde by UV radiation. The extent of antioxidant activity afforded by anthocyanin pigments in a rat liver microsomal system and the scavenging effect of hydroxyl radicals and superoxide anion radicals were influenced by the chemical composition of the anthocyanins present [61].

A white bean extract showed much lower antioxidant activity when studied using a photochemiluminescence technique than the extracts of faba bean, lentil, broad bean, pea, everlasting pea [45]. Orak *et al.* [33], who investigated white bean extracts, reported significant (P < 0.01), strong correlations between the TPCs and FRAP results (r = 0.850) as well as between TEAC and FRAP (r = 0.734) data. For the extracts of red bean, correlation coefficients between TPCs and the results of FRAP, ABTS$^{+\bullet}$, and DPPH radical assays were 0.981, 0.997, and 0.890, respectively [65].

For a β-carotene-linoleate assay with a sample containing white bean extract, after 60 and 120 min of incubation at 50 °C, 58.8 and 32.4% of the β-carotene remained unoxidized, respectively [72].

Addition of an aqueous extract of rajma (*Phaseolus vulgaris* L.) to chicken breast meat enhanced lipid oxidative stability of the cooked product, as monitored by 2-thiobarbituric acid reactive substances (TBARS) [39].

Table 7. Antioxidant capacity/activity of bean seeds or their extracts.

Material	Methods	Unit	Results	Reference
Bean hulls	DPPH β-carotene	TEAC AA	2.1 16.6	[37]

(Table 7) cont.....

Material	Methods	Unit	Results	Reference
Black turtle Eclipse Black turtle T-39 Navy bean Pinto bean Red kidney Pink bean Small red Black turtle Eclipse Black turtle T-39 Navy bean Pinto bean Red kidney Pink bean Small red Black turtle Eclipse Black turtle T-39 Navy bean Pinto bean Red kidney Pink bean Small red	DPPH FRAP ORAC	μmol TE/g seeds mmol Fe^{2+}/100 g seeds μmol TE/g seeds	18.95 ± 0.03 14.49 ± 0.14 1.48 ± 0.04 13.79 ± 0.03 16.81 ± 0.11 15.49 ± 0.17 17.90 ± 0.13 9.70 ± 0.35 6.05 ± 0.20 1.27 ± 0.03 7.24 ± 0.28 7.93 ± 0.47 4.07 ± 0.08 4.53 ± 0.19 92.73 ± 4.99 48.91 ± 2.04 13.30 ± 0.55 51.13 ± 3.64 70.48 ± 6.99 90.85 ± 1.92 70.58 ± 3.24	[42]
Small red bean Red kidney bean Black kidney bean Navy bean Small red bean Red kidney bean Black kidney bean Navy bean	Reducing power Scavenging of ˙OH	Absorbance Scavenging activity %	0.74 ± 0.01 0.63 ± 0.07 0.85 ± 0.07 0.17 ± 0.00 65.44 ± 0.85 66.15 ± 0.01 85.68 ± 2.51 66.82 ± 1.36	[19]
Bean from Brazil	DPPH	μmol Trolox/ g extract	851 ± 9; 852 ± 81; 1930 ± 30	[22]
Bean Negro 8025 flour Bean Madero Flour	DPPH ABTS DPPH ABTS	μmol Trolox/ g flour	1.31; 4.91;1.40 2.77; 5.23; 4.60 1.40; 4.49; 2.02 2.28; 5.19; 4.53	[12]
White bean	ABTS DPPH	mg Trolox/g extract Scavenge of DPPH˙ (%)	14.92 2.96	[58]
Dry common beans from Italy (12 samples)	DPPH	EC_{50}(mg/mg DPPH˙)	39 - 2810	[20]
Dry common beans from Italy (12 samples)	DPPH	EC_{50} (mg extract/ml)	1.57 - 61.6	[21]
Common bean from India (n=23)	FRAP ORAC	μmol TE/g D.M. μmol TE/g D.M.	12 - 58 42 - 223	[4]

(Table 7) cont.....

Material	Methods	Unit	Results	Reference
Black bean Red kidney bean Pinto bean	Copper-mediated LDL oxidation	μmol TE/g seeds	29.43 ± 0.85 22.55 ± 0.59 14.98 ± 0.22	[68]
Snap bean (n=51)	ORAC FRAP	μmol TE/g	42 - 223	[71]
Common bean	β-carotene-linoleate DPPH	AA- antioxidant activity (%) TEAC	60 1.82	[63]
Common bean	FRAP	μmol $FeSO_4$/l	92.98 ± 0.20	[64]
Bean Negro 8025 flour Bean Madero flour Bean Negro 8025 flour Bean Madero flour	ABTS DPPH	μmol TE/g flour Scavenge of DPPH˙ (%)	1.31; 4.91; 3.27 1.40; 4.49; 2.02 2.77; 5.23; 4.60 2.28; 5.19; 4.53	[12]
Bean Methanolic extracts Acetonic extracts	DPPH	Scavenge of DPPH˙ (%)	9.90 - 62.28 32.21 - 67.35	[66]
Seed coats of Brazilian and Peruvian beans (n=28)	ABTS	μmol TE/g seed coat	1.05 - 481	[15]
Red bean: Extract Low molecular fraction Tannin fraction Extract Low molecular fraction Tannin fracti	FRAP ABTS	mmol Fe^{2+}/g extract or fraction mmol TE/g extract or fraction	0.612 ± 0.004 0.213 ± 0.002 11.7 ± 0.05 0.481 ± 0.009 0.093 ± 0.007 4.37 ± 0.05	[53]
White beans from Turkey	ABTS FRAP	μmol TE/g extract μmol TE/g seeds μmol Fe^{2+}/g extract μmol Fe^{2+}/g seeds	27 - 43 3.54 - 5.17 66-86 8.29 - 11.2	[33]

Changes in the Content of Phenolic Compounds in Bean Seeds during Technological Processing and Germination

Soaking beans in water for long periods of time can change the content of the endogenous phenolic compounds. For instance, soaking (12 h at 30 °C) markedly decreased the content of total phenolics and tannins in the seeds of selected beans [36]. On the other hand, the content of phenolic acids and flavonoids in the extract of white beans increased with 10 h of maceration [58].

The effect of cooking on the phenolics of beans has been reported by several researchers. The phenolic content in common bean after pressure cooking (*i.e.*, 121 °C; 103.421 kPa; 5, 7, or 60 min) was drastically reduced: in the case of seed coats the reduction was as much as 90% [43]. For green beans, the TPC increased after cooking, and this is likely due to higher extractability of previously-bound phenolic compounds to the cell walls from the processed material [73]. Cooking of kidney red beans reduced the TPC from 8.9 to 6.3 mg/g D.W., but increased the content of total flavonoids from 0.9 to 1.03 mg/g D.W [38]. Moreover, soaking, boiling, and steaming processes significantly decreased the content of total phenolics in black beans. Steaming resulted in a greater retention of the TPC than boiling, while pressure steaming resulted in significant decrease in the TPC compared to regular steaming [74].

In the experiments of Rocha-Guzmán *et al.* [43], longer cooking times enhanced the diffusion of phenolic compounds out from the seed coats of three common bean cultivars to the cooking water, and from there to the cotyledons. On the other hand, microwave treatment of pinto beans resulted in an increase of the TPC by 20% [26], which is likely *via* stimulation of the pentose phosphate pathway [75].

Extrusion decreased the TPC and tannins content in seeds of selected beans [36]. From the research of Korus *et al.* [13], the effect of extrusion on the TPC of beans depended on the cultivar in question; that is, one variety showed a 14% increase in the amount of phenolic compounds in extrudates compared to the starting raw material, while the other two varieties exhibited a decrease by 21%.

From the experiments of Oomah *et al.* [26], micronization reduced the TPC of red bean and increased it for pinto bean. Apropos black bean, no effect was observed. Reduction in the content of tannins (6%) in red kidney beans was caused by micronization according to the research of Khattab and Arntfield [76]. With respect to pinto and black bean hulls, micronization had little or no effect on the TPC, but increased it for that of red bean hulls [26].

Germination of beans for 24, 48, and 72 h decreased the TPC and tannins content in selected beans [36]. The content of phenolic acids (notably *p*-hydroxybenzoic, vanillic, *p*-coumaric, and ferulic acids) decreased after germination of the bean seeds. The presence of vanillic and *p*-hydroxybenzoic acids was detected only in the germinated seeds: this could have been caused by degradation of the seed's lignin by way of enzymatic oxidation [77]. In bean seeds, the presence of flavonol glycosides (*i.e.,* quercetin-3-*O*-rutinoside, quercetin-3-*O*-ramnoside, kaempferol-3-*O*-rutinoside, and kaempferol-3-*O*-glucoside) was detected only in the germinated products. The highest concentrations of flavonols were observed in the seeds germinated with light [78]. During the germination of kidney beans, the content of flavan-3-ols and anthocyanins was compensated for by a higher content of flavones and flavonols [16].

The increase in various phenolic constituents has also been observed during the fermentation of beans [79]. During fungal fermentation of two varieties of common beans (*Pinto durani* and Bayo) for tempeh, Gamboa-Gómez *et al.* [21] observed a decrease in the TPC. Liquid-state fermentation for 48 and 96 h *via Bacillus subtilis* increased the content of several phenolic acids and flavonoids in a kidney bean extract. Application of natural fermentation with *Lactobacillus plantarum*, however, showed an opposite effect [18].

Changes in the Antioxidant Potential of Bean Seeds during Technological Processing and Germination

Over-night soaking of red kidney beans increased their antioxidant potential, as determined by the $ABTS^{+\bullet}$ and $DPPH^{\bullet}$ assays. An opposite effect was observed from the results of H_2O_2, and hydroxy radical assays [80].

Although the level of phenolics was low for cooked beans, their antiradical activity against $ABTS^{+\bullet}$ was similar to that measured for crude seed coats, and higher than that noted for crude cotyledons [43]. Pressure and microwave cooking decreased the antioxidant potential of red kidney beans, as determined using the $DPPH^{\bullet}$, H_2O_2, and hydroxy radical assays. An opposite effect was noted from results obtained using the $ABTS^{+\bullet}$ assay. Boiling test samples for 15 min reduced their antioxidant potential in all of the *in vitro* assays employed [80]. Boiling and steaming exhibited significantly ($P < 0.05$) lower antioxidant activities than raw beans in DPPH free radical scavenging activity, and oxygen radical absorbing capacity (ORAC) [74]. The percent $DPPH^{\bullet}$ scavenging activity for cooked beans was greater than that for raw beans. The radical-scavenging activity of seed coats was related to their cooking time [43]. Cooking increased the antioxidant capacity (according to the $DPPH^{\bullet}$ assay) of red kidney beans [38]. However, the reducing power of cooked seeds was greater than that of raw seeds. From the study of Ombra *et al.* [28], the effect of cooking on the antioxidant activities of the extracts of twelve bean types only had a marginal impact.

After thermal treatment (121 °C for 10 min), Huber *et al.* [58] observed reduced antioxidant activities for the extract of white bean. The $ABTS^{+\bullet}$ and $DPPH^{\bullet}$ assays were employed in this research. Extrusion of dry beans decreased their antioxidant potential, as determined by electron paramagnetic resonance studies and using a β-carotene/linoleic acid system [13]. Microwave treatment of pinto bean increased the antioxidant potential by 18%, as measured by the $ORAC_{FL}$ method [26].

The result of germination modified the antiradical efficacy of bean phenolics against DPPH• [66]. After a germination period of 4 and 6 days, beans exhibited higher antiradical capacities than those of their raw seeds. Aquilera *et al.* [32] reported that germination led to an increase in antioxidant potential of kidney beans, as determined by the $ORAC_{FL}$ assay.

Compared to commercial antioxidant preparations, an extract of red bean fermented by *Bacillus subtilis* and *Aspergillus oryzae* showed less of a scavenging effect against DPPH• and a weaker reducing power compared to that of α-tocopherol and BHT, but a superior Fe^{2+}-chelating capacity [81, 82]. A decrease in the antioxidant potential of common bean, as affected by fungal fermentation (tempeh), was reported by Gamboa-Gómez *et al.* [21]. These authors used a DPPH• assay and a low-density lipoprotein oxidation method.

Using a mixed culture of tempeh (*Lactobacillus plantarum* and *Rhizopus microspores*), Starzyńska-Janiszewska *et al.* [83] observed an enhanced antioxidant capacity, as determined by the DPPH• assay. In a study by Luzardo-Ocampo *et al.* [84], digested (*i.e.*, mouth, stomach, small intestine, colon) common bean chips exhibited significantly higher antioxidant activities compared to methanolic extract preparations. The *in vitro* digestion increased the antioxidant activity of black bean seed coats. This process was responsible for a reduction in the TPC, PAC content, and monomeric anthocyanins in black bean and red bean seed coats, when measured by the $ORAC_{FL}$ assay [14].

CONCLUDING REMARKS

It is clear that the chemical profiles and levels of phenolic acids, flavonoids, and tannins found in the different types of bean seeds consumed by man are quite varied. Yet, all extracts prepared from these bean seeds using a myriad of solvent systems, seem to impart some degree of antioxidant/anti-radical activity regardless of the model employed. To complicate the story further, it is evident that processing of the bean seeds prior to analysis can greatly impact the phenolics levels detected as well as the observed activities in question. In some cases, processing allows for a superior extraction of antioxidant constituents from the sample matrix, while at other times, processing facilitates the removal or destruction of these compounds. At all times, a detailed accounting of the sample preparation steps is critical in order to permit quantitative comparisons of phenolic constituents and their activities between the bean seed types.

CONSENT FOR PUBLICATION

Not applicable.

CONFLICT OF INTERETS

The author (editor) declares no conflict of interest, financial or otherwise.

ACKNOWLEDGEMENTS

Delacred None.

REFERENCES

[1]   Reynoso-Camacho R, Ramos-Gomez M, Loarca-Pina G. Bioactive Components in Common Beans (Phaseolus vulgaric L.).Advances in Agricultural and Food Biotechnology. Kerala: Research Signpost 2006; pp. 217-36.

[2]   Vaz Patto MC, Amarowicz R, Aryee ANA, *et al.* Achievements and challenges in improving the nutritional quality of food legumes. Crit Rev Plant Sci 2015; 34: 105-43.
[http://dx.doi.org/10.1080/07352689.2014.897907]

[3]   Hayat I, Ahmad A, Masud T, Ahmed A, Bashir S. Nutritional and health perspectives of beans (*Phaseolus vulgaris* L.): an overview. Crit Rev Food Sci Nutr 2014; 54(5): 580-92.
[http://dx.doi.org/10.1080/10408398.2011.596639] [PMID: 24261533]

[4]   Dutta SK, Chatterjee D, Sarkar D, *et al.* Common bean (*Phaseolus vulgaris* L., Fabaceae), landraces of *Lushai* hills in India: Nutrients and antioxidants source for the farmers. Indian J Tradit Knowl 2016; 15: 313-20.

[5]   Amarowicz R, Pegg RB. Legumes as a source of natural antioxidants. Eur J Lipid Sci Technol 2008; 110: 865-78.
[http://dx.doi.org/10.1002/ejlt.200800114]

[6]   Suárez-Martínez SE, Ferriz-Martínez RA, Campos-Vega R, *et al.* Beans seeds: leading nutraceutical source for human nutrition. CYTA J Food 2016; 14: 131-7.
[http://dx.doi.org/10.1080/19476337.2015.1063548]

[7]   Chávez-Mendoza C, Sánchez E. Bioactive compounds from Mexican varieties of the common bean (*Phaseolus vulgaris*): Implications for health. Molecules 2017; 22(8): 1360.
[http://dx.doi.org/10.3390/molecules22081360] [PMID: 28817105]

[8]   Singleton VL, Orthofer R, Lamuela-Raventós RM. Analysis of total phenols and other oxidation substrates and antioxidants by means of Folin-Ciocalteu reagent. Methods Enzymol 1999; 299: 152-78.
[http://dx.doi.org/10.1016/S0076-6879(99)99017-1]

[9]   Christ B, Müller KH. Zur serienmäßigen Bestimmung des Gehaltes an Flavonol-Derivaten in Drogen. Arch Pharm 1960; 293: 1033-42.
[http://dx.doi.org/10.1002/ardp.19602931202]

[10]   Price ML, van Scoyoc S, Butler LG. A critical evaluation of the vanillin reaction as an assay for tannin in sorghum grain. J Agric Food Chem 1978; 26: 1214-8.
[http://dx.doi.org/10.1021/jf60219a031]

[11]   Kyznetsova MY, Makieieva OM, Lavrovska DO, *et al.* Effect of aqueous extract from *Phaseolus vulgaris* pods on lipid peroxidation and antioxidant enzymes activity in the liver and kidney of diabetic rats. J Appl Pharm Sci 2015; 5: 1-6.
[http://dx.doi.org/10.7324/JAPS.2015.50501]

[12]   Ramírez-Jiménez AK, Reynoso-Camacho R, Mendoza-Díaz S, Loarca-Piña G. Functional and technological potential of dehydrated *Phaseolus vulgaris* L. flours. Food Chem 2014; 161: 254-60.
[http://dx.doi.org/10.1016/j.foodchem.2014.04.008] [PMID: 24837948]

[13]   Korus J, Gumul D, Czechowska K. Effect of extrusion on the phenolic composition and antioxidant activity of dry beans of *Phaseolus vulgaris* L. Food Technol Biotechnol 2007; 45: 139-46.

[14]   Sancho RAS, Pavan V, Pastor GM. Effect of *in vitro* digestion on bioactive compounds and antioxidant activity of common bean coat. Food Res Int 2015; 76: 74-8. [http://dx.doi.org/10.1016/j.foodres.2014.11.042]

[15]   Ranilla LG, Genovese MI, Lajolo FM. Polyphenols and antioxidant capacity of seed coat and cotyledon from Brazilian and Peruvian bean cultivars (*Phaseolus vulgaris* L.). J Agric Food Chem 2007; 55(1): 90-8. [http://dx.doi.org/10.1021/jf062785j] [PMID: 17199318]

[16]   Dueñas M, Martínez-Villaluenga C, *et al*. Effect of germination and elicitation on phenolic composition and bioactivity of kidney beans. Food Res Int 2015; 70: 55-63. [http://dx.doi.org/10.1016/j.foodres.2015.01.018]

[17]   Karamać M, Amarowicz R, Weidner S, Shahidi F. Antioxidant activity of phenolic fractions of white bean (*Phaseolus vulgaris*). J Food Lipids 2004; 11: 165-77. [http://dx.doi.org/10.1111/j.1745-4522.2004.tb00268.x]

[18]   Limón RI, Peñas E, Torino MI, Martínez-Villaluenga C, Dueñas M, Frias J. Fermentation enhances the content of bioactive compounds in kidney bean extracts. Food Chem 2015; 172: 343-52. [http://dx.doi.org/10.1016/j.foodchem.2014.09.084] [PMID: 25442563]

[19]   Zhao Y, Du S-K, Wang H, Cai M. *In vitro* antioxidant activity of extracts from common legumes. Food Chem 2014; 152: 462-6. [http://dx.doi.org/10.1016/j.foodchem.2013.12.006] [PMID: 24444962]

[20]   Heimler D, Vignolini P, Dini MG, Romani A. Rapid tests to assess the antioxidant activity of *Phaseolus vulgaris* L. dry beans. J Agric Food Chem 2005; 53(8): 3053-6. [http://dx.doi.org/10.1021/jf049001r] [PMID: 15826058]

[21]   Gamboa-Gómez CI, Muñoz-Martínez A, Rocha-Guzmán NE, *et al*. Changes in phytochemical and antioxidant potential of tempeh common bean flour from two selected cultivars influenced by temperature and fermentation time. J Food Process Preserv 2016; 40: 270-8. [http://dx.doi.org/10.1111/jfpp.12604]

[22]   Veggi PC, Cavalcanti RN, Meireles MAA. Production of phenolic-rich extracts from Brazilian plants using supercritical and subcritical fluid extraction: Experimental data and economic evaluation. J Food Eng 2014; 131: 96-109. [http://dx.doi.org/10.1016/j.jfoodeng.2014.01.027]

[23]   Xu BJ, Chang SKC. A comparative study on phenolic profiles and antioxidant activities of legumes as affected by extraction solvents. J Food Sci 2007; 72(2): S159-66. [http://dx.doi.org/10.1111/j.1750-3841.2006.00260.x] [PMID: 17995858]

[24]   Mojica L, Meyer A, Berhow MA, de Mejía EG. Bean cultivars (*Phaseolus vulgaris* L.) have similar high antioxidant capacity, *in vitro* inhibition of α-amylase and α-glucosidase while diverse phenolic composition and concentration. Food Res Int 2015; 69: 38-48. [http://dx.doi.org/10.1016/j.foodres.2014.12.007]

[25]   Hernández-Salazar M, Osorio-Diaz P, Loarca-Piña G, Reynoso-Camacho R, Tovar J, Bello-Pérez LA. *In vitro* fermentability and antioxidant capacity of the indigestible fraction of cooked black beans (*Phaseolus vulgaris* L.), lentils (*Lens culinaris* L.) and chickpeas (*Cicer arietinum* L.). J Sci Food Agric 2010; 90(9): 1417-22. [http://dx.doi.org/10.1002/jsfa.3954] [PMID: 20549791]

[26]   Oomah BD, Kotzeva L, Allen M, Bassinello PZ. Microwave and micronization treatments affect dehulling characteristics and bioactive contents of dry beans (*Phaseolus vulgaris* L.). J Sci Food Agric 2014; 94(7): 1349-58. [http://dx.doi.org/10.1002/jsfa.6418] [PMID: 24114525]

[27]   Tan Y, Chang SKC, Zhang Y. Comparison of α-amylase, α-glucosidase and lipase inhibitory activity of the phenolic substances in two black legumes of different genera. Food Chem 2017; 214: 259-68. [http://dx.doi.org/10.1016/j.foodchem.2016.06.100] [PMID: 27507474]

[28]   Ombra MN, d'Acierno A, Nazzaro F, *et al.* Phenolic composition and antioxidant and antiproliferative activities of the extracts of twelve common bean (*Phaseolus vulgaris* L.) endemic ecotypes of southern Italy before and after cooking Oxid Med Cell Long. 2016. 1398298

[29]   Wang Y-K, Zhang X, Chen G-L, *et al.* Antioxidant property and their free, soluble conjugate and insoluble-bound phenolic contents in selected beans. J Funct Foods 2016; 24: 359-72. [http://dx.doi.org/10.1016/j.jff.2016.04.026]

[30]   Marathe SA, Deshpande R, Khamesra A, *et al.* Effect of radiation processing on nutritional, functional, sensory and antioxidant properties of red kidney beans. Radiat Phys Chem 2016; 125: 1-8. [http://dx.doi.org/10.1016/j.radphyschem.2016.03.002]

[31]   Preti R, Rapa M, Vinci G. Effect of steaming and boiling on the antioxidant properties and biogenic amines content in green bean (*Phaseolus vulgaris*) varieties of different colours J Food Qual. 2017. 5329070

[32]   Aguilera Y, Mojica L, Rebollo-Hernanz M, Berhow M, de Mejía EG, Martín-Cabrejas MA. Black bean coats: New source of anthocyanins stabilized by β-cyclodextrin copigmentation in a sport beverage. Food Chem 2016; 212: 561-70. [http://dx.doi.org/10.1016/j.foodchem.2016.06.022] [PMID: 27374568]

[33]   Orak HH, Karamać M, Orak A, Amarowicz R. Antioxidant potential and phenolic compounds of some widely consumed Turkish white bean (*Phaseolus vulgaris* L.) varieties. Pol J Food Nutr Sci 2016; 66: 253-60. [http://dx.doi.org/10.1515/pjfns-2016-0022]

[34]   Aquino-Bolaños EN, García-Díaz YD, Chavez-Servia JL, *et al.* Anthocyanin, polyphenol, and flavonoid contents and antioxidant activity in Mexican common bean (*Phaseolus vulgaris* L.) landraces. Emir J Food Agric 2016; 28: 581-8. [http://dx.doi.org/10.9755/ejfa.2016-02-147]

[35]   Chen PX, Zhang H, Marcone MF, *et al.* Anti-inflammatory effects of phenolic-rich cranberry bean (*Phaseolus vulgaris* L.) extracts and enhanced cellular antioxidant enzyme activities in Caco-2 cells. J Funct Foods 2017; 38: 675-85. [http://dx.doi.org/10.1016/j.jff.2016.12.027]

[36]   Alonso R, Aguirre A, Marzo F. Effects of extrusion and traditional processing methods on antinutrients and *in vitro* digestibility of protein and starch in faba and kidney beans. Food Chem 2000; 68: 159-65. [http://dx.doi.org/10.1016/S0308-8146(99)00169-7]

[37]   Cardador-Martínez A, Loarca-Piña G, Oomah BD. Antioxidant activity in common beans (*Phaseolus vulgaris* L.). J Agric Food Chem 2002; 50(24): 6975-80. [http://dx.doi.org/10.1021/jf020296n] [PMID: 12428946]

[38]   Gujral HS, Sharma P, Gupta N, Wani AA. Antioxidant properties of legumes and their morphological fractions as affected by cooking. Food Sci Biotechnol 2013; 22: 187-94. [http://dx.doi.org/10.1007/s10068-013-0026-8]

[39]   Yogesh K, Jha SN, Ahmad T. Antioxidant potential of aqueous extract of some food grain powder in meat model system. J Food Sci Technol 2014; 51(11): 3446-51. [http://dx.doi.org/10.1007/s13197-012-0804-y] [PMID: 26396344]

[40]   Doria E, Campion B, Sparvoli F, *et al.* Anti-nutrient components and metabolites with health implications in seeds of 10 common bean (*Phaseolus vulgaris* L. and *Phaseolus lunatus* L.) landraces cultivated in southern Italy. J Food Compos Anal 2012; 26: 72-80. [http://dx.doi.org/10.1016/j.jfca.2012.03.005]

[41]   Rosales MA, Ocampo E, Rodríguez-Valentín R, Olvera-Carrillo Y, Acosta-Gallegos J, Covarrubias AA. Physiological analysis of common bean (*Phaseolus vulgaris* L.) cultivars uncovers characteristics related to terminal drought resistance. Plant Physiol Biochem 2012; 56: 24-34.
[http://dx.doi.org/10.1016/j.plaphy.2012.04.007] [PMID: 22579941]

[42]   Xu BJ, Yuan SH, Chang SKC. Comparative analyses of phenolic composition, antioxidant capacity, and color of cool season legumes and other selected food legumes. J Food Sci 2007; 72(2): S167-77.
[http://dx.doi.org/10.1111/j.1750-3841.2006.00261.x] [PMID: 17995859]

[43]   Rocha-Guzmán NE, González-Laredo RF, *et al.* Effect of pressure cooking on the antioxidant activity of extracts from three common bean (*Phaseolus vulgaris* L.) cultivars. Food Chem 2007; 100: 31-5.
[http://dx.doi.org/10.1016/j.foodchem.2005.09.005]

[44]   Madhujith T, Naczk M, Shahidi F. Antioxidant activity of common beans (*Phaseolus vulgaris* L.). J Food Lipids 2004; 11: 220-33.
[http://dx.doi.org/10.1111/j.1745-4522.2004.01134.x]

[45]   Amarowicz R, Raab B. Antioxidative activity of leguminous seed extracts evaluated by chemiluminescence methods. Z Naturforsch 1997; 52c: 709-12.
[http://dx.doi.org/10.1515/znc-1997-9-1022]

[46]   Moreno-Jiménez MR, Cervantes-Cardoza V, Gallegos-Infante JA, *et al.* Phenolic composition changes of processed common beans: their antioxidant and anti-inflammatory effects in intestinal cancer cells. Food Res Int 2015; 76: 79-85.
[http://dx.doi.org/10.1016/j.foodres.2014.12.003]

[47]   Takeoka GR, Dao LT, Full GH, *et al.* Characterization of black bean (*Phaseolus vulgaris* L.) anthocyanins. J Agric Food Chem 1997; 45: 3395-400.
[http://dx.doi.org/10.1021/jf970264d]

[48]   Salinas-Moreno Y, Rojas-Herrera L, Sosa-Montes E, Pérez-Herrera P. Anthocyanin composition in black bean (*Phaseolus vulgaris* L.) varieties grown in México. Agrociencia 2005; 39: 385-94.

[49]   de Mejía EG, Guzmán-Maldonado SH, Acosta-Gallegos JA, *et al.* Effect of cultivar and growing location on the trypsin inhibitors, tannins, and lectins of common beans (*Phaseolus vulgaris* L.) grown in the semiarid highlands of Mexico. J Agric Food Chem 2003; 51(20): 5962-6.
[http://dx.doi.org/10.1021/jf030046m] [PMID: 13129302]

[50]   De Mejia EG, Del Carmen Valadez-Vega M, Reynoso-Camacho R, Loarca-Pina G. Tannins, trypsin inhibitors and lectin cytotoxicity in tepary (Phaseolus acutifolius) and common (*Phaseolus vulgaris*) beans. Plant Foods Hum Nutr 2005; 60(3): 137-45.
[http://dx.doi.org/10.1007/s11130-005-6842-0] [PMID: 16187017]

[51]   Aparicio-Fernandez X, Yousef GG, Loarca-Pina G, de Mejia E, Lila MA. Characterization of polyphenolics in the seed coat of Black Jamapa bean (*Phaseolus vulgaris* L.). J Agric Food Chem 2005; 53(11): 4615-22.
[http://dx.doi.org/10.1021/jf047802o] [PMID: 15913334]

[52]   Gu L, Kelm MA, Hammerstone JF, *et al.* Screening of foods containing proanthocyanidins and their structural characterization using LC-MS/MS and thiolytic degradation. J Agric Food Chem 2003; 51(25): 7513-21.
[http://dx.doi.org/10.1021/jf034815d] [PMID: 14640607]

[53]   Amarowicz R, Karamać M, Dueñas M, Pegg RB. Antioxidant activity and phenolic composition of a red bean (*Phaseolus vulgaris*) extract and its fractions. Nat Prod Commun 2017; 12: 541-4.

[54]   Karamać M, Kosińska A, Rybarczyk A, Amarowicz R. Extraction and chromatographic separation of tannin fractions from tannin-rich plant material. Pol J Food Nutr Sci 2007; 57: 471-4.

[55]   Sosulski FW, Dabrowski KJ. Composition of free and hydrolyzable phenolic acids in the flours and hulls of ten legume species. J Agric Food Chem 1984; 32: 131-3.
[http://dx.doi.org/10.1021/jf00121a033]

[56]   Madhujith T, Amarowicz R, Shahidi F. Phenolic antioxidants in beans and their effects on inhibition of radical-induced DNA damage. J Am Oil Chem Soc 2004; 81: 691-6.
[http://dx.doi.org/10.1007/s11746-004-963-y]

[57]   Weidner S, Król A, Karamać M, Amarowicz R. Phenolic compounds and the antioxidant properties in seeds of green- and yellow-podded bean (*Phaseolus vulgaris* L.) varieties. CYTA J Food 2018; 16: 373-80.

[58]   Huber K, Brigide P, Bretas EB, Canniatti-Brazaca SG. Effect of thermal processing and maceration on the antioxidant activity of white beans. PLoS One 2014; 9(7): e99325.
[http://dx.doi.org/10.1371/journal.pone.0099325] [PMID: 24991931]

[59]   Díaz-Batalla L, Widholm JM, Fahey GC Jr, Castaño-Tostado E, Paredes-López O. Chemical components with health implications in wild and cultivated Mexican common bean seeds (*Phaseolus vulgaris* L.). J Agric Food Chem 2006; 54(6): 2045-52.
[http://dx.doi.org/10.1021/jf051706l] [PMID: 16536573]

[60]   Onyeneho SN, Hettiarachchy NS. Effect of navy bean hull extract on the oxidative stability of soy and sunflower oils. J Agric Food Chem 1991; 39: 1701-4.
[http://dx.doi.org/10.1021/jf00010a600]

[61]   Tsuda T, Shiga K, Ohshima K, Kawakishi S, Osawa T. Inhibition of lipid peroxidation and the active oxygen radical scavenging effect of anthocyanin pigments isolated from *Phaseolus vulgaris* L. Biochem Pharmacol 1996; 52(7): 1033-9.
[http://dx.doi.org/10.1016/0006-2952(96)00421-2] [PMID: 8831722]

[62]   Beninger CW, Hosfield GL. Antioxidant activity of extracts, condensed tannin fractions, and pure flavonoids from *Phaseolus vulgaris* L. seed coat color genotypes. J Agric Food Chem 2003; 51(27): 7879-83.
[http://dx.doi.org/10.1021/jf0304324] [PMID: 14690368]

[63]   Cardador-Martínez A, Albores A, Bah M, *et al.* Relationship among antimutagenic, antioxidant and enzymatic activities of methanolic extract from common beans (*Phaseolus vulgaris* L). Plant Foods Hum Nutr 2006; 61(4): 161-8.
[http://dx.doi.org/10.1007/s11130-006-0026-4] [PMID: 17048099]

[64]   Gjorgieva D, Kadifkova Panovska T, Ruskovska T, Bačeva K, Stafilov T. Mineral nutrient imbalance, total antioxidants level and DNA damage in common bean (*Phaseolus vulgaris* L.) exposed to heavy metals. Physiol Mol Biol Plants 2013; 19(4): 499-507.
[http://dx.doi.org/10.1007/s12298-013-0196-0] [PMID: 24431518]

[65]   Orak HH, Karamać M, Amarowicz R. Antioxidant activity of phenolic compounds of red bean (*Phaseolus vulgaris* L.). Oxid Commun 2015; 38: 67-76.

[66]   Rocha-Guzmán NE, Herzog A, González-Laredo RF, *et al.* Antioxidant and antimutagenic activity of phenolic compounds in three different colour groups of common bean cultivars (*Phaseolus vulgaris*). Food Chem 2007a; 103: 521-7.
[http://dx.doi.org/10.1016/j.foodchem.2006.08.021]

[67]   Madhujith T. Beans: A Source of Natural Antioxidants. In: Shahidi F, C-T Ho, Eds. In: Phenolic Compounds in Foods and Natural Health Products. ACS Symposium Series 909; 2005; pp. Washington, DC, American Chemical Society. 83-93.
[http://dx.doi.org/10.1021/bk-2005-0909.ch008]

[68]   Xu BJ, Yuan SH, Chang SKC. Comparative studies on the antioxidant activities of nine common food legumes against copper-induced human low-density lipoprotein oxidation *in vitro.* J Food Sci 2007; 72(7): S522-7.
[http://dx.doi.org/10.1111/j.1750-3841.2007.00464.x] [PMID: 17995667]

[69]   Zhou K, Yu L. Total phenolic contents and its antioxidant properties of commonly consumed vegetables grown in Colorado. Lebensm Wiss Technol 2006; 39: 1155-62.

[http://dx.doi.org/10.1016/j.lwt.2005.07.015]

[70]   Berger M, Küchler T, Maaßen A, *et al.* Correlations of ingredients with sensory attributes in green beans and peas under different storage conditions. Food Chem 2007; 103: 875-84.

[71]   Ou B, Huang D, Hampsch-Woodill M, Flanagan JA, Deemer EK. Analysis of antioxidant activities of common vegetables employing oxygen radical absorbance capacity (ORAC) and ferric reducing antioxidant power (FRAP) assays: A comparative study. J Agric Food Chem 2002; 50(11): 3122-8.
[http://dx.doi.org/10.1021/jf0116606] [PMID: 12009973]

[72]   Amarowicz R, Troszyńska A, Karamać M, Kozłowska H. Antioxidative Properties of Legume Seed Extracts.Agri-Food Quality: An Interdisciplinary Approach. Cambridge, UK: The Royal Society of Chemistry 1996; pp. 376-9.

[73]   Turkmen N, Sari F, Velioglu SY. The effect of cooking methods on total phenolics and antioxidant activity of selected green vegetables Food Chem 2005; 93: 713-8.
[http://dx.doi.org/10.1016/j.foodchem.2004.12.038]

[74]   Xu BJ, Chang SKC. Total phenolic content and antioxidant properties of eclipse black beans (*Phaseolus vulgaris* L.) as affected by processing methods. J Food Sci 2008; 73(2): H19-27.
[http://dx.doi.org/10.1111/j.1750-3841.2007.00625.x] [PMID: 18298732]

[75]   Randhir R, Shetty K. Microwave-induced stimulation of L-DOPA, phenolics and antioxidant activity in fava bean (*Vicia faba*) for Parkinson's diet. Process Biochem 2004; 39: 1775-84.
[http://dx.doi.org/10.1016/j.procbio.2003.08.006]

[76]   Khattab RY, Arntfield SD. Nutritional quality of legume seeds as affected by some physical treatments 2. Antinutritional factors. Lebensm Wiss Technol 2009; 42: 1113-8.
[http://dx.doi.org/10.1016/j.lwt.2009.02.004]

[77]   Dagley S. Catabolism of aromatic compounds by micro-organisms. Adv Microb Physiol 1971; 6(0): 1-46.

[78]   López-Amorós ML, Hernández T, Estrella I. Effect of germination on legume phenolic compounds and their antioxidant activity. J Food Compos Anal 2006; 19: 277-83.
[http://dx.doi.org/10.1016/j.jfca.2004.06.012]

[79]   Murakami H, Asakawa T, Terao J, Matsushita S. Antioxidative stability of tempeh and liberation of isoflavones by fermentation. Agric Biol Chem 1984; 48: 2971-5.

[80]   Chakraborty A, Bhattacharyya S. Thermal processing effects on *in vitro* antioxidant activities of five common Indian pulses. J Appl Pharm Sci 2014; 4: 65-70.

[81]   Chou S-T, Chang C-T, Chao W-W, Chung Y-C. Evaluation of antioxidative and mutagenic properties of 50% ethanolic extract from red beans fermented by *Aspergillus oryzae.* J Food Prot 2002; 65(9): 1463-9.
[http://dx.doi.org/10.4315/0362-028X-65.9.1463] [PMID: 12233859]

[82]   Chung Y-C, Chang C-T, Chao W-W, Lin CF, Chou ST. Antioxidative activity and safety of the 50 ethanolic extract from red bean fermented by *Bacillus* subtilis IMR-NK1. J Agric Food Chem 2002; 50(8): 2454-8.
[http://dx.doi.org/10.1021/jf011369q] [PMID: 11929313]

[83]   Starzyńska-Janiszewska A, Stodolak B, Mickowska B. Effect of controlled lactic acid fermentation on selected bioactive and nutritional parameters of tempeh obtained from unhulled common bean (*Phaseolus vulgaris*) seeds. J Sci Food Agric 2014; 94(2): 359-66.
[http://dx.doi.org/10.1002/jsfa.6385] [PMID: 24037686]

[84]   Luzardo-Ocampo I, Campos-Vega R, Gaytán-Martínez M, Preciado-Ortiz R, Mendoza S, Loarca-Piña G. Bioaccessibility and antioxidant activity of free phenolic compounds and oligosaccharides from corn (*Zea mays* L.) and common bean (*Phaseolus vulgaris* L.) chips during *in vitro* gastrointestinal digestion and simulated colonic fermentation. Food Res Int 2017; 100(Pt 1): 304-11.
[http://dx.doi.org/10.1016/j.foodres.2017.07.018] [PMID: 28873692]

CHAPTER 4

Phytochemicals Content and Health Effects of Cultivated and Underutilized Species of the Cucurbitaceae Family

Nikolaos Tzortzakis[1,*], Antonios Chrysargyris and **Spyridon A. Petropoulos[2]**

[1] *Cyprus University of Technology, Department of Agricultural Sciences, Biotechnology and Food Science, Anexartisias 33, Lemesos, Cyprus*

[2] *University of Thessaly, Department of Agriculture, Crop Production and Rural Environment, 38446 N. Ionia, Magnesia, Greece*

Abstract: Cucurbitaceae represents a large plant family with more than 120 genera and 800 species, among which many significant cultivated vegetable species are included, such as watermelon, melon, cucumber and cucurbits (squash, pumpkin and zucchini). These species are usually consumed for their edible fruits, however several other uses have been reported for the various plant parts, including medicinal and therapeutic ones among others. The present chapter will demonstrate the most common vegetable species in terms of their chemical composition and health effects, as well as their edible, medicinal and industrial uses, based on the phytochemical content of the various plant parts. Special focus will be given on cucurbitacins which are an important group of phytochemicals present in the Cucurbitaceae family, since several studies have confirmed its bioactive properties and multiple health effects. Finally, selected less known species of this family (gourds) will be presented, considering their important health effects and their use in vegetable grafting. In conclusion, future perspectives for further valorization of these species will be highlighted, especially for the ones that are less commonly used.

Keywords: Antioxidant Activity, Anti-Diabetic, Anti-inflammatory, Bioactive Compounds, Bottle Gourd, Cucumber, Cucurbitaceae, Cucurbitacins, Cucurbits, Flavonoids, Melon, Watermelon, Seed Oils.

INTRODUCTION

Cucurbitaceae represents a plant family with more than 800 species, among which many significant cultivated vegetables species are included, such as watermelon,

* **Corresponding author Nikolaos Tzortzakis:** Cyprus University of Technology, Department of Agricultural Sciences, Biotechnology and Food Science, Anexartisias 33, Lemesos, Cyprus; Tel: +357 25002280; E-mail: nikolaos.tzortzakis@cut.ac.cy

melon, cucumber and cucurbits (squash, pumpkin and zucchini) [1]. These vegetables are widely consumed for their edible fruits, which are considered significant sources of proteins, sugars, carotenoids, and vitamins, while in many parts of the world other plant parts such as leaves and young stems, and mature seeds are also found to have culinary uses. However, apart from edible uses all the above-mentioned species have also shown significant health effects and therapeutic properties, such as anti-diabetic, anti-inflammatory, antitumor, anti-viral, hepatoprotective, cardiovascular and immunoregulatory activities [2, 3], and have been extensively used in traditional and folk medicine since centuries [4]. Moreover, except for the widely known cultivated species, Cucurbitaceae family includes several less commonly known species usually classified as gourds which find various uses, including industrial and pharmaceutical ones. Some of these Cucurbitaceae species have been recently introduced as potential rootstocks in vegetable grafting, including *Lagenaria siceraria*, *Benincasa hispida* and interspecific hybrids of *Cucurbita* sp., while others have been traditionally used in the tropics (*e.g. Momordica* sp.).

Most of these health effects have been attributed to various heterogeneous biological active compounds, including carbohydrates, cardiac glycosides, carotenoids, resins, phytosterols, saponins, tannins and terpenoids (especially cucurbitacins), which in most cases act synergistically [5 - 7].

The present chapter will discuss the chemical composition and health related effects of the various phytochemicals for each species, focusing on the main cultivated vegetables, as well as on less common species with regional interest and/or industrial uses.

MELON

Introduction

Melon (*Cucumis melo* L.), also known as Musk melon or Cantaloupe, is considered of great importance across many areas of the world with great diversity in cultivation area as well as in human uses, while the origin and domestication of melon have not been clarified so far. Africa is considered to be the centre of origin of melon, though the recent data identify Asia as the origin of the genus *Cucumis* [8, 9]. Many different melon cultivars are being cultivated worldwide such as reticulatus, cantaloupensis, inodorus (including the honeydew, honeyball, casaba, grenshaw, Persian), flexuosus, dudaim, chito and conomon. Therefore, melon fruit exhibit a great variation in fruit traits such as size, shape, colour, taste, texture, and biochemical composition rendering the species as the most variable within the genus *Cucumis* [10, 11].

The human uses of melon fruit depend on the type of fruit, where sweet types are being consumed as a dessert in the summer period due to its refreshing, sweet and aromatic flesh, while non-sweet types are being used mostly as vegetables, *i.e.* the immature fruits are eaten raw, pickled or cooked [12]. Other uses of melon plants include its ornamental use [13].

C. melo includes the orange flesh cantaloupes, green flesh honeydew, and mixed melons. Cantaloupe and honeydew melons are commonly grown varieties throughout the world and are popular dietary choices [14], since they are low in energy and are excellent sources of nutrients, in particular provitamin A and vitamin C [15]. The melon variety *reticulatus* is highly appreciated and consumed due to its nutritional and organoleptic attributes, as well as to its vitamin C, β-carotene, polyphenol antioxidants and potassium content [16].

Apart from raw consumption, fruit may be used in juices, nectars, pickles or jams, resulting in great amounts of peel and seeds as by-products and waste parts [17]. Edible parts (juice and pulp) represent around 65% of total fruit weight of *Cantaloupes* (being 42% attributed to juice and 23% to pulp), while waste parts (peel and seeds) constitute 32% of total weight (25% to peel and 7% to seeds) [17]. Compositional, nutritional, and functional properties are the significant parameters in determining food quality, where health-related compounds have been classified as phenolics, flavonoids, ascorbic acid, and carotenoids [18].

Fruit Chemical Composition

Melon is a valuable source of **minerals** and antioxidant compounds. Mallek-Ayadi *et al.* reported that melon peels contain significant amounts of calcium (11.53 mg/g), potassium (8.84 mg/g), magnesium (3.89 mg/g) and sodium (1.44 mg/g) on a dry weight basis [19].

Melon peels have relatively low **moisture content** (16.95%), while the **ash content** of melon peels is higher than other curcubits, such as pumpkin and watermelon (3.67% *vs* 2.46% and 3.07%, respectively) [19]. Color of peels depends on different combinations of chlorophyll, carotenoids and flavonoids which gradually change during fruit maturity [20]. **Lipid content** in peels ranges from 1.58-2.12% [21] and total **dietary fiber content** is 41.69%, 37.58% of which is insoluble dietary fiber and 4.38% is soluble dietary fiber [19]. **Protein** content of fresh melon is reported between 0.5% and 1.31% on a fresh weight basis [22], while on a dry weight basis, protein content varies between 5.74% and 16.91% [23].

Juice, pulp and peel are rich sources of vitamin C (323.21, 220.38, and 336.78 µg/g, respectively), whereas Amaro *et al.* [24] reported lower contents

(153.3 µg/g) in *Cantaloupe* fresh-cut tissue. However, different amounts of vitamin C have been reported in other studies for juice (107.59 mg/100 g), pulp (88.08 mg/100 g), peel (42.76 mg/100 g) and seeds (21.50 mg/100 g) [17, 25]. Moreover, Lester and Saftner [26] reported different ascorbic acid content between three different muskmelon cultivars (34.7-44.7 mg/100 g), while Kolayli *et al.* [27] found significantly higher contents in ungrafted melon compared to grafted or hybrid melons. **Total phenolics** content was 100 µg/g in fruit peel [17, 28] while higher values were found in seeds [17]. According to Mallek-Ayadi *et al.* [19] total polyphenol content of melon peel (maazoun cultivar) was 3.32 mg/g extract, while Ismail *et al.* [29] reported that cantaloupe peels contain higher amount of total polyphenols (4.70 mg/g extract) in comparison to seeds (2.85 mg/g extract) and melon flesh (1.68 mg/g extract). Kolayli *et al.* [27] suggested that the main phenolic components in fruit were benzoic, abscisic, vanillic, and trans-cinnamic acids, followed by p-coumaric, ferulic and gallic acids, while protocatechuic acid and catechin were not detected. Both, Mallek-Ayadi *et al.* [19] and Al-Sayed and Ahmed [21] identified 3-Hydroxybenzoic acid as the major phenolic compound in melon peels (3.34 mg/g extract), followed by isovanillic acid (2.37 mg/g extract), m-coumaric acid (1.99 mg/g extract), gallic acid (1.21 mg/g extract), tyrosol (1.13 mg g extract) and hydroxytyrosol (0.91 mg/g extract). Moreover, Díaz-de-Cerio *et al.* [30] reported different amounts of total phenolic compounds when quantified using high performance liquid chromatography method (2.34 mg/g extract) and spectrophotometric analysis (Folin Ciocalteu assay; 3.32 mg/g extract). This is related to the interference of Folin Ciocalteu reagent with other substances (such as sugars and pectins) which result in overestimation of phenolic compounds content.

Several studies reported the **total flavonoids** content of fruit peels and flesh [29, 31, 32] and only a few of them further examined the flavonoids individual compounds [19]. Therefore, melon peels contain notable amounts of apigenin-7-glycoside (0.29 mg/g extract), luteolin-7-glycoside (0.16 mg/g extract), flavone (0.13 mg/g extract) and naringenin (0.16 mg/g extract) [19]. Total flavonoid content of melon peel extract ranged between 0.95 and 1.06 mg/g extract) in fruit from different origin (Tunisia and Brazil, respectively), as reported by Mallek-Ayadi *et al.* [19] and Morais *et al.* [32], indicating a significant effect of genotype, fruit maturity and growing season [20].

Total chlorophylls content ranged between 50 and 750 µg/g, depending on the genus, species, cultivar and environmental factors [20]. For melon fruit, high concentration of **β-carotene** is linked with its nutritional quality and associated with health benefits [16]. Orange-fleshed muskmelons, as *Cantaloupe*, are excellent sources of β-carotene [33], while fruit parts differ in terms of total carotenoids concentration, namely 49.90 mg/g, 30.51 mg/g, 68.92 mg/g, and

23.46 mg/g in juice, seeds, pulp, and peels, respectively. Moreover, Moreira *et al.* [25] reported a value of 40.0 mg/g of total carotenoids in fresh-cut *Cantaloupe* melon, when studying the effect of packaging on the fruit. Additionally, Laur and Tian [34] reported that pigments (carotenoid and chlorophyll) composition is largely different between cantaloupe and honeydew fruits, with cantaloupe fruits do not contain chlorophylls, and β-carotene accounting for the majority of total carotenoids, while in honeydew melons the major pigments are chlorophylls a and b. Moreover, z-carotene (in cantaloupes) and violaxanthin (in both cantaloupe and honeydew melons) are present in small amounts and do not possess provitamin A activities, while cantaloupe melons contain 20.20-2.53 mg/g f.w. β-carotene and 0.099-0.26 mg/g f.w. lutein; honeydew melons contained 0.004-0.30 mg/g f.w. β-carotene and 0.038-0.27 mg/g f.w. lutein [35].

Total soluble solids content is responsible for fruit flavor and depends on genotype and cultivation practices [22]. Kolayli *et al.* [27] reported that the main sugars are mannitol, sucrose, glucose, and fructose, while fructose/glucose ratio (F/G) is a specific indicator for the type of fruit juices (grafted, ungrafted plants and hybrids). **Acidity** in melons is due to the content of several organic acids such as citric, malic, fumaric, acetic, ascorbic or galacturonic [36]. Sweet melons possess low titrable acidity (0.12–0.2%) [37], while Burger *et al.* [38] reported that sugar and acid accumulation are not genetically controlled in sweet melon as it is possible to combine high sugar and acidity in one single genotype. Fundo *et al.* [17] reported seeds are the richest part (11.79 °Brix) in sugars with less acidity (pH 6.58), while peel had the lowest values in sugars (5.67 °Brix) and higher acidity (pH 5.17). The lower pH value obtained for peel than for seeds maybe explained by the presence of stronger organic acids in peel [17].

Seeds and Oils Composition

Seeds of *Cantaloupe* melon are important sources of potassium and total phenolics, and possess significant antioxidant activity [17], while the determination of chemical composition and antioxidant activity would significantly contribute to the valorization of cucurbit oil potential in food, cosmetic and pharmaceutical industries [39]. The high content of **oil** makes *Cucumis* seeds convenient for oil industry application. Regarding seeds of sweet melon varieties, oil content on a dry weight basis varies from 35.6 to 47%, while protein content varies from 23 to 36% [40].

Carbohydrates content ranged from 8.0-39% [41 - 44], while **moisture content** can vary from 4.27 to 7.78% [41, 42, 45 - 47]. **Ash content** of seeds depends on variety and ranges between 1.5% and 4.83% [44, 48]. The **total dietary fiber** content varies from 5.51 to 24.75% [41, 42, 46]. Seeds have higher **K content**

(708 mg/100 g) compared to peel (267 mg/100 g) [17], while its content shows great variability among the different cultivars [44, 49]. Iron, zinc, manganese and copper were present in relatively low amounts in melon seeds.

The total **phenolic content** of melon seeds varied among the studied cultivars with values between 80.0 and 304.1 mg/100 g. Individual phenolic compounds analysis in Maazoun melon seed oil resulted in four phenolic acids (gallic, protocatechuic, caffeic and rosmarinic acids), five flavonoids (luteolin-7-*O*-glycoside, naringenin, apigenin, flavone and amentoflavone), one secoiridoid (oleuropein) and one lignan (pinoresinol) [44]. The **total flavonoid** content of Maazoun melon seeds was 875.2 μg/g, while amentoflavone (32.80 μg/g) and luteolin-7-*O*-glycoside (9.60 μg/g) were the predominant flavones [44]. Other phenolic compounds as naringenin, apigenin and flavones were also identified in melon seed oil which accounted for 4.72 μg/g oil, 3.88 μg/g oil and 1.94 μg/g oil, respectively [44].

Regarding the content of biologically active substances such as sterols, tocopherols, and phospholipids in seed oils, the data are scarce [43]. According to Azhari *et al.* [46] and Mariod and Matthäus [50] sterols content was 0.3-0.8%, while the main phytosterols were campesterol, campestanol, 24-methylene-cholesterol, stigmasterol, chlerosterol, β-sitosterol, sitostanol, Δ^7-campesterol, Δ^7-stigmasterol, Δ^7-avenasterol, Δ^5-avenasterol, Δ^5,23-stigmastadienol, Δ^5,24-stigmastadienol, cholesterol and brassicasterol [43, 44].

The **total tocopherol content** in seed oils of Maazoun and *inodorus* Naudin melon was 27.02 mg/100 g [44, 51], while in Honeydew and Hybrid 1, the detected amounts were 82.8 and 73.1 mg/100 g [43]. The tocopherol composition of Maazoun melon seed oil was 18.13 mg/100 g for β+γ-tocopherols, followed by 6.09 mg/100 g for δ-tocopherol and 2.85 mg/100 g for α-tocopherol [44]. Petkova and Antova [43] reported that the main vitamin E in oils was γ-tocopherol, followed by α-tocopherol. In contrast, Azhari *et al.* [46], reported that δ-toco-pherol was the most abundant vitamer E in seed oil from variety *tibish* where predominated (63.4%), followed by γ-tocopherol (30.3%) and α-tocopherol (6.3%).

The most abundant **fatty acid** in seed oil is linoleic acid with values between 31.0% and 69.0%, followed by oleic acid (12.1% to 31.0%), palmitic acid (7.8% to 39.36%), and stearic acid (4.9% to 10.45%) [41, 43, 45, 46, 50, 52 - 54]. In winter melon (*Benincasa hispida*), locally known as Kundur and very popular especially among Asian communities had high linoleic acid content (63.10-70.64%), followed by palmitic (12.45-17.59%), oleic (8.46-12.87%) and stearic acid (5.13-7.48%) [52].

In addition, melon seed protein is rich in aspartic, glutamic acids and arginine in concentration of 9.07, 12.18 and 19.47 mg/100 g of proteins, respectively [45]. Other amino acids in lower concentrations were methionine, lysine, threonine, valine, leucine, isoleucine, cystine, proline, histidine, phenylalanine, tyrosine, alanine, serine and glycine [45, 55].

Other Uses

Melon presents a high potential for human consumption due to its exceptional taste. During processing of melon fruit, large quantities of by-products are generated. Much of these materials are composed by melon seeds which are generally discarded. The increasing demand for healthy foods has been growing in recent years and for that reason the food industry is constantly looking for new sources of nutritional and healthful components. Therefore, fruit wastes can be used as a source of natural food additives and ingredients, allowing for minimization of wastes during processing, reduction of the environmental impact and profit increasing [17].

Seeds could be considered as renewable resources from which several useful products can be derived. Indeed, due to the growing trend of replacing fats from animal origin with those of vegetables, some industries nowadays are shifting towards the use of natural raw materials for edible oils production [44]. Thus, fruit seeds including melon seeds can be used for the extraction of plant oils which contain a great number of valuable biocomponents and natural antioxidants [56].

Pumpkin and melon seeds are excellent sources of protein and oil and are utilized directly as snacks after salting and roasting mostly in Arabian countries [57, 58]. They are usually dried and used to add flavor to Indian dishes and desserts [59]. Melon seeds are also a good source of natural antioxidants and may serve as food ingredients for maintaining shelf life [60]. Moreover, melon seed oil is frequently used as cooking oil in some countries in Africa and the Middle East [61].

The kernels are sometimes used in sweet meats and in toppings as a substitute for almonds and pistachios in India [62]. Melon seed kernels are major soup ingredients and they are used as a thickener and flavour component of soups [63].

Moreover, leaf and stem extracts of cantaloupe melon are rich sources of antioxidants and can be considered as new sources of natural antioxidants for food and nutraceutical products [29].

Health Effects and Bioactive Properties

The widespread cultivation of melons throughout the world is due to its excellent taste qualities of the fruit, as well as its application in folk medicine as a medicinal plant [64]. Its consumption is recommended in the cases of cardio vascular diseases, liver and kidney diseases, in case of anemia, for patients with atherosclerosis, rheumatism, and gout [65]. In addition, melon seeds have therapeutic effects, such as anti-oxidant, anti-inflammatory, and analgesic effects [66, 67]. Along with their good taste qualities, melons show a great diversity in the various plant parts in terms of chemical composition, which makes them an excellent source of biologically active substances for the human organism.

Fruit

Fruit pulp is the most common edible part and is very refreshing and sweet in taste with a pleasant aroma. Several studies evaluated bioactive compounds in fresh-cut melon tissue [24, 33] although a significant part of bioactive compounds is lost on discarded parts (melon peel and seeds) during processing [68]. Consumption of 1–2 cups of cantaloupe fruit (about 150–170 g per cup) would meet the Recommended Daily Amount (RDA) of vitamins A (ranged from 700-900 mg Retinol activity equivalent (RAE)) and C (75-90 mg RAE) [35, 69].

Lutein is present in both cantaloupe and honeydew melons and is believed to have antioxidant activities in humans [70]. Melon fruit is recommended for the treatment of cardiovascular disorders as diuretic, stomachic and vermifuge [71]. Some reports are also available about the antioxidative, anti-inflammatory effects [72] and urease inhibitory potential [73] of melon fruits. According to Lester and Hodges [18], melons should be included in everyone's diet helping towards the reduction of the risk of cancer and chronic diseases. Moreover, fruit peels are rich in phenolic compounds, flavonoids, carotenoids and other biologically active components which have a positive influence on health [74].

Melon peels are rich sources of carbohydrates which are considered hypoglycaemic agents being known for their ability to modulate the immune system through the stimulation of macrophages and the reduction of inflammation [75].

Dietary fibers have also beneficial effects on human health and body function and its consumption is associated with a reduced incidence of common disorders such as cancer, obesity, diabetes and cardiovascular diseases [76, 77]. The incorporation of Maazoun melon peel powder in some food products (*e.g.* bakery products, yogurt and cornflakes) could increase their fiber contents and consequently enhance the nutritional and functional quality of the final product.

The nutritive value of melons is also related to the significant amounts of minerals as calcium, potassium, magnesium and sodium which are crucial for human health given its intrinsic role in bone and tooth development, preventing hypertension, reducing vasoconstriction and regulating heartbeat [78]. The presence of calcium and potassium in melon peels could enhance the beneficial effects on health of other phytochemicals such as phenolic compounds. Other mineral elements are also present in melon peels for instance magnesium and sodium, which promote a vaso-relaxation and a reduction in blood pressure, in combination with other similar ions such as potassium and calcium. Furthermore, magnesium and potassium are cofactors for many enzyme reactions [79].

Phenolic phytochemicals have the capacity to protect cellular components against free radicals *via* their antioxidant and free radical scavenging effects [80]. Furthermore, polyphenols prevent degenerative diseases such as cardiovascular and cancer diseases [81]. 3-Hydroxybenzoic acid is the predominant phenolic compound in melon peels [19, 21] and posses antifungal, antimutagenic, and antimicrobial activities [82]. Additionally, isovanillic acid exhibits antibacterial and antioxidant activities [82], while chlorogenic and coumaric acids (phenol carboxylic acids) exert beneficial effects on human health through the prevention of degenerative pathologies such as cardiovascular diseases and cancer [83]. Moreover, melon peels contain considerable amounts of gallic acid, tyrosol and hydroxytyrosol as these phenolic compounds prove to have anticancer, anti-inflammatory activities [84], free radical-scavenging and antibacterial properties against intestinal flora [85]. Hydroxytyrosol which is the main product of oleuropein degradation [86] may protect against atherosclerosis and prevent diabetic neuropathies [87].

Flavonoids such as apigenin-7-glycoside and luteolin-7-glycoside are the main flavones in melon peels and have been proved to have antioxidant, free radical scavenging, anti-inflammatory and anti-tumor effects [88, 89]. Since polyphenol components are often used as food stabilizers and natural conservatives due to their antioxidant potential and antibacterial properties [80], melon peels could be considered as a promising raw material for producing food additives and useful bioactive substances.

Seeds and Seed Oils

Melon seeds are a rich source of oil and protein and have been reported to possess medicinal properties and health beneficial properties such as preventing chronic diseases such as cancer and cardiovascular diseases [90 - 92]. *C. melo* seeds are used in Chinese folk medicine as antitussive, digestive, febrifuge and vermifuge [93], while melon seeds extract can be used as an antidiabetic and is also beneficial against chronic eczema [40, 94].

Oils and derived compounds are a main source of tocopherols and alpha-tocopherol in particular which is beneficial to human nutrition due to its higher biological activity compared to other tocopherols [95]. Moreover, high content of total sterols in seed oil makes them a good source for human nutrition. Sterols from plant oils are known to protect against cardiovascular diseases. In fact, phytosterols have an effect on the human body by inhibiting cholesterol absorption from the intestine and decreasing blood levels of low density lipoprotein cholesterol fraction (LDLc) [96, 97]. The main sterol (β-sitosterol) found in melon seed oils exhibited positive effects in the prevention of cell apoptosis and cancer. Ntanios [98] reported that sitosterol contributes to lower blood LDL cholesterol by 10–15% as part of a healthy diet, and therefore, melon seed oil may be used as a new therapeutic agent for the treatment of hypercholesterolemia [44]. Moreover, Δ^5-avenasterol can act as an antioxidant and anti-polymerization agent in frying oils. Sterols with an ethylidene group in the side chain are most effective as antioxidants, and a synergistic effect of the sterols with other antioxidants could be occurring.

The polyunsaturated fatty acids profile of melon seeds, constitutes an appreciable group of phytochemicals thanks to their beneficial health properties, such as the improvement of the immune response [99], the increase of insulin sensitivity [100] and the reduction of both total and LDL cholesterol [101].

Phenolic acids of melon seeds account for almost one-third of dietary phenols and are related with antioxidant, nutritional and organoleptic properties. Gallic acid is known to have anti-inflammatory activity [84] and often being used as additive to prevent the degradation of foods [44]. Oleuropein is also associated with high antioxidant capacity [102].

Bioactive Properties

Anti-Inflammatory and Analgesic Activity

According to Gill *et al.* [103] methanolic extracts of seeds showed significant anti-inflammatory activity against carrageenan-induce paw edema, as well analgesic activity at doses of 300 mg/kg. Similar results have been reported for seed methanolic extracts of *C. melo* var. agrestis at the same doses [104]. Moreover, Vouldoukis *et al.* [72] suggested that cantaloupe melon fruit extracts with high superoxide dismutase activity may exhibit anti-inflammatory properties, mainly through the production of anti-inflammatory cytokines (IL-10).

Antioxidant Activity

Antioxidant activity is associated to the presence of vitamins and polyphenols, and plays an important role in the prevention of oxidative stress responsible for chronic diseases [105]. Ascorbic acid, the most well-known antioxidant, is an important molecule in plant tissues and protects plants against oxidative damage resulting from the oxidant metabolites of photosynthesis and aerobic processes [106]. According to Ismail *et al.* [29], increased antioxidant capacity was found in stem and leaf extracts of cantaloupe fruit, which was suggested to be related to the presence of phenolic compounds and flavonoids in particular.

Other compounds such as chlorophylls have been strongly related to green color characteristic, and antioxidant and antimutagenic activities [107]. Moreover, seed extract showed the lower antioxidant activity (0.41 mg alpha-tocopherol equivalent (Teq)/g extract) towards the bleaching of β-carotene, when compared with other plant parts (leaf, stem *etc.*) [29]. However, according to Contreras-Calderón *et al.* [68] seeds had a considerable antioxidant activity, which highlights its use as an important source of natural antioxidants due to the presence of polyphenol components, mainly phenolic acids and flavonoids [108]. Cucumin S, a phenylethyl chromone isolated from *C. melo* var. *reticulatus* seeds has also shown significant antioxidant properties [109].

Antiulcer Activity

Gill and Bali [110] isolated a new tetracyclic triterpenoid with cucurbitacin structure which exhibited significant antiulcerogenic properties against pyloric ligation, water immersion stress and indomethacin-induced ulcer in rats and mice in a dose dependent manner.

Anticancer Activity

Cucurbitacin B, a compound which is usually present in melon and other species of Cucurbitaceae family is considered a natural anti-cancer agent with significant antiproliferative effects against various human leukemia cells mostly through the inhibition of STAT3 activation and the Raf/MEK/ERK pathway [111]. Moreover, the combined administration of cucurbitacin B and docetaxel resulted in growth inhibition of laryngeal cancer cells, cell cycle arrest and apoptosis induction [112]. Ibrahim *et al.* [113] isolated a new triterpenoid namely cucumol A which was attributed with significant antiproliferative activity against L5178Y and Hela cancer cell lines.

Diuretic-Nephroprotective Effects

Fruit pericarp possess diuretic properties, as reported by Isea Fernández *et al.* [114] who suggested oral administration of 3 and 6 mL/300 g to increase urinary excretion in rats in a dose dependent manner. Ravishankar *et al.* [115] reported that the oral administration of seeds ethanolic extract at doses of 400 mg/kg increased significantly the urinary volume and chloride substance in rats, showing important diuretic effects. According to Naito *et al.* [116] the oral administration of melon fruit extracts (oxykine) prevented diabetes-induced nephropathy in rodent models, through the decrease of urinary albumin and urinary 8-hydroxyde-oxyguanosine (8-OHdG), and renal mesangial injury.

Antidiabetic Activity

The oral administration of fruit peel extracts stimulated thyroid function in propylthiouracil-induced hypothyroid animals through the increase of triiodothyronine (T3) and thyroxin (T4) hormones, while it further decreased lipid peroxidation [117]. Moreover, methanolic and aqueous extracts of fruit peels at doses of 500 mg/kg exhibited significant antihyperlipidemic activity in high cholesterol diet induced hyperlipidemic rats by effectuating reduced body weight gain and improved serum lipid profile [118]. Similarly, fruit peel extracts demonstrated a great potency in amelioration of negative effects in rats with diet-induced dyslipidemia, hypothyroidism and diabetes mellitus, while it was suggested that this potency could be attributed to polyphenols and ascorbic acid content of fruit peels [117].

Cardiovascular Effects

Supplementation of melon concentrates for a 4-week period resulted in alleviation of vascular dysfunctions associated with cardiovascular diseases, probably through the increase of nitric oxide bioavailability and SOD expression [119].

Antimicrobial Activity

Ethanolic extracts of *C. melo* fruit contain various bioactive compounds with antimicrobial activities, as was reported by Sasi Kumar *et al.* [120] who studied their fruit extracts effects against various microorganisms, such as *Salmonella typhi*, *Staphylococcus aureus*, *Bacillus cereus*, *Klebsiella pneumonia* and *Escherichia coli*. Melon seeds have been reported to exhibit antimicrobial and cytotoxic properties, mostly due to their chromones content [121] while peptides from melon seeds also presented antifungal activities against *Fusarium oxysporum* [122]. Moreover, seed oil have been used in formulation of skin creams which also exhibited inhibition properties against various microorganisms [123].

Immunomodulatory Effect

Ethanolic extracts of *C. melo* fruit have been associated with significant *in vivo* immunomodulatory effects in mice, since their administration stimulates cell mediated and humoral immunity [124].

WATERMELON

Introduction

Watermelon (*Citrullus lanatus* L.) is also a member of Cucurbitaceae family and the third most popular fruit vegetable in the world [125], while it accounts for most of the world production of all Cucurbits. Watermelon, native of southern Africa, represents a large portion of the Mediterranean diet, dating to aprroximately 3,000 years ago [126]. Apart from raw consumption, fruit may be used in juice, jam, pickled rind, *etc.*, while the seeds are also edible (seed cultivars). Watermelon flesh color is varied and the sweet interior flesh of the fruit is usually deep red or pink and sometimes orange, yellow, white or even green in non-ripe fruit [127]. Edible parts (flesh) represent around 68% of the total fruit weight, while considerable amounts of plant waste constitute of seeds and rinds (2% and 30%, respectively) [128].

Watermelons are widely appreciated for its refreshing properties and this is related to the high-water content of the pulp and to the mineral salts concentration. The pleasant taste of the fruit is due to the high content of sugars and to the volatile compounds profile. Moreover, the bioactive phytochemicals, such as carotenoids (lycopene and β-carotene) confer health promoting properties [129]. Medicinally, watermelons are mildly diuretic, being effective in the treatment of dropsy and renal stones, in reducing hypertension, in preventing erectile dysfunction, acting as an antioxidant, and treating enlarged liver, jaundice and giardiasis [130 - 132].

Fruit Chemical Composition

Watermelon contains more than 91% water and up to 7% carbohydrates and as such it has low calorie content and is a natural source of minerals like K, P and Mg and antioxidant compounds such as lycopene and vitamin C (ascorbic acid) [133, 134]. The proximate composition of watermelon fruit samples in fresh weight are moisture 90.82-95.00%, protein 0.12-0.60%, fat 0.05-0.10%, fiber 0.20-0.60%, ash 0.28-1.13%, carbohydrates 3.10-8.00%. Mineral composition on a dry weight basis varied between 1.00-4.89 mg/g for K, 0.12-0.14 mg/g for Mg, 0.39-0.43 μg/g for Mn and 0.92-1.88 μg/g for Fe [135].

Fruit pulp redness might be varied by the cultivar at harvest stage, by the storage temperatures (5 °C *versus* 21 °C) [136], by grafting and mycorrhizal presence [137]. **Fruit firmness** is ranged to 15-20 N [129, 138]. The minimum **soluble solids content** (SSC) requirement for ripe watermelons is 8% [139], while

Mantoan *et al.* [140] found a SSC range of 10.6-12.5% for ripe fruit of ten mini-watermelons varieties. In mature watermelons, sucrose and glucose accounted for 20–40% of total sugars content, fructose for 30–50%, while sugars composition is highly dependent on maturity and ripening stage [141 - 143] **Acids** and **esters** have been reported in watermelons too (89). **Vitamin C** is also identified in watermelons, ranging from 38.2 to 576.2 mg/kg f.w. [144], and the content is dependent on fruit ripening stage [139, 145]. Others vitamins also identified in watermelon in low content include vitamin B_6 (0.06-0.07 mg/100 g d.w.) and vitamin E (0.01-0.04 mg/100 g d.w.) [135, 146].

Phenolic compounds are the principal hydrophilic compounds contributing to the hydrophilic antioxidant activity in watermelon [139]. Most studies on Cucurbita species have focused on the total content of phenols and antioxidant activity [29, 139, 147], and limited data are available on the individual phenolic and phyto-components of these plant foods. Vinson *et al.* [148] reported that watermelon total phenolic content was 26 mg/100 g, which includes 9.5% of free and 90.3% of conjugated phenolics [149]. Abu-Reidah *et al.* [150] reported a total of 71 phytochemical compounds tentatively characterized, mainly phenolic compounds: phenolics acids (hydroxybenzoic and hydroxycinnamic acids), flavonoids derivatives, lignans, iridoids, coumarins, stilbenoids, and others. Abu-Hiamed [135] reported rutin and β-sitosterol mean contents of watermelon fruit of 1.66 and 0.460 mg/100 g f.w., respectively.

Watermelon, also contains **polyamines** which are essential for growth and cell proliferation. The most common polyamines are spermidine and spermine, together with their diamine precursor, putrescine [151].

The main **carotenoids** in watermelon are lycopene and β-carotene with lycopene being the prevalent carotenoid in common red-fleshed cultivars [129]. Up to 10–14 different **carotenoids** have been also identified in the flesh of mature watermelon fruits [152, 153] but only seven of them (phytoene, phytofluene ζ-carotene, lycopene, b-carotene, lutein, and violaxanthin), accounted for more than 90% of total carotenoids content [154]. Proietti *et al.* [133] reported that lycopene content increased (by 40%) in grafted watermelons, while in red-fleshed watermelons, the major carotenoid is lycopene [20, 155] in amounts of 48.7 mg/g f.w. [156].

According to Beaulieu and Lea [157], the main **volatile compounds** in watermelons were aldehydes and alcohols with 23 alcohols, 21 aldehydes, eight ketones, seven hydrocarbons, one acid, two lactones, 12 furans, and one oxide being identified. Many of these C9 aldehydes compounds have been attributed to characteristic aroma attributes, for example (*Z*)-3-nonen-1-ol (fresh melon-like odor), (*Z,Z*)-3,6-nonadien-1-ol, (*E,Z*)-2,6-nonadienal, and (*E*)-2-nonenal have been reported as characteristic flavor components in seeded watermelon [89, 158]. The characteristic aroma of watermelon is found to be attributed mainly to the alcohol (*Z,Z*)-3,6-nonaden-1-ol [158]. The aldehyde (*Z*)-6-nonenal was considered flavor important and more abundant in watermelon as compared with muskmelon [159], while both (*Z*)-3-nonenal and (*Z,Z*)-3,6-nonadienal have been considered as important to flavor compounds [158].

L-citrulline is a non-protein **amino acid** that was first identified from watermelon [160] and as such is emerged as an important amino acid both as a product of the NO cycle and as a precursor for arginine [161]. As an organic supplement, citrulline appears to be a powerful pharmaconutrient with therapeutic potential to restore arginine metabolism in critically ill patients with sepsis [162]. **L-citrulline content** in three varieties of watermelon juice varied in the range of 11.25–16.73 mg/g, while content in the rind was between 13.95–28.46 mg/g on dry weight basis [163]. Rimando and Perkins-Veazie [164] reported that the citrulline content in flesh was affected by variety when 14 watermelon varieties examined.

Seeds and Seed Oils Composition

El-Adawy and Taha [58] reported the proximate composition of watermelon seed for *C. lanatus* on a dry weight basis (in parenthesis is the relevant values for *C. vulgaris* seeds), including moisture 8% (9%), protein 34% (35%), fibre 3% (2%), ash 3% (3%), phytic acid 5 mg/g (4 mg/g), oxalate 12% (10%), total phenols 2 mg catechin/g (2 mg catechin/g), hydrocyanic acid 15 mg/100 g (15 mg/100 g), saponin 4% (4%) [58].

According to Johnson *et al.* [165] and Sani [166], saponins, alkaloids, tannins, phenols, and flavonoids were present in the seed and rind of *C. lanatus*. **Fat and protein** together account for almost 75% of seeds weight, which shows its potential as a protein source [167], such as globulin, albumin and glutelin [168].

Watermelon seed oil consisted of 59.6% linoleic acid (18:2n-6) and 78.4% total unsaturated fatty acids. The predominant fatty acid in the oil was linoleic acid, followed by oleic, palmitic, and stearic acids. Linolenic, palmitoleic, and myristic acids were minor constituents [52, 58].

The **phytic acid content** of watermelon seeds meal ranged between 0.99 and 2.63 g/100 g [58, 167], the **tannin content** ranged between 0.24 and 0.61 mg/100 g [58, 169], and the relevant **alkaloid content** was 0.36 mg/100 g [169]. The **phenol** content in seeds was 0.12 mg/100 g [169], while the content for cyanide, phytate and oxalate were 0.79, 0.63 and 0.09 mg/100 g, respectively [169]. The **amino acid** pattern indicated a high concentration of leucine, aromatic amino acids and tryptophan while arginine was found to be higher in watermelon seeds than most of other oilseeds including soya, groundnut and gingelly seed oils [167]. Other amino acids reported in watermelon seeds are aspartic acid, glutamic acid, alanine, serine and proline [58, 167].

The **antinutrients** (expressed in mg/100 g) in rind and seeds, respectively *viz*: saponin (3.0 and 2.31), alkaloid (1.39 and 0.36), tannins (1.33 and 0.61), phenol (0.53 and 0.12) and flavonoid (2.87 and 2.03) were comparably higher in the rind than in the seed [169]. However, seed content (expressed in mg 100/g) in cyanide (0.79), phytate (0.63) and oxalate (0.09) was higher than rind for cyanide (0.00), phytate (0.46) and oxalate (0.08) [169].

The **bioaccessibility of minerals** (bioaccessibility is the fraction of ingested nutrient that is available for utilization in normal physiological functions and for storage) in the oilseeds is a less researched area, but this information is of high significance as oilseeds form major protein sources and are extensively used in complementary foods, supplementary foods and bakery products. Bio-accessible iron in the watermelon seed components was found to be < 1 mg/100 g in whole meal and defatted flour, while it increased by two-fold in seed isolates [167]. Moreover, the same authors reported for the first time on the bioavailable zinc and calcium in the watermelon seed [167].

Other Uses

Harnessing food wastes utilization in diets and drugs could improve food supply, health and the environment while antinutrients composition in food provides an idea of its pharmacological, dietary and toxicity effects [169]. Watermelon wastes such as rinds/peels and seeds, can be used as antinutrient components with pharmacological properties and (especially at higher amounts) could be toxic by forming insoluble complexes with nutrients and reducing their bioavailability and absorption [170].

The rind of watermelon accounts for more than 30–40% of fruit weight and is used in the preparation of pickles and candies due to the presence of vital nutritional components [171]. Watermelon rinds contained 47.87% carbohydrates and 16.46% protein [172]. Watermelon rinds polysaccharides (WMRP) have various functional properties such as water holding capacity (2 g H_2O/g sample), fat-binding capacity (4 g oil/g sample), foam capacity (150%) and high antioxidant activity [172]. Watermelon rinds may be used for the development of novel biologically active components that can be incorporated in cosmetics, food and pharmaceutical products due to their polysaccharides contents, their antioxidant and antihypertensive activities as well as their functional proprieties [172].

Watermelon seeds are consumed as snack food worldwide and are used to prepare edible oil in some countries [52]. In some parts of the world as Nigeria and Middle East countries the seeds are used for extracting cooking oil, while in the majority of the countries its utility is restricted for uses as additive [167]. Seeds are also used as condiments, garnishers, thickeners in soups, fat binders, flavourants and as snacks in most parts of the world [58, 173].

Other seed oil uses include the production of various products such as shoe polishes, foaming agents, paints and creams requiring qualities as high saponification value as well as capacity to flow and spread easily on surfaces [174]. Watermelon seed oil edibility is related to the low molecular weight saturated fatty acids content [175], as the iodine value (28.51 g/100 g) was within the range for the standard oils [174]. Thus, the watermelon seed oil could be nutritionally beneficial and the low iodine value could confer it with greater oxidative stability during storage than most of the standard oils.

Functional properties of watermelon seeds flour (defatted) and isolate are reported by Jyothi lakshmi and Kaul [167] as their water absorption capacity varied from 3.3 to 4.0 mL/g in flour and isolate, respectively, and foam capacity between 68% and 87% in the flour and isolate, respectively.

In Nigeria, *C. lanatus* is an underutilized plant and can be a source of cheap seed oil, while serves as raw feed for animal stocks and for the production of biosurfactant with detergents and cosmetics uses [176]. Adewuyi *et al.* [176] reported that watermelon seed oil biosurfactants inhibited the growth of organisms such as *P. aeruginosa*, *S. aureus*, *K. pneumonia* and *E. coli.*

Watermelon seeds could be used as a non-conventional animal feed because of its availability, low cost and non-human feed [177]. Thus watermelon seed cake is a good source of protein for animal and it is comparable to cotton cake and linseeds cake [178].

Health Effects and Bioactive Properties

Fruit

Lycopene, the main carotenoid in watermelon acts as a powerful *in vitro* free radical scavenger [179] and is effective against oxidative stress, cardiovascular diseases and has a potential role in the prevention of prostate cancer [180 - 182]. Lycopene bioavailability is dependent on the kind of vegetable matrix while the consumption of watermelon juice increases the lycopene concentration in human plasma [183]. **β-carotene** in watermelon, is a potent quencher of singlet oxygen, acts as a free radicals scavenger in human LDL, HDL and in cell membranes and can regenerate the antioxidant form of vitamin E [129].

The **phenols** have high antimicrobial potential, which could be considered in the treatment of typhoid fever and other bacterial infections [184]. The effectiveness of watermelon extracts has been recently described in promoting a laxative effect, this property being due partly to phenolic compounds [185].

Alkaloids (especially at low concentration) are therapeutically significant natural plant products responsible for analgesic, antispasmodic and antibacterial properties [184]. **Saponins** that are present in watermelon rinds have bitter taste which could be associated with pharmacological potential, including hemolytic activities [186] and beneficial effects on blood cholesterol levels, bone health, cancer and the stimulation of the immune system [184]. **Phytates** provide strong antioxidant effects, while the presence of phytate in watermelon rind is suggestive of antioxidant benefits following their consumption [184].

A dietary supplement rich in **L-citrulline** which is usually present in watermelon, seems to improve sexual stamina and erectile functions, however the exact mechanism is not known [187]. L-citrulline is an essential amino acid for young mammals and adults with trauma, burn injury, massive small bowel resection, and renal failure [188]. Extracts of L-citrulline may help in smooth muscle relaxation, which is based on NO producing ability, decrease in intracellular calcium content and inhibition of phosphodiesterase-5A (PDE-5A) and these highlight the importance of potential beneficial role of watermelon extracts at cellular level [163].

Seed and Seed Oils

Watermelon protein-rich seeds are commonly used in traditional medicine with increasing acclaimed efficacy against diabetes mellitus (DM) in Africa [189]. Watermelon **amino acid** composition (high arginine content) is indicative of the possession of medicinal properties [58]. **Tannins** presence in watermelon rind and seeds implied that these tissues may have astringent and antimicrobial properties [184] and may be considered for treating a wide range of ailments, including inflammation, liver injury, kidney problems, arteriosclerosis, hypertension, stomach problems and scavenging of reactive oxygen species [170]. Moreover, the low peroxide value of watermelon seed oil (2.80 mg KOH/g) suggested its high degree of stability and non-susceptibility to oxidative rancidity and could be suitable for human consumption [190].

Bioactive Properties

Antidiabetic Activity

Several authors have reported flavonoids, sterols, tannins, saponins, alkaloids and polyphenols as bioactive antidiabetic compounds [191, 192]. According to Ahn *et al.* [193] the antidiabetic activity of raw *C. lanatus* rind and flesh was studied in streptozotocin-induced diabetic mice and results have shown a reduction in blood glucose level. On top of that, Perkins-Veazie [194] found that 1% supplement with ethanol extract from watermelon rind had lowering effect on blood glucose, while it raised serum insulin concentrations in rats, by using immunohistochemistry as it was determined that this was due to watermelon having a protective effect on the pancreatic β cells [193].

Anti-Hyperglycaemic Activity

Watermelon seeds contain globulins which have been attributed with anti-hyperglycaemic activity and this can be explored further for the development of peptide-drugs and/or phytomedicines incorporating these bioactive proteins which could be used as affordable alternative therapies against diabete mellitus [189]. A high dose of watermelon juice has also been shown to have hypoglycaemic effects in diabetic rats [195].

Antihyperlipidemic Activity

According to El-Razek and Sadeek [195], a high dose of watermelon juice had hypolipidemic effects in diabetic rats. Moreover, oral administration of whole fruit extracts including skin, pulp and seeds exhibited significant anti-hypercholesterolemic effects in hypercholesterolemia-induced atherosclerotic mice, while it also reduced body weight gain and enhanced homeostasis of cytokines [196].

Anti-hypertensive Effects

Romdhane *et al.* [172] reported that watermelon rinds polysaccharides had a very important angiotensin I-converting enzyme (ACE) inhibitory activity, and this activity increased in a concentration-dependent manner.

Antioxidant Activity

Citrulline is also an efficient hydroxyl radical scavenger and is a strong antioxidant [197, 198]. Flesh lycopene content is an important nutritional trait that sets watermelon among lycopene- rich functional foods [199, 200].

Cardiovascular Effects

In an earlier study, it was found that D- or L-citrulline from watermelon rinds significantly attenuates cardiac contractile dysfunction in the isolated perfused rat heart subjected to ischemia/reperfusion *via* non-nitric oxide mediated mechanism [201]. Balderas-Munãoz *et al.* [202] reported that adults with heart failure had improvements in left ventricular ejection fraction, functional class, and endothelial function as assessed by photoplethysmography after oral citrulline treatment for 4 months.

Diuretic Effects

According to Gul *et al.* [203], the high amount of water content of watermelon makes it a powerful diuretic diet.

Anticancer Activity

The potential role of lycopene in watermelon was reported by Rahmat *et al.* [204] in human breast and liver cancer cell lines.

Anti-Inflammatory and Analgesic Activity

Rind, root, seed and leaf extract of watermelon showed analgesic and anti-inflammatory effects which are related with the inhibition of NO production in macrophages without showing toxic effects on these cells [205]. Significant *in vivo* anti-inflammatory activity has been reported for watermelon fruit juice and cucurbitacin E in particular, where according to Abdelwahab *et al.* [206] intraperitoneal injection of cucurbitacin E resulted in significant inhibition of carrageenan-induced paw edema in rats. Moreover, in the same study the mechanisms of action through *in vitro* tests were evaluated and revealed that anti-inflammatory effects are probably associated with the inhibition of COX-1 and COX-2 (cyclooxygenase) enzymes and reactive nitrogen species (RNS).

Antiulcer Activity

Lucky *et al.* [207] presented for the first time that watermelon seeds possess antiulcer properties against acetylsalicylic acid-induced ulceration in animals model. According to Gill *et al.* [208] oral administration of methanolic extract of *C. lanatus* seeds in rats at the dose of 300 mg/kg showed a significant decrease in the gastric volume, free acidity and total acidity in the case of pyloric ligation and showed a significant percentage inhibition of ulcer as indicated by the decrease in ulcerative index.

Hepatoprotective Effects

The effects of the leaf extract of *C. lanatus* on carbon tetrachloride (CCl_4) induced liver damage in rats was investigated by Adebayo *et al.* [209], while Madhavi *et al.* [210] also reported the hepatoprotective activity of watermelon seed oil on CCl_4 induced liver damage in rats. Similar results have been reported in *in vivo* studies with mice models, where methanolic extracts of watermelon seeds reduced lesions, necrotic tissues, lymphocyte infiltration and fatty degeneration [211]. Moreover, oral administration of fruit juice resulted in amelioration of oxidative effects in liver tissues of ethanol-induced intoxicated rats [212].

Antianemia Effects

L-citrulline from watermelon rinds given orally to children and adolescents with sickle cell disease resulted in the improvement of symptoms, raised plasma arginine levels, and reduced elevated total leukocyte and segmented neutrophil counts to within normal limits [213].

Reproduction System Effects

Dietary supplements containing citrulline have been used to improve sexual stamina and erectile function; however, the mode of action for this activity is still not unknown [187].

CUCUMBER

Cucumber (*Cucumis sativus* L.) is the second and third most important cucurbit after watermelon and melon, in terms of total production and harvested area, respectively [214]. The plant is a creeping vine which forms several fruit (pepos) that can be consumed raw as salad vegetables or after processing (pickles) before physiological maturity. Apart from the fruit, other plant parts may have edible uses, such as young shoots and leaves which can be consumed as leafy vegetables, as well as seeds which can be consumed in raw form or being used for oil extraction [7, 215].

Chemical Composition

The phytochemical content of cucumber consist of various compounds, such as alkaloids, flavonoids, carotenoids, phytosterols, saponins, steroids, tannins, and terpenoids [216]. The various plant part differ in terms of chemical composition and although the edible parts of the plant are limited to fruit and to a lesser in tender shoots, all of them have found uses in traditional and folk medicine from time to time due to their phytochemical content. Aqueous fruit extracts contain mostly carbohydrates, flavonoids, glycosides, steroids, and tannins, while leaves and seeds are rich in flavone *C*-glycosides (isovitexin, isoscoparin, saponarin, vicenin-2 and apigenin) and 9-beta-methyl-19-norlanosta-5-ene glycosides, respectively [217, 218]. Carotenoids is another important group of compounds that are present in cucumber fruit, where according to Kowalczyk *et al.* [219] its content is highly affected by cultivation practices (growing season and growth substrate) as also be genotypic factors. In another study, two *Cucumis* species (*C. sativus* L. and *C. trigonus* Roxb.) were compared regarding their fruit phytochemical content and the analyses detected the presence of various compounds such as alkaloids (0.82 mg/kg), total flavonoids (2.63 mg/kg), tannins (0.12 mg/kg), lignin (0.09 mg/kg), glycosides (0.09 mg/kg), serpentines (0.12 mg/kg), terpenoids (0.05 mg/kg), saponins (0.03 mg/kg) and total phenols (0.18 mg/kg) [220]. Moreover, Sotiroudis *et al.* [221] identified fruit pulp as the main source of phenolic compounds (13.8 mg/100 g fruit), comparing to fruit peel (6.2 mg/100 g fruit) and juice (5.4 mg/100 g fruit). In contrast, Sahar *et al.* [222] reported the absence of alkaloids, steroids and anthraquinones in fruit pulps, which indicates that peels are the main source of alkaloids in cucumber fruit. Moreover, Oboh *et al.* [223] have reported the presence of flavonoids such as rutin, quercetin, apigenin, and kaempferol as the main phenolic compounds of cucumber fruit, followed by caffeic, gallic and p-coumaric acid, and luteolin. Abu-Reidah *et al.* [224] have carried out a detailed characterization of phenolic compounds in whole fruit hydromethanolic extracts using a Reversed-Phase High-Performance Liquid Chromatography equipment coupled to Electrospray Ionization Quadrupole Time-of-Flight Mass Spectrometry (RP-HPLC–ESI–Q-TOF-MS), and reported 73 different phenolic compounds, including conjugated and glycosylated forms of caffeic, p-coumaric and ferulic acids, as well as many flavonoids and its derivatives. Chiu *et al.* [225] have also indicated cucumber flesh and peels of fruit as a rich source of chitosan which may have several practical applications due to its biological activities, while Sotiroudis *et al.* [221] identified twenty one different volatile compounds in dichloromethane and methanolic extracts of fruit.

Cucumber leaves are also a good source of a variety of phytochemicals including cucumerin A and B which have been classified as flavone *C*-glycosides, as well as vitexin, isovitexin, orientin, isoorientin and hydroxycinnamic acid [226]. Other important compounds include megastimane, cucumegastimane I and II, megastigmane (+)- dehydrovomifoliol, and 4'-X-*O*-diglucosides of isovitexin and swertiajaponin [227]. The main phenolic compounds detected in flower were quercetin3-*O*-glucoside, kaempferol 3-O-glucoside, isorhamnetin 3-*O*-glucoside, and kaempferol 3-*O*-rhamnoside [228].

Cucumber seeds are also rich in phytochemicals with phenolics (93.5 ± 0.1 mg GAE/g), flavonoids (57.4 ± 0.1 mg QE/g), and β-carotene (19.46 ± 0.4 mg carotenoids/100 g) being the most abundant compounds [229]. Seeds are also rich in oil (23.3-41.07%) and tannins [230, 231], while the main fatty acids were linoleic, oleic and stearic acid [232]. The main detected sterols were 25(27)-dehydroporiferasterol, sitosterol and stigmasterol [233].

An important group of phytochemicals of cucumber belongs to a diverse category of compounds, namely cucurbitacins, which structurally are classified as triterpenes and are usually found in great amounts in both roots and fruit of cucumber plants, as well as of other cucurbits [6], while the presence of cucurbitacins in roots usually indicates their presence in other plant parts [234]. This group of compounds includes several kinds of cucurbitacins (more than 18) with the same cucurbitane skeleton [4], which are distinguished from each other according to oxygens' position, while their glycosidic forms are also present in considerable amounts [2]. The presence of cucurbitacins is associated with bitter taste and especially for cucumber, the main detected cucurbitacins are cucurbitacin B, C, and E, while the existing varieties are divided in three different groups; those which have bitter vegetative parts and fruit that are slightly bitter; those with bitter vegetative parts and not bitter fruit and finally those which have no bitter plant parts [234]. Moreover, Mukherjee *et al.* [235] reported the presence of three more cucurbitacins such as cucurbitacin A, D and I in cotyledons of cucumber seedlings. Apart from the genotype, growing conditions may also affect cucurbitacins production and factors such fruit position on the plant and harvest (earlier harvest) and developmental stage (younger fruit), nitrogen fertilization and low air temperatures may induce cucurbitacins production [234]. Moreover, Qing *et al.* [236] demonstrated that cucurbitacin C is more abundant in leaves and stems of cucumber plants comparing to reproduction organ (flowers and fruit), roots and stalks, while they also suggested a decrease of cucurbitacin C content with the process of leaf senescence due to its transformation to a more stable metabolite, namely 23,24-dihydrocucurbitacin C. Ramezani *et al.* [237] reported that foliar spray of cucumber leaves with chitosan and potassium phosphite may increase cucurbitacin E content and increase bioactive effects.

Health Effects

Cucumber plant parts have been associated with antioxidant properties, with several studies highlighting the *in vitro* radical scavenging activity of fruit (whole fruit or fruit juice) [238 - 241], flesh and peels [221, 242], seeds [230], and cotyledons [243]. In particular, cucumber seeds have been attributed with antioxidative properties, where triterpenoid glycosides showed significant radical scavenging activity comparing to ascorbic acid with two methods (1,1-diphenyl-1-picrylhydrazyl and peroxide method), as well as strong anti-ulcerogenic activity after evaluation with pyrolic ligation and water immersion stress using indomethacin as positive control [218, 230]. Moreover, Nema *et al.* [239] highlighted the skin protective properties of fruit aqueous extracts through the inhibition of hyaluronidase and elastase enzymes that are responsible for skin wrinkling, while Akhtar *et al.* [244] suggested the use of hydroethanolic fruit extracts in skin rejuvenation formulations. Topical application of fruit extracts was highly effective against skin tumor and showed significant potency against tumor incidence, tumor cell proliferation and tumor onset [245]. Plant parts (leaves, stems, roots, whole fruits, calyxes, and fruits without calyxes) of cucumber have also exhibited skin protective properties through the inhibition of melagonenesis in B16 cells, especially methanolic extracts of leaves and stems which reduced significantly tyrosinase expression [246], while Gangale *et al.* [247] suggested the use of cucumber fruit aqueous extracts as natural antisolar agent due to UV absorption properties. According to the same study, the main responsible compounds for such effects was primarily lutein followed by (+)-(1*R*,2*S*,5*R*,6*S*)-2,6-di-(4'-hydroxyphenyl)-3,7-dioxabicyclo[3.3.0]octane. Moreover, the beneficial properties of the proteases of cucumber sap but also provide the scientific basis for the wide use of cucumber in cosmetic industry and as well as in traditional medicine as a skin conditioner [248].

Seed extracts have been attributed with antihyperlipidemic effects through the reduction of total cholesterol, triglycerides, low-density lipoprotein and body mass index and could be considered as potential food supplements for such purposes [249].

Fruit volatiles have been also associated with antimicrobial activities, where according to Sotiroudis *et al.* [221] essential oils isolated from fruit pulp and peels showed significant antibacterial and fungicidal properties and further suggested that (E,Z)-2,6-nonadienal and (E)-2-nonenal were probably the most potent constituents, as previously reported by Cho *et al.* [250]. According to Muruganantham *et al.* [251] cucumber flowers ethanolic extracts exhibited significant antimicrobial activities against bacteria (*Salmonella typhi*, *Escherichia coli*, *E. faecalis*, and *Bacillus cereus*) and fungi (*Candida lunata* and *C. albicans*).

Cucurbitacins are associated with the bitter taste of fruit and originally are involved with plant defense mechanisms against pest and pathogens attacks [252, 253]. However, despite the fact that intake of cucurbitacins and its derivatives in reasonable amounts has been associated with several health effects, including antitumor, antidiabetic, and anti-inflammatory properties among others [6]. In particular, Sharma *et al.* [254] evaluated the potential of using vitexin obtained from cucumber as antiangiogenic agent through molecular docking analysis and reported significant inhibitory effects against Hsp90 protein which is responsible for indirect induction of angiogenesis and metastasis. Moreover, Muruganantham *et al.* [255] reported anticancer activities of ethyl acetate extracts of cucumber flowers against HePG2 cell lines due to its high content in cucurbitacins, lignans and flavonoids. Ethanolic fruit extracts exhibited significant antibacterial properties against *E. coli* and *Pseudomonas aeruginosa* due to its high content in cucurbitacin E [237]. In contrast, there are several studies which report toxicity effects or food poisoning in humans and animals after intake of cucurbitacins [6, 256], however, severity of toxicity symptoms and lethal doses depend on cucurbitacin type [257].

Commercial hybrids and cultivars have been gradually selected through breeding based on their low or no content in such compounds in order to fulfil market demands and consumers' preferences for bitter-less fruit; therefore, nowadays the edible parts of cultivated plants are free of cucurbitacins [234].

According to Chiu *et al.* [225], flesh and peels of cucumber fruit contain significant amounts of chitosan which exhibited several biological activities such as antimutagenic, antiobesity, antimicrobial, antioxidant and dermal protective properties.

Other health effects include hepatoprotective and nephroprotective properties, since according to Bajpai *et al.* [258] the oral administration of heat-treated fruit juice (10 mg/kg) in carbon tetrachloride (CCl_4)-treated rats resulted in significant body weight control and reduced activities of aspartate aminotransferase, alanine aminotransferase, blood urea nitrogen and creatinine content comparing to control treatment. Similar results have been reported by Palanisamy *et al.* [259] who highlighted the hepatoprotective effects of ethanolic extracts of cucumber fruit against paracetamol-induced hepatotoxicity in rats. Aqueous extracts of cucumber fruit have also exhibited *in vitro* inhibitory effects against acetylcholinesterase, butyrylcholinesterase and monoamine oxidase enzymes which are related with neurodegenerative diseases, as well as antioxidant properties against lipid peroxidation [223]. Moreover, Kumar *et al.* [260] and Kumar and Parle [261] have reported that cucumber fruit extracts may improve learning and memory abilities in diazepam-treated aged rodents, and further suggested agmatine, cucurbitacins, and vitamins as the main responsible compounds or these activities. In addition, Foong *et al.* [262] reported significant anticancer activities of aqueous fruit extracts against human non-small cell lung carcinoma cell line (H1299) and human breast adenocarcinoma cell line (MCF-7) and attributed these properties to the presence of flavonoids, saponins, alkaloids and steroids. Aqueous extracts of fruit have been also attributed with laxative effects, with oral administration of doses up to 1000 mg/kg being similarly effective as the reference drug sodium picosulfate and castor oil [263]. Seed hydroethanolic extracts have also exhibited hepatoprotective effects against cisplatin and gentamicin-treated rats [264].

Ethanolic extracts of fruit peels and fruit powder decoctions have also exhibited antidiabetic properties in animal model studies. In particular, Dixit and Kar [265] evaluated the antidiabetic potential of fruit peels and reported the ameliorative effects regarding the regulation of alloxan-induced diabetes mellitus effects on serum lipids and thyroid hormones of mice. Similar results have been reported by Kartiyayini *et al.* [266] who observed that the oral administration of 200-400 mg/kg of fruit powder ethanolic extracts resulted in significant antidiabetic and antihyperlipidemic effects in streptozotocin-induced diabetic rats. According to Park *et al.* [267], the antidiabetic effects of fruit ethanolic extracts are associated with inhibition of α-glucosidase, as well as with the induction of 3T3-L1 cell differentiation into adipocytes. In addition, in the study of Roman-Ramos *et al.* [268] the anti-diabetic properties of cucumber fruit decoctions which have been traditionally used for therapeutic purposes in Mexico were confirmed with glucose tolerance tests in rabbits after the gastric administration of water, tolbudamide and decoctions. Among the various bioactive compounds which are responsible for the antidiabetic properties of cucurbits fruit, polysaccharides are considered as very potent compounds that stimulate hypoglycemic effects and regeneration of β-cells in a similar manner to prescribed anti-diabetic drugs [267]. Seed aqueous extracts exhibited antidiabetic properties through the inhibitory activities against α-amylase and α-glucosidase, while the also presented antiproliferative properties against prostate cancer cells *via* lipoxigenase inhibition [269]. Sudheesh and Vijayalakshmi [270] have reported the antihyperlipidemic effects of fruit pectins, with a dose of 5 g/kg/day resulted in a significant reduction of cholesterol, free fatty acids, phospholipids, and triglycerides in both normal and cholesterol-fed rats, while [271] demonstrated the *in vivo* anti-ulcerogenic properties of seed methanolic extracts in model rats through pyloric ligation and water immersion stress tests, while these effects were further attributed to 9-beta-methyl-19-norlanosta-5-ene cucurbitane glycoside [272]. Anti-ulcerogenic effects have been also exhibited by fruit extracts which ameliorated the acetic acid-induced ulcerative colitis in rats [273]. Similar results have been reported by Sharma *et al.* [274] who suggested the gastroprotective effects of aqueous fruit pulp extracts against indomethacin and pyloric ligation induced ulcers in model rats. Kumar *et al.* [275] suggested the analgesic effects of aqueous fruit extracts at doses of 500 mg/kg and further attributed these effects to its high content in flavonoids and tannins, while Singh Gill *et al.* [276] reported similar analgesic and anti-inflammatory effects for methanolic extracts of seeds at doses of 300 mg/kg. Anti-inflammatory effects through the reduction of proinflammatory cytokines in the cerebellum have been also reported for fruit [277].

Vegetative plant parts have also exhibited cytotoxic and antifungal activities, since according to Das *et al.* [278] ethanolic and chloroform extracts of leaves and stems shown significant *in vitro* cytotoxic activity against brine shrimp nauplii, as well as antifungal effects against *Aspergillus niger, Blastomyces dermatitides, Candida albicans, Pityrosporum ovale, Trichophyton* spp. and *Microsporum* spp. Moreover, Ibrahim *et al.* [279] reported that vegetative parts of cucumbers are usually used as herbs in Egypt due to their antioxidant properties and their bioactive compounds content, especially phenolic acids and chlorogenic acid in particular. Plant stems have also shown antimicrobial properties against phytopathogenic fungi (*Pythium aphanidermatum, Botryosphaeria dothidea, Fusarium oxysporum* f. sp. *cucumerinum* and *Botrytis cinerea*) and Gram-negative and Gram-positive bacteria (*Xanthomonas vesicatoria, Pseudomonas lachrymans* and *Bacillus subtilis*) which were attributed to three sphingolipids isolated from crude stem methanolic extracts [280]. In contrast, Chu *et al.* [241] reported a low content of phenolic compounds in cucumber fruit, as well as low antioxidant and no antiproliferative activity.

Seed extracts have been attributed with antimicrobial properties showing significant inhibitory effects against several bacteria such as *Serratia marcescens, Escherichia coli, Streptococcus thermophilous, Fusarium oxysporum,* and *Trichoderma reesei*, while no inhibition against *Aspergillus niger* was observed [7]. Similar antimicrobial activities against *Bacillus subtilis, Staphylococcus aureus, Streptococcus* sp. and *E. coli* have been reported for fruit pulp extracts by Sahar *et al.* [222], while Al Akeel *et al.* [281] suggested that crude protein extracts of cucumber seeds are responsible for such antimicrobial effects.

CUCURBITA

Cucurbita genus belongs to *Cucurbitaceae* family and includes a group of varieties also known as squash, gourd, zucchini or courgette with fruit that are consumed as winter or summer squashes [282]. The main domesticated species of the genus are *Cucurbita pepo, C. mixta, C. maxima, C. moschata* Durch, *C. argyrosperma* Huber and *C. ficifolia* Bouche, which differ in plant morphology and especially in fruit shape and size [1]. However, apart from differences in visual appearance, significant differences have been also reported in terms of chemical composition of the edible parts of plant (fruit and seeds). In this section, chemical composition and health effects of the main cultivated *Cucurbita* species will be presented.

Chemical Composition of the Main Cultivated Species

Pumpkin plant parts are a good source of phytochemicals with various compounds such as phenolic glycosides, polysaccharides, proteins and 13-hydroxy-9Z, 11E-octadecatrienoic acid been isolated from leaves and seeds so far [283, 284]. Fruit pulp contains significant amounts of carotenoids, chlorophylls, total phenols and vitamins B6, C, E, K, thiamin and rivoflavin, as well as macro and microminerals (K, P, Se and Fe) [285 - 288]. However, a great variation in carotenoids content between *Cucurbita pepo* cultivars exists, while differences are also observed between fruit parts (epicarp and mesocarp) [289]. According to Martínez-Valdivieso *et al.*, epicarp is richer in carotenoids that mesocarp, while the main carotenoids were lutein and β-carotene followed by zeaxanthin [289]. Fruit mesocarp is also rich in dietary fiber that could be extracted for industrial uses [290]. Seeds are considered a rich source of oils and fatty acids, although a great variation exists among the main domesticated species, including *C. pepo, C. maxima, C. moschata* and *C. mixta* [291]. In particular, seed oil content of 12 cultivars of *C. maxima* varied significantly with values ranging between 10.9% and 30.9%, while unsaturated fatty acids content were the most abundant fatty acids class (73.1% to 80.5% of total fatty acids) [292]. Moreover, nutritional value of seeds depends on genotype and apart from fatty acids and oils pumpkin seeds are a good source of proteins (28.0%-40.5%) and polyphenols (34.3-113.0 mg GAE/100 g) [293, 294]. According to Nawirska-Olszańska *et al.* [294], seeds from *C. maxima* contain higher amounts of fatty acids than *C. pepo* cultivars, while fatty acid composition varied between the tested cultivars of both species. In the same study, phenolic composition of seeds was evaluated and the main detected phenolic compounds were caffeic, vanillic, *p*-coumaric, sinapic and hydroxybenzoic acid [294], while Rezig *et al.* reported the presence of protocatechuic, ferulic, caffeic, vanillic, *p*-coumaric and syringic acids in *C. maxima* seeds [295]. Seeds of *Cucrbita* species contain tocopherols (α-, β- and γ-tocopherol) and β-cryptoxanthin in amounts that depend on the species (*C. maxima, C. pepo* and *C. moschata*) and fruit part (flesh, peel and seeds) [296]. According to Andjelkovic *et al.* [297], total phenolic compounds content in pumpkin seed oils ranged between 24.7-50.9 mg GAE/kg oil, while the main detected compounds were luteolin, sinapic acid, tyrosol, vanillic acid, and vanillin. Seed oil has a high content in polyunsaturated fatty acids, namely linoleic and oleic acid (76% of total fatty acids), phytosterols and tocopherols [298, 299], as well as in carotenoids [300]. The main detected tocopherols in seed oils of several *C. maxima* and *C. pepo* varieties were α- and γ-tocopherols, followed by δ-tocopherol, while β-tocopherol was detected in lower amounts [294]. In contrast, this was not always the case in the studies of Stevenson *et al.* and Veronezi and Jorge where tocopherol composition differed between 12 cultivars of *C. maxima* and γ- and δ-tocopherols were the most abundant ones [292, 301].

Other compounds present in *Cucurbita* species include alkaloids and flavonoids, and various antibiotic components such as α- and β-moschins and moschatins which are primarily used for plant defense purposes against pathogens [302 - 304]. Moreover, the mesocarp of *C. moschata* contains cucurmosin, a type 1 ribosome-inactivating protein with significant bioactive properties [305]. Cucurbitacins which are typical triterpens of the Cucurbitaceae family are also present in *Cucurbita* species and exhibit significant bioactive properties. Seeds contain cucurbitacin E and cucurbitacin L 2-*O*-β-glucoside [306], while zucchini fruit also contain cucurbitacin E [307]. Moreover, Wang *et al.* isolated and identified two cucurbitane glucosides, namely cucurbitacin L 2-O-b-D-glucopyranoside and cucurbitacin K 2-O-b-D-glucopyranoside, and two hexan-orcucurbitane glycosides (2,16-dihydroxy-22,23,24,25,26, 27-hexanorcucurbit-5-en-11,20-dione 2-O-b-D-glucopyranoside and 16-hydroxy- 22,23,24,25,26,27-hexanorcucurbit-5-en-11, 20-dione 3-O-a-L-rhamnopyranosyl- (1→2)-b-D-glucopyranoside, from *C. pepo* cv. *dayangua* [308].

Health Effects

Cucurbita species are associated with several therapeutic properties, such as antibacterial, antidiabetic, antihypertension, antitumor, immunomodulatory, antihypercholesterolemia, antiparasitic, antiinflammatory and antalgic among others. Moreover, fruit seeds are the most investigated plant part and have been widely used in traditional and folk medicine for the treatment of several diseases.

Antioxidant Activity

Pumpkin seeds have been reported to possess significant antioxidant properties, which according to Nawirska-Olszańska *et al.* show a varied potency depending on the genotype and the tested assay [294]. Moreover, protein isolates from *C. pepo* seeds have shown significant antioxidant properties against lipid peroxidation and CCl_4 intoxication of rats [309], while further hydrolysis of protein isolate showed the presence of peptides with antioxidant properties [310]. Seed oils of *C. pepo* reported to alleviate toxic effects after subacute aflatoxin poisoning in mice [311]. Seed oils have also significant antioxidant properties which have been associated with phenolic compounds content [297]. Similar results have been reported for seed extracts from different pumpkin seeds, where radical scavenging activity depended on total phenolic compounds content [312]. Water soluble polysaccharides from *C. maxima* fruit have also shown a significant antioxidant potential [313].

Antidiabetic Effects

Pumpkin has been widely used for its medicinal properties, since fruit pulp and seeds have been attributed with antidiabetic effects due to its high content in non-pectin polysaccharides, pectins, hypoglycaemic proteins, tocopherols and oils [314, 315]. Similar results have been reported by Wang *et al.* [316] and Sedigheh *et al.* [317] who suggested that polysaccharides obtained from *C. moschata* fruit pulp have a high potency against diabetes mellitus in alloxan-induced diabetic rats. Moreover, the oral administration of a mixture of pumpkin and flax seeds ameliorated diabetic nephropathy in alloxan-induced diabetic rats [318]. Moreover, extracts of *C. ficifolia* fruit are commonly used as antidiabetic agents in traditional medicine and have been previously confirmed being effective against Type 2 diabetic patients and alloxan and streptozotocin-induced diabetic models without however indications about the involved mechanisms of action [319 - 321]. However, Jessica *et al.* identified five bioactive compounds (p-coumaric acid, p-hydroxybenzoic acid, salicin, stigmast-7,2,2-dien-3-ol and stigmast-7-en-3-ol) in aqueous extracts of *C. ficifolia* fruit and further suggested that hypoglycemic effects should be attributed to glycogen accumulation in liver [322]. In the recent study of Xia and Wang the mechanisms involved for such activity include the renewal of β-cells and the recovery of partially destroyed ones [323], while Jiang and Du reported that two tetrasaccharide glyceroglycolipids from *C. pepo* fruit exhibited significant glucose-lowering effects in diabetic mice [324].

Cardiovascular System Effects

According to Abuelgassim *et al.* [325], the administration of pumpkin seeds in the diet of atherogenic rats resulted in significant decrease of serum total cholesterol and LDL cholesterol, while the authors suggested that this effect was not attributed to high arginine content of seeds. The high content of seeds in unsaturated fatty acids, antioxidants and fibers justifies its anti-atherogenic effects, especially when administered as a mixture with flax seeds [326]. Seed oil is attributed with anti-hypertensive effects and its oral administration resulted in a decrease of arterial blood pressure and protected against histopathological alteration of aorta and heart in l-arginine methyl ester hydrochloride (l-NAME) induced hypertensive rats [315]. Phenolic phytochemicals in various pumpkin cultivars fruit have been also associated with antihypertension activities, where according to Kwon *et al.* traditional foods contain pumpkin fruit have shown significant inhibitory activity against angiotensin I-converting enzyme [327]. Moreover, seed oil which is rich in phytoestrogens showed beneficial effects in HDL cholesterol in postmenopausal women [328], as well as in plasma lipid profile and blood pressure [329].

Anticancer Effects

Anticancer effects of *C. pepo* seed oil may be attributed to several bioactive components, with squalene being an important anticancer agent [299, 330]. Hydroethanolic seed extracts of *Cucurbita pepo* L. subsp. *pepo* var. styriaca showed significant inhibitive properties against hyperplastic and cancer cells, especially against prostate, breast and colon cancer cells [331]. The protective role of pumpkin against prostate cancer has been also reported in *in vitro* and *in vivo* studies where pumpkin seeds were one of the ingredients of polyherbal formulations [332, 333]. Moreover, according to Friedrich *et al.* [334] and Hong *et al.* [335] administration of pumpkin seed oil in patients with benign prostatic hyperplasia (BPH) may significantly improve symptoms based on the international prostate symptom score (IPPS), without however reducing prostate volume and prostate specific antigen. Similar results have been reported by Vahlensieck *et al.* who observed a clinical reduction of IPPS after one year of pumpkin seed oil administration in patients with lower urinary tract symptoms [336]. Other studies, also highlight the beneficial effects of pumpkin seed oil against prostate hyperplasia in testosterone-induced rats [337, 338]. Moreover, moschatin a ribosome inactivating protein isolated from mature seeds of *C. moschata* has been attributed with inhibitive properties against human melanoma cells M_{21} [303], while hexane extracts of *C. ficifolia* fruit showed significant *in vivo* cytotoxicity against human bone marrow-mesenchymal stem cells [339]. Another compound isolated from the mesocarp of *C. moschata*, namely cucurmosin, has shown significant *in vivo* cytotoxicity against lung adenocarcinoma cancer cells (A549), human leukemia cells (K562), murine melanoma cells (B16), human hepatoma HepG2 cells, colon carcinoma cells (CT26) and peripheral blood lymphocyte cells [305, 340, 341]. Moreover, methanolic extracts from the aerial parts of *C. maxima* showed strong *in vivo* antiproliferative activity against Ehrlich ascites carcinoma in rats [342], while Tomar *et al.* identified 2S albumin as the main bioactive compound in *C. maxima* seeds responsible for anticancer activity against breast cancer (MCF-7), ovarian teratocarcinoma (PA-1), prostate cancer (PC-3 and DU-145) and hepatocellular carcinoma (HepG2) cell lines [343]. In the study of Rotimi *et al.* [344] the inhibitory effects of ethanolic extracts of *C. pepo* leaves against NF-κB were evaluated and the authors suggested that this activity could be attributed to octadecadienoic acid derivatives, hexadecanoic acid, methyl strearate and phytol.

Anthelminthic Effects

Cucurbita seeds show significant antiparasitic activities and have been used as veterinary medicines against endoparasites and stomach problems in various domesticated animals [345]. In particular, the oral administration of aqueous and ethanolic extracts of *C. maxima* seeds in *Aspiculuris tetraptera* infected mice resulted in a significant decrease of infection, comparable to invemectin treated models [346]. In addition, Obregón *et al.* [347] performed preclinical studies with aqueous extracts of *C. maxima* seeds against canine tapeworms in rats and reported that administration of extracts at concentrations of 23 g of seeds in 100 mL of distilled water increased helminthic motility. Anthelminthic effects have been also reported for *C. pepo* seeds, with seed meals or seed extracts showing efficacy against human taeniasis [348] and gastrointestinal nematodes in ostriches [349], rodents [350] and ruminants [351]. According to the literature, the anthelminthic properties of *C. pepo* seeds could be attributed to the presence of triterpens (cucurbitacin B, D, E and cucurbitacin L 2-*O*-β-glucoside), cucurbitin (a non-proteic amino acid), sterols and saponins [234, 306, 351].

Hepatoprotective effects

Pumpkin seed oil is also effective against alcohol-induced hepatotoxicity since its oral administration significantly improves liver function as expressed by lipid peroxidation decrease and the increase of catalase and glutathione-S-transferase activities and hepatic glutathione content [352]. In addition, the diet supplementation of hypercholesterolemic rats with a mixture of pumpkin and flax seeds improved significantly liver function, probably due to high content of supplements in unsaturated fatty acids [326]. According to the study of Morrison *et al.* [353], the substitution of saturated fatty acids by polyunsaturated fatty acids from pumpkin seed oil had a protective role against the non-alcoholic fatty liver disease in mice. Moreover, aqueous extracts from *C. moschata* fruit showed cytoprotective effects against hepatocyte cytotoxicity in cumene hydroperoxide- and glyoxal-induced rats, where the administration of extracts (50 μg/mL) reduced hepatocyte lysis, lysosomal damage, and cellular proteolysis among others [354].

Anti-Inflammatory and Immunomodulation Effects

Pumpkin seeds have been suggested as immunonutrients, since according to Immaculata *et al.* [355] a high dose of pumpkin seeds (7.6 g/kg d.w.) showed significant immunomodulatory effects in rats. Moreover, the administration of a dietary mixture of milled seeds (pumpkin, flax and sesame) in hemodialysis patients resulted in improved inflammatory markers (TNF-alpha, IL-6, and hs-CRP) [356]. Seed oil may also have anti-inflammatory activity against adjuvant arthritis in rats due to its high radical scavenging activity [357]. Aqueous extracts of *C. ficifolia* has also modulatory effects against systemic chronic inflammation in monosodium glutamate-induced obese mice, mainly through the reduced expression of TNF-α and IL-6 adipokines and the increase of IFN-γ and IL-10 cytokines [358]. Moreover, ethanolic extracts of *C. maxima* leaves have shown anti-inflammatory potency against TNF-α and are widely used as ingredients in medicinal formulations [359].

Antimicrobial Effects

Pumpkins contain several compounds with antimicrobial properties. Essential oils from seeds of *C. pepo* showed moderate inhibition activity against various micro-organisms [360], while phloem exudates from pumpkin fruit were efficient against animal and plant pathogenic fungi [361]. Wang *et al.* isolated a protein-bound polysaccharide from *C. moschata* seeds which exhibited strong antibacterial activity against *Bacillus subtilis*, *Staphylococcus aureus*, *Pichia fermentans* and *Escherichia coli* [362]. Moreover, pumpkin fruit rinds contain the protein Pr-1 with significant antifungal properties against various pathogenic fungi, including *Botrytis cinerea*, *Fusarium oxysporum*, *F. solani* and *Rhizoctonia solani*, as well as *Candida albicans*, without however showing any potency against *Escherichia coli* and *Staphylococcus aureus* [363]. Methanolic extracts of *C. pepo* leaves were also effective against various Gram negative bacteria, including multi-drag resistant strains [364].

Other Effects

C. pepo seeds are rich sources of cucurbitane type triterpenoids which exhibited significant anti-ulcer activities in rats at doses of 300 μg/mL [351]. Moreover, methanolic extracts of *C. pepo* leaves showed antiobesity activity in progesterone-induced obese mice attribute to various phytochemicals such as flavonoids, steroids, alkaloids, saponins, diterpenes, triterpenes and phenols [365]. Apart from fruit and seeds, seed oil contains several bioactive compounds such as tocopherols, fatty acids and phytosterols which have shown significant healing properties against wounds in animal models [298, 366]. Furthermore, oral administration of pumpkin seed oils was effective against cytotoxic and gynotoxic effects of azathioprine in mice by decreasing DNA fragmentation, reducing the frequencies of micronucleated polychromatic erythrocytes (Mn-PCEs) and sperm abnormalities and increasing total sperm count, the percentage of PCEs and the ratio of PCEs to normochromatic erythrocytes (NCEs) [367]. Oral administration of *C. maxima* seed oil may reduce urinary disorders [368], while a mixture of soy and pumpkin seeds reduces urination and inconsistence frequency [369]. Moreover, consumption of capsules with pumpkin seed oil (PSO) at doses of 400 mg/d^{-1} increased hair count by 40% in patients with androgenetic alopecia, probably due to blocking of 5-α-reductase [370], while pumpkin seed supplementation reduced calcium-oxalate crystal occurrence and calcium level and increased phosphorus levels, thus highlighting its potential use against bladder stone disease [371]. *C. pepo* seed oil has also shown significant healing properties against deep second-degree burns in rats through the increase of collagen production [372].

Toxic Effects

Seeds have been widely used for medicinal purposes and constitute a basic ingredient in various traditional formulations without any toxic effects being observed. Moreover, in a recent study it was reported that toxicity levels for oral administration of seed extracts is extremely high (>5000 mg/kg), rendering its use considerably safe [373]. However, the presence of various antinutrients such as saponins, antimicrobial proteins, raffinose oligosaccharides, trypsin inhibitor, phytic acid and tannins has been reported and high intake of seeds has been associated with several symptoms such as cramps, stomach pain, allergic responses [374]. Another issue regarding pumpkin seeds consumption is correlated with its high phytoestrogens content. Considering that estrogens are involved in several vital functions in human body, overconsumption of estrogens may intervene and result in manipulation of endocrine systems, and the onset of cancers, auto-immune diseases and nephropathies [375 - 378].

Other Uses

Pumpkin seeds have found several uses in the food industry and various food products containing such seeds have been developed, based on its health benefits and functional properties, including beverages, snacks and ready to eat food products, infant food and so forth [374, 379, 380]. Whole and defatted pumpkin seed meals and flour and fiber from fruit have shown good physicochemical and functional properties, while they also contain a high protein content which makes them a promising material for the food industry [381 - 383]. Apart from food and medicinal purposes, pumpkins have been extensively used during the last decades in vegetable grafting of related species (watermelon and melon), since they are considered tolerant in various soil pathogens and soil related problems [384, 385]. However, apart from improvement in agronomic performance the use of *Cucurbita* species as rootstocks has been shown to improve fruit quality, chemical composition and bioactive compounds content of watermelon and melon fruit [133, 386]. Therefore, grafing technique should be used as a means to manipulate chemical composition and bioactive compounds content of fruit of *Cucurbita* species.

OTHER SPECIES (GOURDS)

In the following section chemical composition and health effects of the most common gourds will be presented, including *Lagenaria siceraria* (bottle gourd or bitter melon), *Luffa cylindrica* syn. *L. aegyptiaca* (sponge gourd), and *L. acutangula* (ridge gourd), *Benincasa hispida* (wax or ash gourd), and *Momordica charantia* (bitter melon).

Chemical Composition

Lagenaria siceraria fruit contain considerable amount of phenolic and flavonoid compounds (243.5 and 109.5 μg/mL) which are equally distributed in the mesocarp and epicarp for most of the extraction solvents being used; however, epicarp was richer than mesocarp when ethyl acetate and chloroform were used for phenolic and flavonoid contents extraction, respectively [387]. Seeds of bottle gourd are rich source of lipids (49.0-55.5%) and proteins (25.2-34.8%), while they also contain ash (3.2-4.1%), crude fibers (2.3-5.5%) and carbohydrates (5.8-10.3%) on a dry weight basis [293, 388]. Antinutrients such as phytic acid (4 mg/g f.w.), saponins (4% f.w.), and oxalates (8% f.w.) and toxic compounds such as hydrocyanic acid (12 mg/g 100 mg/g.) have been detected in bottle gourd seed kernels; however, heating treatment may reduce the content of these compounds [389]. Moreover, peels of *Benincasa hispida* and *Luffa acutangula* contain 0.078 mg and 0.111 mg/100 g f.w. of oxalates and 232.44 mg and 256.96 mg 100/g f.w. of phytates, respectively [390].

According to Chiu *et al.* [225], flesh and peels of luffa fruit *(Luffa cylindrica)* are a rich source of chitosan which may have several practical applications as a vegan source of biological active compounds. Bitter melon contains charantin and vicine, compounds with structure similar to insulin which have been attributed with antidiabetic effects [391]. Recently, Singh *et al.* [392] identified two chitotriose-specific lectins in wax gourd fruit with significant antiproliferative activity. Moreover, bottle gourd seeds contain lagenin, a ribosome –inactivating protein with several biological activities [393].

The main detected sterols in bottle gourd (*Lagenaria siceraria*) were sitosterol, codisterol, clerosterol, and stigmasterol, while in luffa seeds (*Luffa cylindrica*) no sterols were detected [233]. Similarly, various sterols have been detected in fruit of bitter melon, namely β-sitosterol-glucoside, stigmast-5,25-dien-3 β-O-glucoside, stigmast-7,25-dien-3 β-ol, and stigmast-7,22,25-trien-3 β-ol [394], while sterols of wax gourd seeds consist mostly of β-sitosterol [395, 396]. Furthermore, fatty acid profile of wax gourd seeds includes mostly unsaturated fatty acids such as linoleic (60.6%), palmitic (15.3%), oleic (14.1%) and stearic acids (7.4%) [397]. Similarly, bottle gourd seeds contain the same fatty acids in different amounts, namely linoleic (69.1%), palmitic (13.0%), oleic (9.0%) and stearic acid (7.9%) [398].

Gourds are rich sources of phenolic compounds with several bioactive properties. According to Doshi *et al.* [399], *B. hispida* fruit are rich sources of flavonoids such as rutin, quercetin and derivatives, while Sheemole *et al.* [396] detected ferulic, syringic and benzoic acid, as well as isovitexin, myricetin, dihydroxyflavan and hydroxyflavan. Other important secondary metabolites which are present in wax gourd seeds, include terpenoids (marasmic acid, hirsuitic acid and β-vetivone), alkaloids (palmatine), and coumarins (umbelliferone) [396].

The main phenolic compounds of bottle gourd fruit were gallic, o-coumaric and caffeic acids [279]. *M. charantia* showed a great variation in phenolic compounds profile between plant parts (leaf, stems and fruit), as well as in fruit of different maturity stages (green and ripe fruit), with leaves having the highest content (474 mg GAE/100 g d.w.), followed by green fruit, stems and ripe fruit (324 mg GAE/100 g d.w., 259 mg GAE/100 g d.w. and 224 mg GAE/100 g d.w., respectively) [400]. Similarly, Horax *et al.* [31] reported significant differences between four bitter lemon varieties in terms of their phenolic acids content, and identified catechin, epicatechin, and gentisic, gallic and chlorogenic acid as the main compounds of fruit pulp, catechin, epicatechin and gallic acid in seeds and gentisic acid, epicatechin and catechin in inner tissue of fruit. Moreover, Madala *et al.* [401] reported that leaves of *Momordica* sp. including *M. charantia* contain kaempferol, quercetin and derivatives and they further suggested a chemotypic variation among three *Momordica* species. Seeds of *B. hispida* differ in phenolic composition, since according to Bimakr *et al.* [397] they contain gallic acid, catechin, myricetin, quercetin and naringenin.

Seeds of *L. siceraria* are a good source of phytochemicals, since they contain significant amounts of phenols (243.2 mg GAE/100 g f.w.), flavonoids (474.9 mg CE/100 g f.w.), anthocyanins (123.3 mg c-3-gE/100 g f.w.) and tannins (23.7 mg LCE/100 g f.w.), while their content in saponins (93.6 mg/100 g f.w.) and alkaloids (1620 mg/g 100 mg/g f.w.) was lower comparing to seeds of other Cucurbitaceae species [230].

Cucurbitacins are also present in these species, while especially for *Momordica* genera there is a unique group of cucurbitacins known as momordicosides, which differ from the rest of the cucurbitacins in C_{19} which is oxidized to an aldehyde [6, 402, 403]. *Momordica charantia* contains apart from cucurbitane triterpene glycosides (momordicosides), several other compounds such as goyaglycosides a, b, c, d, e, f, g, and h; goyasaponins I, II and III; karavilagenins A, B, and C; karavilosides I, II, III, IV, and V [404 - 406]. Moreover, Chen *et al.* [407] identified fourteen cucurbitane triterpenoids from vines and leaves of *M. charantia* which were classified as kuguasins, while Zhang *et al.* [408] reported eight cucurbitane-type glucosides in fruit, namely kuguasaponis A-H. Lin *et al.* [409] have also identified two new compounds in stems and fruit, namely taiwacin A and B, respectively, while Liu *et al.* [410] detected three new triterpenoids in stems of *M. charantia*, one multifloran triterpenoid and two cucurbitane triterpenoids. Finally [411], reported six new cucurbitane-type triterpenoids in leaves and stems of *M. charantia*, which were identified as karavilagenin F, karavilosides XII and XIII, and momordicines VI, VII, and VIII.

Health Effects

Lagenaria siceraria fruit extracts exhibited significant *in vitro* antioxidant activities, regardless of the solvent, the fruit part and the studied assay [387]. Moreover, Ghule *et al.* [412] highlighted the antihyperlipidemic effects of fruit methanolic extracts in high-fat diet-induced hyperlipidemia in rats, through a significant reduction in blood lipids levels. Oral administration of bottle gourd fruit juice in diabetic and healthy patients alleviated diabetes associated disorders through a a decrease of blood glucose, total cholesterol, triglycerides, LDL-c and VLDL-c levels, and an improvement in kidney and liver function [413]. Bottle gourd fruit extracts have also shown significant potency for obesity treatment since they inhibited pancreatic lipase activity, and they also suppressed lipid digestion and intake [414]. Moreover, Rajput *et al.* [415, 416] reported *in vitro* and *in vivo* antithrombotic, antiplatelet and antiatherosclerotic potential for fruit ethanolic extracts. Fruit powder administered in rats with isoprenalin-induced cardiotoxicity at doses of 500 mg/kg alleviated toxic effects by preventing heart rate increase and hypotension [417]. Cardioprotecive effects have been also reported for bottle gourd fruit in rats with isoproterenol-induced myocardial infarction, where oral administration of 400 mg/kg exhibited protective properties against biochemical and histopathological changes [418]. Other health effects include diuretic properties [419], and antioxidant activity [420, 421], anti-stress and adaptogenic [422] among others. Moreover, high concentration seed extracts (100% and 75% of crude extracts) were effective against various microorganisms such as *Serratia marcescens*, *Streptococcus thermophiles*, *Trichoderma reesei*, *Fusarium oxysporum*, *Candida albicans*, and *Aspergillus niger*, while no inhibition was observed against *Escherichia coli* [7]. Seed of *L. siceraria* contain various storage proteins such as globulin, albumin and prolamin which have been associated with significant antidiabetic effects blood glucose reduction [189].

Momordica charantia has been attributed with numerous health effects including antidiabetic, antiviral, immunomodulatory, antiinflammatory, neuroprotective and anticancer properties among others [391, 402, 423 - 427], while it has been traditionally used in folk medicine for remedies against various diseases, such as gastrointestinal, viral, and gynecological diseases, skin problems, malaria, helminthes and so forth [428, 429]. In particular, Tan *et al.* [427] attributed *in vitro* antidiabetic effects to momordicosides Q, R, S and T which induced GLUT4 translocation from cytosol to the cell membrane, as well as they increased the activity of AMP-activated protein kinase. Zhang *et al.* [408] have also reported moderate cytotoxic activities of kuguasaponins from fruit against breast adenocarcinoma (MCF-7), laryngeal carcinoma (HEp-2) and colon carcinoma (WiDr) human cell lines. Moreover, Wang *et al.* [391] suggested that charantin-rich extracts of *M. charantia* fruit exert antidiabetic effects in type 2 diabetes patients through the increase of insulin sensitivity. Similarly, oral administration of bitter gourd fruit freeze-dried powder in rats resulted in a decrease of serum glucose, hypatic cholesterol and triglycerides levels [430], as well as lipid accumulation and adiponectin expression in 3T3-L1 cells [431]. Fruit juice of bottle gourd may also renew β and δ-cells in the pancreas of streptozotocin-induced diabetic rats [432]. Apart from animal model studies, clinical studies have showed significant antidiabetic effects of bitter melon [433].

Ethanolic extracts of bitter gourd seeds also exhibited significant inhibitory effects against cell proliferation of seven T-cell leukaemia-related cell lines, suggesting their use as functional foods [434]. In addition, Patel *et al.* [394] reported *in vivo* analgesic and antipyretic properties through acetic acid-induced writhing and tail-immersion tests in mice, while Gürbüz *et al.* [435] and Alam *et al.* [436] observed anti-ulcerogenic and wound healing properties for fruit ethanolic and methanolic extracts, respectively, in ethanol, diethyldithiocarbamate and HCl+ethanol-induced ulcerogenesis in rats. Kuguasins isolated from vines and leaves of the plant exhibited weak anti-HIV-1 activities [407], while Costa *et al.* [437] reported that leaf extracts showed significant antibacterial properties against *Staphylococcus aureus*, *Escherichia coli* and *Bacillus cereus*. According to Sood *et al.* [7], seed extracts of *M. charantia* with high concentration (100% and 75% of crude extracts) were also effective against various microorganisms such as *Serratia marcescens*, *Streptococcus thermophiles*, *Trichoderma reesei*, *Fusarium oxysporum*, *Escherichia coli*, *Candida albicans*, and *Aspergillus niger*.

Luffa has several therapeutic properties, where according to Chiu *et al.* [225] flesh and peels of luffa fruit contain significant amounts of chitosan which exhibited several biological activities such as antimutagenic, antiobesity, antimicrobial, antioxidant and dermal protective properties. Moreover, luffa seed extracts exhibited protective effects against amyloid β peptide-induced neurotoxicity and neurons cell death [438], while El-Fiky *et al.* [439] noted significant anti-diabetic properties in streptozotocin-induced diabetic rats through decreasing blood glucose levels. Pericarp ethanolic extracts have also significant antioxidant and immunomodulatory potency which was attributed to phenolic compounds content and gallic and hydroxybenzoic acid in particular [440].

Benincasa hispida has been attributed with antidiabetic, anxiolytic, antiobesity and antihyperlipidemic effects [441 - 443]. In a recent study of Singh *et al.* [392], lectins of fruit were attributed with antiproliferative activity against pancreatic cancer cells, as well as with antiangiongenic properties. Fruit peels exhibited antihyperlipidemic effects in high-fat diet-induced hyperlipidemia in mice through the inhibition of peroxisome proliferator-activated receptor γ (PPARγ) and downregulation of HMG-CoA reductase [443]. Moreover, Qadrie *et al.* [444] indicated the antinociceptive and antipyretic effects of seed ethanolic extracts in yeast-induced pyrexia in rats at doses of 250 and 500 mg/kg, while Shetty *et al.* [445] reported the antiulcerogenic effects of fruit extracts in rats with indomethacin-induced ulcers. According to Rachchh *et al.* [446], petroleum ether and methanolic extracts of fruit extracts showed a dose depended anti-inflammatory effect against carrageenan and histamine-induced edema, as well as against cotton pellet-induced granuloma in rats.

Apart from beneficial health effects the abovementioned species have been associated with negative effects on human health. In particular, *Momordica* (β-momorcharin in *M. charantia* seeds) and *Luffa* proteins (luffaculin isolated from *L. acutangula* seeds and luffin-a and luffin-b from *L. cylindrical* seeds), exhibited abortifacient and teratogenic effects in *in vivo* studies with mice, while ethanolic extracts of seeds showed antispermatogenic, antisteroidogenic and androgenic activities in rats [447 - 449]. Moreover, Ho *et al.* [256] reported that although beneficial for health, extremely bitter gourd fruit may have adverse effects such as diarrhea, gastrointestinal bleeding, hypotension and vomiting due to their high content of cucurbitacins.

CONCLUSION REMARKS AND FUTURE PROSPECTS

The Cucurbitaceae family includes very economically important vegetable crops with a world wide interest, as well as less exploited species which are traditionally used for medicinal and food purposes. Considering the polymorphism that this family exhibits, more and more studies continue to uncover the properties of less known species. The bioactive and therapeutic properties of all these species cover a wide spectrum and should be further studied in order to elucidate the mechanisms of action behind these properties. Moreover, more clinical studies should be carried out in order to confirm results from *in vitro* and model studies and further incorporate bioactive compounds from these species in the pharmaceutical industry.

CONSENT FOR PUBLICATION

Not applicable.

CONFLICT OF INTEREST

The author (editor) declares no conflict of interest, financial or otherwise.

ACKNOWLEDGEMENTS

Declare None.

REFERENCES

[1]   Rajasree RS, Sibi PI, Francis F, William H. Phytochemicals of cucurbitaceae family – A review. Int J Pharmacogn Phytochem Res 2016; 8(1): 113-23.

[2]   Saboo SS, Thorat PK, Tapadiya GG, Khadabadi SS. Ancient and Recent Medicinal Uses of Cucurbitaceae Family. Int J Ther Appl 2013; 2013(9): 11-9.

[3]   Wiart C. Terpenes BT - Lead Compounds from Medicinal Plants for the Treatment of Cancer. Massachusetts: Academic Press 2013; pp. 97-265.

[4]   Dhiman K, Gupta A, Sharma DK, Gill NS, Goyal A. A review on the medicinally important plants of the family Cucurbitaceae. Asian J Clin Nutr 2012; 4(1): 16-26.

[5]   De Rosa L, Alvarez-Parrilla E, Gonz GA. Fruit and Vegetable Phytochemicals Chemistry, Nutritional Value and Stability. Indianapolis: Wiley-Blackwell 2010; p. 367.

[6]   Kaushik U, Aeri V, Mir SR. Cucurbitacins - An insight into medicinal leads from nature. Pharmacogn Rev 2015; 9(17): 12-8.
        [PMID: 26009687]

[7]   Sood A, Kaur P, Gupta R. Phytochemical screening and antimicrobial assay of various seeds extract of Cucurbitaceae family. Int J Appl Biol Pharm Technol 2012; 3(3): 401-9.

[8]   Schaefer H, Heibl C, Renner SS. Gourds afloat: a dated phylogeny reveals an Asian origin of the gourd family (Cucurbitaceae) and numerous oversea dispersal events. Proc Biol Sci 2009; 276(1658): 843-51.
        [PMID: 19033142]

[9]   Renner SS, Schaefer H, Kocyan A. Phylogenetics of Cucumis (Cucurbitaceae): cucumber (*C. sativus*) belongs in an Asian/Australian clade far from melon (*C. melo*). BMC Evol Biol 2007; 7: 58.
        [PMID: 17425784]

[10]   Mallick MFR, Masui M. Origin, distribution and taxonomy of melons. Sci Hortic (Amsterdam) 1986; 28(3): 251-61.

[11]   Bates DM, Robinson RW. Cucumber, melons and watermelons, Cucumis and Citrullus (Cucurbitaceae).Evolution of Crop Plants. New York: John Wiley & Sons 1995; pp. 89-111.

[12]   Mc Creight J, Staub J. Report of the cucurbit working group. Washington: DC USDA-ARC 1993.

[13]   Akashi Y, Fukuda N, Wako T, Masuda M, Kato K. Genetic variation and phylogenetic relationships in East and South Asian melons, *Cucumis melo* L., based on the analysis of five isozymes. Euphytica 2002; 125(3): 385-96.

[14]   Boriss H, Brunke H, Kreith M. Commodity Profile : Melons. Agric Issues Cent (AIC), Univ Calif Retrieved Oct 18 2009 (February); http//aic.ucdavis.edu/profiles/Melons-2006.pdf2006 (February); 1-7.

[15]   Lester GE. Environmental Regulation of Human Health Nutrients (Ascorbic Acid, 13-Carotene, and Folic Acid) in Fruits and Vegetables. HortScience 2006; 41: 3-7.

[16]   Maietti A, Tedeschi P, Stagno C, *et al.* Analytical traceability of melon (*Cucumis melo* var reticulatus): proximate composition, bioactive compounds, and antioxidant capacity in relation to cultivar, plant physiology state, and seasonal variability. J Food Sci 2012; 77(6): C646-52.
        [PMID: 22583041]

[17]   Fundo JF, Miller FA, Garcia E, Santos JR, Silva CLM, Brandão TRS. Physicochemical characteristics, bioactive compounds and antioxidant activity in juice, pulp, peel and seeds of Cantaloupe melon. J

Food Meas Charact 2018; 12(1): 292-300.

[18] Lester GE, Hodges DM. Antioxidants associated with fruit senescence and human health: Novel orange-fleshed non-netted honey dew melon genotype comparisons following different seasonal productions and cold storage durations. Postharvest Biol Technol 2008; 48(3): 347-54.

[19] Mallek-Ayadi S, Bahloul N, Kechaou N. Characterization, phenolic compounds and functional properties of *Cucumis melo* L. peels. Food Chem 2017; 221: 1691-7.
       [PMID: 27979149]

[20] Tadmor Y, Burger J, Yaakov I, *et al.* Genetics of flavonoid, carotenoid, and chlorophyll pigments in melon fruit rinds. J Agric Food Chem 2010; 58(19): 10722-8.
       [PMID: 20815398]

[21] Al-Sayed HMA, Ahmed AR. Utilization of watermelon rinds and sharlyn melon peels as a natural source of dietary fiber and antioxidants in cake. Ann Agric Sci 2013; 58(1): 83-95.

[22] Fernandes ALT, Rodrigues GP, Testezlaf R. Mineral and Organomineral Fertirrigation in Relation to Quality of Greenhouse Cultivated Melon. Sci Agric 2003; 60(1): 149-54.

[23] Berdiyev M, Arslan D, Özcan MM. Nutritional composition, microbiological and sensory properties of dried melon: a traditional Turkmen product. Int J Food Sci Nutr 2009; 60(1): 60-8.
       [PMID: 18608557]

[24] Amaro AL, Fundo JF, Oliveira A, Beaulieu JC, Fernández-Trujillo JP, Almeida DPF. 1-methylcyclopropene effects on temporal changes of aroma volatiles and phytochemicals of fresh-cut cantaloupe. J Sci Food Agric 2013; 93(4): 828-37.
       [PMID: 22821412]

[25] Moreira SP, de Carvalho WM, Alexandrino AC, de Paula HCB. Freshness retention of minimally processed melon using different packages and multilayered edible coating containing microencapsulated essential oil. Int J Food Sci Technol 2014; 49(10): 2192-203.

[26] Lester GE, Saftner RA. Marketable quality and phytonutrient concentrations of a novel hybrid muskmelon intended for the fresh-cut industry and its parental lines: Whole-fruit comparisons at harvest and following long-term storage at 1 or 5 °C. Postharvest Biol Technol 2008; 48(2): 248-53.

[27] Kolayli S, Kara M, Tezcan F, *et al.* Comparative study of chemical and biochemical properties of different melon cultivars: standard, hybrid, and grafted melons. J Agric Food Chem 2010; 58(17): 9764-9.
       [PMID: 20715772]

[28] Antonious GF, Turley ET, Hill RR, Snyder JC. Chicken manure enhanced yield and quality of field-grown kale and collard greens. J Environ Sci Health B 2014; 49(4): 299-304.
       [PMID: 24502217]

[29] Ismail HI, Chan KW, Mariod AA, Ismail M. Phenolic content and antioxidant activity of cantaloupe (*Cucumis melo*) methanolic extracts. Food Chem 2010; 119(2): 643-7.

[30] Díaz-de-Cerio E, Gómez-Caravaca AM, Verardo V, Fernández-Gutiérrez A, Segura-Carretero A. Determination of guava (Psidium guajava L.) leaf phenolic compounds using HPLC-DAD-QTOF-MS. J Funct Foods 2016; 22: 376-88.

[31] Horax R, Hettiarachchy N, Islam S. Total Phenolic Contents and Phenolic Acid Constituents in 4 Varieties of Bitter Melons (Momordica charantia) and Antioxidant Activities of their Extracts. J Food Sci 2006; 70(4): C275-80.

[32] Morais DR, Rotta EM, Sargi SC, Schmidt EM, Bonafe EG, Eberlin MN, *et al.* Antioxidant activity, phenolics and UPLC-ESI(-)-MS of extracts from different tropical fruits parts and processed peels. Food Res Int 2015; 77: 392-9.

[33] Fleshman MK, Lester GE, Riedl KM, *et al.* Carotene and novel apocarotenoid concentrations in orange-fleshed *Cucumis melo* melons: determinations of β-carotene bioaccessibility and

bioavailability. J Agric Food Chem 2011; 59(9): 4448-54.
[PMID: 21417375]

[34] Laur LM, Tian L. Provitamin A and vitamin C contents in selected California-grown cantaloupe and honeydew melons and imported melons. J Food Compos Anal 2011; 24(2): 194-201.

[35] USDA National Nutrient Database for Standard Reference 2009.

[36] Flores FB, Martínez-Madrid MC, Sánchez-Hidalgo FJ, Romojaro F. Differential rind and pulp ripening of transgenic antisense ACC oxidase melon. Plant Physiol Biochem 2001; 39(1): 37-43.

[37] Fergany M, Kaur B, Monforte AJ, Pitrat M, Rys C, Lecoq H, *et al.* Variation in melon (*Cucumis melo*) landraces adapted to the humid tropics of southern India. Genet Resour Crop Evol 2011; 58(2): 225-43.

[38] Burger Y, Sa'ar U, Distelfeld A, Katzir N, Yeselson Y, Shen S, *et al.* Development of sweet melon (*Cucumis melo*) genotypes combining high sucrose and organic acid content. J Am Soc Hortic Sci 2003; 128(4): 537-40.

[39] Rezig L, Chouaibi M, Msaada K, Hamdi S. Chemical composition and profile characterisation of pumpkin (Cucurbita maxima) seed oil. Ind Crops Prod 2012; 37(1): 82-7.

[40] Teotia MS, Ramakrishna P. Chemistry and technology of melon seeds. J Food Sci Technol 1984; 21: 332-40.

[41] Yanty NAM, Lai OM, Osman A, Long K, Ghazali HM. Physicochemical properties of *Cucumis melo* var. inodorus (honeydew melon) seed and seed oil. J Food Lipids 2008; 15: 42-55.

[42] Obasi NA, Ukadilonu J, Eze E, Akubugwo EI, Okorie UC. Proximate composition, extraction, characterization and comparative assessment of coconut (*Cocos nucifera*) and melon (*Colocynthis citrullus*) seeds and seed oils. Pak J Biol Sci 2012; 15(1): 1-9.
[PMID: 22530436]

[43] Petkova Z, Antova G. Proximate composition of seeds and seed oils from melon (*Cucumis melo* L.) cultivated in Bulgaria. Cogent Food Agric 2015; 1(1): 1018779.

[44] Mallek-Ayadi S, Bahloul N, Kechaou N. Chemical composition and bioactive compounds of *Cucumis melo* L. seeds: Potential source for new trends of plant oils. Process Saf Environ Prot 2018; 113: 68-77.

[45] de Melo MLS, Narain N, Bora PS. Characterisation of some nutritional constituents of melon (*Cucumis melo* hybrid AF-522) seeds. Food Chem 2000; 68: 411-4.

[46] Azhari S, Xu YS, Jiang QX, Xia WS. Physicochemical properties and chemical composition of Seinat (*Cucumis melo* var. tibish) seed oil and its antioxidant activity. Grasas Aceites 2014; 65(1): e008.http://grasasyaceites.revistas.csic.es/index.php/grasasyaceites/article/view/1473/1504 [Internet].

[47] Ibeto CN, Okoye COB, Ofoefule AU. Comparative Study of the Physicochemical Characterization of Some Oils as Potential Feedstock for Biodiesel Production. ISRN Renew Energy 2012; 2012: 1-5.

[48] Onyeike EN, Acheru GN. Chemical composition of selected Nigerian oil seeds and physicochemical properties of the oil extracts. Food Chem 2002; 77(4): 431-7.

[49] Azhari S, Xu YS, Jiang QX, Xia WS. Chemical and Nutritional Properties of Seinat (*Cucumis melo* var. tibish) seeds. J Acad Ind Res 2014; 2(9): 495-9.

[50] Mariod A, Matthäus B. Fatty acids, tocopherols, sterols, phenolic profiles and oxidative stability of *Cucumis melo* var. agrestis oil. J Food Lipids 2008; 15: 56-67.

[51] da Silva AC, Jorge N. Bioactive compounds of the lipid fractions of agro-industrial waste. Food Res Int 2014; 66: 493-500.

[52] Albishri HM, Almaghrabi OA, Moussa TA. Characterization and chemical composition of fatty acids content of watermelon and muskmelon cultivars in Saudi Arabia using gas chromatography/mass spectroscopy. Pharmacogn Mag 2013; 9(33): 58-66.

[PMID: 23661995]

[53] Mian-Hao H, Yansong A. Characteristics of some nutritional composition of melon (*Cucumis melo* hybrid "ChunLi") seeds. Int J Food Sci Technol 2007; 42(12): 1397-401.

[54] Milovanović M, Pićurić-Jovanović K. Characteristics and Composition of Melon Seed Oil. J Agric Sci 2005; 50(1): 41-7.

[55] Rashwan MRA, El-Syiad SI, Seleim MA. Protein solubility, mineral content, amino acid composition and electrophoretic pattern of some gourd seeds. Acta Aliment 1993; 22(1): 15-24.

[56] Górnaś Paweł and Rudzińska M. Seeds recovered from industry by-products of nine fruit species with a high potential utility as a source of unconventional oil for biodiesel and cosmetic and pharmaceutical sectors. Ind Crops Prod 2016; 83: 329-38.

[57] Al-Khalifa AS. Physicochemical characteristics, fatty acid composition, and lipoxygenase activity of crude pumpkin and melon seed oils. J Agric Food Chem 1996; 44(4): 964-6.

[58] El-Adawy TA, Taha KM. Characteristics and composition of watermelon, pumpkin, and paprika seed oils and flours. J Agric Food Chem 2001; 49(3): 1253-9.
[PMID: 11312845]

[59] Maran JP, Priya B. Supercritical fluid extraction of oil from muskmelon (*Cucumis melo*) seeds. J Taiwan Inst Chem Eng 2015; 47: 71-8.

[60] Zeb A. Phenolic Profile and Antioxidant Activity of Melon (*Cucumis melo* L.) Seeds from Pakistan. Foods 2016; 5(4): 67.
[PMID: 28231162]

[61] Hemavatahy J. Lipid Composition of Melon (*Cucumis melo*) Kernel. J Food Compos Anal 1992; 5: 90-5.

[62] Raw materials. New Delhi: Council of Scientific and Industrial Research 1950; p. 8.

[63] Abiodum OA, Adeleke RO. Comparative studies on nutritional composition of four melon seeds varieties. Pak J Nutr 2010; 9(9): 905-8.

[64] Jeffrey C. Systematics of the Cucurbitaceae: An overview.Biology and utilization of the Cucurbitaceae. Ithaca: Cornell University Press 1990; pp. 3-28.

[65] Ivanova PH. The melons—Raw material for food processing In: 50 years Food RDI International Scientific-Practical Conference —Food, Technologies and Health.l; 2012; pp. Plovdiv, Bulgaria23-6.

[66] Chen L, Kang YH, Suh JK. Roasting processed oriental melon (*Cucumis melo* L. var. makuwa Makino) seed influenced the triglyceride profile and the inhibitory potential against key enzymes relevant for hyperglycemia. Food Res Int 2014; 56: 236-42.

[67] Gill NS, Garg M, Bansal R, Sood S, Muthuraman A, Bali M, *et al.* Evaluation of antioxidant and antiulcer potential of cucumis Sativum L. Seed extract in rats. Asian J Clin Nutr 2009; 1(3): 131-8.

[68] Contreras-Calderón J, Calderón-Jaimes L, Guerra-Hernández E, García-Villanova B. Antioxidant capacity, phenolic content and vitamin C in pulp, peel and seed from 24 exotic fruits from Colombia. Food Res Int 2011; 44(7): 2047-53.

[69] Otten J, Hellwig J, Meyers L. Dietary Reference Intakes: The Essential Guide to Nutrient Requirements. Washington, DC: National Academies Press 2006.

[70] Alves-Rodrigues A, Shao A. The science behind lutein. Toxicol Lett 2004; 150(1): 57-83.
[PMID: 15068825]

[71] Milind P, Kulwant S. Musk melon is eat-must melon: A Review. Int Res J Pharm 2011; 2(8): 52-7.

[72] Vouldoukis I, Lacan D, Kamate C, *et al.* Antioxidant and anti-inflammatory properties of a *Cucumis melo* LC. extract rich in superoxide dismutase activity. J Ethnopharmacol 2004; 94(1): 67-75.
[PMID: 15261965]

[73]   Malhotra OP, Rani I. Occurrence, isolation & characterisation of urease-inhibitor from melon (*Cucumis melo*) seeds. Indian J Biochem Biophys 1978; 15(3): 229-31.
[PMID: 748169]

[74]   Moon JK, Shibamoto T. Antioxidant assays for plant and food components. J Agric Food Chem 2009; 57(5): 1655-66.
[PMID: 19182948]

[75]   Simpson R, Morris GA. The anti-diabetic potential of polysaccharides extracted from members of the cucurbit family: A review. Bioact Carbohydrates Diet Fibre 2014; 3(2): 106-14.

[76]   Johnson IT. New approaches to the role of diet in the prevention of cancers of the alimentary tract. Mutat Res 2004; 551(1-2): 9-28.
[PMID: 15225578]

[77]   Navarro-González I, García-Valverde V, García-Alonso J, Periago MJ. Chemical profile, functional and antioxidant properties of tomato peel fiber. Food Res Int 2011; 44(5): 1528-35.

[78]   Houston MC. Nutraceuticals, vitamins, antioxidants, and minerals in the prevention and treatment of hypertension. Prog Cardiovasc Dis 2005; 47(6): 396-449.
[PMID: 16115519]

[79]   Ghanem N, Mihoubi D, Kechaou N, Mihoubi NB. Microwave dehydration of three citrus peel cultivars: Effect on water and oil retention capacities, color, shrinkage and total phenols content. Ind Crops Prod 2012; 40(1): 167-77.

[80]   Bahloul N, Boudhrioua N, Kouhila M, Kechaou N. Effect of convective solar drying on colour, total phenols and radical scavenging activity of olive leaves (*Olea europaea* L.). Int J Food Sci Technol 2009; 44(12): 2561-7.

[81]   Aydin E, Gocmen D. The influences of drying method and metabisulfite pre-treatment onthe color, functional properties and phenolic acids contents and bioaccessibility of pumpkin flour. Lebensm Wiss Technol 2015; 60(1): 385-92.

[82]   Khadem S, Marles RJ. Monocyclic phenolic acids; hydroxy- and polyhydroxybenzoic acids: occurrence and recent bioactivity studies. Molecules 2010; 15(11): 7985-8005.
[PMID: 21060304]

[83]   Bendini A, Cerretani L, Carrasco-Pancorbo A, *et al.* Phenolic molecules in virgin olive oils: a survey of their sensory properties, health effects, antioxidant activity and analytical methods. An overview of the last decade. Molecules 2007; 12(8): 1679-719.
[PMID: 17960082]

[84]   Soong YY, Barlow PJ. Quantification of gallic acid and ellagic acid from longan (Dimocarpus longan Lour.) seed and mango (*Mangifera indica* L.) kernel and their effects on antioxidant activity. Food Chem 2006; 97(3): 524-30.

[85]   Ismail T, Sestili P, Akhtar S. Pomegranate peel and fruit extracts: a review of potential anti-inflammatory and anti-infective effects. J Ethnopharmacol 2012; 143(2): 397-405.
[PMID: 22820239]

[86]   Benavente-Garcia O, Castillo J, Lorente J, Ortuno A, Del Rio JA. Antioxidant activity of phenolics extracted from *Olea europaea* L. leaves. Food Chem 2000; 68: 457-62.

[87]   Erbay Z, Icier F. Optimization of hot air drying of olive leaves using response surface methodology. J Food Eng 2009; 91(4): 533-41.

[88]   Bhujbal SS, Nanda RK, Deoda RS, Kumar D, Kewatkar SM, More LS, *et al.* Structure elucidation of a flavonoid glycoside from the roots of *Clerodendrum serratum* (L.) Moon, Lamiaceae. Brazilian J Pharmacogn 2010; 20(6): 1001-2.

[89]   Kim KS, Lee HJ, Kim SM. Volatile flavor components in watermelon (*Citrullus vulgaris* S.) and Oriental Melon (*Cucumis melo* L.). Korean J Food Sci 1999; 31(2): 322-8.

[90]   Bellakhdar J, Claisse R, Fleurentin J, Younos C. Repertory of standard herbal drugs in the Moroccan pharmacopoea. J Ethnopharmacol 1991; 35(2): 123-43.
[PMID: 1809818]

[91]   Castelo-Branco VN, Torres AG. Potential application of antioxidant capacity assays to assess the quality of edible vegetable oils. Lipid Technol 2009; 21(7): 152-5.

[92]   Oomah BD, Ladet S, Godfrey DV, Liang J, Girard B. Characteristics of raspberry (*Rubus idaeus* L.) seed oil. Food Chem 2000; 69(2): 187-93.

[93]   Duke JA, Ayensu ES. Medicinal Plants of China. Michigan: Algonac 1985.

[94]   Lal D, Lata K. Plants used by the Bhat community for regulating fertility. Econ Bot 1980; 34(3): 273-5.
[PMID: 12336832]

[95]   Saloua F, Eddine NI, Hedi Z. Chemical composition and profile characteristics of Osage orange Maclura pomifera (Rafin.) Schneider seed and seed oil. Ind Crops Prod 2009; 29(1): 1-8.

[96]   Jones PJ, Raeini-Sarjaz M, Ntanios FY, Vanstone CA, Feng JY, Parsons WE. Modulation of plasma lipid levels and cholesterol kinetics by phytosterol *versus* phytostanol esters. J Lipid Res 2000; 41(5): 697-705.
[PMID: 10787430]

[97]   Tapiero H, Townsend DM, Tew KD. Phytosterols in the prevention of human pathologies. Biomed Pharmacother 2003; 57(8): 321-5.
[PMID: 14568225]

[98]   Ntanios F. Plant sterol-ester-enriched spreads as an example of a new functional food Plant sterol-enriched spreads : a new functional food Mechanisms of action of the plant sterols What are plant sterols? Establishing the efficacy of plant sterol- esters and claim. Eur J Lipid Sci Technol 2001; 103: 102-6.

[99]   Ntambi JM, Choi Y, Park Y, Peters JM, Pariza MW. Effects of conjugated linoleic acid (CLA) on immune responses, body composition and stearoyl-CoA desaturase. Can J Appl Physiol 2002; 27(6): 617-28.
[PMID: 12501000]

[100] Suresh Y, Das UN. Long-chain polyunsaturated fatty acids and chemically induced diabetes mellitus. Effect of omega-3 fatty acids. Nutrition 2003; 19(3): 213-28.
[PMID: 12620523]

[101] Ajayi OB, Ajayi DD. Effect of oilseed diets on plasma lipid profile in albino rats. Pak J Nutr 2009; 8: 116-8.

[102] Rodríguez-Morató J, Xicota L, Fitó M, Farré M, Dierssen M, de la Torre R. Potential role of olive oil phenolic compounds in the prevention of neurodegenerative diseases. Molecules 2015; 20(3): 4655-80.
[PMID: 25781069]

[103] Gill NS, Bajwa J, Dhiman K, Sharma P, Sood S, Sharma PD, *et al.* Evaluation of therapeutic potential of traditionally consumed *Cucumis melo* seeds. Asian J Plant Sci 2011; 10(1): 86-91.

[104] Arora R, Kaur M, Gill NS. Antioxidant activity and pharmacological eval- uation of *Cucumis melo* var. agrestis methanolic seed extract. Res J Phytochem 2011; 5: 146-55.

[105] Gropper SS, Smith JL, Groff JL. Advanced Nutrition and Human Metabolism. Belmont: Thomson Wadsworth Publishing Co 2012.

[106] Smirnoff N. BOTANICAL BRIEFING: The Function and Metabolism of Ascorbic Acid in Plants. Ann Bot 1996; 78(6): 661-9.

[107] Delgado-Pelayo R, Gallardo-Guerrero L. Hornero-M$ı$ndez D. Chlorophyll and carotenoid pigments

in the peel and flesh of commercial apple fruit varieties. Food Res Int 2014; 65: 272-81.

[108] Duda-Chodak A, Tarko T. Antioxidant properties of different fruit seeds and peels. Acta Sci Pol Technol Aliment 2007; 6(3): 29-36.

[109] Ibrahim SRM, Mohamed GA. Cucumin S, a new phenylethyl chromone from *Cucumis melo* var. Reticulatus seeds. Brazilian J Pharmacogn 2015; 25(5): 462-4.

[110] Gill NS, Bali M. Isolation of anti ulcer cucurbitane type triterpenoid from the seeds of *Cucurbita pepo*. Res J Phy 2011; 5(2): 70-9.

[111] Chan KT, Li K, Liu SL, Chu KH, Toh M, Xie WD. Cucurbitacin B inhibits STAT3 and the Raf/MEK/ERK pathway in leukemia cell line K562. Cancer Lett 2010; 289(1): 46-52.
[PMID: 19700240]

[112] Liu T, Zhang M, Zhang H, *et al.* Combined antitumor activity of cucurbitacin B and docetaxel in laryngeal cancer. Eur J Pharmacol 2008; 587(1-3): 78-84.
[PMID: 18442812]

[113] Ibrahim S, Al Haidari R, Mohamed G, Elkhayat E, Moustafa M. Cucumol a: A cytotoxic triterpenoid from *Cucumis melo* seeds. Brazilian J Pharmacogn 2016; 26(6): 701-4.

[114] Isea Fernández GA, Rodríguez Rodríguez IE, Camarillo Sánchez EE, Araujo Gil MA. Dose-response assessment of the diuretic effect in a pericarp aqueous extract of *Cucumis melo* L. var. reticulatus Ser. Rev Cuba Plantas Med 2013; 18(3): 405-11.

[115] Ravishankar K, Vishnu Priya PSV. Evaluation of diuretic effect of ethanolic seed extracts of Macrotyloma uniflorum and *Cucumis melo* in rats. Int J Pharma Bio Sci 2012; 3(3): 251-5.

[116] Naito Y, Akagiri S, Uchiyama K, *et al.* Reduction of diabetes-induced renal oxidative stress by a cantaloupe melon extract/gliadin biopolymers, oxykine, in mice. Biofactors 2005; 23(2): 85-95.
[PMID: 16179750]

[117] Parmar HS, Kar A. Protective role of *Mangifera indica* , *Cucumis melo* and *Citrullus vulgaris* peel extracts in chemically induced hypothyroidism. Chem Biol Interact 2009; 177(3): 254-8.
[PMID: 19059228]

[118] Bidkar JS, Ghanwat DD, Bhujbal MD, Dama GY. Anti-hyperlipidemic activity of *Cucumis melo* fruit peel extracts in high cholesterol diet induced hyperlipidemia in rats. J Complement Integr Med 2012; 9: 22. [A].
[PMID: 23023565]

[119] Carillon J, Jover B, Cristol J-P, Rouanet J-M, Richard S, Virsolvy A. Dietary supplementation with a specific melon concentrate reverses vascular dysfunction induced by cafeteria diet. Food Nutr Res 2016; 60: 32729.
[PMID: 27834185]

[120] Sasi Kumar R, Priyadharshini S, Nandha Kumar KPL, Nivedha S. *In vitro* pharmacognostical studies and evaluation of bioactive constituents from the fruits of *Cucumis melo* L. (Muskmelon). Int J Pharmacogn Phytochem Res 2014; 6(4): 936-41.

[121] Ibrahim SRM. New 2-(2-phenylethyl)chromone derivatives from the seeds of *Cucumis melo* L var. reticulatus. Nat Prod Commun 2010; 5(3): 403-6.
[PMID: 20420317]

[122] Ribeiro SFF, Agizzio AP, Machado OLT, Neves-Ferreira AGC, Oliveira MA, Fernandes KVS, *et al.* A new peptide of melon seeds which shows sequence homology with vicilin: Partial characterization and antifungal activity. Sci Hortic (Amsterdam) 2007; 111(4): 399-405.

[123] Gavarkar P, Thorat S, Adnaik R, Mohite S, Magdum C. Characterisation and formulation of skin cream from seed oil extracted from *Cucumis melo*. Der Pharm Lett 2016; 8(5): 1-4.

[124] Dhasarathan P, Gomathi R, Theriappan P, Paulsi S. Immunomodulatory activity of alcoholic extract of different fruits in mice. J Appl Sci Res 2010; 6(8): 1056-9.

[125] Guner N, Wehner TC. The genes of watermelon. HortScience 2004; 39(6): 1175-82.

[126] Vaughan JG, Geissler CA. The new oxford book of food plants. Oxford: University Press 2009.

[127] Arshiya S. The antioxidant effect of certain fruits: - A review. J Pharm Sci Res 2013; 5(12): 265-8.

[128] Kumar P. Watermelon-utilization of peel waste for pickle processing. Indian Food Pack 1985; 39: 49-52.

[129] Bianchi G, Rizzolo A, Grassi M, Provenzi L, Lo Scalzo R. External maturity indicators, carotenoid and sugar compositions and volatile patterns in "Cuoredolce®" and "Rugby" mini-watermelon (*Citrullus lanatus* (Thunb) Matsumura & Nakai) varieties in relation of ripening degree at harvest. Postharvest Biol Technol 2018; 136: 1-11.

[130] Erukainure OL, Oke OV, Daramola AO, Adenekan SO, Umanhonlen EE. Improvement of the biochemical properties of watermelon rinds subjected to Saccharomyces cerevisae solid media fermentation. Pak J Nutr 2010; 9: 806-9.

[131] Figueroa A, Sanchez-Gonzalez MA, Wong A, Arjmandi BH. Watermelon extract supplementation reduces ankle blood pressure and carotid augmentation index in obese adults with prehypertension or hypertension. Am J Hypertens 2012; 25(6): 640-3.
[PMID: 22402472]

[132] Hassan LEA, Koko WS, Osman E-BE, Dahab MM, Sirat HM. *In vitro* antigiardial activity of *Citrullus lanatus* Var. citroides extracts and cucurbitacins isolated compounds. J Med Plants Res 2011; 5(15): 3338-46.

[133] Proietti S, Rouphael Y, Colla G, Cardarelli M, De Agazio M, Zacchini M, *et al.* Fruit quality of mini-watermelon as affected by grafting and irrigation regimes. J Sci Food Agric 2008; 88: 1107-14.

[134] Desamero NV, Adelberg JW, Hale A, Young RE, Rhodes BB, Horticulture I, *et al.* Nutrient utilization in liquid / membrane system for watermelon micropropagation. Plant Cell Tissue Organ Cult 1993; 33: 265-71.

[135] Abu-Hiamed HA. Chemical Composition, Flavonoids and β-sitosterol Contents of Pulp and Rind of Watermelon (*Citrullus lanatus*) Fruit. Pak J Nutr 2017; 16(7): 502-7.

[136] Perkins-Veazie P, Collins JK. Carotenoid changes of intact watermelons after storage. J Agric Food Chem 2006; 54(16): 5868-74.
[PMID: 16881688]

[137] Miceli A, Romano C, Moncada A, Piazza G, Torta L, D'Anna F, *et al.* Yield and quality of mini-watermelon as affected bygrafting and mycorrhizal inoculum. J Agric Sci Technol 2016; 18(2): 505-16.

[138] Shiu JW, Slaughter DC, Boyden LE, Barrett DM. Correlation of Descriptive Analysis and Instrumental Puncture Testing of Watermelon Cultivars. J Food Sci 2016; 81(6): S1506-14.
[PMID: 27105291]

[139] Tlili I, Hdider C, Lenucci MS, Ilahy R, Jebari H, Dalessandro G. Bioactive compounds and antioxidant activities during fruit ripening of watermelon cultivars. J Food Compos Anal 2011; 24(7): 923-8.

[140] Mantoan C. Yield and quality of mini- and midi-watermelons. Inf Agrar 2009; 65(3): 44-6. [Rese e qualità di mini e midi cocomeri].

[141] Yativ M, Harary I, Wolf S. Sucrose accumulation in watermelon fruits: genetic variation and biochemical analysis. J Plant Physiol 2010; 167(8): 589-96.
[PMID: 20036442]

[142] Guo S, Liu J, Zheng Y, *et al.* Characterization of transcriptome dynamics during watermelon fruit development: sequencing, assembly, annotation and gene expression profiles. BMC Genomics 2011; 12: 454.

[PMID: 21936920]

[143] Soteriou GA, Kyriacou MC, Siomos AS, Gerasopoulos D. Evolution of watermelon fruit physicochemical and phytochemical composition during ripening as affected by grafting. Food Chem 2014; 165: 282-9.
[PMID: 25038677]

[144] Leskovar DI, Bang H, Crosby KM, Maness N, Franco JA, Perkins-Veazie P. Lycopene, carbohydrates, ascorbic acid and yield components of diploid and triploid watermelon cultivars are affected by deficit irrigation. J Hortic Sci Biotechnol 2004; 79(1): 75-81.

[145] Vanderslice JT, Higgs DJ, Hayes JM, Block G. Ascorbic acid and dehydroascorbic acid content of foods-as-eaten. J Food Compos Anal 1990; 3(2): 105-18.

[146] Chun J, Lee J, Ye L, Exler J, Eitenmiller RR. Tocopherol and tocotrienol contents of raw and processed fruits and vegetables in the United States diet. J Food Compos Anal 2006; 19(2–3): 196-204.

[147] Tarazona-Díaz MP, Viegas J, Moldao-Martins M, Aguayo E. Bioactive compounds from flesh and by-product of fresh-cut watermelon cultivars. J Sci Food Agric 2011; 91(5): 805-12.
[PMID: 21384347]

[148] Vinson JA, Su X, Zubik L, Bose P. Phenol antioxidant quantity and quality in foods: fruits. J Agric Food Chem 2001; 49(11): 5315-21.
[PMID: 11714322]

[149] Fabiani A, Versari A, Parpinello GP, Castellari M, Galassi S. High-performance liquid chromatographic analysis of free amino acids in fruit juices using derivatization with 9-fluorenylmethyl-chloroformate. J Chromatogr Sci 2002; 40(1): 14-8.
[PMID: 11866381]

[150] Abu-Reidah IM, Arráez-Román D, Segura-Carretero A, Fernández-Gutiérrez A. Profiling of phenolic and other polar constituents from hydro-methanolic extract of watermelon (*Citrullus lanatus*) by means of accurate-mass spectrometry (HPLC-ESI-QTOF-MS). Food Res Int 2013; 51(1): 354-62.

[151] Kumar A, Altabella T, Taylor MA, Tiburcio AF. Recent advances in polyamine research. Trends Plant Sci 1997; 2(4): 124-30.

[152] Liu C, Zhang H, Dai Z, Liu X, Liu Y, Deng X, *et al.* Volatile chemical and carotenoid profiles in watermelons [*Citrullus vulgaris* (Thunb.) Schrad (Cucurbitaceae)] with different flesh colors. Food Sci Biotechnol 2012; 21(2): 531-41.

[153] Yoo KS, Bang H, Lee EJ, Crosby K, Patil BS. Variation of carotenoid, sugar, and ascorbic acid concentrations in watermelon genotypes and genetic analysis. Hortic Environ Biotechnol 2012; 53(6): 552-60.

[154] Lv P, Li N, Liu H, Gu H, Zhao WE. Changes in carotenoid profiles and in the expression pattern of the genes in carotenoid metabolisms during fruit development and ripening in four watermelon cultivars. Food Chem 2015; 174: 52-9.
[PMID: 25529651]

[155] Bang H. Environmental and genetic strategies to improve carotenoids and quality in watermelon. USA: Texas A&M University 2005.

[156] Holden JM, Eldridge AL, Beecher GR, Marilyn Buzzard I, Bhagwat S, Davis CS, *et al.* Carotenoid Content of U.S. Foods: An Update of the Database. J Food Compos Anal 1999; 12(3): 169-96.

[157] Beaulieu JC, Lea JM. Characterization and semiquantitative analysis of volatiles in seedless watermelon varieties using solid-phase microextraction. J Agric Food Chem 2006; 54(20): 7789-93.
[PMID: 17002453]

[158] Yajima I, Sakakibara H, Ide J, Yanai T, Hayash K. Volatile flavor components of watermelon (*Citrullus vulgaris*). Agric Biol Chem 1985; 49(11): 3145-50.

[159] Kemp TR, Knavel DE, Stoltz LP, Lundin RE. 3,6-Nonadien-1-ol from *Citrullus Vulgaris* and *Cucumis melo*. Phytochemistry 1974; 13: 1167-70.

[160] Kasting R, Delwiche CC. Ornithine, Citrulline, and Arginine Metabolism in Watermelon Seedlings. Plant Physiol 1958; 33(5): 350-4.
[PMID: 16655146]

[161] Romero MJ, Platt DH, Caldwell RB, Caldwell RW. Therapeutic use of citrulline in cardiovascular disease. Cardiovasc Drug Rev 2006; 24(3-4): 275-90.
[PMID: 17214603]

[162] Luiking YC, Poeze M, Ramsay G, Deutz NE. Reduced citrulline production in sepsis is related to diminished *de novo* arginine and nitric oxide production. Am J Clin Nutr 2009; 89(1): 142-52.
[PMID: 19056593]

[163] Jayaprakasha GK, Chidambara Murthy KN, Patil BS. Rapid HPLC-UV method for quantification of l-citrulline in watermelon and its potential role on smooth muscle relaxation markers. Food Chem 2011; 127(1): 240-8.

[164] Rimando AM, Perkins-Veazie PM. Determination of citrulline in watermelon rind. J Chromatogr A 2005; 1078(1-2): 196-200.
[PMID: 16007998]

[165] Johnson JT, Iwang EU, Hemen JT, Odey MO, Efiong EE. Evaluation of anti-nutrient contents of watermelon citrallus lanatus. Biol Res 2012; 3(11): 5145-50.

[166] Sani UM. Phytochemical screening and antidiabetic effect of extracts of the seeds of *Citrullus lanatus* in alloxan-induced diabetic albino mice. J Appl Pharm Sci 2015; 5(3): 51-4.

[167] Jyothi lakshmi A, Kaul P. Nutritional potential, bioaccessibility of minerals and functionality of watermelon (*Citrullus vulgaris*) seeds. Lebensm Wiss Technol 2011; 44(8): 1821-6.

[168] Wani AA, Sogi DS, Singh P, Wani IA, Shivhare US. Characterisation and functional properties of watermelon (*Citrullus lanatus*) seed proteins. J Sci Food Agric 2011; 91(1): 113-21.
[PMID: 20824684]

[169] Egbuonu ACC. Comparative Investigation of the Proximate and Functional Properties of Watermelon (*Citrullus lanatus*) Rind and Seed. Res J Environ Toxicol 2015; 9(3): 160-7.

[170] Zhu M, Phillipson JD, Greengrass PM, Bowery NE, Cai Y. Plant polyphenols: biologically active compounds or non-selective binders to protein? Phytochemistry 1997; 44(3): 441-7.
[PMID: 9014370]

[171] Simonne A, Carter M, Fellers R, *et al*. Chemical, physical and sensory characterization of watermelon rind pickles. J Food Process Preserv 2003; 26(6): 415-31.

[172] Romdhane MB, Haddar A, Ghazala I, Jeddou KB, Helbert CB, Ellouz-Chaabouni S. Optimization of polysaccharides extraction from watermelon rinds: Structure, functional and biological activities. Food Chem 2017; 216: 355-64.
[PMID: 27596431]

[173] King RD, Onuora JO. Aspects of melon seed protein characteristics. Food Chem 1984; 14(1): 65-77.

[174] Egbuonu ACC, Aguguesi RG, Samuel R, Ojunkwu O, Onyenmeri F, Uzuegbu U. Some physicochemical properties of the petroleum ether-extracted watermelon (*Citrullus lanatus*) seed oil. Asian J Sci Res 2015; 8(4): 519-25.

[175] Osagie AU, Okoye WF, Oluwayose BO, Dawudu OA. Chemical quality parameters and fatty acid composition of oil of some under exploited tropical seeds. Niger J Appl Sci 1986; 4: 154-62.

[176] Adewuyi A, Oderinde RA, Ademisoye AO. Antibacterial activities of nonionic and anionic surfactants from *Citrullus lanatus* seed oil. Jundishapur J Microbiol 2013; 6(3): 205-8.

[177] Mustafa AB, Alamin AAM. Chemical composition and Protein Degradability of Watermelon

(*Citrullus lanatus*) Seeds Cake grown in Western Sudan. Asian J Anim Sci 2012; 6(1): 33-7.

[178] Pal RN, Mahadevan V. Chemical composition and nutritive value of Bijada cake (*Citrulus vulgaris*). Indian Vet J 1968; 45(5): 433-9.
[PMID: 5756747]

[179] Di Mascio P, Kaiser S, Sies H. Lycopene as the most efficient biological carotenoid singlet oxygen quencher. Arch Biochem Biophys 1989; 274(2): 532-8.
[PMID: 2802626]

[180] Rao AV. Tomatoes, Lycopene and Human Health Preventing Chronic Diseases Ardersier. Caledonian Science Press Ltd 2007.

[181] Hong MY, Hartig N, Kaufman K, Hooshmand S, Figueroa A, Kern M. Watermelon consumption improves inflammation and antioxidant capacity in rats fed an atherogenic diet. Nutr Res 2015; 35(3): 251-8.
[PMID: 25631716]

[182] Han GM, Meza JL, Soliman GA, Islam KMM, Watanabe-Galloway S. Higher levels of serum lycopene are associated with reduced mortality in individuals with metabolic syndrome. Nutr Res 2016; 36(5): 402-7.
[PMID: 27101758]

[183] Edwards AJ, Vinyard BT, Wiley ER, *et al.* Consumption of watermelon juice increases plasma concentrations of lycopene and beta-carotene in humans. J Nutr 2003; 133(4): 1043-50.
[PMID: 12672916]

[184] Adeolu AT, Enesi DO. Assessment of proximate, mineral, vitamin and phytochemical compositions of plantain (Musa paradisiaca) bract – an agricultural waste. Int Res J Plant Sci 2013; 4(7): 192-7.

[185] Sharma S, Dwivedi J, Tilak A. First report on laxative activity of *Citrullus lanatus*. Pharmacologyonline 2011; 2: 790-7.

[186] Sodipo OA, Akiniyi JA, Ogunbamosu JU. Studies on certain characteristics of extracts of bark of Pausinystalia johimbe and Pausinystalia macroceras (K Schum) Pierre ex Beille. Glob J Pure Appl Sci 2000; 6: 83-7.

[187] Drewes SE, George J, Khan F. Recent findings on natural products with erectile-dysfunction activity. Phytochemistry 2003; 62(7): 1019-25.
[PMID: 12591255]

[188] Collins JK, Wu G, Perkins-Veazie P, *et al.* Watermelon consumption increases plasma arginine concentrations in adults. Nutrition 2007; 23(3): 261-6.
[PMID: 17352962]

[189] Teugwa CM, Boudjeko T, Tchinda BT, Mejiato PC, Zofou D. Anti-hyperglycaemic globulins from selected Cucurbitaceae seeds used as antidiabetic medicinal plants in Africa. BMC Complement Altern Med 2013; 13(63): 63.
[PMID: 23506532]

[190] Othman OC, Ngassapa FN. Physicochemical Characteristics of Some Imported Edible Vegetable Oils and Fat Marketed in Dar es Salaam. Tanzania. J Sci 2001; 27: 49-58.

[191] Sikarwar MS, Patil MB. Antidiabetic activity of Pongamia pinnata leaf extracts in alloxan-induced diabetic rats. Int J Ayurveda Res 2010; 1(4): 199-204.
[PMID: 21455444]

[192] Kunyanga CN, Imungi JK, Okoth M, Momanyi C, Biesalski HK, Vadivel V. Antioxidant and antidiabetic properties of condensed tannins in acetonic extract of selected raw and processed indigenous food ingredients from Kenya. J Food Sci 2011; 76(4): C560-7.
[PMID: 22417336]

[193] Ahn J, Choi W, Kim S, Ha T. Anti-diabetic effect of watermelon (*Citrullus vulgaris* Schrad) on

streptozotocin-induced diabetic mice. Food Sci Biotechnol 2011; 20(1): 251-4.

[194] Perkins-Veazie P. Cucurbits, watermelon, and benefits to human health. Acta Hortic 2010; (871): 25-32.

[195] El-Razek FHA, Sadeek EA. Dietary supplementation with watermelon (*Citrullus Ianatus*) juice enhances arginine availability and modifies hyperglycemia, hyperlipidemia and oxidative stress in diabeticrats. Aust J Basic Appl Sci 2011; 5: 1284-95.

[196] Poduri A, Rateri DL, Saha SK, Saha S, Daugherty A. *Citrullus lanatus* 'sentinel' (watermelon) extract reduces atherosclerosis in LDL receptor-deficient mice. J Nutr Biochem 2013; 24(5): 882-6.
[PMID: 22902326]

[197] Akashi K, Miyake C, Yokota A. Citrulline, a novel compatible solute in drought-tolerant wild watermelon leaves, is an efficient hydroxyl radical scavenger. FEBS Lett 2001; 508(3): 438-42.
[PMID: 11728468]

[198] Fang Y-Z, Yang S, Wu G. Free radicals, antioxidants, and nutrition. Nutrition 2002; 18(10): 872-9.
[PMID: 12361782]

[199] Fish WW, Davis AR. The effects of frozen storage conditions on lycopene stability in watermelon tissue. J Agric Food Chem 2003; 51(12): 3582-5.
[PMID: 12769528]

[200] Perkins-Veazie P, Collins JK, Pair SD, Roberts W. Lycopene content differs among red-fleshed watermelon cultivars. J Sci Food Agric 2001; 81(10): 983-7.

[201] Ikeda Y, Young LH, Scalia R, Lefer AM. Cardioprotective effects of citrulline in ischemia/reperfusion injury *via* a non-nitric oxide-mediated mechanism. Methods Find Exp Clin Pharmacol 2000; 22(7): 563-71.
[PMID: 11196344]

[202] Balderas-Munoz K, Castillo-Martínez L, Orea-Tejeda A, *et al.* Improvement of ventricular function in systolic heart failure patients with oral L-citrulline supplementation. Cardiol J 2012; 19(6): 612-7.
[PMID: 23224924]

[203] Gul S, Rashid Z, Sarwer G. *Citrullus lanatus* (Watermelon) as Diuretic Agent: An *in vivo* Investigation on Mice. Am J Drug Deliv Ther 2014; 1(4): 89-92.

[204] Rahmat A, Rosli R, Zain W, Endrini S, Sani HA. Antiproliferative Activity of Pure Lycopene Compared to Both Extracted Lycopene and Juices. J Med Sci 2002; 2(2): 55-8.

[205] Peters RR, Farias MR, Ribeiro-do-Valle RM. Anti-inflammatory and analgesic effects of cucurbitacins from Wilbrandia ebracteata. Planta Med 1997; 63(6): 525-8.
[PMID: 9434604]

[206] Abdelwahab SI, Hassan LE, Sirat HM, *et al.* Anti-inflammatory activities of cucurbitacin E isolated from *Citrullus lanatus* var. citroides: role of reactive nitrogen species and cyclooxygenase enzyme inhibition. Fitoterapia 2011; 82(8): 1190-7.
[PMID: 21871542]

[207] Lucky OO, John UO, Kate ie, Peter OO, Jude OE. Quantitative determination, metal analysis and antiulcer evaluation of methanol seed extract of Citrillus lanatus Thunb (Cucurbitaceae) in rats. Asian Pac J Trop Biomed 2012; 2(3): 1261-5.

[208] Gill NS, Sood S, Muthuraman A, Bali M, Sharma DP, Singh Gill N, *et al.* Evaluation of Antioxidant and Anti-ulcerative Potential of *Citrullus lanatus* Seed Extract in Rats. Lat Am J Pharm 2011; 30(3): 429-34.

[209] Adebayo AH, Yakubu OF, Balogun TM. Protective Properties of *Citrullus lanatus* on Carbon Tetrachloride Induced Liver Damage in Rats. European J Med Plants 2014; 4(8): 979-89.

[210] Madhavi P, Vakati K, Rahman H. Hepatoprotective Activity of *Citrullus lanatus* Seed Oil on CCl 4 Induced Liver Damage in Rats. Sch Acad J Pharm 2012; 1(1): 30-3.

[211] Zhang F, Lin L, Xie J. A mini-review of chemical and biological properties of polysaccharides from Momordica charantia. Int J Biol Macromol 2016; 92(235): 246-53.
[PMID: 27377459]

[212] Oyenihi OR, Afolabi BA, Oyenihi AB, Ogunmokun OJ, Oguntibeju OO. Hepato- and neuro-protective effects of watermelon juice on acute ethanol-induced oxidative stress in rats. Toxicol Rep 2016; 3: 288-94.
[PMID: 28959549]

[213] Waugh WH, Daeschner CW III, Files BA, McConnell ME, Strandjord SE. Oral citrulline as arginine precursor may be beneficial in sickle cell disease: early phase two results. J Natl Med Assoc 2001; 93(10): 363-71.
[PMID: 11688916]

[214] FAO Statistics Division. Production and trade statistics. Rome, Italy 2016.

[215] Patel S, Rauf A. Edible seeds from Cucurbitaceae family as potential functional foods: Immense promises, few concerns. Biomed Pharmacother 2017; 91: 330-7.
[PMID: 28463796]

[216] Babajide JM, Olaluwoye AA, Taofik Shittu TA, Adebisi MA. Physicochemical Properties and Phytochemical Components of Spiced Cucumber-Pineapple Fruit Drink. Niger Food J 2013; 31(1): 40-52.

[217] Abou-Zaid MM, Lombardo DA, Kite GC, Grayer RJ, Veitch NC. Acylated flavone C-glycosides from *Cucumis sativus*. Phytochemistry 2001; 58(1): 167-72.
[PMID: 11524127]

[218] Singh N, Bali M. Evaluation of antioxidant, antiulcer activity of 9-beta-methyl-19-norlanosta-5-ene type glycosides from *Cucumis sativus* seeds. Res J Med Plant 2012; 6: 309-17.

[219] Kowalczyk K, Gajc-Wolska J, Bujalski D, Radzanowska J. The effect of cultivation term, substrates and cultivars on fruit sensory value of cucumber (*Cucumis sativus* L.) in greenhouse production. Acta Hortic 2010; (877): 235-8.

[220] Gopalakrishnan SB, Kalaiarasi T, Subramanian R. Comparative DFT Study of Phytochemical Constituents of the Fruits of Cucumis trigonus Roxb. and *Cucumis sativus* Linn J Comput Methods Phys. 1-6. 2014; pp. Article ID 623235

[221] Sotiroudis G, Melliou E, Sotiroudis TG, Chinou I. Chemical analysis, antioxidant and antimicrobial activity of three Greek cucumber (*Cucumis sativus*) cultivars. J Food Biochem 2010; 34 (Suppl. 1): 61-78.

[222] Sahar A, Naqvi SAR, Hussain Z, Nosheen S, Khan ZA, Ahmad M, *et al.* Screening of Phytoconstituents, Investigation of Antioxidant and Antibacterial Activity of Methanlic and Aqueous Extracts of *Cucumis sativus* L. J Chem Soc Pak 2013; 35(2): 456-62.

[223] Oboh G, Ademiluyi AO, Ogunsuyi OB, Oyeleye SI, Dada AF, Boligon AA. Cabbage and cucumber extracts exhibited anticholinesterase, antimonoamine oxidase and antioxidant properties. J Food Biochem 2017; 41(3): 1-7.

[224] Abu-Reidah IM, Arráez-Román D, Quirantes-Piné R, Fernández-Arroyo S, Segura-Carretero A, Fernández-Gutiérrez A. HPLC-ESI-Q-TOF-MS for a comprehensive characterization of bioactive phenolic compounds in cucumber whole fruit extract. Food Res Int 2012; 46(1): 108-17.

[225] Chiu HF, Huang SR, Lu YY, Han YC, Shen YC, Venkatakrishnan K, *et al.* Antimutagenicity, antibacteria, and water holding capacity of chitosan from Luffa aegyptiaca Mill and *Cucumis sativus* L. J Food Biochem 2017; 41(3): 1-10.

[226] McNally DJ, Wurms KV, Labbé C, Bélanger RR. Synthesis of C-glycosyl flavonoid phytoalexins as a site-specific response to fungal penetration in cucumber. Physiol Mol Plant Pathol 2003; 63(6): 293-303.

[227] Kai H, Baba M, Okuyama T. Two new megastigmanes from the leaves of *Cucumis sativus*. Chem Pharm Bull (Tokyo) 2007; 55(1): 133-6.
[PMID: 17202717]

[228] Krauze-Baranowska M, Cisowski W. Flavonoids from some species of the genus Cucumis. Biochem Syst Ecol 2001; 29(3): 321-4.
[PMID: 11152951]

[229] Azhagu Saravana Babu P, Vajiha Aafrin B, Archana G, Sabina K, Sudharsan K, Radha Krishnan K, *et al.* Polyphenolic and phytochemical content of *Cucumis sativus* seeds and study on mechanism of preservation of nutritional and quality outcomes in enriched mayonnaise. Int J Food Sci Technol 2016; 51(6): 1417-24.

[230] Kamda AGS, Fokou E, Loh MBA, Raducanu D, Kansci G, Ifrim I, *et al.* Protective effect of edible Cucurbitaceae seed extract from Cameroon against oxidative stress. Environ Eng Manag J 2014; 13(7): 1721-7.

[231] Ngure JW, Cheng C, Yang S, Lou Q, Li J, Qian C, *et al.* Cultivar and seasonal effects on seed oil content and fatty acid composition of cucumber as a potential industrial crop. J Am Soc Hortic Sci 2015; 140(4): 362-72.

[232] Alnadif AM, Mirghani MES, Hussein I. Unconventional Oilseeds and Oil Sources London. Academic Press 2017; pp. 1-365.

[233] Garg VK, Nes WR. Occurrence of Δ5-sterols in plants producing predominantly Δ7-sterols: Studies on the sterol compositions of six cucurbitaceae seeds. Phytochemistry 1986; 25(11): 2591-7.

[234] Gry J, Andersson HC. Cucurbitacins in plant food. TemaNord 2006; 1-68.

[235] Mukherjee PK, Nema NK, Maity N, Sarkar BK. Phytochemical and therapeutic potential of cucumber. Fitoterapia 2013; 84(1): 227-36.
[PMID: 23098877]

[236] Qing ZX, Zhou Y, Liu XB, Cheng P, Zeng JG. 23,24-Dihydrocucurbitacin C: a new compound regarded as the next metabolite of cucurbitacin C. Nat Prod Res 2014; 28(15): 1165-70.
[PMID: 24896808]

[237] Ramezani M, Rahmani F, Dehestani A. Comparison between the effects of potassium phosphite and chitosan on changes in the concentration of Cucurbitacin E and on antibacterial property of *Cucumis sativus*. BMC Complement Altern Med. BMC Complement Altern Med 2017; 17(1): 1-6.
[PMID: 28049463]

[238] Miller HE, Rigelhof F, Marquart L, Prakash A, Kanter M. Antioxidant Content of Whole Grain Breakfast Cereals, Fruits and Vegetables J Am Coll Nutr 2000; 19 ((sup3)): 312S-9S.

[239] Nema NK, Maity N, Sarkar B, Mukherjee PK. *Cucumis sativus* fruit-potential antioxidant, anti-hyaluronidase, and anti-elastase agent. Arch Dermatol Res 2011; 303(4): 247-52.
[PMID: 21153830]

[240] Kaur C, Kapoor HC. Antioxidant activity and total phenolic content of some Asian vegetables. Int J Food Sci Technol 2002; 37: 153-61.

[241] Chu YF, Sun J, Wu X, Liu RH. Antioxidant and antiproliferative activities of common vegetables. J Agric Food Chem 2002; 50(23): 6910-6.
[PMID: 12405796]

[242] Pellegrini N, Serafini M, Colombi B, *et al.* Total antioxidant capacity of plant foods, beverages and oils consumed in Italy assessed by three different *in vitro* assays. J Nutr 2003; 133(9): 2812-9.
[PMID: 12949370]

[243] Jain K, Kataria S, Guruprasad KN. Effect of UV-B radiation on antioxidant enzymes and its modulation by benzoquinone and α-tocopherol in cucumber cotyledons. Curr Sci 2004; 87(1): 87-90.

[244] Akhtar N, Mehmood A, Khan BA, Mahmood T, Khan HMS, Saeed T. Exploring cucumber extract for skin rejuvenation. Afr J Biotechnol 2011; 10(7): 1206-16.

[245] Villaseñor IM, Simon MKB, Villanueva AMA. Comparative potencies of nutraceuticals in chemically induced skin tumor prevention. Nutr Cancer 2002; 44(1): 66-70.
[PMID: 12672643]

[246] Kai H, Baba M, Okuyama T. Inhibitory effect of *Cucumis sativus* on melanin production in melanoma B16 cells by downregulation of tyrosinase expression. Planta Med 2008; 74(15): 1785-8.
[PMID: 19009501]

[247] Gangale A, Ambore S, Gavit M, Lokhande S. A pilot study on evaluation of *Cucumis sativus* L. As a natural antisolar agent. Res J Pharm Technol 2016; 9(3): 212-4.

[248] Manjula B, Mahadeswara Swamy YH, Devaraja S, Girish KS, Kemparaju K. Clot promoting and dissolving properties of cucumber (*Cucumis sativus*) sap, validating its use in traditional medicine. Int J Pharm Pharm Sci 2015; 7: 104-11.

[249] Soltani R, Hashemi M, Farazmand A, *et al.* Evaluation of the Effects of *Cucumis sativus* Seed Extract on Serum Lipids in Adult Hyperlipidemic Patients: A Randomized Double-Blind Placebo-Controlled Clinical Trial. J Food Sci 2017; 82(1): 214-8.
[PMID: 27886382]

[250] Cho MJ, Buescher RW, Johnson M, Janes M. Inactivation of pathogenic bacteria by cucumber volatiles (E,Z)-2,6-nonadienal and (E)-2-nonenal. J Food Prot 2004; 67(5): 1014-6.
[PMID: 15151242]

[251] Muruganantham N, Solomon S, Senthamilselvi M. Antimicrobial Activity of Cucumis sativas (Cucumber) Flowers. Int J Pharm Sci Rev Res 2016; 36(1): 97-100.

[252] Izawa K, Amino Y, Kohmura M, Ueda Y, Kuroda M. 4.16 - Human–Environment Interactions – Taste BT - Comprehensive Natural Products II.Oxford. Elsevier 2010; pp. 631-71.

[253] Ujváry I. Chapter 3 - Pest Control Agents from Natural Products A2 New York: Academic PressKrieger, Robert BT - Hayes' Handbook of Pesticide Toxicology (Third Edition) 2010; pp. 119-229.

[254] Sharma N, Akhtar S, Jamal QMS, *et al.* Elucidation of antiangiogenic potential of vitexin obtained from *Cucumis sativus* targeting Hsp90 Protein: A novel multipathway targeted approach to restrain angiogenic phenomena. Med Chem 2017; 13(3): 282-91.
[PMID: 27834134]

[255] Muruganantham N, Solomon S, Senthamilselvi MM. Anti-cancer activity of *Cucumis sativus* (cucumber) flowers against human liver cancer. Int J Pharm Clin Res 2016; 8(1): 39-41.

[256] Ho CH, Ho MG, Ho SP, Ho HH. Bitter bottle gourd (*Lagenaria siceraria*) toxicity. J Emerg Med 2014; 46(6): 772-5.
[PMID: 24360122]

[257] Lueangamornnara U, Jiratchariyakul W. Immunosuppressive effects of Cucurbitacin B on human peripheral blood lymphocytes. J Med Plants Res 2010; 4(22): 2340-7.

[258] Bajpai VK, Kim J-E, Park Y-H, Kang SC. *In vivo* Pharmacological Effectiveness of Heat-treated Cucumber (*Cucumis sativus* L.) Juice against CCI4- induced Detoxification in a Rat Model. Indian J Pharm Educ Res 2017; 51(2): 280-7.

[259] Palanisamy V, Shanmugam S, Balakrishnan S. Hepato protective activity of *Cucumis sativus* L. Int J Res Pharm Sci 2015; 6(2): 85-8.

[260] Kumar M, Garg A, Parle M. Amelioration of diazepam induced memory impairment by fruit of *Cucumis sativus* L. in aged mice by using animal models of alzheimer's disease. Int J Pharmacogn Phytochem Res 2014; 6(4): 1015-23.

[261] Kumar M, Parle M. Pharmacological evaluation of cucumber for cognition enhancing effect on brain of mice. Pharmacogn J 2014; 6(3): 100-7.

[262] Foong F, Aqeelah M, Ichwan S. Biological Properties of Cucumber (*Cucumis sativus* L.) Extracts. Malays J Anal Sci 2015; 19(6): 1218-22.

[263] Sharma S, Yadav S, Singh G, Paliwal S, Dwivedi J. First report on laxative activity of *Cucumis sativus*. Int J Pharm Sci Rev Res 2012; 12(2): 129-31.

[264] Prasanthi D, Adikay S. Amelioration of cisplatin and gentamicin -induced nephrotoxicity by seeds of *Cucumis sativus*. Int J Pharma Bio Sci 2016; 7(4): 245-53.

[265] Dixit Y, Kar A. Protective role of three vegetable peels in alloxan induced diabetes mellitus in male mice. Plant Foods Hum Nutr 2010; 65(3): 284-9.
[PMID: 20614191]

[266] Karthiyayini T, Kumar R, Kumar KLS, Sahu RK, Roy A. Evaluation of antidiabetic and hypolipidemic effect of *Cucumis sativus* fruit in streptozotocin-induced-diabetic rats. Biomed Pharmacol J 2009; 2(2): 351-5.

[267] Park J-H, Kim R-Y, Park E. Antidiabetic activity of fruits and vegetables commonly consumed in Korea: Inhibitory potential against α-glucosidase and insulin-like action *in vitro*. Food Sci Biotechnol 2012; 21(4): 1187-93.

[268] Roman-Ramos R, Flores-Saenz JL, Alarcon-Aguilar FJ. Anti-hyperglycemic effect of some edible plants. J Ethnopharmacol 1995; 48(1): 25-32.
[PMID: 8569244]

[269] Packirisamy ASB, Basheer VA, Kasirajan S, Sukumar M. Assessment of antioxidant and *in vitro* inhibitory potential against key enzymes catalyse for hyperglycemia and prostate inflammation. Res J Biotechnol 2016; 11(4): 42-7.

[270] Sudheesh S, Vijayalakshmi NR. Lipid-lowering action of pectin from *Cucumis sativus*. Food Chem 1999; 67(3): 281-6.

[271] Gill N, Garg M, Bansal R, Sood S, Muthuraman A, Bali M, *et al.* Evaluation of Antioxidant and Antiulcer Potential of Cucumis sativum L. seeds Extracts in rats. Asian J Clin Nutr 2009; 1(3): 131-8.

[272] Gill N, Bali M. Evaluation of Antioxidant, Antiulcer Activity of 9-beta-methyl-19-norlanosta-5-ene Type Glycosides from *Cucumis sativus* Seeds. Res J Med Plant 2012; 6(4): 309-17.

[273] Patil MVK, Kandhare AD, Bhise SD. Effect of aqueous extract of *Cucumis sativus* Linn. fruit in ulcerative colitis in laboratory animals. Asian Pac J Trop Biomed 2012; 2(2): S962-9.

[274] Sharma S, Dwivedi J, Agrawal M, Paliwal S. Cytoprotection mediated antiulcer effect of aqueous fruit pulp extract of *Cucumis sativus*. Asian Pac J Trop Dis 2012; 2(2): S61-7.

[275] Kumar D, Kumar S, Singh J, *et al.* Free Radical Scavenging and Analgesic Activities of *Cucumis sativus* L. Fruit Extract. J Young Pharm 2010; 2(4): 365-8.
[PMID: 21264095]

[276] Singh Gill N, Sood S, Muthuraman A, Garg M, Kumar R, Bali M, *et al.* Antioxidant, anti-inflammatory and analgesic potential of *Cucumis sativus* seed extract. Lat Am J Pharm 2010; 29(6): 927-32.

[277] Gemma C, Mesches MH, Sepesi B, Choo K, Holmes DB, Bickford PC. Diets enriched in foods with high antioxidant activity reverse age-induced decreases in cerebellar β-adrenergic function and increases in proinflammatory cytokines. J Neurosci 2002; 22(14): 6114-20.
[PMID: 12122072]

[278] Das J, Chowdhury A, Biswas SK, Karkamar UK, Sharif SR, Raihan SZ, *et al.* Cytotoxicity and Antifungal Activities of Ethanolic and Chloroform Extract of *Cucumis sativus* L. Leaves and Stems. Reasearch J Phytochem 2012; 1: 25-30.

[279] Ibrahim TA, El-Hefnawy HM, El-Hela AA. Antioxidant potential and phenolic acid content of certain cucurbitaceous plants cultivated in Egypt. Nat Prod Res 2010; 24(16): 1537-45.
[PMID: 20835955]

[280] Tang J, Meng X, Liu H, *et al.* Antimicrobial activity of sphingolipids isolated from the stems of cucumber (*Cucumis sativus* L.). Molecules 2010; 15(12): 9288-97.
[PMID: 21160453]

[281] Al Akeel R, Al-Sheikh Y, Mateen A, Syed R, Janardhan K, Gupta VC. Evaluation of antibacterial activity of crude protein extracts from seeds of six different medical plants against standard bacterial strains. Saudi J Biol Sci 2014; 21(2): 147-51.
[PMID: 24600307]

[282] Paris HS. History of the Cultivar-Groups of *Cucurbita pepo*. Hortic Rev (Am Soc Hortic Sci) 2001; 25: 71-170.

[283] Adams GG, Imran S, Wang S, *et al.* Extraction, isolation and characterisation of oil bodies from pumpkin seeds for therapeutic use. Food Chem 2012; 134(4): 1919-25.
[PMID: 23442639]

[284] Bang MH, Han JT, Kim HY, *et al.* 13-Hydroxy-9Z,11E,15E-octadecatrienoic acid from the leaves of Cucurbita moschata. Arch Pharm Res 2002; 25(4): 438-40.
[PMID: 12214851]

[285] Adubofuor J, Amoah I, Agyekum PB. Physicochemical Properties of Pumpkin Fruit Pulp and Sensory Evaluation of Pumpkin-Pineapple Juice Blends. Am J od. Food Sci Technol 2016; 4(4): 89-96.

[286] Dinu M, Soare R, Hoza G, Becherescu A. Biochemical composition of some local pumpkin population. Agric Agric Sci Procedia 2016; 10: 185-91.

[287] Imaeda N, Tokudome Y, Ikeda M, Kitagawa I, Fujiwara N, Tokudome S. Foods contributing to absolute intake and variance in intake of selected vitamins, minerals and dietary fiber in middle-aged Japanese. J Nutr Sci Vitaminol (Tokyo) 1999; 45(5): 519-32.
[PMID: 10683805]

[288] Blanco-Díaz MT, Font R, Martínez-Valdivieso D, Del Río-Celestino M. Diversity of natural pigments and phytochemical compounds from exocarp and mesocarp of 27 *Cucurbita pepo* accessions. Sci Hortic (Amsterdam) 2015; 197: 357-65.

[289] Martínez-Valdivieso D, Gómez P, Font R, Alonso-Moraga A, Río-Celestino MD. Physical and chemical characterization in fruit from 22 summer squash (*Cucurbita pepo* L.) cultivars. Lebensm Wiss Technol 2015; 64: 1225-33.

[290] Fissore EN, Ponce NM, Stortz CA, Rojas AM, Gerschenson LN. Characterisation of fiber obtained from Pumpkin (Cucumis moschata Duch.) mesocarp through enzymatic treatment. Food Sci Technol Int 2007; 13(2): 141-51.

[291] Applequist WL, Avula B, Schaneberg BT, Wang YH, Khan IA. Comparative fatty acid content of seeds of four Cucurbita species grown in a common (shared) garden. J Food Compos Anal 2006; 19(6–7): 606-11.

[292] Stevenson DG, Eller FJ, Wang L, Jane J-L, Wang T, Inglett GE. Oil and tocopherol content and composition of pumpkin seed oil in 12 cultivars. J Agric Food Chem 2007; 55(10): 4005-13.
[PMID: 17439238]

[293] Achu MB, Fokou E, Tchiégang C, Fotso M, Tchouanguep MF. Nutritive value of some Cucurbitaceae oilseeds from different regions in Cameroon. Afr J Biotechnol 2005; 4(11): 1329-34.

[294] Nawirska-Olszańska A, Kita A, Biesiada A, Sokół-Łętowska A, Kucharska AZ. Characteristics of antioxidant activity and composition of pumpkin seed oils in 12 cultivars. Food Chem 2013; 139(1-4): 155-61.
[PMID: 23561092]

[295] Rezig L, Chouaibi M, Msaada K, Hamdi S. Chemical composition and profile characterisation of pumpkin (Cucurbita maxima) seed oil. Ind Crops Prod 2012; 37(1): 82-7.

[296] Achilonu MC, Nwafor IC, Umesiobi DO, Sedibe MM. Biochemical proximates of pumpkin (Cucurbitaeae spp.) and their beneficial effects on the general well-being of poultry species. J Anim Physiol Anim Nutr (Berl) 2018; 102(1): 5-16.
[PMID: 28158900]

[297] Andjelkovic M, Van Camp J, Trawka A, Verhé R. Phenolic compounds and some quality parameters of pumpkin seed oil. Eur J Lipid Sci Technol 2010; 112: 208-17.

[298] Bardaa S, Ben Halima N, Aloui F, *et al.* Oil from pumpkin (*Cucurbita pepo* L.) seeds: evaluation of its functional properties on wound healing in rats. Lipids Health Dis 2016; 15: 73.
[PMID: 27068642]

[299] Rabrenovi BB, Dimi EB, Novakovi MM, Te VV. The most important bioactive components of cold pressed oil from different pumpkin (*Cucurbita pepo* L.) seeds. Lebensm Wiss Technol 2014; 55: 521-7.

[300] Siano F, Straccia MC, Paolucci M, Fasulo G, Boscaino F, Volpe MG. Physico-chemical properties and fatty acid composition of pomegranate, cherry and pumpkin seed oils. J Sci Food Agric 2016; 96(5): 1730-5.
[PMID: 26033409]

[301] Veronezi CM, Jorge N. Bioactive compounds in lipid fractions of pumpkin (Cucurbita sp) seeds for use in food. J Food Sci 2012; 77(6): C653-7.
[PMID: 22671521]

[302] Yadav M, Jain S, Tomar R, Prasad GBKS, Yadav H. Medicinal and biological potential of pumpkin: an updated review. Nutr Res Rev 2010; 23(2): 184-90.
[PMID: 21110905]

[303] Xia HC, Li F, Li Z, Zhang ZC. Purification and characterization of Moschatin, a novel type I ribosome-inactivating protein from the mature seeds of pumpkin (Cucurbita moschata), and preparation of its immunotoxin against human melanoma cells. Cell Res 2003; 13(5): 369-74.
[PMID: 14672560]

[304] Ng TB, Parkash A, Tso WW. Purification and characterization of moschins, arginine-glutamate-rich proteins with translation-inhibiting activity from brown pumpkin (Cucurbita moschata) seeds. Protein Expr Purif 2002; 26(1): 9-13.
[PMID: 12356464]

[305] Hou X, Meehan EJ, Xie J, Huang M, Chen M, Chen L. Atomic resolution structure of cucurmosin, a novel type 1 ribosome-inactivating protein from the sarcocarp of Cucurbita moschata. J Struct Biol 2008; 164(1): 81-7.
[PMID: 18652900]

[306] Elhadi IM, Koko WS, Dahab MM, Mohamed Y, Imam E, Elmonem MA, *et al.* Antigiardial Activity of some Cucurbita Species Antigiardial Activity of some Cucurbita Species and *Lagenaria siceraria*. J For Prod Ind 2014; 2(4): 43-7.

[307] Hutt TF, Herrington ME. The Determination of Bitter Principles in Zucchinis. J Sci Food Agric 1985; 36: 1107-12.

[308] Wang DC, Pan HY, Deng XM, *et al.* Cucurbitane and hexanorcucurbitane glycosides from the fruits of *Cucurbita pepo* cv dayangua. J Asian Nat Prod Res 2007; 9(6-8): 525-9.
[PMID: 17885839]

[309] Nkosi CZ, Opoku AR, Terblanche SE. Antioxidative effects of pumpkin seed (*Cucurbita pepo*) protein isolate in CCl4-induced liver injury in low-protein fed rats. Phytother Res 2006; 20(11): 935-40.
[PMID: 16909447]

[310] Fan S, Hu Y, Li C, Liu Y. Optimization of preparation of antioxidative peptides from pumpkin seeds using response surface method. PLoS One 2014; 9(3): e92335.
[PMID: 24637721]

[311] Eraslan G, Kanbur M, Aslan Ö, Karabacak M. The antioxidant effects of pumpkin seed oil on subacute aflatoxin poisoning in mice. Environ Toxicol 2013; 28(12): 681-8.
[PMID: 24591108]

[312] Xanthopoulou MN, Nomikos T, Fragopoulou E, Antonopoulou S. Antioxidant and lipoxygenase inhibitory activities of pumpkin seed extracts. Food Res Int 2009; 42(5–6): 641-6.

[313] Nara K, Yamaguchi A, Maeda N, Koga H, Duchesne C. Antioxidative activity of water soluble polysaccharide in pumpkin fruits (Cucurbita maxima Duchesne). Biosci Biotechnol Biochem 2009; 73(6): 1416-8.
[PMID: 19502750]

[314] Adams GG, Imran S, Wang S, Mohammad A, Kok S, Gray DA, *et al.* The hypoglycaemic effect of pumpkins as anti-diabetic and functional medicines. Food Res Int 2011; 44(4): 862-7.

[315] El-Mosallamy AEMK, Sleem AA, Abdel-Salam OME, Shaffie N, Kenawy SA. Antihypertensive and cardioprotective effects of pumpkin seed oil. J Med Food 2012; 15(2): 180-9.
[PMID: 22082068]

[316] Wang S, Lu A, Zhang L, *et al.* Extraction and purification of pumpkin polysaccharides and their hypoglycemic effect. Int J Biol Macromol 2017; 98: 182-7.
[PMID: 28153462]

[317] Sedigheh A, Jamal MS, Mahbubeh S, Somayeh K, Mahmoud R, Azadeh A, *et al.* Hypoglycaemic and hypolipidemic effects of pumpkin (*Cucurbita pepo* L.) on alloxan-induced diabetic rats. Afr J Pharm Pharmacol 2011; 5(23): 2620-6.

[318] Makni M, Sefi M, Fetoui H, *et al.* Flax and Pumpkin seeds mixture ameliorates diabetic nephropathy in rats. Food Chem Toxicol 2010; 48(8-9): 2407-12.
[PMID: 20570704]

[319] Acosta-Patiño JL, Jiménez-Balderas E, Juárez-Oropeza MA, Díaz-Zagoya JC. Hypoglycemic action of Cucurbita ficifolia on Type 2 diabetic patients with moderately high blood glucose levels. J Ethnopharmacol 2001; 77(1): 99-101.
[PMID: 11483384]

[320] Alarcon-Aguilar FJ, Hernandez-Galicia E, Campos-Sepulveda AE, *et al.* Evaluation of the hypoglycemic effect of Cucurbita ficifolia Bouché (Cucurbitaceae) in different experimental models. J Ethnopharmacol 2002; 82(2-3): 185-9.
[PMID: 12241994]

[321] Xia T, Wang Q. Antihyperglycemic effect of Cucurbita ficifolia fruit extract in streptozotocin-induced diabetic rats. Fitoterapia 2006; 77(7-8): 530-3.
[PMID: 16905276]

[322] Jessica GG, Mario GL, Alejandro Z, *et al.* Chemical characterization of a hypoglycemic extract from Cucurbita ficifolia Bouche that induces liver glycogen accumulation in diabetic mice. Afr J Tradit Complement Altern Med 2017; 14(3): 218-30.
[PMID: 28480434]

[323] Xia T, Wang Q. Hypoglycaemic role of Cucurbita ficifolia (Cucurbitaceae) fruit extract in streptozotocin-induced diabetic rats. J Sci Food Agric 2007; 87: 1753-7.

[324] Jiang Z, Du Q. Glucose-lowering activity of novel tetrasaccharide glyceroglycolipids from the fruits of Cucurbita moschata. Bioorg Med Chem Lett 2011; 21(3): 1001-3.
[PMID: 21215628]

[325] Abuelgassim AO, Al-showayman SI. The effect of pumpkin (*Cucurbita pepo* L) seeds and L-arginine

supplementation on serum lipid concentrations in atherogenic rats. Afr J Tradit Complement Altern Med 2011; 9(1): 131-7.
[PMID: 23983330]

[326]   Makni M, Fetoui H, Gargouri NK, *et al.* Hypolipidemic and hepatoprotective effects of flax and pumpkin seed mixture rich in omega-3 and omega-6 fatty acids in hypercholesterolemic rats. Food Chem Toxicol 2008; 46(12): 3714-20.
[PMID: 18938206]

[327]   Kwon Y-I, Apostolidis E, Kim Y-C, Shetty K. Health benefits of traditional corn, beans, and pumpkin: *in vitro* studies for hyperglycemia and hypertension management. J Med Food 2007; 10(2): 266-75.
[PMID: 17651062]

[328]   Gossell-Williams M, Hyde C, Hunter T, *et al.* Improvement in HDL cholesterol in postmenopausal women supplemented with pumpkin seed oil: pilot study. Climacteric 2011; 14(5): 558-64.
[PMID: 21545273]

[329]   Gossell-Williams M, Lyttle K, Clarke T, Gardner M, Simon O. Supplementation with pumpkin seed oil improves plasma lipid profile and cardiovascular outcomes of female non-ovariectomized and ovariectomized Sprague-Dawley rats. Phytother Res 2008; 22(7): 873-7.
[PMID: 18567058]

[330]   330. Dessì M, Deiana M, Day B, Rosa A, Banni S, Corongiu F. Oxidative stability of polyunsaturated fatty acids : effect of squalene. Eur J Lipid Sci Technol 2002; 104: 506-12.

[331]   Medjakovic S, Hobiger S, Ardjomand-Woelkart K, Bucar F, Jungbauer A. Pumpkin seed extract: Cell growth inhibition of hyperplastic and cancer cells, independent of steroid hormone receptors. Fitoterapia 2016; 110: 150-6.
[PMID: 26976217]

[332]   Jiang J, Eliaz I, Sliva D. Suppression of growth and invasive behavior of human prostate cancer cells by ProstaCaid™: mechanism of activity. Int J Oncol 2011; 38(6): 1675-82.
[PMID: 21468543]

[333]   Jiang J, Loganathan J, Eliaz I, Terry C, Sandusky GE, Sliva D. ProstaCaid inhibits tumor growth in a xenograft model of human prostate cancer. Int J Oncol 2012; 40(5): 1339-44.
[PMID: 22293856]

[334]   Friederich M, Theurer C, Schiebel-Schlosser G. Prosta Fink Forte -kapseln in der behandlung der benignen prostatahyperplasie. Eine multizentrische Anwendungsbeobachtung an 2245 patienten]. Forsch Komplementarmed Klass Naturheilkd 2000; 7(4): 200-4.
[PMID: 11025395]

[335]   Hong H, Kim CS, Maeng S. Effects of pumpkin seed oil and saw palmetto oil in Korean men with symptomatic benign prostatic hyperplasia. Nutr Res Pract 2009; 3(4): 323-7.
[PMID: 20098586]

[336]   Vahlensieck W, Theurer C, Pfitzer E, Patz B, Banik N, Engelmann U. Effects of pumpkin seed in men with lower urinary tract symptoms due to benign prostatic hyperplasia in the one-year, randomized, placebo-controlled GRANU study. Urol Int 2015; 94(3): 286-95.
[PMID: 25196580]

[337]   Tsai Y-S, Tong Y-C, Cheng J-T, Lee C-H, Yang F-S, Lee H-Y. Pumpkin seed oil and phytosterol-F can block testosterone/prazosin-induced prostate growth in rats. Urol Int 2006; 77(3): 269-74.
[PMID: 17033217]

[338]   Gossell-Williams M, Davis A, O'Connor N. Inhibition of testosterone-induced hyperplasia of the prostate of sprague-dawley rats by pumpkin seed oil. J Med Food 2006; 9(2): 284-6.
[PMID: 16822218]

[339]   Aristatile B, Alshammari GM. *In vitro* biocompatibility and proliferative effects of polar and non-polar extracts of cucurbita ficifolia on human mesenchymal stem cells. Biomed Pharmacother 2017;

89: 215-20.
[PMID: 28231542]

[340] Xie J, Que W, Liu H, Liu M, Yang A, Chen M. Anti-proliferative effects of cucurmosin on human hepatoma HepG2 cells. Mol Med Rep 2012; 5(1): 196-201.
[PMID: 21964700]

[341] Shokrzadeh M, Azadbakht M, Ahangar N, Hashemi A, Saeedi Saravi SS. Cytotoxicity of hydro-alcoholic extracts of Cucurbitapepo and Solanum nigrum on HepG2 and CT26 cancer cell lines. Pharmacogn Mag 2010; 6(23): 176-9.
[PMID: 20931075]

[342] Saha P, Mazumder UK, Haldar PK, Naskar S, Kundu S, Bala A, *et al.* Anticancer activity of methanol extract of Cucurbita maxima against Ehrlich as-cites carcinoma. Int J Res Pharm Sci 2011; 2(1): 52-9.

[343] Tomar PP, Nikhil K, Singh A, Selvakumar P, Roy P, Sharma AK. Characterization of anticancer, DNase and antifungal activity of pumpkin 2S albumin. Biochem Biophys Res Commun 2014; 448(4): 349-54.
[PMID: 24814706]

[344] Rotimi SO, Rotimi OA, Obembe OO. In silico analysis of compounds characterized from ethanolic extract of *Cucurbita pepo* with NF- κ B-inhibitory potential. Bangladesh J Pharmacol 2014; 9: 551-6.

[345] Lans C, Turner N, Khan T, Brauer G. Ethnoveterinary medicines used to treat endoparasites and stomach problems in pigs and pets in British Columbia, Canada. Vet Parasitol 2007; 148(3-4): 325-40.
[PMID: 17628343]

[346] Ayaz E, Gökbulut C, Co H, Türker A, Özsoy Ş. Evaluation of the anthelmintic activity of pumpkin seeds (Cucurbita maxima) in mice naturally infected with Aspiculuris tetraptera. J Pharmacogn Phytother 2015; 7(9): 189-93.

[347] Díaz Obregón D, Lloja Lozano L, Carbajal Zúñiga V. [Preclinical studies of cucurbita maxima (pumpkin seeds) a traditional intestinal antiparasitic in rural urban areas]. Rev Gastroenterol Peru 2004; 24(4): 323-7.
[PMID: 15614300]

[348] Li T, Ito A, Chen X, *et al.* Usefulness of pumpkin seeds combined with areca nut extract in community-based treatment of human taeniasis in northwest Sichuan Province, China. Acta Trop 2012; 124(2): 152-7.
[PMID: 22910218]

[349] Feitosa TF, Vilela VL, Athayde AC, *et al.* Anthelmintic efficacy of pumpkin seed (*Cucurbita pepo* Linnaeus, 1753) on ostrich gastrointestinal nematodes in a semiarid region of Paraíba State, Brazil. Trop Anim Health Prod 2013; 45(1): 123-7.
[PMID: 22684690]

[350] Grzybek M, Kukula-Koch W, Strachecka A, *et al.* Evaluation of Anthelmintic Activity and Composition of Pumpkin (*Cucurbita pepo* L.) Seed Extracts-*In Vitro* and *in vivo* Studies. Int J Mol Sci 2016; 17(9): 1-21.
[PMID: 27598135]

[351] Marie-Magdeleine C, Hoste H, Mahieu M, Varo H, Archimede H. *In vitro* effects of Cucurbita moschata seed extracts on Haemonchus contortus. Vet Parasitol 2009; 161(1-2): 99-105.
[PMID: 19135803]

[352] Sayed H, Seif A. Ameliorative effect of pumpkin oil (*Cucurbita pepo* L.) against alcohol-induced hepatotoxicity and oxidative stress in albino rats. Beni-Suef Univ J Basic Appl Sci 2014; 3(3): 178-85.

[353] Morrison MC, Mulder P, Stavro PM, *et al.* Replacement of Dietary Saturated Fat by PUFA-Rich Pumpkin Seed Oil Attenuates Non-Alcoholic Fatty Liver Disease and Atherosclerosis Development, with Additional Health Effects of Virgin over Refined Oil. PLoS One 2015; 10(9): e0139196.
[PMID: 26405765]

[354] Shayesteh R, Kamalinejad M, Adiban H, Kardan A, Keyhanfar F, Eskandari MR. Cytoprotective Effects of Pumpkin (Cucurbita Moschata) Fruit Extract against Oxidative Stress and Carbonyl Stress. Drug Res (Stuttg) 2017; 67(10): 576-82.
[PMID: 28586926]

[355] Immaculata M, Insanu M, Anne C, Dass S. Development of Immunonutrient from Pumpkin (Cucurbita moschata Duchense Ex. Lamk.) Seed. Procedia Chem 2014; 13: 105-11.

[356] Ristic-Medic D, Perunicic-Pekovic G, Rasic-Milutinovic Z, *et al.* Effects of dietary milled seed mixture on fatty acid status and inflammatory markers in patients on hemodialysis. Sci World J 2014; 2014: 563576.
[PMID: 24578648]

[357] Fahim AT, Abd-el Fattah AA, Agha AM, Gad MZ. Effect of pumpkin-seed oil on the level of free radical scavengers induced during adjuvant-arthritis in rats. Pharmacol Res 1995; 31(1): 73-9.
[PMID: 7784309]

[358] Fortis-Barrera Á, García-Macedo R, Almanza-Perez JC, *et al.* Cucurbita ficifolia (Cucurbitaceae) modulates inflammatory cytokines and IFN-γ in obese mice. Can J Physiol Pharmacol 2017; 95(2): 170-7.
[PMID: 27918843]

[359] Tufts HR, Harris CS, Bukania ZN, Johns T. Antioxidant and Anti-Inflammatory Activities of Kenyan Leafy Green Vegetables , Wild Fruits , and Medicinal Plants with Potential Relevance for Kwashiorkor Evidence-Based Complement Altern Med. 2015; pp. 1-9. Article ID

[360] Hammer KA, Carson CF, Riley TV. Antimicrobial activity of essential oils and other plant extracts. J Appl Microbiol 1999; 86(6): 985-90.
[PMID: 10438227]

[361] Macgibbon DB, Mann JD. Inhibition of Animal and Pathogenic Fungal Proteases by Phloem Exudate from Pumpkin Fruits (Cucurbitaceae). J Sci Food Agric 1986; 37: 515-22.

[362] Wang L, Liu F, Wang A, Yu Z, Xu Y, Yang Y. Purification, characterization and bioactivity determination of a novel polysaccharide from pumpkin (Cucurbita moschata) seeds. Food Hydrocoll 2017; 66: 357-64.

[363] Park SC, Lee JR, Kim JY, *et al.* Pr-1, a novel antifungal protein from pumpkin rinds. Biotechnol Lett 2010; 32(1): 125-30.
[PMID: 19760117]

[364] Noumedem JAK, Mihasan M, Lacmata ST, Stefan M, Kuiate JR, Kuete V. Antibacterial activities of the methanol extracts of ten Cameroonian vegetables against Gram-negative multidrug-resistant bacteria. BMC Complement Altern Med 2013; 13(26): 26.
[PMID: 23368430]

[365] Nderitu KW, Mwenda NS, Macharia NJ, Barasa SS, Ngugi MP. Antiobesity Activities of Methanolic Extracts of Amaranthus dubius, *Cucurbita pepo*, and Vigna unguiculata in Progesterone-Induced Obese Mice Evidence-Based Complement Altern 2017; pp. 1-10. Article ID

[366] Bahramsoltani R, Farzaei MH, Abdolghaffari AH, *et al.* Evaluation of phytochemicals, antioxidant and burn wound healing activities of *Cucurbita moschata* Duchesne fruit peel. Iran J Basic Med Sci 2017; 20(7): 798-805.
[PMID: 28852445]

[367] Elfiky SA, Elelaimy IA. Protective effect of pumpkin seed oil against genotoxicity induced by azathioprine. J Basic Appl Zool 2012; 65(5): 289-98.

[368] Nishimura M, Ohkawara T, Sato H, Takeda H, Nishihira J. Pumpkin Seed Oil Extracted From Cucurbita maxima Improves Urinary Disorder in Human Overactive Bladder. J Tradit Complement Med 2014; 4(1): 72-4.
[PMID: 24872936]

[369] Shim B, Jeong H, Lee S, Hwang S. A randomized double-blind placebo-controlled clinical trial of a product containing pumpkin seed extract and soy germ extract to improve overactive bladder-related voiding dysfunction. J Funct Foods 2014; 8: 111-7.

[370] Cho YH, Lee SY, Jeong DW, Choi EJ, Kim YJ, Lee JG, *et al.* Effect of Pumpkin Seed Oil on Hair Growth in Men with Androgenetic Alopecia : A Randomized, Double-Blind, Placebo-Controlled Trial 2014; pp. 1-7. Article ID

[371] Suphakarn VS, Yarnnon C, Ngunboonsri P. The effect of pumpkin seeds on oxalcrystalluria and urinary compositions of children in hyperendemic area. Am J Clin Nutr 1987; 45(1): 115-21. [PMID: 3799495]

[372] Bardaa S, Moalla D, Ben Khedir S, Rebai T, Sahnoun Z. The evaluation of the healing proprieties of pumpkin and linseed oils on deep second-degree burns in rats. Pharm Biol 2016; 54(4): 581-7. [PMID: 26186459]

[373] Cruz RCB, Meurer CD, Silva EJ, Schaefer C, Santos ARS, Cruz AB, *et al.* Toxicity Evaluation of Cucurbita maxima. Seed Extract in Mice Toxicity Evaluation of Cucurbita maxima Seed Extract in Mice. Pharm Biol 2006; 44(4): 301-3.

[374] Patel S, Rauf A. Edible seeds from Cucurbitaceae family as potential functional foods: Immense promises, few concerns. Biomed Pharmacother 2017; 91: 330-7. [PMID: 28463796]

[375] Patisaul HB, Jefferson W. The pros and cons of phytoestrogens. Front Neuroendocrinol 2010; 31(4): 400-19. [PMID: 20347861]

[376] Vrtačnik P, Ostanek B, Mencej-Bedrač S, Marc J. The many faces of estrogen signaling. Biochem Med (Zagreb) 2014; 24(3): 329-42. [PMID: 25351351]

[377] Chandrareddy A, Muneyyirci-Delale O, McFarlane SI, Murad OM. Adverse effects of phytoestrogens on reproductive health: a report of three cases. Complement Ther Clin Pract 2008; 14(2): 132-5. [PMID: 18396257]

[378] Tung N, Wang Y, Collins LC, *et al.* Estrogen receptor positive breast cancers in BRCA1 mutation carriers: clinical risk factors and pathologic features. Breast Cancer Res 2010; 12(1): R12. [PMID: 20149218]

[379] El-Adawy TA, Taha KM. Characteristics and composition of watermelon, pumpkin, and paprika seed oils and flours. J Agric Food Chem 2001; 49(3): 1253-9. [PMID: 11312845]

[380] Pla MDE, Rojas AM, Gerschenson LN. Effect of Butternut (Cucurbita moschata Duchesne ex Poiret) Fibres on Bread Making, Quality and Staling. Food Bioprocess Technol 2013; 6: 828-38.

[381] Rodríguez-Miranda J, Hernández-Santos B, Herman-Lara E, Vivar-Vera MA, Carmona-García R, Go'mez-Aldapa CA, *et al.* Physicochemical and functional properties of whole and defatted meals from Mexican (*Cucurbita pepo*) pumpkin seeds. Int J Food Sci Technol 2012; 47: 2297-303.

[382] Promsakha P, Jangchud K, Jangchud A. Comparisons of physicochemical properties and antioxidant activities among pumpkin (Cucurbita moschata L.) flour and isolated starches from fresh pumpkin or flour. Int J Food Sci Technol 2017; 52: 2436-44.

[383] Przetaczek-Rożnowska I. Physicochemical properties of starches isolated from pumpkin compared with potato and corn starches. Int J Biol Macromol 2017; 101: 536-42. [PMID: 28322952]

[384] Petropoulos SA, Khah EM, Passam HC. Evaluation of rootstocks for watermelon grafting with reference to plant development, yield and fruit quality. Int J Plant Prod 2012; 6(4): 481-92.

[385] Petropoulos SA, Olympios C, Ropokis A, Vlachou G, Ntatsi G. Fruit volatiles, quality, and yield of

watermelon as affected by grafting. J Agric Sci Technol 2014; 16(4): 873-85.

[386] Rouphael Y, Schwarz D, Krumbein A, Colla G. Impact of grafting on product quality of fruit vegetables. Sci Hortic (Amsterdam) 2010; 127(2): 172-9.

[387] Ahmed D, Fatima M, Saeed S. Phenolic and flavonoid contents and anti-oxidative potential of epicarp and mesocarp of *Lagenaria siceraria* fruit: a comparative study. Asian Pac J Trop Med 2014; 7S1(S1): S249-55.
[PMID: 25312131]

[388] Edith DMJ, Dimitry MY, Richard NM, Leopold TN, Nicolas NY. Physico-chemical characteristics and rehydration kinetics of five species of cucurbitacae seeds. J Food Meas Charact 2017; 11(2): 736-45.

[389] Badifu GIO. Effect of Processing on Proximate Composition, Antinutritional and Toxic Contents of Kernels from Cucurbitaceae Species Grown in Nigeria. J Food Compos Anal 2001; 14(2): 153-61.

[390] Nagarajaiah SB, Prakash J. Chemical composition and bioactive potential of dehydrated peels of benincasa hispida, luffa acutangula, and sechium edule. J Herbs Spices Med Plants 2015; 21(2): 193-202.

[391] Wang HY, Kan WC, Cheng TJ, Yu SH, Chang LH, Chuu JJ. Differential anti-diabetic effects and mechanism of action of charantin-rich extract of Taiwanese Momordica charantia between type 1 and type 2 diabetic mice. Food Chem Toxicol 2014; 69: 347-56.
[PMID: 24751968]

[392] Singh R, Nawale L, Sarkar D, Suresh CG. Two chitotriose-specific lectins show anti-angiogenesis, induces caspase-9-mediated apoptosis and early arrest of pancreatic tumor cell cycle. PLoS One 2016; 11(1): e0146110.
[PMID: 26795117]

[393] Wang HX, Ng TB. Lagenin, a novel ribosome-inactivating protein with ribonucleolytic activity from bottle gourd (*Lagenaria siceraria*) seeds. Life Sci 2000; 67(21): 2631-8.
[PMID: 11104364]

[394] Patel R, Mahobia N, Upwar N, Waseem N, Talaviya H, Patel Z. Analgesic and antipyretic activities of Momordica charantia Linn. fruits. J Adv Pharm Technol Res 2010; 1(4): 415-8.
[PMID: 22247882]

[395] Doshi GM, Chaskar PK, Une HD. Elucidation of β-sitosterol from Benincasa hispida Seeds, Carissa congesta Roots and Polyalthia longifolia Leaves by High Performance Thin Layer Chromatography. Pharmacogn J 2015; 7(4): 221-7.

[396] Sheemole MS, Antony VT. K K, Saji A. Phytochemical Analysis of Methanolic Extract of Benincasa hispida(Thunb.)Cogn. Fruit Using LC-MS Technique. Int J Pharm Sci Rev Res 2016; 36(1): 244-9.

[397] Bimakr M, Rahman RA, Ganjloo A, Taip FS, Adzahan NM, Sarker MZI. Characterization of Valuable Compounds from Winter Melon (Benincasa hispida (Thunb.) Cogn.) Seeds Using Supercritical Carbon Dioxide Extraction Combined with Pressure Swing Technique. Food Bioprocess Technol 2016; 9(3): 396-406.

[398] Fokou E, Achu M, Kansci G, Ponka R, Fotso M, Tchiégang C, *et al.* Chemical properties of some Cucurbitaceae oils from Cameroon. Pak J Nutr 2009; 8(9): 1325-34.

[399] Doshi GM, Nalawade VV, Mukadam AS, *et al.* Elucidation of flavonoids from Carissa congesta, Polyalthia longifolia, and Benincasa hispida plant extracts by hyphenated technique of liquid chromatography-mass spectroscopy. Pharmacognosy Res 2016; 8(4): 281-6.
[PMID: 27695269]

[400] Kubola J, Siriamornpun S. Phenolic contents and antioxidant activities of bitter gourd (Momordica charantia L.) leaf, stem and fruit fraction extracts *in vitro*. Food Chem 2008; 110(4): 881-90.
[PMID: 26047274]

[401] Madala NE, Piater L, Dubery I, Steenkamp P. Distribution patterns of flavonoids from three Momordica species by ultra-high performance liquid chromatography quadrupole time of flight mass spectrometry: A metabolomic profiling approach. Brazilian J Pharmacogn 2016; 26(4): 507-13.

[402] Raina K, Kumar D, Agarwal R. Promise of bitter melon (Momordica charantia) bioactives in cancer prevention and therapy. Semin Cancer Biol 2016; 40-41: 116-29.
[PMID: 27452666]

[403] Jiang Y, Peng XR, Yu MY, Wan LS, Zhu GL, Zhao GT, *et al.* Cucurbitane-type triterpenoids from the aerial parts of Momordica charantia L. Phytochem Lett 2016; 16(132): 164-8.

[404] Murakami T, Emoto A, Matsuda H, Yoshikawa M. Medicinal foodstuffs. XXI. Structures of new cucurbitane-type triterpene glycosides, goyaglycosides-a, -b, -c, -d, -e, -f, -g, and -h, and new oleanane-type triterpene saponins, goyasaponins I, II, and III, from the fresh fruit of Japanese Momordica charantia L. Chem Pharm Bull (Tokyo) 2001; 49(1): 54-63.
[PMID: 11201226]

[405] Nakamura S, Murakami T, Nakamura J, Kobayashi H, Matsuda H, Yoshikawa M. Structures of new cucurbitane-type triterpenes and glycosides, karavilagenins and karavilosides, from the dried fruit of Momordica charantia L. in Sri Lanka. Chem Pharm Bull (Tokyo) 2006; 54(11): 1545-50.
[PMID: 17077551]

[406] Kimura Y, Akihisa T, Yuasa N, *et al.* Cucurbitane-type triterpenoids from the fruit of Momordica charantia. J Nat Prod 2005; 68(5): 807-9.
[PMID: 15921438]

[407] Chen JC, Liu WQ, Lu L, *et al.* Kuguacins F-S, cucurbitane triterpenoids from Momordica charantia. Phytochemistry 2009; 70(1): 133-40.
[PMID: 19041990]

[408] Zhang LJ, Liaw CC, Hsiao PC, Huang HC, Lin MJ, Lin ZH, *et al.* Cucurbitane-type glycosides from the fruits of Momordica charantia and their hypoglycaemic and cytotoxic activities. J Funct Foods 2014; 6(1): 564-74.

[409] Lin KW, Yang SC, Lin CN. Antioxidant constituents from the stems and fruits of Momordica charantia. Food Chem 2011; 127(2): 609-14.
[PMID: 23140707]

[410] Liu C-H, Yena M-H, Tsang S-F, Gana K-H, Hsub H-Y, Lin C-N. Antioxidant triterpenoids from the stems of Momordica charantia Chiung-Hui. Food Chem 2010; 118: 751-6.

[411] Zhao GT, Liu JQ, Deng YY, *et al.* Cucurbitane-type triterpenoids from the stems and leaves of Momordica charantia. Fitoterapia 2014; 95: 75-82.
[PMID: 24631764]

[412] Ghule BV, Ghante MH, Saoji AN, Yeole PG. Antihyperlipidemic effect of the methanolic extract from *Lagenaria siceraria* Stand. fruit in hyperlipidemic rats. J Ethnopharmacol 2009; 124(2): 333-7.
[PMID: 19397976]

[413] Katare C, Saxena S, Agrawal S, Prasad GBKS. Alleviation of diabetes induced dyslipidemia by *Lagenaria siceraria* fruit extract in human type 2 diabetes. J Herb Med 2013; 3(1): 1-8.

[414] Maqsood M, Ahmed D, Atique I, Malik W. Lipase inhibitory activity of *Lagenaria siceraria* fruit as a strategy to treat obesity. Asian Pac J Trop Med 2017; 10(3): 305-10.
[PMID: 28442115]

[415] Rajput MS, Balekar N, Jain DK. Inhibition of ADP-induced platelet aggregation and involvement of non-cellular blood chemical mediators are responsible for the antithrombotic potential of the fruits of *Lagenaria siceraria*. Chin J Nat Med 2014; 12(8): 599-606.
[PMID: 25156285]

[416] Rajput MS, Balekar N, Jain DK. *Lagenaria siceraria* ameliorates atheromatous lesions by modulating

HMG–CoA reductase and lipoprotein lipase enzymes activity in hypercholesterolemic rats. J Acute Dis 2014; 3(1): 14-21.

[417] Mali VR, Bodhankar SL. Cardioprotective effect of *Lagenaria siceraria* (LS) fruit powder in isoprenalin-induced cardiotoxicity in rats. Eur J Integr Med 2010; 2(3): 143-9.

[418] Upaganlawar A, Balaraman R. Cardioprotective Effects of *Lagenaria siceraria* Fruit Juice on Isoproterenol-induced Myocardial Infarction in Wistar Rats: A Biochemical and Histoarchitecture Study. J Young Pharm 2011; 3(4): 297-303.
[PMID: 22224036]

[419] Harini K, Jayasree T. Evaluation of diuretic activity of *Lagenaria siceraria* in Albino rats. IP Int J Compr Adv Pharmacol 2017; 2(3): 99-103.

[420] Mayakrishnan V, Veluswamy S, Sundaram KS, Kannappan P, Abdullah N. Free radical scavenging potential of *Lagenaria siceraria* (Molina) Standl fruits extract. Asian Pac J Trop Med 2013; 6(1): 20-6.
[PMID: 23317881]

[421] Mohan R, Birari R, Karmase A, Jagtap S, Bhutani KK. Antioxidant activity of a new phenolic glycoside from *Lagenaria siceraria* Stand. fruits. Food Chem 2012; 132(1): 244-51.
[PMID: 26434287]

[422] Lakshmi BVS, Sudhakar M. Adaptogenic activity of *Lagenaria siceraria*: An experimental Study using acute stress models on rats. J Pharmacol Toxicol 2009; 4(8): 300-6.

[423] Norfaizatul SO, Zetty Akmal CZ, Noralisa AK, Then SM, Wan Zunnah WN, Musalmah M. Dual effects of plant antioxidants on neuron cell viability. Faslnamah-i Giyahan-i Daruyi 2010; 9 (Suppl. 6.): 113-23.

[424] Nerurkar P, Ray RB. Bitter melon: antagonist to cancer. Pharm Res 2010; 27(6): 1049-53.
[PMID: 20198408]

[425] Nagarani G, Abirami A, Siddhuraju P. A comparative study on antioxidant potentials, inhibitory activities against key enzymes related to metabolic syndrome, and anti-inflammatory activity of leaf extract from different Momordica species. Food Sci Hum Wellness 2014; 3(1): 36-46.

[426] Joseph B, Jini D. Antidiabetic effects of Momordica charantia (bitter melon) and its medicinal potency. Asian Pacific J Trop Dis Asian Pacific Tropical Medicine Press 2013; 3(2): 93-102.

[427] Tan MJ, Ye JM, Turner N, *et al.* Antidiabetic activities of triterpenoids isolated from bitter melon associated with activation of the AMPK pathway. Chem Biol 2008; 15(3): 263-73.
[PMID: 18355726]

[428] Beloin N, Gbeassor M, Akpagana K, *et al.* Ethnomedicinal uses of Momordicacharantia (Cucurbitaceae) in Togo and relation to its phytochemistry and biological activity. J Ethnopharmacol 2005; 96(1-2): 49-55.
[PMID: 15588650]

[429] Fang EF, Ng TB. Bitter gourd (Momordica charantia) is a cornucopia of health: a review of its credited antidiabetic, anti-HIV, and antitumor properties. Curr Mol Med 2011; 11(5): 417-36.
[PMID: 21568930]

[430] Jayasooriya AP, Sakono M, Yukizaki C, Kawano M, Yamamoto K, Fukuda N. Effects of Momordica charantia powder on serum glucose levels and various lipid parameters in rats fed with cholesterol-free and cholesterol-enriched diets. J Ethnopharmacol 2000; 72(1-2): 331-6.
[PMID: 10967491]

[431] Popovich DG, Li L, Zhang W. Bitter melon (Momordica charantia) triterpenoid extract reduces preadipocyte viability, lipid accumulation and adiponectin expression in 3T3-L1 cells. Food Chem Toxicol 2010; 48(6): 1619-26.
[PMID: 20347917]

[432] Ahmed I, Adeghate E, Sharma AK, Pallot DJ, Singh J. Effects of Momordica charantia fruit juice on islet morphology in the pancreas of the streptozotocin-diabetic rat. Diabetes Res Clin Pract 1998; 40(3): 145-51.
[PMID: 9716917]

[433] Tan SP, Kha TC, Parks SE, Roach PD. Bitter melon (Momordica charantia L.) bioactive composition and health benefits: A review. Food Rev Int 2016; 32(2): 181-202.

[434] Kai H, Akamatsu E, Torii E, *et al.* Inhibition of proliferation by agricultural plant extracts in seven human adult T-cell leukaemia (ATL)-related cell lines. J Nat Med 2011; 65(3-4): 651-5.
[PMID: 21293936]

[435] Gürbüz I, Akyüz C, Yeşilada E, Şener B. Anti-ulcerogenic effect of Momordica charantia L. fruits on various ulcer models in rats. J Ethnopharmacol 2000; 71(1-2): 77-82.
[PMID: 10904148]

[436] Alam S, Asad M, Asdaq SMB, Prasad VS. Antiulcer activity of methanolic extract of Momordica charantia L. in rats. J Ethnopharmacol 2009; 123(3): 464-9.
[PMID: 19501279]

[437] Costa JGM, Nascimento EMM, Campos AR, Rodrigues FFG. Antibacterial activity of Momordica charantia (Curcubitaceae) extracts and fractions. J Basic Clin Pharm 2010; 2(1): 45-51.
[PMID: 24826002]

[438] Okada Y, Okada M. Protective effects of plant seed extracts against amyloid β-induced neurotoxicity in cultured hippocampal neurons. J Pharm Bioallied Sci 2013; 5(2): 141-7.
[PMID: 23833520]

[439] El-Fiky FK, Abou-Karam MA, Afify EA. Effect of Luffa aegyptiaca (seeds) and Carissa edulis (leaves) extracts on blood glucose level of normal and streptozotocin diabetic rats. J Ethnopharmacol 1996; 50(1): 43-7.
[PMID: 8778506]

[440] Kalaskar MG, Surana SJ. Free Radical Scavenging, Immunomodulatory Activity and Chemical Composition of Luffa acutangula var. Amara (Cucurbitaceae) pericarp. J Chil Chem Soc 2014; 59(1): 4-7.

[441] Nimbal SK. VenkatraoN LS, Pujar B. Anxiolytic behavioural model for Benincasa hispida eIJPPR 2011; 1: 96-101.

[442] Nadhiya K, Vijayalakshmi K, Gaddam Aadinath Reddy G. Antiobesity effect of Benincasa hispida fruit extract in high fat diet fed wistar albino rats. Int J Pharm Clin Res 2016; 8(12): 1590-9.

[443] Gu M, Fan S, Liu G, Guo L, Ding X, Lu Y, *et al.* Extract of wax gourd peel prevents high-fat diet-induced hyperlipidemia in C57BL/6 mice *via* the inhibition of the PPAR γ pathway Evide nce-based Complement Altern Med 2013; 1: pp. 1-11.

[444] Qadrie ZL, Hawisa NT, Khan MWA, Samuel M, Anandan R. Antinociceptive and anti-pyretic activity of Benincasa hispida (thunb.) cogn. in Wistar albino rats. Pak J Pharm Sci 2009; 22(3): 287-90.
[PMID: 19553176]

[445] Shetty BV, Arjuman A, Jorapur A, *et al.* Effect of extract of Benincasa hispida on oxidative stress in rats with indomethacin induced gastric ulcers. Indian J Physiol Pharmacol 2008; 52(2): 178-82.
[PMID: 19130862]

[446] Rachchh MA, Yadav PN, Gokani RH, Jain SM. Anti-inflammatory activity of Benincasa hispida fruit. Int J Pharma Bio Sci 2011; 2(3): 98-106.

[447] Ng TB, Chan WY, Yeung HW. Proteins with abortifacient, ribosome inactivating, immunomodulatory, antitumor and anti-AIDS activities from Cucurbitaceae plants. Gen Pharmacol 1992; 23(4): 579-90.
[PMID: 1397965]

[448] Naseem MZ, Patil SR, Patil SR, Ravindra , Patil RS. Antispermatogenic and androgenic activities of Momordica charantia (Karela) in albino rats. J Ethnopharmacol 1998; 61(1): 9-16.
[PMID: 9687077]

[449] Grover JK, Yadav SP. Pharmacological actions and potential uses of Momordica charantia: a review. J Ethnopharmacol 2004; 93(1): 123-32.
[PMID: 15182917]

CHAPTER 5

Phytochemicals in *Asteraceae* Leafy Vegetables

Maria Gonnella[a,*], Massimiliano Renna[a,b], Massimiliano D'Imperio[a], Giulio Testone[c] and **Donato Giannino[c]**

[a] *Institute of Sciences of Food Production, CNR, Bari, Italy*

[b] *Department of Agricultural and Environmental Science, Univ. of Bari Aldo Moro, Bari, Italy*

[c] *Institute of Agricultural Biology and Biotechnology, CNR, Unit of Rome, Italy*

Abstract: Knowledge about quantitative and qualitative characterization of phytochemical contents in *Asteraceae* leafy vegetables is continuously increasing due to the strong interest from scientists and consumers. This chapter deals with leafy vegetables of *Lactuca sativa*, *Cichorium intybus* and *C. endivia* species, given their relevance within the botanical family. It gives an overview of the wide differences occurring in genotype panels, in bioactive compound types and levels, focusing on phenolics, carotenoids and sesquiterpenes lactones. It also provides information on their biosynthesis pathways in plants together with health effects, bioaccessibility and bioavailability of diverse phytonutrient classes. Finally, it outlines the effects of the main pre-harvest and post-harvest factors affecting the amount and status of final products intended for dietary intake, and reports on some genetic aspects and biotechnologies aimed to biofortification of these species.

Keywords: Bioavailability, Bioaccessibility, Biosynthesis Pathways, Carotenoids, Chicory, Endive, Genetics and Biotechnological Tools, Health impacts, Lettuce, Phenolics, Phytonutrient Contents, Pre- and post-Harvest Factors, Sesquiterpene Lactones.

INTRODUCTION

The previous notion that vegetables could provide 14 vitamins and 16 minerals essential for human health has been overcome by the assessment that they provide hundreds of phytochemicals, a heterogeneous group of secondary metabolites with biological activity [1], including terpenoids, phenolics, alkaloids and glucosinolates [2]. Naturally, phytochemicals are synthesized in defense mechanisms against biotic and abiotic stress (photosynthetic stress, reactive oxygen species, herbivores, wounds and damage by ultraviolet light) [2].

---

[*] **Corresponding author Maria Gonnella:** Institute of Sciences of Food Production, National Research Council, Bari, Italy; Tel: +39 080 5929306, +39 080 4732974; E-mail: maria.gonnella@ispa.cnr.it

In leafy vegetables the phytochemicals of interest are those contained in leaves and stems. The strong link between plant dietary nutrients and human disease prevention has been established and includes risk reduction of cardiovascular disease, cancer, diabetes, Alzheimer, cataracts and age-related decline [3]. Phytochemicals can act through overlapping mechanisms (detoxification enzymes, scavenging oxidative agents, stimulation of the immune system and hormone metabolism, antibacterial and antiviral effects) [4]. Health benefits of vegetables-rich diets derive from additive and synergistic effects of a mixture of phytochemicals, not equivalent to those obtained from dietary supplements of purified molecules, such as β-carotene or vitamin C [4].

This chapter focuses on leafy vegetables *Lactuca sativa*, *Cichorium intybus* and *endivia* (*Cichorieae* tribe), given their relevance within the *Asteraceae* family. The ample lettuce cultivar panel offers wide differences in types and levels of bioactive compounds; for instance, romaine had the highest content of β-carotene, lutein and ascorbic acid, followed by the loose-leaf group (intermediate contents except for γ-tocopherol) and the crisphead/iceberg type (with the lowest contents of vitamin E, ascorbic acid, β-carotene and lutein, due to the abundance of etiolated leaves) [5, 6]. Red types are rich in anthocyanins (responsible for the red color), chlorophylls, total phenolics, ascorbic acid [6, 7], caffeic acid derivates, flavones and flavonols [8]. Similarly, leaf red sectors of radicchio chicory have higher antioxidant, cytoprotective and antiproliferative activities compared to the whole leaf, attributed to higher anthocyanin content [9]. Phytochemical contents can vary with different factors. For example, external leaves have higher contents of flavonoids than the internal ones, *i.e.* the edible portion. The inner leaves of lettuce can show from 10 to 50% the quercetin content of the outer leaves, the variation due to a different genotypic plant habit (10% referred to red butterhead lettuce, 50% to red Lollo loose-leaf lettuce) [10]. The difference in the flavonoid concentration between outer and inner leaves reflects the impact of light exposure (48 h light treatment on inner leaves increases the quercetin content from 3 to 100% the concentration of the outer leaves) [11]. Notably, external portions are mostly discarded and represents the byproducts of the harvested mass [6]. When referring to ready-to-eat leafy vegetables, the byproducts, namely rich in bioactive compounds, can reach up to 50% of the harvested weight [8].

The focus falls on phenolics, carotenoids and sesquiterpene lactones (STL). Phenolics can be ascribed to two major classes [12], phenolic acids (PA) and flavonoids (FLV). Three groups of compounds are mostly represented in lettuce, chicory and endive: hydroxycinnamic acids (HA) within the PA, flavonols, flavones (and isoflavones to less extent), and anthocyanins (ANT) within the FLV. Carotenoids are essential pigments and classified into xanthophylls and carotenes groups. The former contains yellow pigments including lutein, zeaxanthin, neoxanthin, violaxanthin, flavoxanthin, α- and β-cryptoxanthin; the latter includes mainly α- and β-carotene. STL are 15 carbon atom terpenoids bearing lactone rings, typical of *Asteraceae* [13 - 15].

PHYTOCHEMICALS

Phenolics

Table (1) reports some content ranges distinguished into red and green cultivars. PA constitute the major fraction in lettuce, over 80%, compared to FLV [16, 17], though exceptions include Iceberg where FLV was comparable or greater than PA contents [18]. Chicoric, chlorogenic, and caftaric were the most abundant HA [18 - 20], maximal in chicory, followed by lettuce and endive with comparable contents. As for FLV, the most frequent flavonol is quercetin and its conjugate/derivate forms (mainly 3-O-glucuronide 3-O-6"-malonyl-glucoside), whilst kaempferol occurs as traces in lettuce and chicory, and detectable in endives. As for flavones, luteolin (mostly as glucuronide or glucoside conjugates) is the most abundant, whereas isoflavones traces are measured [18]. Anthocyanins are mainly cyanidin conjugates representing 12-15% of total phenol contents [6] and abundance ranges were overviewed in biotech and conventional red cultivars [21].

As for chicory and endive, a large-scale survey revealed total phenol content differences between the two species and within cultivars [22]. Indeed, chicory was three fold richer than endive and curly endives had 40% higher amounts than smooth ones. Radicchio chicories had maximal phenolic concentration followed by witloof and sugarloaf. The HA (80 and 67% of total phenols in endives and chicory) included mono- and di-caffeoyl tartaric, caffeic, chicoric and chlorogenic acids with sugar-conjugated or derivate forms. Chicoric and chlorogenic acids were the most abundant in chicory and endives (30% of total phenolics). Endives contained double amounts of chicoric acid than chicories, while these latter doubled the content of mono-caffeoyl tartaric acid *versus* endives. Looking at chicory cultigroups, chicoric acid mono-caffeoyltartaric acids levels peaked in sugarloaf, followed by witloof and radicchio [22]. The FLV (20% of total phenolics in both species) were mostly kaempferol and quercetin forms (flavonols), and luteolin and apigenin sugar conjugate forms (flavones). Kaempferol was more abundant in endives than green chicories, while luteolin abounded in chicories. Kaempferol prevailed in curly endives *vs* smooth types; quercetin just occurred as traces in curly cultivars, while luteolin was undetected in smooth ones. Traces of apigenins glucoside- and glucuronide-forms occur in lettuce [18, 23] and endives [22], respectively. Finally, the highest level of anthocyanins occurs in red chicories, followed by variegated and red spotted cultigroups, and cyanidins with conjugated forms prevailed (over 60% of total phenolics) [24].

Table 1. Phenolic contents in edible leaves.

Compound	Lettuce			Chicory			Endive		
	[mg/100g] FW		**Ref.**	**[mg/100g] FW**		**Ref.**	**[mg/100g] FW**		**Ref.**
	Green	**Red**		**Green**	**Red**		**Curly**	**Smooth**	
Total phenol content	18-125	322-571	[16]	43.2-180.7	218-1010	[25]	11.8-95.1	10.2-52.5	[26]
	44-66	88	[27]	124-146	134-300	[28]	243-274		[16]
	21-203	74-175	[27, 29]	118-470		[23, 27]			
Hydroxycinnamic acids	77-198	165-176	[23]	160-207	101-346	[28]	9.06-76.4	9.4-41.8	[26]
		146-203	[16]	258-436		[23]			
Chlorogenic acid	0.3-4.7		[19]	43.2-158.6	51.3-432.9	[24]	4.3-39.6	2.7-17.9	[26]
		ca. 45	[30]	44-72	28-89	[28]			
				1.6-224	15.1-56.6	[25]			
Chicoric acid	0.2-13.8		[19]	21.8-61.1	14.2-59.7	[24]	1.2-19.6	3.8-12.3	[26]
	2.1-3.2		[31]	71-103	58-186	[28]			
		ca. 15	[30]	17.8-90.6	33-153	[25]			
Caftaric acid	0.4-12.8		[19]	4.8	1.8-9.1	[28]	0.3-6.8	0.6-2.4	[26]
	0.6-0.9		[31]	4.9-22.2		[25]			
		ca.15	[30]						
Flavonoids	15-44	36	[23]	28-33		[23]	2.4-23.8	0.8-10.6	[26]
	<9.6	7.6-20.7	[6]	17.9-38.1		[32]			
	4.6	19.0	[33]						
Luteolin	<0.9	0.8-2.3	[6]	3.5-12.8	35.4-164.8	[25]	0.6		[33]
	0.2	5.7	[33]	<7.8		[32]			
	0.4-1.0		[31]						
Quercetins	3.2	10.6	[33]	3.7-25.2		[32]	<5.9	n.d.	[26]
	<8.8	5.5-15.7	[6]				<0.1		[34]
3-O-glucuronide	<0.4		[19]	8.1-32.7	18.3-106.5	[25]			
3-O-(6"-malonyl-glucoside)	0.1-0.9		[19]						
Kaempferols		2.4	[33]	3.6-11.1		[32]			
3-O-glucoside	<0.2		[34]	<0.2		[34]	1.5-9.5		[34]
3-O-glucuronide							1.9-21.2	0.8-8.9	[26]
3-O-(6"-malonyl-glucoside)							<3.6	<1.7	[26]
Apigenins				1.8-2.8		[32]			

(Table 1) cont.....

Compound	Lettuce		Ref.	Chicory		Ref.	Endive		Ref.
	[mg/100g] FW			[mg/100g] FW			[mg/100g] FW		
	Green	Red		Green	Red		Curly	Smooth	
7-O-glucoside	<0.1		[18]						
7-O-glucuronide							0.0-1.3	<0.1	[26]
Anthocyanins		0.3-7	[35]		102-119	[9]			
					129.7-548.5	[25]			

Regarding the phenolic biosynthetic pathway, extensive reviews have been issued [36, 37]. We address those branches that lead to beneficial compounds (paragraph 2.3) of lettuce and chicories. The phenylpropanoid pathway (Fig. **1**) produces a wide range of HA and esters, which are further diversified through the activity of reductases, oxygenases, and transferases, eventually leading to organ-, developmental- and species- specific metabolites. Phenylalanine from the shikimate pathway is converted into cinnamic acid by phenylalanine amino lyase (**PAL**), followed by the two-step synthesis of *p*-Coumaric acid and of *p*-Coumaroyl-CoA *via* cinnammate-4-hyroxylase (**C4H**) and 4-(hydroxyl) cinnamoyl-CoA ligase (**4CL**). *p*-Coumaroyl-CoA is a key substrate that opens the branches of other HA synthesis, *via* hydroxycinnamoyl-CoA shikimate/quinate hydroxycinnamoyl transferase (**HCT**), of flavonoids, *via* chalcone synthases (**CHS**), and of several other phenylpropanoids. The biosynthesis of chlorogenic acid (CGA; syn.: caffeoyl quinic acid) relies on three pathways. Fig. (**1**) depicts two of them, one indicates the CGA origin from caffeoyl-CoA and quinic acid by the hydroxycinnamoyl-CoA quinate hydroxycinnamoyl transferase (**HQT**), and the other proposes the hydroxylation of p-Coumaroyl quinic acid by the p-coumarate 3'-hydroxylase (**C3'H**) to form CGA [38]. The route of CGA synthesis in artichoke (*Asteraceae*) preferentially involves HCT, C3'H, and HQT1 and 2. These two enzymes are HCT with stronger preference for quinate than shikimate and might be involved in different steps in the synthesis of CGA. Namely, HQT1 preferentially catalyses the step leading directly to CGA from caffeoyl-CoA and quinic acid, whereas HQT2 is more likely involved in the synthesis of p-coumaroylquinate [39]. Consistently, in chicory two HCT and three HQT are redundantly responsible for isochlorogenic acid and CGA synthesis [40]. Chicoric acid (CHA, syn.: di-caffeoyltartaric acid) is mostly found as L-chicoric form of which the biosynthesis is still unravelled because specific enzymes have not been found [41]. As for FLV, the p-Coumaroyl-CoA - naringenin chalcone - narigenin metabolic branch is achieved by the sequential action of **CHS** and chalcone-flavanon isomerase (**CHI**) and the flavanone narigenin is precursor of flavones, flavonols and anthocyanins. The flavone biosynthesis comprises the narigenin conversion into apigenin by a flavone synthase (**FNSI**) and into eriodictyol *via* flavonoid 3'-monooxygenase (**F3'H**); these flavones can be independently converted into luteolin. The apigenin-luteolin switch occurs *via* either flavonoid 3',5'-hydroxylase (**CYP75**) or F3'H, while the eriodictyol-luteolin conversion is catalysed by the FNSI. The flavonoid biosynthesis branch comprises the narigenin-dihydrokaempferol step by the naringenin 3-dioxygenase (**F3H**) followed by the kaempferol synthesis *via* flavonol synthase 1 (**FLS1**). Quercetin either directly derives from kaempferol through the flavonoid 3'-monooxygenase (**F3'H**) or from the eriodictyol-dihydroquercetin sequence, catalysed by the action of naringenin, 2-oxoglutarate 3-dioxygenase (**F3H**) and FLS1. The flavonoid modifications provide three precursors of anthocyanidins (aglycones) and anthocyanins (glycosides): cyanidin, delphinidin, pelargonidin that has different hydroxyl group numbers in the phenyl group. Focussing on cyanidin, the two-step conversion of dihydroquercetin and leucocyanidin precursors is catalysed by the dihydroflavonol 4-reductase (**DFR**) and leucoanthocyanidin dioxygenase (**LDOX**). Subsequently, various glycosyltransferases (**BZ1**, **GT1**) modify cyanidins acting on type, number and positions of additional glucosides and the derived anthocyanins are further conjugated to aromatic groups by anthocyanin aromatic acyltransferases (**3AT**, **5AT**). Delphinidins, which are detected in radicchio [25], derive from the dihydrotricetin (flavanone) - dihydromyricetin (flavonol) -leucodelphinidin (flavandiol) branch (not in Fig. (**1**)) and variants are generated by glycosyltransferases and acyltransferases. As for regulatory aspects of the pathways, the initial three steps catalysed by PAL, C4H and 4CL provide the precursors for all downstream branches and metabolites, and a tight degree of coordination occurs at both the mRNA and protein levels [42]. PAL-genes (*PAL*) form families encoding multiple isoforms [43]. Each *PAL* respond to various stimuli during organ development *via* multifaceted regulatory mechanisms acting at the transcriptional and post translational levels, and including metabolite feedbacks [44]. Consistently, lettuce *PAL* genes and enzymes respond to heat shock, chilling, highlight intensity and CO_2 concentration [45, 46]. C4H is encoded by single genes in several species [47, 48] and mutations dramatically affect development causing accumulation of unusual derivate [49]. 4CL ligases are also encoded by gene families [50] as well as HCT in lettuce [39], this latter belonging to BAHD-like acyltransferases under complex regulation at multiple levels [51]. Regarding transcriptional control of flavonoids and anthocyanin, major studies derive from the model Arabidopsis [52]. Upstream biosynthetic genes (CHS, CHI, F3H, and FLS1) are controlled by transcription factors (MYB11, MYB12, and MYB111) related to the R2R3-MYB proteins [53, 54]. Downstream biosynthetic genes are regulated by a complex made of MYB-bHLH-WD40 (MBW) factors [55], belonging to the types R2R3-MYBs, bHLHs, and Transparent Testa Glabrous1 (TTG1). Finally, genes of the anthocyanin pathway are regulated by the MBW complex made of TTG1, one R2R3-MYB protein from production of anthocyanin pigments (PAP1), PAP2, MYB113, or MYB114, as well as one bHLH protein from TT8, Glabrous3 (GL3), or enhancer of Glabra3 [55].

Fig. (1). Schematic representation of phenylpropanoid biosynthesis pathway. Compounds with assessed beneficial effects in *Cichorieae* are in green. Abbreviation of enzyme names are provided in the text.

Terpenoids

Carotenoids

In lettuce, β-carotene and lutein are the major fractions among eight carotenoids (Table **2**) found in butterhead, batavia, and oak cultivars [56]. A survey based on over 50 genotypes showed that carotenoid amounts (β-carotene + lutein) were maximal in green leaf and romaine cultivars followed by red leaf, butterhead and crisphead cultivars. Moreover, most of them had higher levels of β-carotene than lutein except for crisphead. Finally, β-carotene and lutein contents were highly correlated with each other and with chlorophylls [57]. Chicories and endives have β-carotene and lutein as major fractions and lack lactucaxanthin [58 - 60]. β-cryptoxanthin and antheraxanthin characterize chicory landraces [61]. Carotenoid content (β-carotene + lutein) was higher in curly- than smooth-leafed endives [62]. Finally, lutein content mostly exceeds that of β-carotene in endives [63].

Table 2. Carotenoid contents[a] in edible leaves.

Compound	Lettuce		Chicory		Endive	
	[mg/100g]	Ref.	[mg/100g]	Ref.	[mg/100g]	Ref.
β-carotene	0.32-4.03	[57]	3.6	[58]	2.5-3.1	[60]
	1.7-2.5	[58]	0.3-0.5	[61]	2.4	[63]
	2.3-3.8	[64]	3.9-7.3	[59]		
			2.3	[63]		
Lutein	0.37-3.98	[57]	5.7	[58]	3.5-4.3	[60]
	1.5-2.3	[58]	0.8-0.9	[61]	8.7	[63]
	3.0-4.5	[64]	3.9-5.9	[59]		
			9.1	[63]		
Neoxanthin	0.6-1.1	[58]	1.5	[58]	1.2-1.6	[60]
Violaxanthin	1.4-2.0	[58]	2.1	[58]	1.8-2.3	[60]
			0.7-1.7	[59]		
Lactucaxanthin	0.8-1.2	[58]				
β-Cryptoxanthin			<0.1	[61]		
Antheraxanthin			0.4-0.5	[59]		
Zeaxanthin[b]	0.75-1.04	[56]	<0.1	[59]		

a, values refer to fresh weight. b, values from the original source were 15.0-20.9 mg/100g dry weight and were converted in fresh weigh based on literature [57] that calculated dry/fresh weight ratio of leaves of ca. 5% as the mean of over 50 lettuce genotypes.

Biosynthesis of carotenoid pigments (C40 isoprenoids) occur in the plastids (Fig. **2**) *via* the 2-C-methyl-D-erythritol 4-phosphate (**MEP**) pathway [65]. Downstream the MEP, two key enzymes act sequentially, the isopentenyl diphosphate isomerase (**IPI**), which catalyzes the mutual conversion of isopentenylpyrophosphate (**IPP**) and dimethylallyl pyrophosphate (**DMAPP**), and the geranylgeranyl pyrophosphate synthase (**GGPS**), which condenses three IPP and one DMAPP units into geranylgeranyl pyrophosphate (**GGPP**), precursor of carotenoids, gibberellins, quinones and chains of chlorophylls and tocopherols. The phytoenesynthase (**PSY**) opens the carotenoid pathway and fuses two GGPP units into 15-*cis*-phytoene, which is converted into all-*trans*-lycopene by a poly-*cis* route. This includes the sequential cascade of two phytoene desaturases (**PDS**), z-carotene isomerase (**ZISO**), two z-carotene desaturases (**ZDS**) and the prolycopene isomerase (**CrtISO**). All-*trans*-lycopene is targeted by both lycopene β- and lycopene ε-cyclases (**LCYB** and **LCYE**) which add respectively β- and ε-rings (at the 5,6 and 4,5 positions) generating the two α- and β- carotene branches (vitamin A precursors are specifically carotenoids that have β-rings without any substitution). Lutein derives from α-carotene by the action of carotenoid ε-hydr-oxylases (**CYP97**) that introduces hydroxyl groups in the rings. In the β-branch, β-carotene turns into zeaxanthin by carotenoid β hydroxylases (**BCH1/2**), then zeaxanthin and violaxanthin are converted into each other, respectively by enzyme-mediated epoxidation (**ZEP**) and de-epoxidation (**VDE**). Finally, lettuce produces lactucaxanthin, a peculiar ε-ε-carotene accomplished with two ε-rings added by a specific **LCYE** that cyclizes both all-*trans*-lycopene ends (not in Fig. (**2**)). This LCYE harbors specific amino acid stretches that confer ability for adding one or two ε-rings [66].

Carotenoid synthesis (CS) genes are regulated by light, epigenetic mechanisms and redox state of signaling molecules. Chlorophyll and CS genes are coordinately governed by light during the whole plant life. The light induced up-regulation of CS genes is mediated by light receptors such as phytochromes (red and far-redranges), which interact with transcription factors named PIFs [67], and cryptochromes (blue and UV ranges) [68]. Epigenetic control of *CrtISO* gene derives from promoter methylation grade caused by chromatin-modifying histone methyltransferase [69]. The *ZEP* expression is controlled by the redox state of plastoquinones, which are molecules involved the electron transport chain during the light-dependent photosynthesis [70].

Post transcriptional regulation include feedback mechanisms at the level of MEP enzyme accumulation due to PSY activity [71] in order to control the metabolic flow of isoprenoid precursors conveyed to carotenoid pathway. Another mechanism is the reduction of MEP specific disulfide groups mediated by the ferredoxin/thioredoxin system [72]. PDS, ZDS, CrtISO, LCYB, LCYE, ZEP enzymes have domains that bind to the flavin adenine dinucleotide cofactor, suggesting that they all are under redox control [73] and both PDS and ZDS are associated to plastoquinone redox state [74]. Finally, steady-state carotenoid levels can be controlled by active degradation *via* enzymatic oxidation (non-specific peroxidases and lipoxygenases), non-enzymatic photo-oxidation [75], and specific cleavage dioxygenases [76].

Studies on carotenoid genetic pathways of *Cichorieae* crops regard violaxanthin de-epoxidase (VDE) and lactucaxanthin-LYCE characterization which dates back to nineteens [77] and interest has arisen again after *Asteraceae* genome sequencing (paragraph 4).

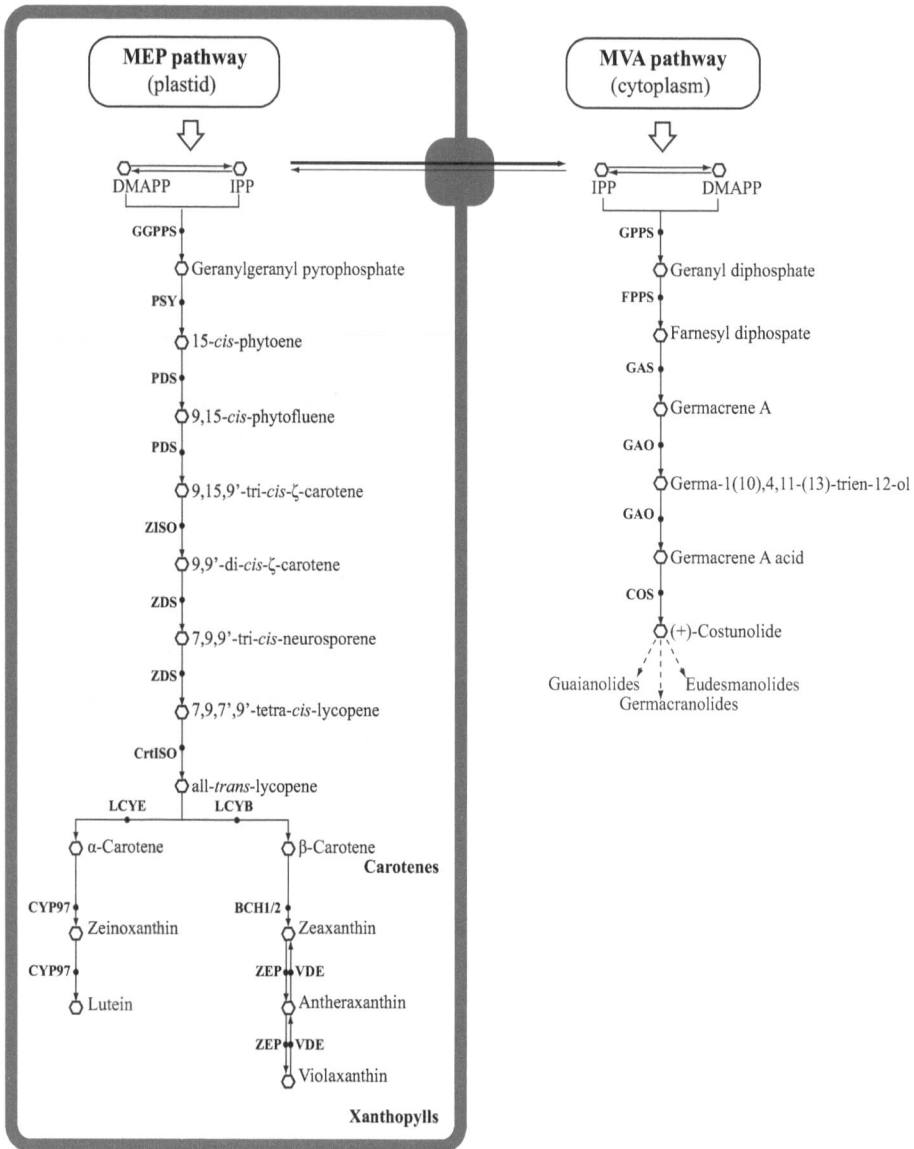

Fig. (2). Carotenoid and sesquiterpenoid biosynthesis pathways.

Sesquiterpene Lactones

Sesquiterpene lactones (STL) play roles in plant defense, contribute to bitter flavor in food and exert various effects on human health [78, 79]. In both lettuce and chicories the most abundant STLs are lactucin, 8-deoxylactucin, lactucopicrin and the respective 11β,13-dihydroderivatives (Table **3**). Conjugated forms include glycoside, oxalate and sulphate groups [22, 79 - 81]. Less frequent leaf types are chicorioside B, 8-deacetylmatricarin-8-O-sulfate and lettucenin A, the last two being specific for cultivated lettuce [80]. Finally, several other STL are found in lettuce wild species [82]. Higher total STL contents were observed in red *vs* green cultivars in Korean lettuce cultivars [83]. Moreover, lettuce contents were much lower than that of chicory [84]. In Korean lettuce germplasm STL amounts were maximal in crisphead followed by leafy type, cos and butter head [85]. The STL content of chicories was higher than endives; within chicory cultigroups, sugarloaf had maximal levels followed by leafy radicchio, witloof and head radicchio; within endives, curly-leafed types had greater amounts than smooth ones. Chicory STL ranges did not correlate with the red color. The 11β,13-dihydrolactucin is usually the most abundant fraction in both chicory and endive, and lactucopicrin, which impacts on bitterness, is maximal in sugarloaf and minimal in head radicchio, while twice higher in curly than smooth endives [22]. Finally, STL contents of chicory leaves are much higher than edible stems [86].

Sesquiterpenes are terpenes made of three isoprene units and, so far, their biosynthesis has been partially unraveled [78]. The isopentenyl pyrophosphate (IPP) backbone can originate from both the plastidial MEP (paragraph 2.2.1) and the cytoplasmic mevalonic acid (MVA) pathways [87] (Fig. **2**). IPP and dimethylallyl pyrophosphate (**DMAPP**), mutually converted by isopentenyl diphosphate isomerase, are substrates of the geranyl diphosphate synthase (**GPPS**) that produces the geranyl diphosphate (**GPP**), rapidly converted into farnesyl diphosphate (**FPP**) by the farnesyl diphosphate synthases (**FPPS**). FPP is the sesquiterpenoid backbone of 15 carbon atoms and undergoes a series of modifications (cyclizations, hydroxylations, oxidations, and lactonization) that open the branches of various sesquiterpenoids (*e.g.* acyclic-, bisabolene-, germacrene-, humulene- and cadinyl types). STL from chicory and lettuce leaves belong to the germacrene A type sesquiterpenoids. The biosynthesis starts with the FPP cyclization by germacrene A synthase (**GAS**) to form (C)-germacrene A. The methyl group at C-12 of this latter undergoes a three-step hydroxylation-oxidation by the cytochrome P450 germacrene A oxidase (**GAO**) bringing to germacrene A acid (**GAA**). The costunolide synthase (**COS**) performs GAA hydroxylation to yield 6α- and 8α-hydroxy-GAA intermediates; the former undergoes spontaneous non-enzymatic lactonization and eventually originates the core STL (+)-costunolide [88 - 92]. Studies support that STLs can originate from both cytoplasm and plastids [78], though it appears that capitate glandular trichomes are major biosynthesis sites in *Asteraceae* and STL are secreted into extracellular and subcuticular spaces in defense process [93]. However, STLs are not restricted to leaf trichrome or aerial parts because lactucopicrin, lactucin, and 8-deoxylactucin occur in laticifers and roots of chicory and lettuce [89, 94].

GAS, GAO and COS enzymes have been characterized in lettuce [89, 90] and chicory [88, 91, 92]. Phylogenetic analyses in *Asteraceae* assigned GAS and GAO to two clades, the clade I was retained by ancestors (*Barnadesioideae*) and the evolutionary related *Cichorioideae* subfamily, the clade II included the ancestral basis common to many other species. Retention of the GAS/GAO clades I in the *Cichorioideae* may be associated with the evolution of specialized cell types such as laticifers not present in other subfamilies. As for COS, chicory and lettuce COS [90, 91] share high identity and function. So far, the enzymes that modify costunolide to yield lactucin and lactucopicrin have been unknown as well as the regulatory mechanisms at the protein levels on GAS/GAO/COS module. However, GAS/GAO synchronized action and STL biosynthesis are consistent with the co-existence of terpene synthase/P450 sets in sequenced plant genomes [95]. Analogies with mechanisms common to other terpene synthase can be assumed and include end-product mediated feedback control, light, and methyl-jasmonate [96]. The STL biosynthesis mostly occurs in the cytosol and the variable localization of GAS (cytoplasm) and GAO/COS (endoplasmatic reticulum) suggests that protein modifications for transport [97].

Lettuce genome sequence reveals that *GAS*, *GAO* and *COS* genes belong to families located on different chromosomes; members of each gene are predicted to encode isoforms by alternative splicing. The cloning of multiple transcripts for each gene type in chicory and phylogenetic analyses [86, 92] support that *Cichorium* spp. may share genomic similarities with *L. sativa*. The expression of *GAS/GAO/COS* genes is thought to be coordinated, reflecting the synchronism of the enzyme module. Moreover, positive correlation of gene expression and sesquiterpene levels have been reported in chicory [86, 88] and artichoke [97]. GAS transcription levels vary with tissue type, organ development stages, and responses to pathogen infections. Relatedly, two lettuce *GAS* genes were expressed constitutively in roots, hypocotyls and true leaves but not in cotyledons, in which transcription was induced during mildew disease [89]. Both chicory *GAS* short and long forms are down regulated in leaves compared to roots, though the long form was more abundant in leaves than the short one [88]. Trichome-specific *GAS* expression may occur in *Cichorieae* similarly to artichoke [97] and it is likely that the *GAS/GAO/COS* module of these crops respond to methyl jasmonate and salicylic acid signals [98]. The knowledge on transcription factors (TF) involved in sesquiterpenoids biosynthesis has been fragmentary. Recently, transcriptome based studies in chicory [86] reported that different classes of TF showed strong positive (MYB factors) and negative (bHLH factors) correlations with both *GAS/GAO* expression and total STL levels, consistently with MYB and bHLH roles acting on terpene synthase genes [99]. Relatedly, MYB factors have been found to affect sesquiterpene production, though direct gene targets need further investigation [100] and some WRKY transcription factor members exert control on sesquiterpene cyclases [101]. Lastly, *GAS/GAO/COS* genomic sequence availability will allow the identification of gene motifs targeted by TF and enhancing knowledge on regulatory mechanisms.

Table 3. Sesquiterpene lactone contents[a] in leaves.

Compound	Lettuce		Chicory		Endive	
	[mg/100g]	Ref.	[mg/100g]	Ref.	[mg/100g]	Ref.
Lactucin	1.5-30.4	[84]	15.8-57.3	[84]	4.0-6.5	[22]
	0.3-1.7	[83]	9.7-26.2	[22]		
	0.8-6.2	[82]	<1.3	[81]		
	<1.6	[79]	0.5-0.7	[86]		
11ß,13-Dihydrolactucin	<0.5	[82]	34.5-46.9	[22]	10.1-17.0	[22]
			0.3-1.3	[81]		
			1.0-2.9	[86]		
Lactucin 15-oxalate	n.q.	[102]				
15-p-Hydroxy-phenylacetyllactucin-8-sulfate	<3.9	[79]				
11ß,13-Dihydroxy-lactucin-8-O-sulfate	<1.0	[80]	<10.0	[80]	1.1-10.0	[80]
8-Deoxylactucin	<15.2	[84]	2.2-418.5	[84]	0.4-9.0	[22]
	0.3-1.7	[83]	12.7-36.1	[22]		
	2.6-4.4	[82]	0.4-1.1	[81]		
			1.1-3.2	[86]		
8-Deoxylactucin 15-sulphate	<68.2	[79]				
8-Deoxylactucin 15-oxalate	<7.8	[79]				
11ß,13-Dihydro-8-deoxylactucin (jaquinellin)	<0.9	[82]	15.1-38.8	[22]	1.1-11.7	[22]
			<1.0	[81]		
11ß,13-Dihydro-8-deoxy-lactucin 15-glycoside (syn. lactuside C; jaquinellin glycoside)	n.q.	[102]	<3.6	[81]		
15-deoxylactucin-8-sulfate	0.8	[102]				
Lactucopicrin	2.9-17.6	[84]	22.0-148.8	[84]	2.3-13.8	[22]
	0.9-3.6	[83]	6.2-25.3	[22]		
	8.0-11.4	[82]	1.3-9.9	[81]		
	0.5-3.6	[79]	1.1-1.5	[86]		
Lactucopicrin 15-oxalate	0.7-8.6	[79]				
11(S),13-Dihydrolactucopicrin	n.q.	[103]	2.4-4.7	[22]	0.8-2.2	[22]
			0.4-1.2	[81]		
			0.4-0.7	[86]		

(Table 3) cont.....

Compound	Lettuce		Chicory		Endive	
	[mg/100g]	Ref.	[mg/100g]	Ref.	[mg/100g]	Ref.
11ß,13-Dihydro-lactucopicrin oxalate			<0.2	[81]		
Lettucenin A	<0.4	[104]	<0.1	[104]	<0.1	[104]
Chicorioside B	1.1-10.0	[80]	1.1-100.0	[80]	1.1-10.0	[80]
8-Deacetylmatricarin-8-O-sulfate	<10.0	[80]	<10.0	[80]	1.1-10.0	[80]

a, white and grey cells refer to dry and fresh weights, respectively. Based on literature, the average dry/fresh weight ratio of leaves ranged between 5-10% for lettuce and 6-9% for chicory [23] and 5-6% for endives [105]. Underlined values refer to edible stems; n.q., assigned but not quantified.

Nutritional and Health Effects

The intake of a rich antioxidant diet is inversely associated with the risk to develop some pathologies [106] and bioavailable antioxidants from vegetables may compensate the inability of human antioxidants to counteract the oxidative stress [107]. Several chemical assays have been developed to measure food total antioxidant capacity in a context of lack of standardized quantification methods [108]. Assays based on scavenging activity *versus* free radicals (DPPH, ABTS), reduction of metal ions (FRAP) and competitive reactions (ORAC) were applied to extracts of lettuce and chicories [8, 23, 27, 105, 109]. Several works report linear correlation between total phenol contents (TPC) and TAC [8, 23, 27], although they can also fail to correlate [23, 105, 110] due to the interference of molecules other than polyphenols in a given assay [111]. In radicchio, peroxyl radical trapping capacity reflected the TPC content and red cultivars showed highest indexes [24]. Consistently, red cultivars had greater ORAC indexes than green ones [27], likely due to higher TPC contents and the anthocyanin contribution. Higher ORAC values of curly *vs* smooth endives were inferred to higher flavonol contents [62, 105]. Chemical assays do not necessarily reflect *in vitro* antioxidant effects because they exclude biological parameters and their significance as direct health indicators is questionable; the accurate and careful use was recommended just for food quality assessment [112]. Alternatively, cellular-based assays (CAA) represent sustainable options to animal/human models [113]. For instance, antioxidant activity of lettuce and endive extracts was measured in erythrocyte systems and CAA-indexes placed these species in middle-ranking among vegetables [62, 114]. Finally, CAA-index values were confirmed to be higher in red sectors than the whole leaf of radicchio [9].

Referring to phenolics, CGA exerts anti-diabetic/obesity/hypertension effects together with antioxidant and anti-microbial/virus properties [115]. Extracts from chicory leaves rich in CGA reinstated the insulin sensitivity in diabetic rats [116]. CHA has anti-cancer [117] and anti-diabetic [118] properties though additional research is needed to better understand CHA benefits [41]. As for flavonoids, luteolin (including that of lettuce) has beneficial properties against several tumor diseases, such as lung [119] cancers, and in cardiovascular recovery [120], together with anti-viral effects [121]. Quercetins represent important drugs in cancer therapy [122] but they have also beneficial effects against other human pathologies such as arthritis [123] and rethinopathy [124]. Therapeutic effects of kaempferol forms have been exhaustively reviewed, and include anti-angiogenesis/metastasis/inflammatory properties [125]. Specifically, endive extracts rich in kaempferol had high antioxidant and protective activity in mice with injured liver [126]. Cyanindins, anthocyanins and delphinidins act efficiently against several tumor diseases [127]. Several other anthocyanin functions span cardio/thrombosis/kidney protection [128 - 130], survival extension in amyotrophic lateral sclerosis [131]. High anthocyanin red lettuce reduced hyperglycemia and improved insulin sensitivity in high fat diet-induced obese hyperglycemic mice [21].

Carotenoids play beneficial roles in human health and disease [132]. Specifically, β-carotene (together with α-/γ-carotene) and β-cryptoxanthin are precursors of vitamin A (retinol) necessary for human growth and development, preservation of good eyesight and immune system equilibrium [133]. Regarding cancer, β-carotene pro-apoptotic activity was reported in several human cancer cell lines [134, 135], but meta-analysis studies concluded that it has not clear roles in cancer and the administration needs rigorous evaluation before acceptance in chemotherapy [136]. β-carotene content of *Cichorieae* leaves can exceed 4 mg/100g FW Table (**2**), following carrot, spinach and kale, top rankers among vegetables. Amounts of 15-50 grams of leaves are sufficient to reach 25% of the vitamin A recommended daily allowance at 25% vitamin A equivalency ratios [65]. Lutein and zeaxanthin accumulate in the fovea of human retina exerting protection against light and dietary intake is important for eyesight [137]. Both isomers are associated with a reduced risk of age-related macular degeneration [138]. Both compounds improve visual performance and sleep quality in people with long time screen exposure [139], have neuroprotective and anti-inflammatory function in diabetic retinopathy [140] and coronary artery disease [141]. Neoxanthincan inhibit crucial carcinogenesis stages of mouth [142], while violaxanthin has anti-inflammatory effects in macrophage cells [143]. Finally, lactucaxanthin, peculiar of lettuce, has anti-diabetic properties by inhibition of α-amylase and α-glucosidase activities [144].

Sesquiterpene lactones have a wide range of pharmacological activities [15] tested in clinical trials [145] and include surveys on potential toxicity [146]. Lactucin, lactucopicrin and 11β,13-dihydrolactucin of chicory have analgesic properties comparable to ibuprofen treated mice [147] and 8-deoxylactucin from chicory is the major anti-inflammatory effector in human colon carcinoma [148].

Bioaccessibility and Bioavailability

Nutrients undergo modifications following gastro-intestinal digestion and gut microflora processing [149] and the assessment of beneficial effects of leafy vegetables require knowledge on bioaccessibility and bioavailability of bioactive compounds (Fig. **3**). Bioaccessibility is the percentage of nutrient released from the food matrix and accessible for absorption during gastrointestinal digestion [150]. It is measured by *in vitro* methods [109, 151] that reproduce the chemio-physical conditions in the gastrointestinal digestion process and simulate gut microbiome effects [152, 153]. Bioaccessibility of a given compound may vary with species, *e.g.* β-carotene in parsley is 6.6% *vs* 55% of lettuce [154, 155] and with food matrix (raw or cooked foods) and processing method [156]. Bioavailability defines the fraction of a nutrient absorbed in the intestinal tract and available for physiological functions. It is assessed by *in vitro*, *ex vivo* and *in vitro* models based on measures of nutrient concentrations in plasma, in tissue (storage), and in excretions after ingestion of vegetables and/or purified/standard compounds [150].

Bioaccessibility and bioavailability of polyphenols have been intensely studied from ingestion to excretion [152, 157]. Disruption of cell components and dissolution of conjugated forms are necessary for polyphenols release from the food matrix. Release is low in mouth, but high in the gastric phase. In the small intestine flavonoid and isoflavonoid aglycones are further converted into mycelles, while antocyanins are largely degraded causing bioaccessibility drop to less than 10%. Polyphenols that are not absorbed in the small intestine function as substrates for the gut microbial communities, which modify bioavailability [157]. At the same time, polyphenols can alter the gut microbiota community, resulting in a greater abundance of beneficial microbes, and a consequent indirect increase in bioavailability. As for lettuce and chicories, bioavailability is quite variable (Table **4**) ranging from 0.22% (apigenin) to 95% (caffeic acid); caftaric acid and chicoric acid of chicory are degraded by gut microflora before absorption and metabolization [158].

Reviews reporting detailed analyses on factors affecting bioaccessibility and bioavailability of carotenoids have been issued [159, 160]. Generally, carotenoids are poorly bioaccessible because their chemical structure deeply interacts with macromolecules within the leafy vegetables matrix. Indeed, they are organized in pigment-protein complexes in cellular chloroplasts and/or chromoplasts, often fused to lipid or protein-bound droplets; chloroplasts are less efficiently destroyed in the intestinal tract than chromoplasts [161]. The release from vegetables matrix starts in the stomach and continues in the intestinal tract, where the bile salts emulsify carotenoids and favour the absorption [162]. The process requires dietary fat for the formation of micelles containing lipids, emulsified by bile salts. Once internalized into the enterocytes, tocopherols and tocotrienols are packaged into chylomicrons and enter the circulation *via* lymph. During circulation, chylomicron triglycerides are hydrolyzed by endothelial bound lipoprotein lipase [163]. Nonetheless, carotenoids have low bioavailability in humans Table (**4**), ranging from 30.4 to 780 nmol/L, in relation to the model used. Bioavailability can be improved acting on bioaccessibility by applying processing treatments, different cooking methods, enzymatic processes that can enhance cell wall break and the protein-carotenoid complex disruption [164].

Regarding STL, the available literature has addressed bioavailability of α-humulene and artemisin [165, 166] that are not found in lettuce and chicories, prompting the urge to address studies on guainolides.

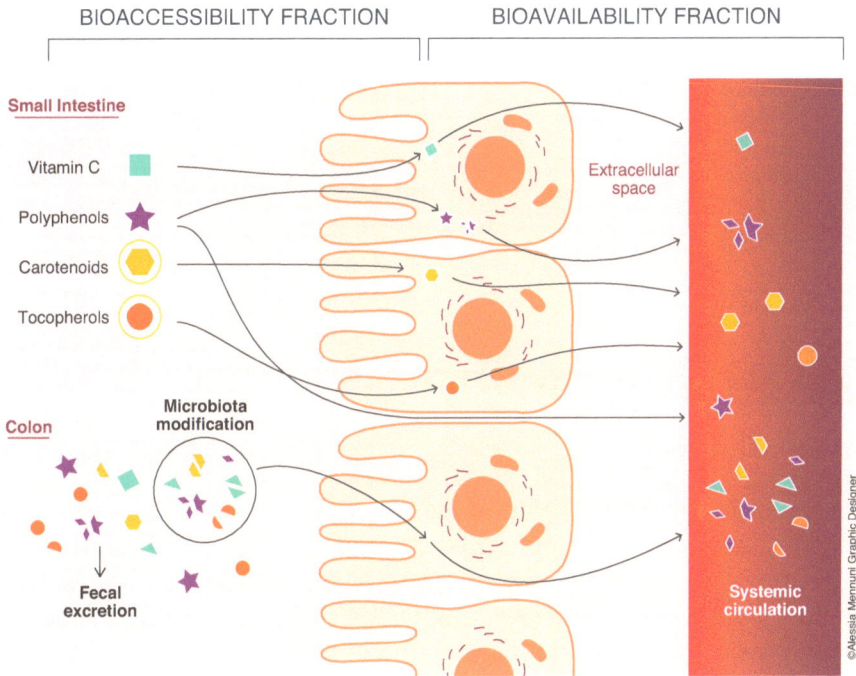

Fig. (3). Bioaccessibility and bioavailability fraction of nutrients from leafy vegetables during digestion in the small intestine and colon tract.

Table 4. Bioaccessibility and bioavailability of bioactive compounds from *Chicorieae* leafy vegetables.

Compound	Bioaccessibility	Ref.	Bioavailability[a]	Ref.
HA	Chicory: 3 caffeoylquinic acid, 206-638 µg/g 5 caffeoylquinic acid, 1060-3920 µg/g	[167]	Chlorogenic acid, 113 µmol/L (animal)	[168]
			Chlorogenic acid, 33% (human)	[169]
			Caffeic acid, 7.6 - 108 µM (animal) Caffeic acid 95 % (human)	[170]
Flavones	Chicory: Apigenin-7-O-glucoside, 0.34-0.42 µg/g Quercetin-3-O-(6"-malonyl-glucoside), 280-521 µg/g	[167]	Quercetin, 0.29-2.26 µM (human)	[171]
			Apigenin, 0.22% (human)	[172]
			Apigenin, 1.45 µg/ml (plasma)	[173]
Flavonols	Chicory: Kaempferol-7-O-glucoside, 6.55-7.00 µg/g	[167]	Kaempferol 0.1-0.5 µg/ml (plasma)	[174]
Anthocyanins	Chicory: Cyanidin 3-O-glucoside, 61-14.6 µg/g	[167]	Cyanidin-3-O-glucoside, 88 % (in vivo), Cyanidin-3-O-glucoside, 0.14 µmol/L (human)	[175]
Carotenoids	Lettuce: Lutein, 13.7% Zeaxanthin, 5.6% β-carotene, 55%	[155]	Trans- and cis-β-carotene, 0.78 and 0.02 µmol/L	[176]
			α-, β-carotene, lutein 28-46% (Caco-2 cell);	[177]
			Lutein, 47.6 nmol/L(plasma)	[56]
			β-carotene, 30.4- 47.6 /L (plasma)	[56]

a,expressed as maximal concentration or percentage; model used is in brackets

FACTORS AFFECTING THE PHYTOCHEMICAL CONTENTS

Pre-Harvest Treatments

Secondary metabolite composition in leafy crops is influenced by plant age, light and temperature levels, and nutrients. In wide terms, both abiotic and biotic stresses (*e.g.* water, nutrients, temperature, light stresses or pathogen attack) can activate the synthesis of secondary compounds. All stresses able to cause a lower photosynthetic rate contribute to increase the need for energy dissipation, which is facilitated by carotenoids and specifically by xanthophylls. Indeed, in a wide variety of plant species the xanthophyll cycle-dependent energy dissipation acts to lose excitation energy, coming from the excess of light, in the form of heat [178]. Other compounds involved in the generation of active oxygen species can be induced in response to pathogen attacks or chemical stressors [2].

As for growth stage, carotenoid and vitamin E, C, and K contents of microgreens (edible seedlings of vegetables) are much higher than those of mature leaves [179]. Moreover, the leaf total phenol levels and antioxidant capacity are 2-4 times higher at early than mature stages in radicchio varieties [24].

Decrease of root-zone temperature (10 *vs* 20 °C) results in higher leaf levels of anthocyanins, chlorophylls, ascorbic acid and total antioxidant power in lettuce [180]. Temperature lowering (from 30 to 20 °C in controlled environment) brings about the increase of pigment (anthocyanins and chlorophyll-b) concentrations in Lollo Rosso lettuce. Consistently, low temperature trigger transcription of genes encoding enzymes involved in anthocyanin biosynthesis [181]. One day chilling stress (4 °C) can increase phenolic content in lettuce, even after three days of recovery from the stress treatment, showing that the enhanced nutritional quality can persist for some days [45]. Temperature and light conditions seem to affect the carotenoid content in lettuce more than the growing system. In fact, lower carotenoid levels of lettuce in hydroponic *versus* open-air conventional cultivation were attributed to polyethylene covering effects rather than to the growing system [58]. Lettuce grown in open-air results richer in flavonoids than plants coming from greenhouse. Moreover, transplant stress is associated to high values of caffeic acid derivates and flavonoids in lettuce grown both in greenhouse and open field at 5 days after transplanting [17]. Growing season (as temperature and light conditions) affects differently each bioactive compound class in lettuce, for instance, winter cultivation is associated to increase of carotenoid and anthocyanin levels, decrease of phenolics and unmodified chlorophyll content [182].

Anthocyanin concentration is affected by light intensity and wavelength as observed in the anthocyanin increase in red leaf lettuce exposed to ambient light intensity compared to 50% shading. In a commercial greenhouse a natural low light intensity-induced bleaching of red color is usually observed [183]. Application of UV filter plastic coverings has demonstrated not only that there was an inhibitory effect of ambient UV radiation on lettuce growth, but also that UV radiation was responsible for the synthesis of anthocyanins and chlorophyll-a, functioning as UV absorbing compounds to protect plant from the specific damage [184]. Specifically, a reduction of anthocyanins and chlorophyll-a, which occurred when UV-B or both UV-A and UV-B radiations were excluded, has been found together with the progressive loss of red coloration. Leaves turned into deep red under the not-filtering UV light covering, while they were light red under the UV-B filter, and green under both UV radiations filter [184]. The decrease of bioactive compound content measured in greenhouse-grown vegetables may be caused by the UV light-filter properties of covering film, after assessing that the UV component of natural light reduces lettuce yield (as a consequence of UV damage to photosynthetic apparatus). On the contrary, there is a high increase in the flavonoid and phenolic acid contents under UV-transparent films compared to plastics with complete or partial block of UV light [185]. Lettuce plants exposed to blue+red LED irradiation showed raised amount of chlorophylls and carotenoids, while higher vitamin C content occurred in plants grown under blue LED light [186]. Moreover, supplemental blue LED enhances anthocyanin and carotenoid contents in lettuce, while far-red LED decreases the same compounds plus chlorophylls compared to white light control [186]. Moreover, supplemental blue LED enhances anthocyanin and carotenoid contents in lettuce, while far-red LED decreases the same compounds plus chlorophylls compared to white light control [187]. Finally, the effect of short-term supplemental red LED is species-dependent with respect to ascorbic acid, total phenols and anthocyanins [188].

Several environmental stresses were applied to lettuce to stimulate production of health-promoting phytochemicals [45]. Briefly, high light stress (800 μmol/m/s for 1 day) produced the highest increment of total phenolics (threefold higher than in control plants), followed by heat shock (40 °C for 10 minutes) and chilling (4 °C for 1 day). After 3 days of recovery from stress treatments, the major phenolic compounds responsible for the total phenolic content variation were chicoric and chlorogenic acids (2.0 and 1.5 mg/g FW) followed by caffeic acid (0.0065 mg/g FW). Moreover, quercetin-3-O-glucoside and luteolin-7-O-glucoside, absent in the control plants, were detected at a significant level (0.12 and 0.012 mg/g FW, respectively) after exposure to high light stress.

Regarding nutritive availability, nitrogen deficiency is associated with greater secondary metabolites and antioxidant capacity in lettuce [180]. Low levels of electrical conductivity (0.5 dS/m) and sulfur (S-SO_4) in the nutrient solution caused the lowest contents of sesquiterpene lactones in lettuce. On the other hand, the highest level of P-PO_4 (48 mg/L, the normal level of P supplied in hydroponic cultivation) determined the highest synthesis of the same compounds [83].

Short term induced salt stress enhanced the content of chlorophylls, carotenoids and anthocyanins in two differently pigmented lettuces, raising the nutritional value without decreasing biomass production [7]. Similarly, phenolic acids and flavonoids increased under a ten-day salt treatment at a moderate rate (50 mmol/L) applied to soilless lettuce [189]. After supplying different macronutrient (NO_3, NH_4, K, Ca) concentrations in a lettuce soilless system (hence increasing electrical conductivity of the nutrient solution), the highest content of ascorbic acid was observed with the intermediate ion concentration in pigmented and green genotypes [190]. Generally, the phenolic compounds also increase at the medium ion concentration. Lettuce plants subjected to high cadmium levels (up to 12 mg/L) show increased contents of total phenolic compounds, flavonoids, and antioxidant activity, probably due to plant defence mechanisms against physiological and structural injuries.

Inoculation with arbuscular mycorrhizal fungi caused significant increase in antioxidant activity and phenolic content in red and green lettuce together with a raise of anthocyanins in red lettuce [191]. At the same time, the effect of mycorrizhal inoculation on carotenoid content in lettuce is dependent on cultivar, leaf position (higher in the inner ones) and growing season [182].

Knowledge about the influence of growing and environmental conditions on the phytochemical contents of plants may help to manage controlled environmental stresses aiming to enhance the synthesis of health-promoting compounds in plants.

Post-Harvest Treatments

Similarly to genetic and environmental conditions, storage and processing techniques are also important factors affecting phytochemical content in leafy vegetable products [192]. Lettuce is consumed as fresh and minimally processed product. It is distributed on the market in wraps or loose bags, as mono-reference or mixed with other vegetables in fresh-cut products. During the postharvest period, both appearance and compositional changes take place and are greatly affected by temperature. For instance, chlorophyll and ascorbic acid contents of iceberg lettuce decline very quickly with a respective loss of circa 60% and 20% after 2 weeks at 1 °C [193]. In 'Butter Crunch' lettuce, the ascorbic acid retention reaches 40% and 16% respectively after 1 and 2 weeks at 7 °C storage [194]. In dark green lettuce stored for 14 days at 5 °C, ascorbic acid content decreases by 39% and 66% after 7 and 14 days, respectively [195]. It was observed that the total content of phenolic acids and flavonoids was preserved throughout storage in different lettuce cultivars, contrary to anthocyanin concentration [30]. Both anthocyanin presence and higher contents of flavonoids and caffeic acid derivates are associated to better visual quality of red cultivars compared to green ones after cutting and during storage [190]. Phenolic metabolites are important in lettuce as their oxidation causes tissue browning during processing. However, it is important to highlight that the shredding step can increase the antioxidant capacity associated with wound-induced phenolic compound synthesis [196]. Relatedly, it was reported that wounded tissues increased concentration of soluble phenolic compounds in the midribs of iceberg lettuce [197], butter leaf and romaine lettuce [198] and in stems [199]. Biochemically, phenol accumulation by the action of phenylalanine ammonia lyase (PAL) and the subsequent oxidization by polyphenoloxidase (PPO) contribute to accelerate tissue browning [200]. PAL activity increases in response to several stress types, including wounding [197], and is correlated to russet spotting development in ethylene-treated lettuce midribs, which show significant increase of phenolic compounds in the affected sectors [201]. In order to prevent browning and other alterations, the use of modified atmospheres with low oxygen and high carbon dioxide composition can be useful in the storage of both intact and fresh-cut iceberg and other lettuce cultivars [46, 202]. Modified atmosphere packaging (MAP) can also exert various effect on phenolic compound metabolism and key enzymes of biosynthesis pathway [203]. As an example, the increment of phenolic level together with brown stain occurred in wounded lettuce exposed to high CO_2 atmosphere and then transferred to air [46]. The event was attributed to PAL activity restoration triggered by CO_2-induced acidity increase [203]. In a study aimed to evaluate phytochemical modifications of minimally processed "Lollo Rosso" lettuce during storage under MAP (2-3% O_2 and 12-14% CO_2), the flavonoid and anthocyanin concentrations did not vary in midribs and green tissues, while they significantly decreased in red tissues [204].

The four cultivar groups of leafy chicory (witloof, pain de sucre, radicchio, catalogna) are used raw as salads or as cooked vegetables in many types of dishes. In the first case, fresh-cut processing can enhance the antioxidant activity associated with total phenol content increase in "Radicchio" hybrids during storage [205]. Storage temperature can affect the compositional characteristics of Witloof chicory. For example, total flavonol glycoside content was reduced of about 55% after 7 day storage at 1 °C [6], while total phenolic content and antioxidant capacity in witloof chicory dropped of 25% after 2 day storage at 4 °C [206].

Similarly to fresh-cut processing, cooking treatments can affect the phenol content in leafy chicory. Boiling, steaming and microwaving are well-known conventional methods to cook vegetables and were shown to alter the phytochemical content in leafy chicories depending on the genotype. Interestingly, cooking by microwaving caused the increase of antioxidant activity and total phenols content in Catalogna-type chicories of about 25% and 19%, respectively, and independently of genotype [207]. The antioxidant activity enhancement was attributed to several factors such as production of strong radical-scavenging antioxidants by thermal chemical reaction, suppression of the oxidation capacity of antioxidants by thermal inactivation of oxidative enzymes and synthesis of novel compounds such as Maillard reaction products with antioxidant activity [208]. Simultaneously, total phenol increase after microwaving could be due to cell tissue disruption during heat treatment and, consequently, to enhanced release and accessibility of compounds in the extraction [209, 210]. The effects of steaming and boiling on phenolics in leafy chicory may dependon genotype. For instance, steaming and boiling did not affect the total phenol content in 'Molfettese' chicory, while boiling caused a 46% decrease in 'Galatina' chicory [207]. The decrease occurring in Catalogna chicory after boiling may be due to leaching and dissolution of phenolic compounds into the cooking water as a consequence of differences in tissue structural properties [210 - 212]. *Sous vide* is an emerging cooking technique thought to provide superior quality owing to the low amount of oxygen inside the pack and to its capacity to better retain the nutritive value of vegetables compared with other conventional methods. *Sous vide* (from French 'under vacuum') requires cooking raw food, preliminarily vacuum-sealed in heat-stable plastic pouches, by dipping in hot water, steaming or microwaving. Evaluating Catalogna chicories under several cooking methods, the *sous vide*-microwaving combination maintained unchanged the antioxidant activity and total phenols content independently of the genotype [207].

GENETICS AND BIOTECHNOLOGIES FOR BIOFORTIFICATION

Lactuca spp. are classified into gene pools according to their cross compatibility, and *L. serriola*, *L. saligna* and *L. virosa* are breeding sources of useful traits transferred to cultivated *L. sativa* [213]. Lettuce is a self-pollinating diploid species (2n=2x=18) bearing a genome size of ca. 2.5 Gb, classified into seven types (crisphead, romaine, butterhead, leaf, latin, stem, and oilseed) which are highly inbred with extensive homozygosity. Major genomic and genetic resources include the high quality and well-validated genome assembly [214], a reference ultra-densetranscript-based genetic map (anchored to genome) of *L. sativa* Salinas × *L. serriola* US96UC23 [215] and an integrated genetic map [216]. *Cichorium* genus comprises six species of which *C. endivia* L. and *C. intybus* L. include cultivated vegetables [217]. *C. intybus* is a perennial species with extensive phenotypic diversity, mainly allogamous for incompatibility systems that imply natural hybridization [218]. Breeding strategies rely on populations maintained by local farmers and on the production of synthetic and F_1 hybrid varieties [219]. Genetic analyses classify cultivated chicory into witloof, root chicory and leaf chicory cultigroups and these latter include the radicchio, sugarloaf and catalogna subgroups [217]. *C. intybus* is diploid (2n=2x=18), genome size is in the range of 0.76-1.3 Gb [220]. *C. endivia* is annual, mainly autogamous and high inbreeding level. Breeding strategies are narrowed to mass or individual selections, pedigree breeding and back-crossing [221]. Chicory and endive share the same chromosome number [222]; endive genome has not been sequenced yet and transcriptomes for both species are available [223]. A linkage consensus map was developed [224] and enriched with nuclear male-sterility and sporophytic self-incompatibility markers [225]. In lettuce, quantitative trait loci (QTL) for chlorophylls and antioxidant capacity fell in the same region of linkage group 1 (LG1) and co-located with a QTL for head formation [226]. Moreover, 96 QTLs for antioxidant potential, total chlorophyll, carotenoids, and phenolic compounds were located in the lettuce reference map. Specifically, LG3 included a key QTL accounting for 30% of the variation in antioxidant potential, which harbored several genes for FLV biosynthesis (*MYB114*; *anthocyanin pigment 2*, *PAP2*; *F3H*), carotenoids (*zeaxanthin epoxidase* and *GGPS*) and anthocyanins (*ferulate-5-hydroxylase*, *F5H*) [227]. As for chicory, QTLs for various agro- and flavour-traits have been identified [228], whilst the genetics of nutrients has been increasing. Relatedly, transcriptome guided studies allowed the identification of markers and expression patterns of key genes (*GAS* and *GAO*) involved in STL content differences and bitter flavour between two stem chicory ecotypes [86]. Finally, *GAS* genes occur in the LG3 and 9 of the chicory consensus map [224].

Successful plant *in vitro* manipulation and tissue culture rely on efficient regeneration. This latter, in lettuce, is strongly genotype-dependent [229]; recalcitrance can be obstructive and overcome by activated charcoal-containing media that favor shoot induction [230]. Tissue culture induced somaclonal variation led to a novel red lettuce cultivar highly enriched with polyphenols [21]. Regeneration from *Cichorium* spp. tissues can occur *via* direct organogenesis or somatic embryogenesis [231]; recently, leaf explant based technology allowed direct multiplication of *C. intybus* producing lines with significant increase of esculin contents compared to controls [232]. Callus and cell suspension can undergo reiterative culture aimed to increase phytochemicals as reported for the hydroxycinnamate content in *Cichorieae* crops [233 - 235]. Alternatively, *in vitro* culturing of lettuce and chicory hairy roots, transgenic for *Agrobacterium rhizogenes* rol genes, showed accumulation of several phytochemicals including phenolics, flavonoids, STL and antioxidant activity compared to controls [233, 235, 236]. Providing plant cell and tissue cultures with exogenous elicitors can enhance phytochemical contents as observed in the case of lettucenin A increment, triggered by methyl-jasmonate in lettuce cell suspension cultures [237], and of lactuside A and crepidiaside B accumulation in lettuce hairy roots, treated with pectinase [238]. Methyl-jasmonate addition can also enhance STL abundance in chicory hairy roots similarly to salicylic acid supply [98]. Moreover, methyl-jasmonate led to doubled chlorogenic and isochlorogenic acid production in chicory cell cultures [40]. Zinc treatment of chicory hairy roots caused accumulation of phenols, flavonoids and chicoric acid, and enhanced antioxidant enzyme activity [239]. Exogenous gibberellin hormones stimulated the production of coumarins in chicory hairy roots [240].

Gene transfer is an effective technology to generate bio-fortified plants [241]; so far, reliable *Agrobacterium*-mediated transformation protocols have been available for *Lactuca* and *Cichorium* spp [229, 230, 242, 243]. Regarding lettuce, a recurrent and effective strategy was the over-expression of single biosynthetic enzymes, such as homogentisate phytyltransferase (HPT) and γ-tocopherol methyltransferase (γ-TMT), to alter vitamin E content [244, 245]. Moreover, co-overexpression of both *γ-TMT* and *HPT* produced far superior tocopherol amounts [245] to lettuce that only over-expressed *γ –TMT* [246]. As for phenolics, total phenol and flavonoid enrichment occurred in lettuce transgenic for the brassica chalcone isomerase gene [247], while β-carotene levels raised in transgenic lettuce for Arabidopsis *PSY* gene [248]. Gene transfer targeted to plastids is an alternative to nuclear transformation, ensuring high and correct transgene expression and minimal transgene dispersion [249]. Plastid engineering was successful to produce bio-fortified lettuce lines and is yet unexplored in chicory. Transplastomic lettuce lines for tocopherol synthase showed higher vitamin E levels and activity [250] and those transformed with bacterial carotenoid synthesis genes mainly accumulated free astaxanthin and its esters [251].

Genome editing of lettuce was optimized to generate transgene-free and gene edited plants at the same time. Briefly, a mix of endonuclease Cas9 protein and *in vitro*-transcribed single guide RNA was transfected into protoplasts to disrupt a gene of the brassinosteroid signaling pathway, achieving a 46% mutation rate without off-targets and stable inheritance of the edited gene [252].

CONCLUSION

Knowledge about factors influencing phytochemicals in the most relevant *Asteraceae* vegetables has been expanding. The available knowledge allows the quantitative and qualitative increase of phytochemicals and their preservation in plant food. Starting from the concept that quality is determined in the field, biofortification will rely on integrative approaches (from genetic/biotechnological tools to agro-practices) to enhance and finely modulate contents, types and properties of phytonutrients in the primary produce. Concurrently, processing technologies will be applied to both improve and preserve phytonutrients in the fresh-cut segment. Considered that at least five portions of vegetables and fruits per day are recommended in the human diet, the efforts of research will converge with policy guidelines to build consumers' conscious consumption by enhancing information on phytonutrient contents and properties for the labeling of vegetable products, in an ideal context of science-driven market communication.

CONSENT FOR PUBLICATION

Not applicable.

CONFLICT OF INTEREST

The author (editor) declares no conflict of interest, financial or otherwise.

ACKNOWLEDGMENT

Declare None.

REFERENCES

[1]   Slavin JL, Lloyd B. Health benefits of fruits and vegetables. Adv Nutr 2012; 3(4): 506-16.
[http://dx.doi.org/10.3945/an.112.002154] [PMID: 22797986]

[2]   Hounsome N, Hounsome B. Biochemistry of vegetables: major classes of primary (carbohydrates, amino acids, fatty acids, vitamins, and organic acids) and secondary metabolites (terpenoids, phenolics, alkaloids, and sulfur-containing compounds) in vegetables Handbook of Vegetables and Vegetable Processing. Indianapolis: Wiley-Blackwell 2010; pp. 23-58.

[3]   Prior RL, Cao G. Antioxidant Phytochemicals in Fruits and Vegetables: Diet and Health Implications. HortScience 2000; 35: 588-92.

[4]   Liu RH. Health benefits of fruit and vegetables are from additive and synergistic combinations of phytochemicals. Am J Clin Nutr 2003; 78(3) (Suppl.): 517S-20S.
[http://dx.doi.org/10.1093/ajcn/78.3.517S] [PMID: 12936943]

[5]   Simonne A, Simonne E, Eitenmiller R, Coker CH. Bitterness and Composition of Lettuce Varieties Grown in the Southeastern United States. Horttechnology 2002; 12: 721-6.

[6]   DuPont MS, Mondin Z, Williamson G, Price KR. Effect of variety, processing, and storage on the flavonoid glycoside content and composition of lettuce and endive. J Agric Food Chem 2000; 48(9): 3957-64.
[http://dx.doi.org/10.1021/jf0002387] [PMID: 10995297]

[7]   Pérez-López U. Growth and nutritional quality improvement in two differently pigmented lettuce cultivars grown under elevated CO2 and/or salinity. Sci Hortic (Amsterdam) 2015; 195: 56-66.
[http://dx.doi.org/10.1016/j.scienta.2015.08.034]

[8]   Llorach R, Tomás-Barberán FA, Ferreres F. Lettuce and chicory byproducts as a source of antioxidant phenolic extracts. J Agric Food Chem 2004; 52(16): 5109-16.
[http://dx.doi.org/10.1021/jf040055a] [PMID: 15291483]

[9]   D'evoli L, Morroni F, Lombardi-Boccia G, *et al.* Red chicory (*Cichorium intybus* L. cultivar) as a potential source of antioxidant anthocyanins for intestinal health. Oxid Med Cell Longev 2013; 2013: 704310.
[http://dx.doi.org/10.1155/2013/704310] [PMID: 24069504]

[10]   Crozier A, Lean MEJ, McDonald MS, Black C. Quantitative Analysis of the Flavonoid Content of Commercial Tomatoes, Onions, Lettuce, and Celery. J Agric Food Chem 1997; 45: 590-5.
[http://dx.doi.org/10.1021/jf960339y]

[11]   Hohl U, Neubert B, Pforte H, Schonhof I, Böhm H. Flavonoid concentrations in the inner leaves of head lettuce genotypes. Eur Food Res Technol. 2001; 213: pp. 205-11.
[http://dx.doi.org/10.1007/s002170100361]

[12]   Vermerris W, Nicholson R. Phenolic Compound Biochemistry New York. Springer 2006; p. 276.

[13]   Zidorn C. Sesquiterpene lactones and their precursors as chemosystematic markers in the tribe *Cichorieae* of the *Asteraceae.* Phytochemistry 2008; 69(12): 2270-96.
[http://dx.doi.org/10.1016/j.phytochem.2008.06.013] [PMID: 18715600]

[14]   Scotti MT, Emerenciano V, Ferreira MJ, *et al.* Self-organizing maps of molecular descriptors for sesquiterpene lactones and their application to the chemotaxonomy of the *Asteraceae* family. Molecules 2012; 17(4): 4684-702.
[http://dx.doi.org/10.3390/molecules17044684] [PMID: 22522398]

[15]   Chadwick M, Trewin H, Gawthrop F, Wagstaff C. Sesquiterpenoids lactones: benefits to plants and

people. Int J Mol Sci 2013; 14(6): 12780-805.
[http://dx.doi.org/10.3390/ijms140612780] [PMID: 23783276]

[16]   Llorach R, Martínez-Sánchez A, Tomás-Barberán FA, Gil MI, Ferreres F. Characterisation of polyphenols and antioxidant properties of five lettuce varieties and escarole. Food Chem 2008; 108(3): 1028-38.
[http://dx.doi.org/10.1016/j.foodchem.2007.11.032] [PMID: 26065768]

[17]   Romani A, Pinelli P, Galardi C, Sani G, Cimato A, Heimler D. Polyphenols in greenhouse and open-air-grown lettuce. Food Chem 2002; 79: 337-42.
[http://dx.doi.org/10.1016/S0308-8146(02)00170-X]

[18]   Alarcón-Flores MI, Romero-González R, Martínez Vidal JL, Garrido Frenich A. Multiclass determination of phenolic compounds in different varieties of tomato and lettuce by ultra high performance liquid chromatography coupled to tandem mass spectrometry. Int J Food Prop 2016; 19: 494-507.
[http://dx.doi.org/10.1080/10942912.2014.978010]

[19]   Ribas-Agustí A, Gratacós-Cubarsí M, Sárraga C, García-Regueiro JA, Castellari M. Analysis of eleven phenolic compounds including novel p-coumaroyl derivatives in lettuce (*Lactuca sativa* L.) by ultra-high-performance liquid chromatography with photodiode array and mass spectrometry detection. Phytochem Anal 2011; 22(6): 555-63.
[http://dx.doi.org/10.1002/pca.1318] [PMID: 21433163]

[20]   Mai F, Glomb MA. Isolation of phenolic compounds from iceberg lettuce and impact on enzymatic browning. J Agric Food Chem 2013; 61(11): 2868-74.
[http://dx.doi.org/10.1021/jf305182u] [PMID: 23473017]

[21]   Cheng DM, Pogrebnyak N, Kuhn P, Krueger CG, Johnson WD, Raskin I. Development and phytochemical characterization of high polyphenol red lettuce with anti-diabetic properties. PLoS One 2014; 9(3): e91571.
[http://dx.doi.org/10.1371/journal.pone.0091571] [PMID: 24637790]

[22]   Ferioli F, Manco MA, D'Antuono LF. Variation of sesquiterpene lactones and phenolics in chicory and endive germplasm. J Food Compos Anal 2015; 39: 77-86.
[http://dx.doi.org/10.1016/j.jfca.2014.11.014]

[23]   Heimler D, Isolani L, Vignolini P, Tombelli S, Romani A. Polyphenol content and antioxidative activity in some species of freshly consumed salads. J Agric Food Chem 2007; 55(5): 1724-9.
[http://dx.doi.org/10.1021/jf0628983] [PMID: 17279769]

[24]   Rossetto M, Lante A, Vanzani P, Spettoli P, Scarpa M, Rigo A. Red chicories as potent scavengers of highly reactive radicals: a study on their phenolic composition and peroxyl radical trapping capacity and efficiency. J Agric Food Chem 2005; 53(21): 8169-75.
[http://dx.doi.org/10.1021/jf051116n] [PMID: 16218660]

[25]   Innocenti M, Gallori S, Giaccherini C, Ieri F, Vincieri FF, Mulinacci N. Evaluation of the phenolic content in the aerial parts of different varieties of *Cichorium intybus* L. J Agric Food Chem 2005; 53(16): 6497-502.
[http://dx.doi.org/10.1021/jf050541d] [PMID: 16076140]

[26]   Filippo D'Antuono L, Ferioli F, Manco MA. The impact of sesquiterpene lactones and phenolics on sensory attributes: An investigation of a curly endive and escarole germplasm collection. Food Chem 2016; 199: 238-45.
[http://dx.doi.org/10.1016/j.foodchem.2015.12.002] [PMID: 26775966]

[27]   Ninfali P, Mea G, Giorgini S, Rocchi M, Bacchiocca M. Antioxidant capacity of vegetables, spices and dressings relevant to nutrition. Br J Nutr 2005; 93(2): 257-66.
[http://dx.doi.org/10.1079/BJN20041327] [PMID: 15788119]

[28]   Sinkovič L, Demšar L, Žnidarčič D, Vidrih R, Hribar J, Treutter D. Phenolic profiles in leaves of chicory cultivars (*Cichorium intybus* L.) as influenced by organic and mineral fertilizers. Food Chem

2015; 166: 507-13.
[http://dx.doi.org/10.1016/j.foodchem.2014.06.024] [PMID: 25053087]

[29] Wu X, Beecher GR, Holden JM, Haytowitz DB, Gebhardt SE, Prior RL. Lipophilic and hydrophilic antioxidant capacities of common foods in the United States. J Agric Food Chem 2004; 52(12): 4026-37.
[http://dx.doi.org/10.1021/jf049696w] [PMID: 15186133]

[30] Ferreres F, Gil MI, Castañer M, Tomás-Barberán FA. Phenolic metabolites in red pigmented lettuce (*Lactuca sativa*). Changes with minimal processing and cold storage. J Agric Food Chem 1997; 45: 4249-54.
[http://dx.doi.org/10.1021/jf970399j]

[31] Luna MC, Tudela JA, Martinez-Sanchez A, Allende A, Marin A, Gil MI. Long-term deficit and excess of irrigation influences quality and browning related enzymes and phenolic metabolism of fresh-cut iceberg lettuce (*Lactuca sativa* L.). Postharvest Biol Technol 2012; 73: 37-45.
[http://dx.doi.org/10.1016/j.postharvbio.2012.05.011]

[32] Arabbi PR, Genovese MI, Lajolo FM. Flavonoids in vegetable foods commonly consumed in Brazil and estimated ingestion by the Brazilian population. J Agric Food Chem 2004; 52(5): 1124-31.
[http://dx.doi.org/10.1021/jf0499525] [PMID: 14995109]

[33] Cao J, Chen W, Zhang Y, Zhang YQ, Zhao XJ. Content of Selected Flavonoids in 100 Edible Vegetables and Fruits. Food Sci Technol Res 2010; 16: 395-402.
[http://dx.doi.org/10.3136/fstr.16.395]

[34] Hertog MGL, Hollman PCH, Katan MB. Content of Potentially Anticarcinogenic Flavonoids of 28 Vegetables and 9 Fruits Commonly Consumed in the Netherlands. J Agric Food Chem 1992; 40: 2379-83.
[http://dx.doi.org/10.1021/jf00024a011]

[35] Chon SU, Boo HO, Heo BG, Gorinstein S. Anthocyanin content and the activities of polyphenol oxidase, peroxidase and phenylalanine ammonia-lyase in lettuce cultivars. Int J Food Sci Nutr 2012; 63(1): 45-8.
[http://dx.doi.org/10.3109/09637486.2011.595704] [PMID: 21718113]

[36] Hichri I, Barrieu F, Bogs J, Kappel C, Delrot S, Lauvergeat V. Recent advances in the transcriptional regulation of the flavonoid biosynthetic pathway. J Exp Bot 2011; 62(8): 2465-83.
[http://dx.doi.org/10.1093/jxb/erq442] [PMID: 21278228]

[37] Cheynier V, Comte G, Davies KM, Lattanzio V, Martens S. Plant phenolics: recent advances on their biosynthesis, genetics, and ecophysiology. Plant Physiol Biochem 2013; 72: 1-20.
[http://dx.doi.org/10.1016/j.plaphy.2013.05.009] [PMID: 23774057]

[38] Comino C, Lanteri S, Portis E, *et al.* Isolation and functional characterization of a cDNA coding a hydroxycinnamoyltransferase involved in phenylpropanoid biosynthesis in *Cynara cardunculus* L. BMC Plant Biol 2007; 7: 14.
[http://dx.doi.org/10.1186/1471-2229-7-14] [PMID: 17374149]

[39] Sonnante G, D'Amore R, Blanco E, *et al.* Novel hydroxycinnamoyl-coenzyme A quinate transferase genes from artichoke are involved in the synthesis of chlorogenic acid. Plant Physiol 2010; 153(3): 1224-38.
[http://dx.doi.org/10.1104/pp.109.150144] [PMID: 20431089]

[40] Legrand G, Delporte M, Khelifi C, *et al.* Identification and Characterization of Five BAHD Acyltransferases Involved in Hydroxycinnamoyl Ester Metabolism in Chicory. Front Plant Sci 2016; 7: 741.
[http://dx.doi.org/10.3389/fpls.2016.00741] [PMID: 27375627]

[41] Lee J, Scagel CF. Chicoric acid: chemistry, distribution, and production. Front Chem 2013; 1: 40.
[http://dx.doi.org/10.3389/fchem.2013.00040] [PMID: 24790967]

[42]   Biochemistry of Plant Secondary Metabolism. Petersen M, Hans J, Matern U. Biosynthesis of Phenylpropanoids and Related Compounds. Annual Plant ReviewsLondon: Wiley-Blackwell 2010; 40: pp. 182-257.
[http://dx.doi.org/10.1002/9781444320503.ch4]

[43]   Hamberger B, Ellis M, Friedmann M, de Azevedo Souza C, Barbazuk B, Douglas CJ. Genome-wide analyses of phenylpropanoid-related genes in Populus trichocarpa, Arabidopsis thaliana, and Oryza sativa: the Populus lignin toolbox and conservation and diversification of angiosperm gene families. This article is one of a selection of papers published in the Special Issue on Poplar Research in Canada. Can J Bot 2007; 85: 1182-201.
[http://dx.doi.org/10.1139/B07-098]

[44]   Zhang X, Liu CJ. Multifaceted regulations of gateway enzyme phenylalanine ammonia-lyase in the biosynthesis of phenylpropanoids. Mol Plant 2015; 8(1): 17-27.
[http://dx.doi.org/10.1016/j.molp.2014.11.001] [PMID: 25578269]

[45]   Oh MM, Carey EE, Rajashekar CB. Environmental stresses induce health-promoting phytochemicals in lettuce. Plant Physiol Biochem 2009; 47(7): 578-83.
[http://dx.doi.org/10.1016/j.plaphy.2009.02.008] [PMID: 19297184]

[46]   Mateos M, Ke D, Cantwell M, Kader AA. Phenolic metabolism and ethanolic fermentation of intact and cut lettuce exposed to CO_2-enriched atmospheres. Postharvest Biol Technol 1993; 3: 225-33.
[http://dx.doi.org/10.1016/0925-5214(93)90058-B]

[47]   Frank MR, Deyneka JM, Schuler MA. Cloning of wound-induced cytochrome P450 monooxygenases expressed in pea. Plant Physiol 1996; 110(3): 1035-46.
[http://dx.doi.org/10.1104/pp.110.3.1035] [PMID: 8819874]

[48]   Bell-Lelong DA, Cusumano JC, Meyer K, Chapple C. *Cinnamate-4-hydroxylase* expression in Arabidopsis. Regulation in response to development and the environment. Plant Physiol 1997; 113(3): 729-38.
[http://dx.doi.org/10.1104/pp.113.3.729] [PMID: 9085570]

[49]   Schilmiller AL, Stout J, Weng JK, Humphreys J, Ruegger MO, Chapple C. Mutations in the cinnamate 4-hydroxylase gene impact metabolism, growth and development in Arabidopsis. Plant J 2009; 60(5): 771-82.
[http://dx.doi.org/10.1111/j.1365-313X.2009.03996.x] [PMID: 19682296]

[50]   Cao Y, Han Y, Li D, Lin Y, Cai Y. Systematic Analysis of the 4-Coumarate:Coenzyme A Ligase (4CL) Related Genes and Expression Profiling during Fruit Development in the Chinese Pear. Genes (Basel) 2016; 7(10): 89.
[http://dx.doi.org/10.3390/genes7100089] [PMID: 27775579]

[51]   D'Auria JC. Acyltransferases in plants: a good time to be BAHD. Curr Opin Plant Biol 2006; 9(3): 331-40.
[http://dx.doi.org/10.1016/j.pbi.2006.03.016] [PMID: 16616872]

[52]   Li S. Transcriptional control of flavonoid biosynthesis: fine-tuning of the MYB-bHLH-WD40 (MBW) complex. Plant Signal Behav 2014; 9(1): e27522.
[http://dx.doi.org/10.4161/psb.27522] [PMID: 24393776]

[53]   Mehrtens F, Kranz H, Bednarek P, Weisshaar B. The Arabidopsis transcription factor MYB12 is a flavonol-specific regulator of phenylpropanoid biosynthesis. Plant Physiol 2005; 138(2): 1083-96.
[http://dx.doi.org/10.1104/pp.104.058032] [PMID: 15923334]

[54]   Stracke R, Ishihara H, Huep G, *et al.* Differential regulation of closely related R2R3-MYB transcription factors controls flavonol accumulation in different parts of the *Arabidopsis thaliana* seedling. Plant J 2007; 50(4): 660-77.
[http://dx.doi.org/10.1111/j.1365-313X.2007.03078.x] [PMID: 17419845]

[55]   Gou J-Y, Felippes FF, Liu C-J, Weigel D, Wang J-W. Negative regulation of anthocyanin biosynthesis

in Arabidopsis by a miR156-targeted SPL transcription factor. Plant Cell 2011; 23(4): 1512-22.
[http://dx.doi.org/10.1105/tpc.111.084525] [PMID: 21487097]

[56]   Nicolle C, Cardinault N, Gueux E, *et al.* Health effect of vegetable-based diet: lettuce consumption improves cholesterol metabolism and antioxidant status in the rat. Clin Nutr 2004; 23(4): 605-14.
[http://dx.doi.org/10.1016/j.clnu.2003.10.009] [PMID: 15297097]

[57]   Mou B. Genetic Variation of Beta-carotene and Lutein Contents in Lettuce. J Am Soc Hortic Sci 2005; 130: 870-6.

[58]   Kimura M, Rodriguez-Amaya DB. Carotenoid composition of hydroponic leafy vegetables. J Agric Food Chem 2003; 51(9): 2603-7.
[http://dx.doi.org/10.1021/jf020539b] [PMID: 12696944]

[59]   Znidarčič D, Ban D, Šircelj H. Carotenoid and chlorophyll composition of commonly consumed leafy vegetables in Mediterranean countries. Food Chem 2011; 129(3): 1164-8.
[http://dx.doi.org/10.1016/j.foodchem.2011.05.097] [PMID: 25212352]

[60]   de Azevedo-Meleiro CH, Rodriguez-Amaya DB. Carotenoids of endive and New Zealand spinach as affected by maturity, season and minimal processing. J Food Compos Anal 2005; 18: 845-55.
[http://dx.doi.org/10.1016/j.jfca.2004.10.006]

[61]   Montefusco A, Semitaio G, Marrese PP, *et al.* Antioxidants in Varieties of Chicory (*Cichorium intybus* L.) and Wild Poppy (*Papaver rhoeas* L.) of Southern Italy. J Chem 2015; 2015: 8.
[http://dx.doi.org/10.1155/2015/923142]

[62]   Giannino D, Gonnella M, Russo R, *et al.* Antioxidant properties of minimally processed endives and escaroles vary as influenced by the cultivation site, cultivar and storage time. Postharvest Biol Technol 2018; 138: 82-90.
[http://dx.doi.org/10.1016/j.postharvbio.2017.12.004]

[63]   Su Q, Rowley KG, Itsiopoulos C, O'Dea K. Identification and quantitation of major carotenoids in selected components of the Mediterranean diet: green leafy vegetables, figs and olive oil. Eur J Clin Nutr 2002; 56(11): 1149-54.
[http://dx.doi.org/10.1038/sj.ejcn.1601472] [PMID: 12428183]

[64]   Cruz R, Baptista P, Cunha S, Pereira JA, Casal S. Carotenoids of lettuce (*Lactuca sativa* L.) grown on soil enriched with spent coffee grounds. Molecules 2012; 17(2): 1535-47.
[http://dx.doi.org/10.3390/molecules17021535] [PMID: 22314378]

[65]   Giuliano G. Provitamin A biofortification of crop plants: a gold rush with many miners. Curr Opin Biotechnol 2017; 44: 169-80.
[http://dx.doi.org/10.1016/j.copbio.2017.02.001] [PMID: 28254681]

[66]   Cunningham FX Jr, Gantt E. One ring or two? Determination of ring number in carotenoids by lycopene epsilon-cyclases. Proc Natl Acad Sci USA 2001; 98(5): 2905-10.
[http://dx.doi.org/10.1073/pnas.051618398] [PMID: 11226339]

[67]   Leivar P, Monte E. PIFs: systems integrators in plant development. Plant Cell 2014; 26(1): 56-78.
[http://dx.doi.org/10.1105/tpc.113.120857] [PMID: 24481072]

[68]   Li F, Vallabhaneni R, Yu J, Rocheford T, Wurtzel ET. The maize phytoene synthase gene family: overlapping roles for carotenogenesis in endosperm, photomorphogenesis, and thermal stress tolerance. Plant Physiol 2008; 147(3): 1334-46.
[http://dx.doi.org/10.1104/pp.108.122119] [PMID: 18508954]

[69]   Cazzonelli CI, Roberts AC, Carmody ME, Pogson BJ. Transcriptional control of SET DOMAIN GROUP 8 and CAROTENOID ISOMERASE during Arabidopsis development. Mol Plant 2010; 3(1): 174-91.
[http://dx.doi.org/10.1093/mp/ssp092] [PMID: 19952001]

[70]   Woitsch S, Römer S. Expression of xanthophyll biosynthetic genes during light-dependent chloroplast differentiation. Plant Physiol 2003; 132(3): 1508-17.

[http://dx.doi.org/10.1104/pp.102.019364] [PMID: 12857831]

[71] Rodríguez-Villalón A, Gas E, Rodríguez-Concepción M. Phytoene synthase activity controls the biosynthesis of carotenoids and the supply of their metabolic precursors in dark-grown Arabidopsis seedlings. Plant J 2009; 60(3): 424-35.
[http://dx.doi.org/10.1111/j.1365-313X.2009.03966.x] [PMID: 19594711]

[72] Balmer Y, Koller A, del Val G, Manieri W, Schürmann P, Buchanan BB. Proteomics gives insight into the regulatory function of chloroplast thioredoxins. Proc Natl Acad Sci USA 2003; 100(1): 370-5.
[http://dx.doi.org/10.1073/pnas.232703799] [PMID: 12509500]

[73] Yu Q, Ghisla S, Hirschberg J, Mann V, Beyer P. Plant carotene cis-trans isomerase CRTISO: a new member of the FAD(RED)-dependent flavoproteins catalyzing non-redox reactions. J Biol Chem 2011; 286(10): 8666-76.
[http://dx.doi.org/10.1074/jbc.M110.208017] [PMID: 21209101]

[74] Carol P, Kuntz M. A plastid terminal oxidase comes to light: implications for carotenoid biosynthesis and chlororespiration. Trends Plant Sci 2001; 6(1): 31-6.
[http://dx.doi.org/10.1016/S1360-1385(00)01811-2] [PMID: 11164375]

[75] Beisel KG, Jahnke S, Hofmann D, Köppchen S, Schurr U, Matsubara S. Continuous turnover of carotenes and chlorophyll a in mature leaves of Arabidopsis revealed by 14CO2 pulse-chase labeling. Plant Physiol 2010; 152(4): 2188-99.
[http://dx.doi.org/10.1104/pp.109.151647] [PMID: 20118270]

[76] Havaux M. Carotenoid oxidation products as stress signals in plants. Plant J 2014; 79(4): 597-606.
[http://dx.doi.org/10.1111/tpj.12386] [PMID: 24267746]

[77] Bugos RC, Yamamoto HY. Molecular cloning of violaxanthin de-epoxidase from romaine lettuce and expression in *Escherichia coli.* Proc Natl Acad Sci USA 1996; 93(13): 6320-5.
[http://dx.doi.org/10.1073/pnas.93.13.6320] [PMID: 8692813]

[78] Padilla-Gonzalez GF, dos Santos FA, Da Costa FB. Sesquiterpene Lactones: More Than Protective Plant Compounds With High Toxicity. CRC Crit Rev Plant Sci 2016; 35: 18-37.
[http://dx.doi.org/10.1080/07352689.2016.1145956]

[79] Chadwick M, Gawthrop F, Michelmore RW, Wagstaff C, Methven L. Perception of bitterness, sweetness and liking of different genotypes of lettuce. Food Chem 2016; 197(Pt A): 66-74.
[http://dx.doi.org/10.1016/j.foodchem.2015.10.105] [PMID: 26616925]

[80] Mai F, Glomb MA. Structural and Sensory Characterization of Novel Sesquiterpene Lactones from Iceberg Lettuce. J Agric Food Chem 2016; 64(1): 295-301.
[http://dx.doi.org/10.1021/acs.jafc.5b05128] [PMID: 26727458]

[81] Graziani G, Ferracane R, Sambo P, Santagata S, Nicoletto C, Fogliano V. Profiling chicory sesquiterpene lactones by high resolution mass spectrometry. Food Res Int 2015; 67: 193-8.
[http://dx.doi.org/10.1016/j.foodres.2014.11.021]

[82] Beharav A, Stojakowska A, Ben-David R, Malarz J, Michalska K, Kisiel W. Variation of sesquiterpene lactone contents in *Lactuca georgica* natural populations from Armenia. Genet Resour Crop Evol 2015; 62: 431-41.
[http://dx.doi.org/10.1007/s10722-014-0171-9]

[83] Seo MW, Yang DS, Kays SJ, Lee GP, Park KW. Sesquiterpene Lactones and Bitterness in Korean Leaf Lettuce Cultivars. HortScience 2009; 44: 246-9.

[84] Price KR, Dupont MS, Shepherd R, Chan HWS, Fenwick GR. Relationship between the chemical and sensory properties of exotic salad crops—coloured lettuce (*Lactuca sativa*) and chicory (*Cichorium intybus*). J Sci Food Agric 1990; 53: 185-92.
[http://dx.doi.org/10.1002/jsfa.2740530206]

[85] Sung JS, Hur OS, Ryu KY, *et al.* Variation in phenotypic characteristics and contents of sesquiterpene lactones in Lettuce (*Lactuca sativa* L.) Germplasm. Kor J Plant Res 2016; 29: 679-89.

[http://dx.doi.org/10.7732/kjpr.2016.29.6.679]

[86] Testone G, Mele G, Di Giacomo E, *et al.* Insights into the sesquiterpenoid pathway by metabolic profiling and *de novo* transcriptome assembly of stem-chicory (*Cichorium intybus* cultigroup "Catalogna"). Front Plant Sci 2016; 7: 1676.
[http://dx.doi.org/10.3389/fpls.2016.01676] [PMID: 27877190]

[87] Dubey VS, Bhalla R, Luthra R. An overview of the non-mevalonate pathway for terpenoid biosynthesis in plants. J Biosci 2003; 28(5): 637-46.
[http://dx.doi.org/10.1007/BF02703339] [PMID: 14517367]

[88] Bouwmeester HJ, Kodde J, Verstappen FW, Altug IG, de Kraker JW, Wallaart TE. Isolation and characterization of two *germacrene A synthase* cDNA clones from chicory. Plant Physiol 2002; 129(1): 134-44.
[http://dx.doi.org/10.1104/pp.001024] [PMID: 12011345]

[89] Bennett MH, Mansfield JW, Lewis MJ, Beale MH. Cloning and expression of sesquiterpene synthase genes from lettuce (*Lactuca sativa* L.). Phytochemistry 2002; 60(3): 255-61.
[http://dx.doi.org/10.1016/S0031-9422(02)00103-6] [PMID: 12031443]

[90] Liu Q, Majdi M, Cankar K, *et al.* Reconstitution of the costunolide biosynthetic pathway in yeast and *Nicotiana benthamiana*. PLoS One 2011; 6(8): e23255.
[http://dx.doi.org/10.1371/journal.pone.0023255] [PMID: 21858047]

[91] Ikezawa N, Göpfert JC, Nguyen DT, *et al.* Lettuce costunolide synthase (CYP71BL2) and its homolog (CYP71BL1) from sunflower catalyze distinct regio- and stereoselective hydroxylations in sesquiterpene lactone metabolism. J Biol Chem 2011; 286(24): 21601-11.
[http://dx.doi.org/10.1074/jbc.M110.216804] [PMID: 21515683]

[92] Nguyen TD, Faraldos JA, Vardakou M, Salmon M, O'Maille PE, Ro DK. Discovery of *germacrene A* synthases in *Barnadesia spinosa*: The first committed step in sesquiterpene lactone biosynthesis in the basal member of the *Asteraceae*. Biochem Biophys Res Commun 2016; 479(4): 622-7.
[http://dx.doi.org/10.1016/j.bbrc.2016.09.165] [PMID: 27697527]

[93] Majdi M, Liu Q, Karimzadeh G, *et al.* Biosynthesis and localization of parthenolide in glandular trichomes of feverfew (*Tanacetum parthenium* L. Schulz Bip.). Phytochemistry 2011; 72(14-15): 1739-50.
[http://dx.doi.org/10.1016/j.phytochem.2011.04.021] [PMID: 21620424]

[94] Franssen MC, Konig WA, Bouwmeester HJ, Konig WA, Bouwmeester HJ. (+)-Germacrene A biosynthesis. The committed step in the biosynthesis of bitter sesquiterpene lactones in chicory. Plant Physiol 1998; 117(4): 1381-92.
[http://dx.doi.org/10.1104/pp.117.4.1381] [PMID: 9701594]

[95] Boutanaev AM, Moses T, Zi J, *et al.* Investigation of terpene diversification across multiple sequenced plant genomes. Proc Natl Acad Sci USA 2015; 112(1): E81-8.
[http://dx.doi.org/10.1073/pnas.1419547112] [PMID: 25502595]

[96] Tholl D. Biosynthesis and Biological Functions of Terpenoids in Plants.Biotechnology of Isoprenoids. Cham: Springer International Publishing 2015; pp. 63-106.
[http://dx.doi.org/10.1007/10_2014_295]

[97] Eljounaidi K, Comino C, Moglia A, *et al.* Accumulation of cynaropicrin in globe artichoke and localization of enzymes involved in its biosynthesis. Plant Sci 2015; 239: 128-36.
[http://dx.doi.org/10.1016/j.plantsci.2015.07.020] [PMID: 26398797]

[98] Malarz J, Stojakowska A, Kisiel W. Effect of methyl jasmonate and salicylic acid on sesquiterpene lactone accumulation in hairy roots of *Cichorium intybus*. Acta Physiol Plant 2007; 29: 127-32.
[http://dx.doi.org/10.1007/s11738-006-0016-z]

[99] Hong GJ, Xue XY, Mao YB, Wang LJ, Chen XY. Arabidopsis MYC2 interacts with DELLA proteins in regulating sesquiterpene synthase gene expression. Plant Cell 2012; 24(6): 2635-48.

[http://dx.doi.org/10.1105/tpc.112.098749] [PMID: 22669881]

[100] Matías-Hernández L, Jiang W, Yang K, Tang K, Brodelius PE, Pelaz S. AaMYB1 and its orthologue AtMYB61 affect terpene metabolism and trichome development in Artemisia annua and Arabidopsis thaliana. Plant J 2017; 90(3): 520-34.
[http://dx.doi.org/10.1111/tpj.13509] [PMID: 28207974]

[101] Schluttenhofer C, Yuan L. Regulation of specialized metabolism by WRKY transcription factors. Plant Physiol 2015; 167(2): 295-306.
[http://dx.doi.org/10.1104/pp.114.251769] [PMID: 25501946]

[102] Sessa RA, Bennett MH, Lewis MJ, Mansfield JW, Beale MH. Metabolite profiling of sesquiterpene lactones from *Lactuca* species. Major latex components are novel oxalate and sulfate conjugates of lactucin and its derivatives. J Biol Chem 2000; 275(35): 26877-84.
[PMID: 10858433]

[103] Abu-Reidah IM, Contreras MM, Arráez-Román D, Segura-Carretero A, Fernández-Gutiérrez A. Reversed-phase ultra-high-performance liquid chromatography coupled to electrospray ionization-quadrupole-time-of-flight mass spectrometry as a powerful tool for metabolic profiling of vegetables: *Lactuca sativa* as an example of its application. J Chromatogr A 2013; 1313: 212-27.
[http://dx.doi.org/10.1016/j.chroma.2013.07.020] [PMID: 23891214]

[104] Mai F, Glomb MA. Lettucenin sesquiterpenes contribute significantly to the browning of lettuce. J Agric Food Chem 2014; 62(20): 4747-53.
[http://dx.doi.org/10.1021/jf500413h] [PMID: 24818869]

[105] D'Acunzo F, Giannino D, Longo V, *et al.* Influence of cultivation sites on sterol, nitrate, total phenolic contents and antioxidant activity in endive and stem chicory edible products. Int J Food Sci Nutr 2017; 68(1): 52-64.
[http://dx.doi.org/10.1080/09637486.2016.1221386] [PMID: 27575665]

[106] Hung HC, Joshipura KJ, Jiang R, *et al.* Fruit and vegetable intake and risk of major chronic disease. J Natl Cancer Inst 2004; 96(21): 1577-84.
[http://dx.doi.org/10.1093/jnci/djh296] [PMID: 15523086]

[107] Jones DP. Redefining oxidative stress. Antioxid Redox Signal 2006; 8(9-10): 1865-79.
[http://dx.doi.org/10.1089/ars.2006.8.1865] [PMID: 16987039]

[108] Apak R, Gorinstein S, Böhm V, Schaich Karen M, Özyürek M, Güçlü K. Methods of measurement and evaluation of natural antioxidant capacity/activity (IUPAC Technical Report). Pure Appl Chem 2013; 85(5): 957-8.
[http://dx.doi.org/10.1351/PAC-REP-12-07-15]

[109] Sahan Y, Gurbuz O, Guldas M, Degirmencioglu N, Begenirbas A. Phenolics, antioxidant capacity and bioaccessibility of chicory varieties (*Cichorium* spp.) grown in Turkey. Food Chem 2017; 217: 483-9.
[http://dx.doi.org/10.1016/j.foodchem.2016.08.108] [PMID: 27664662]

[110] Schaffer S, Schmitt-Schillig S, Muller WE, Eckert GP. Antioxidant properties of Mediterranean food plant extracts: Geographical differences Journal of physiology and pharmacology: an official journal of the Pol Physiol Soc 2005; 56 (Suppl 1): 115-24.

[111] Stratil P, Klejdus B, Kubán V. Determination of total content of phenolic compounds and their antioxidant activity in vegetables--evaluation of spectrophotometric methods. J Agric Food Chem 2006; 54(3): 607-16.
[http://dx.doi.org/10.1021/jf052334j] [PMID: 16448157]

[112] Pompella A, Sies H, Wacker R, *et al.* The use of total antioxidant capacity as surrogate marker for food quality and its effect on health is to be discouraged. Nutrition 2014; 30(7-8): 791-3.
[http://dx.doi.org/10.1016/j.nut.2013.12.002] [PMID: 24984994]

[113] López-Alarcón C, Denicola A. Evaluating the antioxidant capacity of natural products: a review on chemical and cellular-based assays. Anal Chim Acta 2013; 763: 1-10.

[http://dx.doi.org/10.1016/j.aca.2012.11.051] [PMID: 23340280]

[114] Song W, Derito CM, Liu MK, He X, Dong M, Liu RH. Cellular antioxidant activity of common vegetables. J Agric Food Chem 2010; 58(11): 6621-9.
[http://dx.doi.org/10.1021/jf9035832] [PMID: 20462192]

[115] Naveed M, Hejazi V, Abbas M, *et al.* Chlorogenic acid (CGA): A pharmacological review and call for further research. Biomed Pharmacother 2018; 97: 67-74.
[http://dx.doi.org/10.1016/j.biopha.2017.10.064] [PMID: 29080460]

[116] Muthusamy VS, Saravanababu C, Ramanathan M, *et al.* Inhibition of protein tyrosine phosphatase 1B and regulation of insulin signalling markers by caffeoyl derivatives of chicory (*Cichorium intybus*) salad leaves. Br J Nutr 2010; 104(6): 813-23.
[http://dx.doi.org/10.1017/S0007114510001480] [PMID: 20444318]

[117] Tsai YL, Chiu CC, Yi-Fu Chen J, Chan KC, Lin SD. Cytotoxic effects of Echinacea purpurea flower extracts and cichoric acid on human colon cancer cells through induction of apoptosis. J Ethnopharmacol 2012; 143(3): 914-9.
[http://dx.doi.org/10.1016/j.jep.2012.08.032] [PMID: 22971663]

[118] Tousch D, Lajoix AD, Hosy E, *et al.* Chicoric acid, a new compound able to enhance insulin release and glucose uptake. Biochem Biophys Res Commun 2008; 377(1): 131-5.
[http://dx.doi.org/10.1016/j.bbrc.2008.09.088] [PMID: 18834859]

[119] Lee YJ, Lim T, Han MS, *et al.* Anticancer effect of luteolin is mediated by downregulation of TAM receptor tyrosine kinases, but not interleukin-8, in non-small cell lung cancer cells. Oncol Rep 2017; 37(2): 1219-26.
[http://dx.doi.org/10.3892/or.2016.5336] [PMID: 28035396]

[120] Zhu S, Xu T, Luo Y, *et al.* Luteolin Enhances Sarcoplasmic Reticulum Ca2+-ATPase Activity through p38 MAPK Signaling thus Improving Rat Cardiac Function after Ischemia/Reperfusion. Cell Physiol Biochem 2017; 41(3): 999-1010.
[http://dx.doi.org/10.1159/000460837] [PMID: 28222421]

[121] Cui XX, Yang X, Wang HJ, *et al.* Luteolin-7-O-Glucoside present in lettuce extracts inhibits hepatitis B surface antigen production and viral replication by human hepatoma cells *in vitro*. Front Microbiol 2017; 8: 2425.
[http://dx.doi.org/10.3389/fmicb.2017.02425] [PMID: 29270164]

[122] Chikara S, Nagaprashantha LD, Singhal J, Horne D, Awasthi S, Singhal SS. Oxidative stress and dietary phytochemicals: Role in cancer chemoprevention and treatment. Cancer Lett 2018; 413: 122-34.
[http://dx.doi.org/10.1016/j.canlet.2017.11.002] [PMID: 29113871]

[123] Borghi SM, Mizokami SS, Pinho-Ribeiro FA, *et al.* The flavonoid quercetin inhibits titanium dioxide (TiO$_2$)-induced chronic arthritis in mice. J Nutr Biochem 2018; 53: 81-95.
[http://dx.doi.org/10.1016/j.jnutbio.2017.10.010] [PMID: 29197723]

[124] Gao FJ, Zhang SH, Xu P, *et al.* Quercetin declines apoptosis, ameliorates mitochondrial function and improves retinal ganglion cell survival and function in *in vitro* model of glaucoma in rat and retinal ganglion cell culture *in vitro*. Front Mol Neurosci 2017; 10: 285.
[http://dx.doi.org/10.3389/fnmol.2017.00285] [PMID: 28936163]

[125] Kashyap D, Sharma A, Tuli HS, Sak K, Punia S, Mukherjee TK. Kaempferol – A dietary anticancer molecule with multiple mechanisms of action: Recent trends and advancements. J Funct Foods 2017; 30: 203-19.
[http://dx.doi.org/10.1016/j.jff.2017.01.022]

[126] Chen CJ, Deng AJ, Liu C, Shi R, Qin HL, Wang AP. Hepatoprotective activity of *Cichorium endivia* L. extract and its chemical constituents. Molecules 2011; 16(11): 9049-66.
[http://dx.doi.org/10.3390/molecules16119049] [PMID: 22033140]

[127] Munagala R, Aqil F, Jeyabalan J, *et al*. Exosomal formulation of anthocyanidins against multiple cancer types. Cancer Lett 2017; 393: 94-102.
[http://dx.doi.org/10.1016/j.canlet.2017.02.004] [PMID: 28202351]

[128] Huang PC, Wang GJ, Fan MJ, *et al*. Cellular apoptosis and cardiac dysfunction in STZ-induced diabetic rats attenuated by anthocyanins *via* activation of IGFI-R/PI3K/Akt survival signaling. Environ Toxicol 2017; 32(12): 2471-80.
[http://dx.doi.org/10.1002/tox.22460] [PMID: 28856781]

[129] Yao Y, Chen Y, Adili R, *et al*. Plant-based Food Cyanidin-3-Glucoside Modulates Human Platelet Glycoprotein VI Signaling and Inhibits Platelet Activation and Thrombus Formation. J Nutr 2017; 147(10): 1917-25.
[http://dx.doi.org/10.3945/jn.116.245944] [PMID: 28855423]

[130] Qi ZL, Wang Z, Li W, *et al*. Nephroprotective Effects of Anthocyanin from the Fruits of Panax ginseng (GFA) on Cisplatin-Induced Acute Kidney Injury in Mice. Phytother Res 2017; 31(9): 1400-9.
[http://dx.doi.org/10.1002/ptr.5867] [PMID: 28731262]

[131] Winter AN, Ross EK, Wilkins HM, *et al*. An anthocyanin-enriched extract from strawberries delays disease onset and extends survival in the hSOD1(G93A) mouse model of amyotrophic lateral sclerosis. Nutr Neurosci 2018; 21(6): 414-26.
[PMID: 28276271]

[132] Bahonar A, Saadatnia M, Khorvash F, Maracy M, Khosravi A. Carotenoids as Potential Antioxidant Agents in Stroke Prevention: A Systematic Review. Int J Prev Med 2017; 8: 70.
[http://dx.doi.org/10.4103/ijpvm.IJPVM_112_17] [PMID: 28983399]

[133] von Lintig J. Provitamin A metabolism and functions in mammalian biology. Am J Clin Nutr 2012; 96(5): 1234S-44S.
[http://dx.doi.org/10.3945/ajcn.112.034629] [PMID: 23053549]

[134] Park Y, Choi J, Lim JW, Kim H. β-Carotene-induced apoptosis is mediated with loss of Ku proteins in gastric cancer AGS cells. Genes Nutr 2015; 10(4): 467.
[http://dx.doi.org/10.1007/s12263-015-0467-1] [PMID: 25981694]

[135] Gloria NF, Soares N, Brand C, Oliveira FL, Borojevic R, Teodoro AJ. Lycopene and beta-carotene induce cell-cycle arrest and apoptosis in human breast cancer cell lines. Anticancer Res 2014; 34(3): 1377-86.
[PMID: 24596385]

[136] Druesne-Pecollo N, Latino-Martel P, Norat T, *et al*. Beta-carotene supplementation and cancer risk: a systematic review and metaanalysis of randomized controlled trials. Int J Cancer 2010; 127(1): 172-84.
[http://dx.doi.org/10.1002/ijc.25008] [PMID: 19876916]

[137] Abdel-Aal SM, Akhtar H, Zaheer K, Ali R. Dietary sources of lutein and zeaxanthin carotenoids and their role in eye health. Nutrients 2013; 5(4): 1169-85.
[http://dx.doi.org/10.3390/nu5041169] [PMID: 23571649]

[138] Ma L, Dou HL, Wu YQ, *et al*. Lutein and zeaxanthin intake and the risk of age-related macular degeneration: a systematic review and meta-analysis. Br J Nutr 2012; 107(3): 350-9.
[http://dx.doi.org/10.1017/S0007114511004260] [PMID: 21899805]

[139] Stringham JM, Stringham NT, O'Brien KJ. Macular Carotenoid Supplementation Improves Visual Performance, Sleep Quality, and Adverse Physical Symptoms in Those with High Screen Time Exposure. Foods 2017; 6(7): 6.
[http://dx.doi.org/10.3390/foods6070047] [PMID: 28661438]

[140] Neelam K, Goenadi CJ, Lun K, Yip CC, Au Eong KG. Putative protective role of lutein and zeaxanthin in diabetic retinopathy. Br J Ophthalmol 2017; 101(5): 551-8.
[http://dx.doi.org/10.1136/bjophthalmol-2016-309814] [PMID: 28232380]

[141] Chung RWS, Leanderson P, Lundberg AK, Jonasson L. Lutein exerts anti-inflammatory effects in patients with coronary artery disease. Atherosclerosis 2017; 262: 87-93.
[http://dx.doi.org/10.1016/j.atherosclerosis.2017.05.008] [PMID: 28527371]

[142] Chang JM, CHen WC, Hong D, Lin JK. The inhibition of DMBA-induced carcinogenesis by neoxanthin in hamster buccal pouch. Nutr Cancer 1995; 24(3): 325-33.
[http://dx.doi.org/10.1080/01635589509514421] [PMID: 8610051]

[143] Soontornchaiboon W, Joo SS, Kim SM. Anti-inflammatory effects of violaxanthin isolated from microalga Chlorella ellipsoidea in RAW 264.7 macrophages. Biol Pharm Bull 2012; 35(7): 1137-44.
[http://dx.doi.org/10.1248/bpb.b12-00187] [PMID: 22791163]

[144] Gopal SS, Lakshmi MJ, Sharavana G, Sathaiah G, Sreerama YN, Baskaran V. Lactucaxanthin - a potential anti-diabetic carotenoid from lettuce (*Lactuca sativa*) inhibits α-amylase and α-glucosidase activity *in vitro* and in diabetic rats. Food Funct 2017; 8(3): 1124-31.
[http://dx.doi.org/10.1039/C6FO01655C] [PMID: 28170007]

[145] Ren Y, Yu J, Kinghorn AD. Development of Anticancer Agents from Plant-Derived Sesquiterpene Lactones. Curr Med Chem 2016; 23(23): 2397-420.
[http://dx.doi.org/10.2174/0929867323666160510123255] [PMID: 27160533]

[146] Amorim MH, Gil da Costa RM, Lopes C, Bastos MM. Sesquiterpene lactones: adverse health effects and toxicity mechanisms. Crit Rev Toxicol 2013; 43(7): 559-79.
[http://dx.doi.org/10.3109/10408444.2013.813905] [PMID: 23875764]

[147] Wesołowska A, Nikiforuk A, Michalska K, Kisiel W, Chojnacka-Wójcik E. Analgesic and sedative activities of lactucin and some lactucin-like guaianolides in mice. J Ethnopharmacol 2006; 107(2): 254-8.
[http://dx.doi.org/10.1016/j.jep.2006.03.003] [PMID: 16621374]

[148] Cavin C, Delannoy M, Malnoe A, *et al.* Inhibition of the expression and activity of cyclooxygenase-2 by chicory extract. Biochem Biophys Res Commun 2005; 327(3): 742-9.
[http://dx.doi.org/10.1016/j.bbrc.2004.12.061] [PMID: 15649409]

[149] Fernández-Navarro T, Salazar N, Gutiérrez-Díaz I, *et al.* Bioactive compounds from regular diet and faecal microbial metabolites. Eur J Nutr 2018; 57(2): 487-97.
[http://dx.doi.org/10.1007/s00394-016-1332-8] [PMID: 27744545]

[150] La Frano MR, de Moura FF, Boy E, Lönnerdal B, Burri BJ. Bioavailability of iron, zinc, and provitamin A carotenoids in biofortified staple crops. Nutr Rev 2014; 72(5): 289-307.
[http://dx.doi.org/10.1111/nure.12108] [PMID: 24689451]

[151] Courraud J, Berger J, Cristol J-P, Avallone S. Stability and bioaccessibility of different forms of carotenoids and vitamin A during *in vitro* digestion. Food Chem 2013; 136(2): 871-7.
[http://dx.doi.org/10.1016/j.foodchem.2012.08.076] [PMID: 23122139]

[152] Ozdal T, Sela DA, Xiao J, Boyacioglu D, Chen F, Capanoglu E. The Reciprocal Interactions between Polyphenols and Gut Microbiota and Effects on Bioaccessibility. Nutrients 2016; 8(2): 78.
[http://dx.doi.org/10.3390/nu8020078] [PMID: 26861391]

[153] Juániz I, Ludwig IA, Bresciani L, *et al.* Bioaccessibility of (poly)phenolic compounds of raw and cooked cardoon (*Cynara cardunculus* L.) after simulated gastrointestinal digestion and fermentation by human colonic microbiota. J Funct Foods 2017; 32: 195-207.
[http://dx.doi.org/10.1016/j.jff.2017.02.033]

[154] Daly T, Jiwan MA, O'Brien NM, Aherne SA. Carotenoid content of commonly consumed herbs and assessment of their bioaccessibility using an *in vitro* digestion model. Plant Foods Hum Nutr 2010; 65(2): 164-9.
[http://dx.doi.org/10.1007/s11130-010-0167-3] [PMID: 20443063]

[155] Granado-Lorencio F, Olmedilla-Alonso B, Herrero-Barbudo C, Pérez-Sacristan B, Blanco-Navarro I, Blazquez-García S. Comparative *in vitro* bioaccessibility of carotenoids from relevant contributors to

carotenoid intake. J Agric Food Chem 2007; 55(15): 6387-94.
[http://dx.doi.org/10.1021/jf070301t] [PMID: 17595101]

[156] Reboul E, Richelle M, Perrot E, Desmoulins-Malezet C, Pirisi V, Borel P. Bioaccessibility of carotenoids and vitamin E from their main dietary sources. J Agric Food Chem 2006; 54(23): 8749-55.
[http://dx.doi.org/10.1021/jf061818s] [PMID: 17090117]

[157] Bohn T. Dietary factors affecting polyphenol bioavailability. Nutr Rev 2014; 72(7): 429-52.
[http://dx.doi.org/10.1111/nure.12114] [PMID: 24828476]

[158] Bel-Rhlid R, Pagé-Zoerkler N, Fumeaux R, *et al.* Hydrolysis of chicoric and caftaric acids with esterases and *Lactobacillus johnsonii in vitro* and in a gastrointestinal model. J Agric Food Chem 2012; 60(36): 9236-41.
[http://dx.doi.org/10.1021/jf301317h] [PMID: 22920606]

[159] Kopec RE, Failla ML. Recent advances in the bioaccessibility and bioavailability of carotenoids and effects of other dietary lipophiles. J Food Compos Anal 2017.

[160] Borel P, Preveraud D, Desmarchelier C. Bioavailability of vitamin E in humans: an update. Nutr Rev 2013; 71(6): 319-31.
[http://dx.doi.org/10.1111/nure.12026] [PMID: 23731443]

[161] van Het Hof KH, West CE, Weststrate JA, Hautvast JGAJ. Dietary factors that affect the bioavailability of carotenoids. J Nutr 2000; 130(3): 503-6.
[http://dx.doi.org/10.1093/jn/130.3.503] [PMID: 10702576]

[162] Alves-Rodrigues A, Shao A. The science behind lutein. Toxicol Lett 2004; 150(1): 57-83.
[http://dx.doi.org/10.1016/j.toxlet.2003.10.031] [PMID: 15068825]

[163] Lodge JK. Vitamin E bioavailability in humans. J Plant Physiol 2005; 162(7): 790-6.
[http://dx.doi.org/10.1016/j.jplph.2005.04.012] [PMID: 16008106]

[164] Fernández-García E, Carvajal-Lérida I, Jarén-Galán M, Garrido-Fernández J, Pérez-Gálvez A, Hornero-Méndez D. Carotenoids bioavailability from foods: From plant pigments to efficient biological activities. Food Res Int 2012; 46: 438-50.
[http://dx.doi.org/10.1016/j.foodres.2011.06.007]

[165] Chaves JS, Leal PC, Pianowisky L, Calixto JB. Pharmacokinetics and tissue distribution of the sesquiterpene alpha-humulene in mice. Planta Med 2008; 74(14): 1678-83.
[http://dx.doi.org/10.1055/s-0028-1088307] [PMID: 18951339]

[166] Qigui L, Peter JW, Wilbur KM. Pharmacokinetic and Pharmacodynamic Profiles of Rapid-Acting Artemisinins in the Antimalarial Therapy. Curr Drug Ther 2007; 2: 210-23.
[http://dx.doi.org/10.2174/157488507781695649]

[167] Bergantin C, Maietti A, Cavazzini A, *et al.* Bioaccessibility and HPLC-MS/MS chemical characterization of phenolic antioxidants in Red Chicory (*Cichorium intybus*). J Funct Foods 2017; 33: 94-102.
[http://dx.doi.org/10.1016/j.jff.2017.02.037]

[168] Gonthier MP, Verny MA, Besson C, Rémésy C, Scalbert A. Chlorogenic acid bioavailability largely depends on its metabolism by the gut microflora in rats. J Nutr 2003; 133(6): 1853-9.
[http://dx.doi.org/10.1093/jn/133.6.1853] [PMID: 12771329]

[169] Olthof MR, Hollman PCH, Katan MB. Chlorogenic acid and caffeic acid are absorbed in humans. J Nutr 2001; 131(1): 66-71.
[http://dx.doi.org/10.1093/jn/131.1.66] [PMID: 11208940]

[170] Adam A, Crespy V, Levrat-Verny M-A, *et al.* The bioavailability of ferulic acid is governed primarily by the food matrix rather than its metabolism in intestine and liver in rats. J Nutr 2002; 132(7): 1962-8.
[http://dx.doi.org/10.1093/jn/132.7.1962] [PMID: 12097677]

[171] Kaushik D, O'Fallon K, Clarkson PM, Dunne CP, Conca KR, Michniak-Kohn B. Comparison of

quercetin pharmacokinetics following oral supplementation in humans. J Food Sci 2012; 77(11): H231-8.
[http://dx.doi.org/10.1111/j.1750-3841.2012.02934.x] [PMID: 23094941]

[172] Meyer H, Bolarinwa A, Wolfram G, Linseisen J. Bioavailability of apigenin from apiin-rich parsley in humans. Ann Nutr Metab 2006; 50(3): 167-72.
[http://dx.doi.org/10.1159/000090736] [PMID: 16407641]

[173] Ding SM, Zhang ZH, Song J, Cheng XD, Jiang J, Jia XB. Enhanced bioavailability of apigenin *via* preparation of a carbon nanopowder solid dispersion. Int J Nanomedicine 2014; 9: 2327-33.
[http://dx.doi.org/10.2147/IJN.S60938] [PMID: 24872695]

[174] Barve A, Chen C, Hebbar V, Desiderio J, Saw CL, Kong AN. Metabolism, oral bioavailability and pharmacokinetics of chemopreventive kaempferol in rats. Biopharm Drug Dispos 2009; 30(7): 356-65.
[http://dx.doi.org/10.1002/bdd.677] [PMID: 19722166]

[175] Felgines C, Krisa S, Mauray A, *et al.* Radiolabelled cyanidin 3-O-glucoside is poorly absorbed in the mouse. Br J Nutr 2010; 103(12): 1738-45.
[http://dx.doi.org/10.1017/S0007114510000061] [PMID: 20187984]

[176] Tang G, Serfaty-Lacrosniere C, Camilo ME, Russell RM. Gastric acidity influences the blood response to a beta-carotene dose in humans. Am J Clin Nutr 1996; 64(4): 622-6.
[http://dx.doi.org/10.1093/ajcn/64.4.622] [PMID: 8839509]

[177] Garrett DA, Failla ML, Sarama RJ. Development of an *in vitro* digestion method to assess carotenoid bioavailability from meals. J Agric Food Chem 1999; 47(10): 4301-9.
[http://dx.doi.org/10.1021/jf9903298] [PMID: 10552806]

[178] Demmig-Adams B, Adams Iii WW, Barker DH, Logan BA, Bowling DR, Verhoeven AS. Using chlorophyll fluorescence to assess the fraction of absorbed light allocated to thermal dissipation of excess excitation. Physiol Plant 1996; 98: 253-64.
[http://dx.doi.org/10.1034/j.1399-3054.1996.980206.x]

[179] Xiao Z, Lester GE, Luo Y, Wang Q. Assessment of vitamin and carotenoid concentrations of emerging food products: edible microgreens. J Agric Food Chem 2012; 60(31): 7644-51.
[http://dx.doi.org/10.1021/jf300459b] [PMID: 22812633]

[180] Bumgarner NR, Scheerens JC, Mullen RW, Bennett MA, Ling PP, Kleinhenz MD. Root-zone temperature and nitrogen affect the yield and secondary metabolite concentration of fall- and spring-grown, high-density leaf lettuce. J Sci Food Agric 2012; 92(1): 116-24.
[http://dx.doi.org/10.1002/jsfa.4549] [PMID: 21842529]

[181] Gazula A, Kleinhenz MD, Streeter JG, Miller AR. Temperature and Cultivar Effects on Anthocyanin and Chlorophyll b Concentrations in Three Related Lollo Rosso Lettuce Cultivars. HortScience 2005; 40: 1731-3.

[182] Baslam M, Garmendia I, Goicoechea N. The arbuscular mycorrhizal symbiosis can overcome reductions in yield and nutritional quality in greenhouse-lettuces cultivated at inappropriate growing seasons. Sci Hortic (Amsterdam) 2013; 164: 145-54.
[http://dx.doi.org/10.1016/j.scienta.2013.09.021]

[183] Kleinhenz MD, Gazula A, Scheerens JC, French DG. Variety, Shading, and Growth Stage Effects on Pigment Concentrations in Lettuce Grown under Contrasting Temperature Regimens. Horttechnology 2003; 13: 677-83.

[184] Krizek DT, Britz SJ, Mirecki RM. Inhibitory effects of ambient levels of solar UV-A and UV-B radiation on growth of cv. New Red Fire lettuce. Physiol Plant 1998; 103: 1-7.
[http://dx.doi.org/10.1034/j.1399-3054.1998.1030101.x]

[185] García-Macías P, Ordidge M, Vysini E, *et al.* Changes in the flavonoid and phenolic acid contents and antioxidant activity of red leaf lettuce (Lollo Rosso) due to cultivation under plastic films varying in ultraviolet transparency. J Agric Food Chem 2007; 55(25): 10168-72.

[http://dx.doi.org/10.1021/jf071570m] [PMID: 18001028]

[186] Amoozgar A, Mohammadi A, Sabzalian MR. Impact of light-emitting diode irradiation on photosynthesis, phytochemical composition and mineral element content of lettuce cv. Grizzly. Photosynthetica 2017; 55: 85-95.
[http://dx.doi.org/10.1007/s11099-016-0216-8]

[187] Li Q, Kubota C. Effects of supplemental light quality on growth and phytochemicals of baby leaf lettuce. Environ Exp Bot 2009; 67: 59-64.
[http://dx.doi.org/10.1016/j.envexpbot.2009.06.011]

[188] Giedre S, Ausra B, Ramunas S, Algirdas N, Pavelas D. Supplementary red-LED lighting affects phytochemicals and nitrate of baby leaf lettuce. J Food Agric Environ 2011; 9: 271-4.

[189] Garrido Y, Tudela JA, Marín A, Mestre T, Martínez V, Gil MI. Physiological, phytochemical and structural changes of multi-leaf lettuce caused by salt stress. J Sci Food Agric 2014; 94(8): 1592-9.
[http://dx.doi.org/10.1002/jsfa.6462] [PMID: 24170602]

[190] Luna MC, Martínez-Sánchez A, Selma MV, Tudela JA, Baixauli C, Gil MI. Influence of nutrient solutions in an open-field soilless system on the quality characteristics and shelf life of fresh-cut red and green lettuces (*Lactuca sativa* L.) in different seasons. J Sci Food Agric 2013; 93(2): 415-21.
[http://dx.doi.org/10.1002/jsfa.5777] [PMID: 22806347]

[191] Avio L, Sbrana C, Giovannetti M, Frassinetti S. Arbuscular mycorrhizal fungi affect total phenolics content and antioxidant activity in leaves of oak leaf lettuce varieties. Sci Hortic (Amsterdam) 2017; 224: 265-71.
[http://dx.doi.org/10.1016/j.scienta.2017.06.022]

[192] Rodriguez-Amaya DB, Kimura M, Godoy HT, Amaya-Farfan J. Updated Brazilian database on food carotenoids: Factors affecting carotenoid composition. J Food Compos Anal 2008; 21: 445-63.
[http://dx.doi.org/10.1016/j.jfca.2008.04.001]

[193] He SY, Feng GP, Yang HS, Wu Y, Li YF. Effects of pressure reduction rate on quality and ultrastructure of iceberg lettuce after vacuum cooling and storage. Postharvest Biol Technol 2004; 33: 263-73.
[http://dx.doi.org/10.1016/j.postharvbio.2004.03.006]

[194] Albrecht JA. Ascorbic acid content and retention in lettuce. J Food Qual 1993; 16: 311-6.
[http://dx.doi.org/10.1111/j.1745-4557.1993.tb00116.x]

[195] Rivera JRE, Stone MB, Stushnoff C, Pilon-Smits E, Kendall PA. Effects of Ascorbic Acid Applied by Two Hydrocooling Methods on Physical and Chemical Properties of Green Leaf Lettuce Stored at 5°C. J Food Sci 2006; 71: S270-6.
[http://dx.doi.org/10.1111/j.1365-2621.2006.tb15653.x]

[196] Alarcón-Flores MI, Romero-González R, Martínez Vidal JL, Egea González FJ, Garrido Frenich A. Monitoring of phytochemicals in fresh and fresh-cut vegetables: a comparison. Food Chem 2014; 142: 392-9.
[http://dx.doi.org/10.1016/j.foodchem.2013.07.065] [PMID: 24001857]

[197] Ke D, Saltveit ME. Wound-induced ethylene production, phenolic metabolism and susceptibility to russet spotting in iceberg lettuce. Physiol Plant 1989; 76: 412-8.
[http://dx.doi.org/10.1111/j.1399-3054.1989.tb06212.x]

[198] Tomás-Barberán FA, Loaiza-Velarde J, Bonfanti A, Saltveit ME. Early Wound- and Ethylene-induced Changes in Phenylpropanoid Metabolism in Harvested Lettuce. Am Soc Hortic Sci 1997; 122: 399-404.

[199] Tomás-Barberán FA, Gil MI, Castañer M, Artés F, Saltveit ME. Effect of Selected Browning Inhibitors on Phenolic Metabolism in Stem Tissue of Harvested Lettuce. J Agric Food Chem 1997; 45: 583-9.
[http://dx.doi.org/10.1021/jf960478f]

[200] Ke D, Saltveit MEJ. Effects of calcium and auxin on russet spotting and phenylalanine ammonia-lyase activity in iceberg lettuce. J Am Soc Hortic Sci 1986; 21(5): 1169-71.

[201] Hyodo H, Kuroda H, Yang SF. Induction of phenylalanine ammonia-lyase and increase in phenolics in lettuce leaves in relation to the development of russet spotting caused by ethylene. Plant Physiol 1978; 62(1): 31-5.
[http://dx.doi.org/10.1104/pp.62.1.31] [PMID: 16660463]

[202] López-Gálvez G, Saltveit M, Cantwell M. The visual quality of minimally processed lettuces stored in air or controlled atmosphere with emphasis on romaine and iceberg types. Postharvest Biol Technol 1996; 8: 179-90.
[http://dx.doi.org/10.1016/0925-5214(95)00002-X]

[203] Siriphanich J, Kader AA. Effects of CO2 on total phenolics, phenylalanine ammonia lyase, and polyphenol oxidase in lettuce tissue. J Am Soc Hortic Sci 1985; 110(2): 149-53.

[204] Gil MI, Castaner M, Ferreres F, Artes F, Tomas-Barberan FA. Modified-atmosphere packaging of minimally processed "Lollo Rosso" (*Lactuca sativa*) - Phenolic metabolites and quality changes. Z Lebensm Unters F A 1998; 206: 350-4.
[http://dx.doi.org/10.1007/s002170050271]

[205] Cefola M. Phenolic profiles and postharvest quality changes of fresh-cut radicchio (*Cichorium intybus* L.): nutrient value in fresh *vs.* stored leaves. Subt Plant Sci 2016; 51: 76-84.

[206] Mayer-Miebach E, Gärtner U, Großmann B, Wolf W, Spieß WEL. Influence of low temperature blanching on the content of valuable substances and sensory properties in ready-to-use salads. J Food Eng 2003; 56: 215-7.
[http://dx.doi.org/10.1016/S0260-8774(02)00254-6]

[207] Renna M, Gonnella M, Giannino D, Santamaria P. Quality evaluation of cook-chilled chicory stems (*Cichorium intybus* L., Catalogna group) by conventional and sous vide cooking methods. J Sci Food Agric 2014; 94(4): 656-65.
[http://dx.doi.org/10.1002/jsfa.6302] [PMID: 23847094]

[208] Jiménez-Monreal AM, García-Diz L, Martínez-Tomé M, Mariscal M, Murcia MA. Influence of cooking methods on antioxidant activity of vegetables. J Food Sci 2009; 74(3): H97-H103.
[http://dx.doi.org/10.1111/j.1750-3841.2009.01091.x] [PMID: 19397724]

[209] Schweiggert U, Schieber A, Carle R. Effects of blanching and storage on capsaicinoid stability and peroxidase activity of hot chili peppers (*Capsicum frutescens* L.). Innov Food Sci Emerg Technol 2006; 7: 217-24.
[http://dx.doi.org/10.1016/j.ifset.2006.03.003]

[210] Effect of cooking on the capsaicinoids and phenolics contents of Mexican peppers. Food Chem 2010; 119: 1619-25.
[http://dx.doi.org/10.1016/j.foodchem.2009.09.054]

[211] Bunea A, Andjelkovic M, Socaciu C, *et al.* Total and individual carotenoids and phenolic acids content in fresh, refrigerated and processed spinach (*Spinacia oleracea* L.). Food Chem 2008; 108(2): 649-56.
[http://dx.doi.org/10.1016/j.foodchem.2007.11.056] [PMID: 26059144]

[212] Zhang D, Hamauzu Y. Phenolics, ascorbic acid, carotenoids and antioxidant activity of broccoli and their changes during conventional and microwave cooking. Food Chem 2004; 88: 503-9.
[http://dx.doi.org/10.1016/j.foodchem.2004.01.065]

[213] Lebeda A, Ryder EJ, Grube R, Dolezalova I, Kristkova E. Lettuce (Asteraceae; Lactuca spp.). Genetic resources, chromosome engineering, and crop improvement: vegetable crops. CRC press - Taylor & Francis group 2006; pp. 377-472.

[214] Reyes-Chin-Wo S, Wang Z, Yang X, *et al.* Genome assembly with *in vitro* proximity ligation data and whole-genome triplication in lettuce. Nat Commun 2017; 8: 14953.

[http://dx.doi.org/10.1038/ncomms14953] [PMID: 28401891]

[215] Truco MJ, Ashrafi H, Kozik A, *et al.* An ultra-high-density, transcript-based, genetic map of lettuce. G3 (Bethesda) 2013; 3(4): 617-31.
[http://dx.doi.org/10.1534/g3.112.004929] [PMID: 23550116]

[216] Truco MJ, Antonise R, Lavelle D, *et al.* A high-density, integrated genetic linkage map of lettuce (*Lactuca* spp.). Theor Appl Genet 2007; 115(6): 735-46.
[http://dx.doi.org/10.1007/s00122-007-0599-9] [PMID: 17828385]

[217] Raulier P, Maudoux O, Notte C, Draye X, Bertin P. Exploration of genetic diversity within *Cichorium endivia* and *Cichorium intybus* with focus on the gene pool of industrial chicory. Genet Resour Crop Evol 2016; 63: 243-59.
[http://dx.doi.org/10.1007/s10722-015-0244-4]

[218] Kiær LP, Felber F, Flavell A, *et al.* Spontaneous gene flow and population structure in wild and cultivated chicory, *Cichorium intybus* L. Genet Resour Crop Evol 2009; 56: 405-19.
[http://dx.doi.org/10.1007/s10722-008-9375-1]

[219] Barcaccia G, Ghedina A, Lucchin M. Current advances in genomics and breeding of leaf chicory (*Cichorium intybus* L.). Agriculture 2016; 6: 50.
[http://dx.doi.org/10.3390/agriculture6040050]

[220] Galla G, Ghedina A, Tiozzo SC, Barcaccia G. Toward a First High-quality Genome Draft for Marker-assisted Breeding in Leaf Chicory, Radicchio (Cichorium intybus). Rijeka, Plant Genomics 2016.

[221] Lucchin M, Varotto S, Barcaccia G, Parrini P. Chicory and Endive.Vegetables I: Asteraceae, Brassicaceae, Chenopodicaceae, and Cucurbitaceae. New York, NY: Springer New York 2008; pp. 3-48.
[http://dx.doi.org/10.1007/978-0-387-30443-4_1]

[222] Bernardes EC, Benko-Iseppon AM, Vasconcelos S, Carvalho R, Brasileiro-Vidal AC. Intra- and interspecific chromosome polymorphisms in cultivated *Cichorium* L. species (*Asteraceae*). Genet Mol Biol 2013; 36(3): 357-63.
[http://dx.doi.org/10.1590/S1415-47572013005000025] [PMID: 24130443]

[223] Hodgins KA, Lai Z, Oliveira LO, *et al.* Genomics of *Compositae* crops: reference transcriptome assemblies and evidence of hybridization with wild relatives. Mol Ecol Resour 2014; 14(1): 166-77.
[http://dx.doi.org/10.1111/1755-0998.12163] [PMID: 24103297]

[224] Cadalen T, Morchen M, Blassiau C, *et al.* Development of SSR markers and construction of a consensus genetic map for chicory (*Cichorium intybus* L.). Mol Breed 2010; 25: 699-722.
[http://dx.doi.org/10.1007/s11032-009-9369-5]

[225] Gonthier L, Blassiau C, Mörchen M, *et al.* High-density genetic maps for loci involved in *nuclear male sterility* (*NMS1*) and s*porophytic self-incompatibility* (*S-locus*) in chicory (*Cichorium intybus* L., *Asteraceae*). Theor Appl Genet 2013; 126(8): 2103-21.
[http://dx.doi.org/10.1007/s00122-013-2122-9] [PMID: 23689744]

[226] Hayashi E, Aoyama N, Still DW. Quantitative trait loci associated with lettuce seed germination under different temperature and light environments. Genome 2008; 51(11): 928-47.
[http://dx.doi.org/10.1139/G08-077] [PMID: 18956026]

[227] Damerum A, Selmes SL, Biggi GF, *et al.* Elucidating the genetic basis of antioxidant status in lettuce (*Lactuca sativa*). Hortic Res 2015; 2: 15055.
[http://dx.doi.org/10.1038/hortres.2015.55] [PMID: 26640696]

[228] Van Stallen N, Vandenbussche B, Londers E, Noten V, De Proft M. QTL analysis of production and taste characteristics of chicory (*Cichorium intybus* var. *foliosum*). Plant Breeding 2005; 124: 59-62.
[http://dx.doi.org/10.1111/j.1439-0523.2004.01043.x]

[229] Curtis IS, Power JB, Blackhall NW, Delaat AMM, Davey MR. Genotype-Independent Transformation of Lettuce Using *Agrobacterium-Tumefaciens.* J Exp Bot 1994; 45: 1441-9.

[http://dx.doi.org/10.1093/jxb/45.10.1441]

[230] Armas I, Pogrebnyak N, Raskin I. A rapid and efficient *in vitro* regeneration system for lettuce (*Lactuca sativa* L.). Plant Methods 2017; 13: 58.
[http://dx.doi.org/10.1186/s13007-017-0208-0] [PMID: 28736573]

[231] Vasseur J, Dubois J, Hilbert JL, Couillerot JP. Somatic embryogenesis in chicory (Cichorium Species).Somatic Embryogenesis and Synthetic Seed II. Berlin, Heidelberg: Springer Berlin Heidelberg 1995; pp. 125-37.
[http://dx.doi.org/10.1007/978-3-642-78643-3_11]

[232] Rafsanjani MS, Alvari A, Mohammad A, Abdin MZ, Hejazi MA. *in vitro* propagation of *Cichorium intybus* L. and quantification of enhanced secondary metabolite (esculin). Recent Pat Biotechnol 2011; 5(3): 227-34.
[http://dx.doi.org/10.2174/187220811797579123] [PMID: 22360470]

[233] Malarz J, Stojakowska A, Kisiel W. Long-term cultured hairy roots of chicory-a rich source of hydroxycinnamates and 8-deoxylactucin glucoside. Appl Biochem Biotechnol 2013; 171(7): 1589-601.
[http://dx.doi.org/10.1007/s12010-013-0446-1] [PMID: 23975347]

[234] Stojakowska A, Malarz J. Bioactive phenolics from *in vitro* cultures of *Lactuca aculeata* Boiss. et Kotschy. Phytochem Lett 2017; 19: 7-11.
[http://dx.doi.org/10.1016/j.phytol.2016.11.003]

[235] Ismail H, Dilshad E, Waheed MT, Mirza B. Transformation of Lettuce with rol ABC Genes: Extracts Show Enhanced Antioxidant, Analgesic, Anti-Inflammatory, Antidepressant, and Anticoagulant Activities in Rats. Appl Biochem Biotechnol 2017; 181(3): 1179-98.
[http://dx.doi.org/10.1007/s12010-016-2277-3] [PMID: 27734289]

[236] Ismail H, Dilshad E, Waheed MT, Sajid M, Kayani WK, Mirza B. Transformation of Lactuca sativa L. with rol C gene results in increased antioxidant potential and enhanced analgesic, anti-inflammatory and antidepressant activities *in vivo*. 3 Biotech 2016; 6

[237] Gundlach H, Müller MJ, Kutchan TM, Zenk MH. Jasmonic acid is a signal transducer in elicitor-induced plant cell cultures. Proc Natl Acad Sci USA 1992; 89(6): 2389-93.
[http://dx.doi.org/10.1073/pnas.89.6.2389] [PMID: 11607285]

[238] Malarz J, Kisiel W. Effect of pectinase on the production of sesquiterpene lactones in the hairy root culture of *Lactuca virosa* L. Acta Soc Bot Pol 2000; 69: 115-7.
[http://dx.doi.org/10.5586/asbp.2000.014]

[239] Azarmehr B, Karimi F, Taghizadeh M, Gargari SLM. Secondary metabolite contents and antioxidant enzyme activities of *Cichorium intybus* hairy roots in response to zinc. J Med Plants By-Prod 2013; 2: 131-8.

[240] Bais HP, Sudha G, George J, Ravishankar GA. Influence of exogenous hormones on growth and secondary metabolite production in hairy root cultures of Cichorium intybus L cv. Lucknow Local. In Vitro Cell Dev-Pl 2001; 37: pp. 293-9.

[241] Karuppusamy S. A review on trends in production of secondary metabolites from higher plants by *in vitro* tissue, organ and cell cultures. J Med Plants Res 2009; 3: 1222-39.

[242] Matvieieva NA, Vasylenko MY, Shahovsky AM, Bannykova MO, Kvasko OY, Kuchuk NV. Effective *Agrobacterium*-mediated transformation of chicory (*Cichorium intybus* L.) with *Mycobacterium tuberculosis* antigene ESAT6. Cytol Genet 2011; 45: 7-12.
[http://dx.doi.org/10.3103/S0095452711010038]

[243] Giannino D, Nicolodi C, Testone G, *et al.* The overexpression of *asparagine synthetase A* from *E-coli* affects the nitrogen status in leaves of lettuce (*Lactuca sativa* L.) and enhances vegetative growth. Euphytica 2008; 162: 11-22.
[http://dx.doi.org/10.1007/s10681-007-9506-3]

[244] Tang Y, Fu X, Shen Q, Tang K. Roles of MPBQ-MT in promoting alpha/gamma-tocopherol production and photosynthesis under high light in lettuce. PLoS One 2016; 11(2): e0148490.
[http://dx.doi.org/10.1371/journal.pone.0148490] [PMID: 26867015]

[245] Li Y, Wang G, Hou R, *et al.* Engineering tocopherol biosynthetic pathway in lettuce. Biol Plant 2011; 55: 453-60.
[http://dx.doi.org/10.1007/s10535-011-0110-y]

[246] Cho EA, Lee CA, Kim YS, Baek SH, de los Reyes BG, Yun SJ. Expression of gamma-tocopherol methyltransferase transgene improves tocopherol composition in lettuce (*Latuca sativa* L.). Mol Cells 2005; 19(1): 16-22.
[PMID: 15750335]

[247] Han EH, Lee JS, Lee JW, Chung IS, Lee YH. Transgenic lettuce expressing *chalcone isomerase* gene of chinese cabbage increased levels of flavonoids and polyphenols. Kor J Hortic Sci 2011; 29: 467-73.

[248] Fu X, Guo X, You L, *et al.* Co-transformation of *Atpsy* and *folE* genes elevated carotenoids and folic acid contents in Romanie lettuce. J STGU Agric Sci 2012; 30: 22-35.

[249] Adem M, Beyene D, Feyissa T. Recent achievements obtained by chloroplast transformation. Plant Methods 2017; 13: 30.
[http://dx.doi.org/10.1186/s13007-017-0179-1] [PMID: 28428810]

[250] Yabuta Y, Tanaka H, Yoshimura S, *et al.* Improvement of vitamin E quality and quantity in tobacco and lettuce by chloroplast genetic engineering. Transgenic Res 2013; 22(2): 391-402.
[http://dx.doi.org/10.1007/s11248-012-9656-5] [PMID: 22990376]

[251] Harada H, Maoka T, Osawa A, *et al.* Construction of transplastomic lettuce (*Lactuca sativa*) dominantly producing astaxanthin fatty acid esters and detailed chemical analysis of generated carotenoids. Transgenic Res 2014; 23(2): 303-15.
[http://dx.doi.org/10.1007/s11248-013-9750-3] [PMID: 24287848]

[252] Woo JW, Kim J, Kwon SI, *et al.* DNA-free genome editing in plants with preassembled CRISPR-Cas9 ribonucleoproteins. Nat Biotechnol 2015; 33(11): 1162-4.
[http://dx.doi.org/10.1038/nbt.3389] [PMID: 26479191]

CHAPTER 6

Headspace Analysis of Volatile Compounds From Fruits of Selected Vegetable Species of Apiaceae Family

Milica G. Aćimović[1,*], Mirjana T. Cvetković[2], Jovana M. Stanković[2], Vele V. Tešević[3] and Marina M. Todosijević[3]

[1] *Institute of Field and Vegetable Crops Novi Sad, Maksima Gorkog 30, 21000 Novi Sad, Serbia*

[2] *Institute of Chemistry, Technology and Metallurgy, University of Belgrade, Njegoševa 12, 11000 Belgrade, Serbia*

[3] *Faculty of Chemistry, University of Belgrade, Studentski trg 12-16, 11000 Belgrade, Serbia*

Abstract: Parsley (*Petroselinum crispum* L.), celery (*Apium graveolens* L.), celeriac (*Apium graveolens* var. *rapaceum*), carrot (*Daucus carota* L.), parsnip (*Pastinaca sativa* L.), lovage (*Levisticum officinale* Koch.) and angelica (*Angelica archangelica* L.) are vegetable plants belonging to the Apiaceae family. They are often used as spices due to their characteristic aroma, originating from the volatile compounds present in the plant tissues. Mainly, all parts of the plant *i.e.* roots, leaves and fruit are used in nutrition. However, the focus of this chapter is plant fruit (*i.e.* seed), which is mostly used as spice. The contemporary method used for the analysis of volatiles compounds is called headspace and it is widely applied in flavor chemistry. The dominant compounds in *P. crispum* are α-pinene (46.2-49.0%) and β-pinene (33.5-35.4%), while in *A. graveolens,* it is limonene (84.1-94.4%). In *D. carota,* the main components are sabinene (28.3%) and α-pinene (25.0%), while in *P. sativa* fruit, it is octyl ester of butanoic acid (53.8%) and 1-octanol (27.6%). In *L. officinale* and *A. archangelica,* the dominant component in fruit is *β*-phellandrene (77.1% and 84.7%, respectively).

Keywords: Angelica, Apiaceae, Aroma, Celery, Celeriac, Carrot, Headspace, Lovage, Parsley, Parsnip, Terpenes, Volatile compounds.

INTRODUCTION

The analysis of the headspace (HS) volatiles above food stuffs by gas chromatography or gas chromatography-mass spectrometry (GC-MS) has been widely applied in flavor chemistry [1]. Since characteristic aroma is an important

---
[*] **Corresponding author Milica G. Aćimović:** Institute of Field and Vegetable Crops, Novi Sad, Serbia; Tel: +021780-365111; E-mail: acimovicbabicmilica@gmail.com

Spyridon A. Petropoulos, Isabel C.F.R. Ferreira and Lillian Barros (Eds.)

quality criterion for many vegetables and spices, application of HS is interesting because this method is a non-destructive and solvent-free method for collecting volatile compounds produced in plants. However, in everyday nutrition aromatic spices and vegetables are consumed directly from food (the reason why HS profile of volatile compounds is important), while extracts and essential oils are used in the food industry.

When consuming fresh plants, it should be noted that the chemical composition of extracted essential oil is not identical to the composition in the plant prior to extraction. The plant itself in its natural habitat doesnot have the same smell as the extracted essential oil. During the hydro-distillation, the compounds found in the plant will undergo certain chemical changes: water vapor can destroy aldehydes, esters of different acids can be created during evaporation, while certain soluble molecules can be trapped in the liquid phase. All of these change the profile of the essential oil and hence, the aroma of the extracted oil.

Parsley (*Petroselinum crispum* L.), celery (*Apium graveolens* L.), celeriac (*Apium graveolens* var. *rapaceum*), carrot (*Daucus carota* L.), parsnip (*Pastinaca sativa* L.), lovage (*Levisticum officinale* Koch.) and angelica (*Angelica archangelica* L.) are vegetable plants from Apiaceae family. They are often used as spices due to their characteristic aroma, originating from the volatile compounds present in the various plant tissues. Although they are biennial plants, they are commercially grown as annual because of their thick root and leaves. During the second year, these plant species form rosette stalks, topped with divided leaves, and inflorescences with flowers formed in umbels. Fruits are schizocarps split into two mericarps (seeds).

Parsley (*Petroselinum crispum* L.)

There are two parsley cultivars: root (ssp. *tuberosum*) and leaf vegetable (ssp. *foliosum*). The root in root parsley (*P. crispum* var. *tuberosum*) is thick, tapered in shape, with different lengths and shapes depending on the variety. The root in leaf parsley has rootlets. The root in both cultivars is white or yellowish. The leaf rosette in root parsley consists of 15-20 leaves, while in the leaf parsley it forms up to 180 leaves during vegetation. The leaf is complex, made up of two or more leaflets with slightly ridged edges. The leaves can be smooth and flat as in Italian parsley (*P. crispum* var. *neapolitanum*) or ruffled as in French parsley (*P. crispum* var. *crispum*) [2].

Parsley leaf is used as a garnish (for salads, soups, boiled potatoes and egg dishes), blended in dips, cooked sauces and stews, while the root is used as a vegetable to enhance soup flavor, in stews and condiments [3]. Essential oil can be found in all plant parts: fruit 3-7%, leaves 0.16-0.3% and root 0.1% [4].

Parsley essential oil (*Petroselini aetheroleum*) obtained from grounded ripe fruits (*Petroselini fructus*) is a colorless, yellowish or chartreuse liquid, denser than water (specific density 1.043-1.110), with a parsley like odor and taste originating from apiole and myristicine, the dominant constituents in root and leaf essential oil [5]. Parsley fruit is schizocarp, oval, flat on the side with five dark ribs. It consists of two equal mericarps. The seed is small, up to 3.5 mm long and up to 2 mm wide. The 1000 seed mass is around 1 g [6].

Headspace analysis of parsley seed var. *neapolitanum* ("Domaći lišćar", NS seme, Serbia) detected 35 volatile compounds (Table **1**), among which α-pinene (49.0%), β-pinene (35.4%), β-phellandrene (8.8%) and myristicine (1.2%) as the dominant compounds accounting for 94.4% of total detected oils. Other compounds were present in less than 1% or in traces, among them six were not identified (NI). According to the results, the most dominant class among volatile compounds were monoterpenes [7].

Table 1. Volatile compounds from *Petroselinum crispum* var. *neapolitanum* fruits.

No	Compound name	R.t. [min]	RI exp.	RI lit.	[%]
1	NI	5.172	909	/	tr
2	α-thujene	5.605	925	924	0.4
3	α-pinene	5.975	939	932	48.9
4	camphene	6.201	947	946	1.1
5	sabinene	6.896	971	969	0.6
6	β-pinene	7.161	982	974	35.4
7	myrcene	7.352	989	988	0.8
8	α-phellandrene	7.779	1004	1002	0.3
9	δ-3-karene	7.973	1009	1008	tr
10	α-terpinene	8.198	1015	1014	0.1
11	p-cymene	8.447	1022	1020	0.3
12	β-phellandrene	8.715	1029	1025	8.8
13	cis-β-ocimene	8.892	1034	1032	tr
14	trans-β-ocimene	9.245	1044	1044	0.1
15	γ-terpinene	9.646	1055	1054	0.5
16	terpinolene	10.772	1086	1086	0.1
17	NI	11.139	1096	/	tr
18	α-campholenal	12.281	1123	1122	tr
19	trans-pinocarveole	12.781	1135	1134	tr

(Table 1) cont.....

No	Compound name	R.t. [min]	RI exp.	RI lit.	[%]
20	NI	13.063	1141	/	tr
21	NI	13.713	1160	/	tr
22	pinocarvone	13.800	1163	1161	tr
23	myrtenal	15.199	1190	1195	0.3
24	daucene	23.354	1376	1380	tr
25	NI	23.607	1381	/	tr
26	ethyl ester of butanoic acid	23.810	1386	1394	0.1
27	trans-caryophyllene	25.062	1415	1417	tr
28	trans-α-bergamotene	25.799	1433	1432	tr
29	trans-β-farnesene	26.740	1455	1454	tr
30	dauca-5.8-dien	27.265	1468	1471	tr
31	myristicine	29.470	1521	1517	1.2
32	elemicin	30.853	1554	1555	0.2
33	octyl ester of hexanoic acid	31.969	1581	1575	tr
34	6-metoksielemicin	32.532	1594	1595	0.3
35	apiole	35.797	1678	1677	0.4

R.t. – Retention time, RI – Retention index experimental, RI lit. – Retention index literature reported, NI– not identified, tr – detected in traces (less than 0.1%), / – not detected

Headspace analysis of parsley var. *tuberosum* ("NS Molski", NS seme, Serbia) detected 19 volatile compounds among which one was not identified (NI) (Table 2). The main compounds were: α-pinene (46.2%), β-pinene (33.5%), apiole (8.0%), γ-terpinene (2.2%), myristicine (2.1%), β-phellandrene (2.0%). As observed, the most abundant compounds were monoterpenes.

Table 2. Volatile compounds from *Petroselinum crispum* var. *tuberosum* fruits.

No	Compound name	R.t. [min]	RI exp.	RI lit.	[%]
1	tricyclene	5,533	922	921	tr
2	α-thujene	5,607	926	924	0.4
3	α-pinene	5,808	933	932	46.2
4	camphene	6,200	946	946	1.1
5	sabinene	6,842	970	969	1.7
6	β-pinene	6,968	975	974	33.5
7	myrcene	7,346	990	988	0.1
8	α-phellandrene	7,799	1004	1002	0.1

(Table 2) cont.....

No	Compound name	R.t. [min]	RI exp.	RI lit.	[%]
9	α-terpinene	8,212	1016	1014	0.1
10	p-cymene	8,429	1022	1020	0.7
11	β-phellandrene	8,583	1026	1025	2.0
12	γ-terpinene	9,638	1055	1054	2.2
13	terpinolene	10,827	1088	1086	tr
14	myrtenal	15,200	1190	1195	1.5
15	NI	22,092	1347	/	tr
16	myristicine	29,397	1519	1517	2.1
17	elemicine	30,944	1556	1555	0.1
18	6-metoxyelemicine	32,548	1595	1594	0.2
19	apiole	35,837	1680	1677	8.0

R.t. – Retention time, RI – Retention index experimental, RI lit. – Retention index literature reported, NI– not identified, tr – detected in traces (less than 0.1%), / – not detected

It can be seen from the tables (Table **2**), that in both varieties, the dominant compounds are α-pinene (49.0% in var. *neapolitanum*, and 46.2% in var. *tuberosum*), and β-pinene (35.4% and 33.5% in var. *neapolitanum* and var. *tuberosum*, respectively). The chemical compositions of parsley seed essential oils from 11 cultivars (three of leaf parsley and eight of root parsley) obtained from different commercial sources, marketed in Poland, were also compared. There were significant differences between quantities of the oil components between cultivars. The content of phenylpropene derivatives in the seed essential oils permits distinction between convar. *crispum* (leaf parsley) with main constituent being myristicine, and convar. *radicosum* (root parsley) with main constituent being apiole [8]. Investigations show that parsley seed from Iran contains mainly myristicine (42.65%), β-phellandrene (21.83%), p-1,3,8-menthatriene (9.97%), and β-myrcene (4.25%) [9], while main compounds from Egyptian parsley are myristicine (25.2%), apiole (18.23%) and α-pinene (16.16%) [10].

A wide range of pharmacological activity has been reported for this plant, in traditional medicine, as well as in modern phytotherapy [11]. In the past few years, the focus was on antimicrobial and antioxidative activity. The investigations of antimicrobial activities of parsley essential oil against foodborne diseases and opportunistic pathogens show that parsley essential oil had bacteriostatic activity against: *Staphylococcus aureus*, *Listeria monocytogenes* and *Salmonella enterica*, as well as fungistatic activity against fungi: *Penicillium ochrochloron* and *Trichoderma viride*. This indicated the antimicrobial activity of parsley essential oil as an alternative natural antimicrobial agent [12]. Apart from this, parsley has health-promoting properties with the potential to prevent oxidative stress-related diseases and can be developed into functional food [13]. Parsley oil and its two major components (myristicine and apiole) can be potential alternative natural antioxidants [14].

Parsley is used for the treatment of urinary tract problems, such as urinary tract infections [15]. Additionally, parsley inhibits urinary tract stone formation [16], reduces the dysfunction in kidney caused by prostadin-induced abortion and could have beneficial effect in reducing the progression of prostaglandin-induced edema [17]. Parsley has beneficial effects on liver diseases [3]. Parsley showed hepatoprotective effect against acute liver injury induced by carbon tetra chloride [18, 19]. The hepatoprotective effect of parsley was proven in a few studies conducted on liver injury caused by either paracetamol [20] or sodium valproate [21], as well as injuries caused by complication of diabetes [22]. Furthermore, myristicin from the essential oil induced the activity of glutathione S-transferase enzyme in the liver [23]. Parsley can be used to treat gastrointestinal disorders, such as spasms, because it relaxes the ileum [24]. It is also effective against reducing stress-induced gastric injury by supporting the cellular antioxidant defence system [25].

Celery (*Apium graveolens* L.)

Similar to parsley, there are three celery cultivars: celeriac (*A. graveolens* var. *rapaceum*), with its thick root being used as vegetable, sweet celery (*A. graveolens* var. *dulce*), with thick leaf stalks which are consumed as salad vegetables or in stews, and leaf celery (*A. graveolens* var. *secalinum*), where its fresh or dried leaves are used as spice [26]. The root is white, well developed, and thick in the upper part with a great number of rootlets in the lower part. This gives it the bearded appearance (celeriac) or shallowroot(leaf celery and sweet celery) [2]. Celery leaves are bigger than parsley leaves; they are smooth and shiny, pinnate to bipinnate with rhombic leaflets. Celery fruit (*Apii fructus*) is a brown schizocarp, 1.0-1.5 mm long, 1.0 mm in diameter. It spontaneously divides itself into two parts. Thousand seed mass is 0.2-0.5 g [27].

Celery root, leaf and leaf stalk are used as a vegetable, fresh for salads, preparation of juices, in meat dishes, snacks, gravies, sauces, while the seed is used in pickling vegetables, salad dressings, baked goods, biscuits, soups, celery salt and bouquet garnish [3]. The entire plant contains essential oil (*Apii aetheroleum*), however the chemical composition differs depending on the plant part, influencing the flavor as well [28]. Fresh celery root contains under 0.02% of essential oil, leaves about 0.1%, while the seed contains 2.5–3% [29, 30]. Celery root has a distinctive odor and taste, sweet at the beginning, then pungent and a little bit bitter, similar to pepper, due to phtalidesfrom essential oil [31].

Sedanolide, sedanonicanhydride and 3-n-butil phthalideare the main components of the root essential oil [32]. Celery fruit and leaves have a fresh citrusy odor and a spicy nutty flavor. The dominant components in the leaves are limonene and β-pinene, while limonene and β-selineneare dominant in the fruit [33].

Twenty six components were isolated from the leaf celery fruit ("Tall Utah 52-70", Agrounikum, Serbia), the dominant ones being: limonene (84.1%), β-selinene (9.1%), β-pinene (1.6%) and pentyl-benzene (1.0%) (Table **3**). These four constituents comprise 95.8% of total oil, while other components comprise less than 1%, among which three components were unidentified [7].

An investigation conducted on celeriac fruit ("Praški", NS seme, Serbia) by applying headspace analysis determined 17 volatile compounds (Table **4**), among which the dominant one was limonene with 94.4%, followed by myrcene (2.2%) and β-pinene (1.6%), while other compounds were present in small amounts (<1%). The dominant class were monoterpenes, whereas sedanenolide was detected in low amounts (0.2%), and 3-n-butilphthalide and sedanolide were present in traces [28].

As it can be seen from the Tables (**1**) and (**4**), our investigations showed that in fruit of both celery varieties the most abundant component was limonene (with 94.4% and 84.1% in var. *rapaceum* and var. *dulce*, respectively), followed by myrcene and β-pinene in var. *rapaceum*, while β-selinene, β-pinene and pentyl-benzene were detected in higher amounts in var. *dulce*.

Table 3. Volatile compounds from *Apium graveolens* var. *dulce* fruits.

	Compound name	R.t. [min]	RI exp.	RI lit.	[%]
1	α-pinene	5.780	932	932	0.1
2	camphene	6.194	947	946	tr
3	sabinene	6.844	971	969	tr
4	β-pinene	6.933	974	974	1.6
5	myrcene	7.313	988	988	0.8
6	α-phellandrene	7.848	1004	1002	tr
7	α-terpinene	8.265	1017	1014	0.3
8	limonene	8.750	1031	1024	84.1
9	γ-terpinene	9.740	1057	1054	0.1
10	terpinolene	10.802	1087	1086	0.4
11	trans-p-menta-2,8-dien-1-ol	12.029	1117	1119	0.1
12	cis-p-menta-2,8-dien-1-ol	12.618	1131	1133	0.1

(Table 3) cont.....

	Compound name	R.t. [min]	RI exp.	RI lit.	[%]
13	trans-d-limonene oxide	12.760	1134	1137	0.1
14	pentyl-benzene	13.538	1152	1152	1.0
15	NI	15.318	1193	/	tr
16	trans-pinocarvyl-acetate	19.827	1296	1298	0.1
17	β-elemene	23.921	1389	1389	tr
18	trans-caryophyllene	25.070	1415	1417	tr
19	β-selinene	28.007	1485	1489	9.1
20	δ-selinene	28.157	1489	1492	0.1
21	α-selinene	28.316	1493	1498	1.2
22	NI	29.642	1524	/	0.1
23	NI	29.981	1533	/	tr
24	butyl phthalide	34.719	1652	1647	0.2
25	apiole	35.835	1680	1677	tr
26	sedanenolide	37.249	1719	1719	0.2

R.t. – Retention time, RI – Retention index experimental, RI lit. – Retention index literature reported, NI– not identified, tr – detected in traces (less than 0.1%), / – not detected

Table 4. Volatile compounds from *Apium graveolens* var. *rapaceum* fruits.

No	Compound name	R.t. [min]	RI exp.	RI lit.	[%]
1	α-thujene	5.609	925	924	tr
2	α-pinene	5.789	932	932	0.2
3	sabinene	6.861	966	969	0.1
4	β-pinene	6.970	974	974	1.6
5	myrcene	7.344	988	988	2.2
6	limonene	8.744	1025	1024	94.4
7	linalool	11.260	1094	1095	tr
8	1-octen-3-yl acetate	11.759	1106	1110	tr
9	cis-limonene oxide	12.673	1131	1132	tr
10	trans-limonene oxide	12.822	1135	1137	0.1
11	pentyl-benzene	13.643	1154	1152	0.2
12	pentyl cyclohexa-1,3-diene	13.661	1155	1156	0.5
13	β-selinene	28.025	1486	1489	0.2
14	α-selinene	28.401	1494	1498	tr
15	3-n-butylphthalide	34.724	1651	1647	tr
16	sedanenolide	37.295	1719	1719	0.2

(Table 4) cont.....

No	Compound name	R.t. [min]	RI exp.	RI lit.	[%]
17	sedanolide	37.503	1725	1722	tr

R.t. – Retention time, RI – Retention index experimental, RI lit. – Retention index literature reported, tr – detected in traces (less than 0.1%)

Investigations conducted in India, showed that limonene (with about 60%) was the main constituent, followed by selinene (20%) [34]. However, the important flavor constituents of the oil responsible for the specific aroma are 3-*n*-butyl-4- 5-dihydrophthalide (sedanenolide), 3-*n*-butyl phthalide, sedanolide, and sedanonic anhydride present in very low levels (1-3%). Phthalides, especially sedanenolide, possess many health benefits.

Experiments in Egypt with celery seed essential oil obtained by hydro-distillation using a Clevenger-type apparatus, also show that limonene (58.38%) is the main constituent, followed by β-selinene (27.03%), α-selinene (2.67%) and myrcene (2.33%) [33].

Celery seed has the potential to be developed into dietary supplements and nutraceuticals [35]. It possesses many nutraceutical properties, such as antioxidant [36], as well as antimicrobial potential [33]. Many studies indicated that celery exhibited good anti-oxidant activity [37 - 41]. Because of this it has a potent role in ameliorating stressful complications accompanied by diabetes mellitus [42]. Furthermore, it is shown that celery possesses anticonvulsant activity which is accompanied by the antioxidant effect in the brain. Because of this, celery can be used for prevention of epileptic seizures [43]. Antimicrobial activity of celery can be used to treat urinary tract infections caused by antibiotic resistant bacteria such as Escherichia coli and *Pseudomonas aeruginosa* [44].

Apart from this, celery has hypolipidemic [45], hypoglycemic [46], anti-inflammatory [47] and anti-platelet aggregation properties [48]. It also expresses significant hepatoprotective [3], antiulcerogenic [32] and spasmolytic activity [49]. Celery has been known to inhibit the growth of a variety of cancer tissues. The antiproliferative effect exerted by a celery ethanolic extract is triggered by induction of apoptosis. It is also demonstrated that vascular endothelial growth factor (signal protein produced by cells that stimulates the formation of blood vessels) expression was regulated with celery extract treatment [50]. Chemoprevention, which includes the use of celery seeds, has been assessed. Celery seed possess cytotoxicity and provokes DNA fragmentation, a sign of induction of apoptosis. Therefore, this plant may be a potential candidate in the field of anticancer drug discovery [51].

Carrot (*Daucus carota* L.)

Cultivated carrot (*D. carota* L. ssp. *sativus* (Hoffm.) Arcang.) originates from the wild carrot (*Daucus carota* L. ssp. *carota*). The color of the root in the cultivated varieties ranges from white, yellow, orange, light purple or deep red to deep violet. The shape varies from short stumps to tapering cones. Leaves are finely dissected, twice or thrice-pinnate, segments are linear to lanceolate, 0.5-3 cm long. Upper leaves are reduced, with a sheathing petiole. Stem is striate or ridged, glabrous to hispid, up to 1 m tall. Flowers are borne in compound, more or less globose, up to 7 cm in diameter umbels. Rays are numerous, bracts 1-2 pinnated, lobes linear, with 7-10 bracteoles similar to bracts. Flowers are white or yellowish; the outer are usually the largest. Sepals are minute or absent, there are five petals and stamens, ovary inferior with two cells and one ovule per cell, two styles. Fruit are oblong, with bristly hairs along ribs, 2-4 mm long [52].

Raw carrot root is used for juices, salads, cakes, or for pickling, while cooked root is used in casseroles, soups and stews. Seed is mainly used for essential oil distillation, and the obtained oil is used in beverages, baked goods, condiments, relishes, meat products. Apart from seed, essential oil can also be obtained from aerial parts of carrot, *i.e.* when carrot is in the flowering stage. This oil is sometimes used as a flavoring agent in food products and in the cosmetics industry.

The composition of volatile compounds of carrot fruit ("Nantes", NS seme, Serbia) is shown in Table (**5**), obtained by headspace extraction. The percentage of each constituent is also shown in the same table. Headspace method has proven itself to be good for isolating monoterpenes as they have a lower boiling point, *i.e.* lower partition coefficients and higher concentration in gas phase. In accordance with this, the main components are sabinene (28.3%), α-pinene (25.0%), myrcene (5.9%), limonene (4.6%), β-pinene (3.5%) and carotol (3.4%). Carotol is the only sesquiterpene among these components.

Table 5. Volatile compounds from *Daucus carota* ssp. *sativus* fruits.

No	Compound name	R.t. [min]	RI exp.	RI lit.	[%]
1	tricyclene	5,525	922	921	tr
2	α-thujene	5,617	926	924	1.2
3	α-pinene	5,824	933	932	25.0
4	camphene	6,185	947	946	1.8
5	2,4 (10)-thujadien	6,317	951	953	0.3
6	sabinene	6,891	972	989	28.3

(Table 5) cont.....

No	Compound name	R.t. [min]	RI exp.	RI lit.	[%]
7	β-pinene	6,968	975	974	3.5
8	myrcene	7,330	989	988	5.9
9	α-phellandrene	7,804	1005	1002	0.1
10	α-terpinene	8,185	1015	1014	0.4
11	p-cymene	8,429	1022	1020	0.3
12	limonene	8,584	1026	1024	4.6
13	γ-terpinene	9,652	1055	1054	0.6
14	cis-sabinene hydrate	9,916	1062	1065	0.5
15	NI	10,537	1080	/	0.1
16	terpinolene	10,778	1086	1086	0.2
17	NI	11,103	1095	/	0.4
18	linalool	11,175	1097	1095	0.8
19	NI	11,438	1104	/	0.8
20	β-thujone	11,895	1114	1112	tr
21	NI	12,039	1118	/	tr
22	α-campholenal	12,241	1122	1122	0.2
23	cis-p-mentha-2,8-dien-1-ol	12,611	1131	1133	tr
24	NI	12,683	1133	/	tr
25	trans-pinokarveol	12,754	1134	1135	0.4
26	trans-verbenol	13,016	1140	1137	1.4
27	NI	13,456	1150	/	0.1
28	sabina ketone	13,532	1151	1154	0.5
29	pinocarvone	13,750	1157	1160	0.2
30	cis-sabinol	14,026	1163	*	0.1
31	Terpinene-4-ol	14,419	1172	1174	0.1
32	α-thujenal	14,739	1180	1181	tr
33	α-terpyneol	15,013	1186	1186	tr
34	myrtenal	15,213	1190	1195	0.4
35	verbenone	15,754	1203	1204	0.5
36	NI	16,271	1215	/	tr
37	carvone	17,409	1241	1239	tr
38	bornyl acetate	19,173	1280	1283	0.7
39	pinokarvyl acetate trans	19,826	1295	1298	tr
40	α-terpynil acetate	22,019	1345	1346	0.1

(Table 5) cont.....

No	Compound name	R.t. [min]	RI exp.	RI lit.	[%]
41	α-copaene	23,188	1372	1374	tr
42	daucene	23,355	1376	1380	1.0
43	geranyl acetate	23,544	1380	1379	1.5
44	NI	23,835	1386	/	0.1
45	NI	24,525	1402	/	0.1
46	cis-α-bergamotene	24,917	1412	1411	1.1
47	trans-caryophyllene	25,094	1416	1417	3.4
48	trans-α-bergamotene	25,794	1432	1432	1.4
49	NI	26,097	1440	/	0.4
50	NI	26,185	1442	/	0.1
51	epi-β-santalene	26,282	1444	1457	tr
52	α-humulene	26,523	1450	1452	0.1
53	trans-β-farnesene	26,727	1455	1457	3.9
54	NI	27,267	1468	/	0.2
55	γ-muurolen	27,710	1478	1478	0.1
56	β-selinene	27,942	1484	1489	2.2
57	α-selinene	28,307	1493	1498	0.2
58	isodaucene	28,447	1496	1500	0.1
59	β-bisabolene	28,888	1506	1505	0.7
60	β-sesquiphellandrene	29,524	1522	1521	0.2
61	caryophyllene oxide	31,882	1578	1582	0.2
62	carotol	32,486	1593	1594	3.4
63	daucol	34,130	1635	1641	tr
64	apiole	35,818	1680	1677	0.1

R.t. – Retention time, RI – Retention index experimental, RI lit. – Retention index literature reported, NI – not identified, tr – detected in traces (less than 0.1%), / – not detected

The major constituents of essential oil obtained by hydro-distillation from cultivated carrot seed ("Nantes", NS seme, Serbia) are carotol (22.0%), sabinene (19.6%) and α-pinene (13.2%). The mixture of aromadendrene, β-farnesane and sesquisabinene comprise 8.2%, the content of trans-caryophyllene is 5.7% while that of myrcene amounts to 4.7% [53].

The main compound in carrot seed essential oil cultivated in Turkey was carotol (66.78%), followed by daucene (8.74%) and α-farnesene (5.86%) [54]. Analysis of essential oil from different cultivars of carrot in Poland showed that the content of carotol in umbels depends on the cultivar and varies between 23 and 48% [55].

However, in Chinese *D. carota* ssp. *sativa* seed the dominant compounds are β-bisabolene (80.49%), α-asarone (8.8%) and *cis-α*-bergamoten (5.51%) [56].

Carrot is used in the traditional medicine for the treatment of variety of ailments and its beneficial health effects have been proven in phytotherapy. It has been used to treat hypertension [57]. However, results indicate that carrot may be acting through blockade of calcium channels and this effect may be responsible for the lowering blood pressure [58]. Carrot also expresses anti-inflammatory effects [59], antinociceptive properties [60] and anti-ulcerogenic potentials [61]. Hepatoprotective effects of carrot are also reported. The carrot could provide significant protection against carbon tetra chloride, thioacetamide, lindane, paracetamol, isoniazid and alcohol induced hepatocellular injury in animal models [3].

Strong antioxidant potential of carrot is also proven [62, 63], as well as its promising anticancer activity [64]. Carrot is one of the most commonly used functional foods, used on a daily basis by patients who suffer from prostate, breast and colorectal cancer [65]. Investigation results show that carrot possesses anti-proliferative effects and stimulates the differentiation of HL-60 cells – cells which exhibit phagocytic activity and responsiveness to chemotactic stimulus [66]. In addition, bioactive compounds from carrot could be considered in reaching a potential cancer therapy targeting proliferation and apoptosis, as well as cancer motility and metastasis [67]. Apart from this, the antioxidant potential of carrot seeds has contributed to the reduction of oxidative stress and lipid levels in experimental rats [68]. Application of carrot in hypercholesterolemic experimental animals' diet can significantly decrease total lipid, total cholesterol, triglycerides, low density lipoprotein cholesterol and liver enzymes [69]. Moreover, this plant possesses strong antimicrobial activity against fungi such as *Candida albicans* and *Alternaria alternate*, as well as bacteria (*Staphylococcus aureus*) [56, 70].

Parsnip (*Pastinaca sativa* L.)

Parsnip (*Pastinaca sativa* L., syn. *Peucedanum sativum* Benth. and Hook.) is a herbaceous biennial plant. During the first year, it forms a thick root and a leaf rosette. The alternate leaves are oddlypinnate, consisting of up to nine pairs of long leaflets coarsely toothed. The leaflets have soft hairs. The leaves can be up to 40 cm long. During the second year, a hollow stem with tiny hair develops. It is 50-150 cm tall with umbels and golden yellow flowers. The umbel diameter can range up to 15 cm. The plant blooms from July to August. The fruit is flat and chartreuse. Different cultivars are grown. Depending on the cultivar, the root can be conical, cylindrical or bulbous. It is cream-colored, rough and without shine.

The flesh and the heart are white. The presence of rootlets and eyes on the root depends on the cultivar. Rootlets are mostly located on the lower half, while the eyes are located on the upper half [2]. Domestic parsnip originates from the wild parsnip, still widespread throughout Europe and Asia. Prior to the potato being introduced to Europe from Middle America, parsnip root was the main source of carbohydrates in nutrition. Parsnip root consists of: galactose, mannose, arabinose, xylose, ramnose, and starch [71]. Apart from the root (*Pastinacae radix*), the fruit (*Pastinacae fructus*) and leaf (*Pastinacae folium*) are also used. The entire plant has the characteristic aromatic odor and flavorful sweet taste. The leaf and thick root are mainly used in nutrition in stews, as spice, fresh, dried and frozen. Parsnip fruit is also used as spice similar to other plants from Apiaceae family.

Table (6) shows the content of volatile compounds of parsnip fruit ("Dugi beli glatki" NS seme, Serbia) obtained by headspace extraction, as well as the percentage of each constituent. Out of 15 isolated components, one was not identified (Table 6). The most abundant compounds were: octyl ester of butanoic acid (53.8%), 1-octanol (27.6%), 1-octanol acetate (7.2%) and hexyl ester of butanoic acid (5.8%).

The investigation of *P. sativa* subsp. *urens* seed essential oil from Turkey detected 18 components comprising 95%, the most abundant being octylbutyrate (79.5%) and octylhexanoate (5.3%) [72]. According to GC-MS and GC-FID analyses, β-pinene was one of the major components of the root and stem headspace volatiles of the endemic Balkan parsnip *Pastinaca hirsuta* Pančić (50.6% and 24.1%, respectively). Trans-β-ocimene was found in a significant percentage in the stem and flowers volatiles (31.6% and 57.3%, respectively). The most abundant constituent of the fruit, flower and fruit essential oils and both extracts was hexyl butanoate (70.5%, 31.1%, 80.4%, 47.4% and 52.7%, respectively). Apiole, accompanied by myristicin and cis-falcarinol, make up over 70% of the root essential oils. γ-Palmitolactone was the major component of the stem essential oils (51.9% at the flowering stage and 45.7% at the fruiting stage) [73].

Table 6. Volatile compounds from *Pastinaca sativa* fruits.

No	Compound name	R.t. [min]	RI exp.	RI lit.	[%]
1	ethyl ester of butanoic acid	3,303	804	802	0.2
2	1-hexanol	4,282	863	863	1.4
3	butyl ester of butanoic acid	7,442	993	993	0.3
4	octanal	7,670	1001	998	0.9
5	hexyl ester of acetic acid	8,052	1011	1007	0.2

(Table 6) cont.....

No	Compound name	R.t. [min]	RI exp.	RI lit.	[%]
6	limonene	8,598	1027	1024	0.2
7	(Z) -3-octen-1-ol	9,761	1058	*	0.3
8	1-octanol	10,141	1069	1063	27.6
9	hexyl ester of butanoic acid	15,079	1187	1191	5.8
10	decanal	15,683	1201	1201	0.2
11	1-octanol acetate	15,949	1207	1211	7.2
12	NI	23,244	1373	/	0.7
13	hexyl ester of hexanoic acid	23,687	1383	1382	0.2
14	octyl ester of butanoic acid	23,973	1389	*	53.8
15	octyl ester of hexanoic acid	31,931	1580	*	1.0

R.t. – Retention time, RI – Retention index experimental, RI lit. – Retention index literature reported, NI – not identified, tr – detected in traces (less than 0.1%), * – Compound retention index not found in the literature.

Essential oil of areal parts of parsnip from Kopaonik was obtained by hydro-distillation with the most abundant constituents being: cis-β-ocimene (10.8%), hexyl-butyrate (10.4%), trans-β-farnesene (6.1%) and lavandulolacetate (5.2%) [74, 75]. By applying the modified dynamic headspace method on sorbent (PorapakQ), it was determined that the main components in parsnip umbels were: cis-ocimene(31.0%) and trans-ocimene (35.0%), metilantranilat (12.0%), trans-3-farnesene (5.5%) and γ-kadinen (5,0%) [76].

In traditional medicine, the parsnip fruit is used as a remedy, an appetite stimulant, for digestion, against bowel cramps and as a diuretic [77]. In addition to the common use of parsnip, traditional medicine validates its use as it reduces cramps [78]. Latest research shows that parsnip is a natural source of antimicrobic ageneses. Parsnip essential oil is effective against the most common human gastrointestinal pathogenic microbial strains: *Escherichia coli*, *Pseudomonas aeruginosa*, *Salmonella enteritidis*, *Bacillus cereus*, *Listeria monocytogenes*, *Staphylococcus aureus* and yeast *Candida albicans* [75]. Moreover, parsnip essential oils are used to reduce and eliminate *Candida* spp. population in human patients [79]. Apart from this, investigations show that parsnip is a good antioxidant agent. Therefore, it is suggested that parsnip, as a natural herb, could be used to extend the shelf life of meat products, providing the consumer with food containing natural additives, which might be seen as healthier than those of synthetic origin. In addition, parsnip could effectively reduce lipid oxidation, maintain or improve sensory attributes [80].

Lovage (*Levisticum officinale* Koch.)

Lovage is a perennial herbaceous plant with a well-developed spindle 40-50 cm long. The root is brown to light grey. A leaf rosette is formed during the first year. The leaves are smooth and shiny. There are three types of leaves on the plant. The basal leaves (leaves and leaf rosettes) are tripinnate, with long stalks and can be up to 60 cm long and 50 cm wide. The stem leaves are alternately located, less divided with shorter stalks; the top leaves are simple, sessile (epetiolate). Leaves wither after the first autumn frost. From the second year on, vegetation starts early in March. Stem formation begins at the end of May and the stalk can be up to 2 m tall. Infloresence is a complex umbel consisting of 10-15 secondary umbels with 3-5 yellow flowers. Lovage flowers from July to August. Fruit matures slow and unevenly during September and shed easily. Fruit is a flat yellow-brown schizocarp 4-7 mm long, consisting of two mericarps, each with five distinct ribs on the lower side. Thousand seed mass is 3-7g [81].

Lovage leaf is used for seasoning soups, sauces and meat dishes. The root is used for producing soup seasonings, finished flavorings in liqueurs and tobacco, while the seed is used as spice, for flavoring cakes, soups, salads and for pickled vegetables (especially cabbage and cucumbers) [3]. All parts of lovage plant contain essential oil (*Levistici aethroleum*) with a characteristic aroma similar to celery, described as sweet (like caramel or honey) [82]. Root (*Levistici radix*) contains 0.5-1% essential oil, leaves (*Levistici folium*) between 0.08 and 0.24%, fruits (*Levistici fructus*) 0.8-1.5%, while aboveground parts with unripe fruits contain 0.15-0.45% essential oil [81].

Headspace analysis of lovage fruits shows 21 compounds (Table **5**), the dominant was *β*-phellandrene with 77.1%, followed by *α*-pinene (4.7%), *α*-phellandrene (4.3%), sabinene (3.3%), myrcene (3.3%) and *cis-β*-ocimene (2.7%). Other compounds were present with less than 1%. The most abundant class were monoterpenes, while only one compound from phthalide class was present, *cis*-ligustilide, with 0.3% [28].

Table 7. Volatile compounds from *Levisticum officinale* fruits.

No	Compound name	R.t. [min]	RI exp.	RI lit.	[%]
1	α-thujene	5.609	925	924	0.5
2	α-pinene	5.789	932	932	4.7
3	camphene	6.189	941	946	0.6
4	sabinene	6.861	966	969	3.3
5	β-pinene	6.970	974	974	0.9

(Table 7) cont.....

No	Compound name	R.t. [min]	RI exp.	RI lit.	[%]
6	myrcene	7.344	988	988	3.3
7	α-phellandrene	7.800	1003	1002	4.3
8	δ-3-carene	7.999	1008	1008	0.1
9	α-terpinene	8.221	1012	1014	0.1
10	p-cymene	8.482	1022	1020	0.3
11	β-phellandrene	8.745	1025	1025	77.1
12	cis-β-ocymene	8.917	1030	1032	2.7
13	trans-β-ocymene	9.294	1040	1044	0.4
14	γ-terpinene	9.708	1056	1054	0.6
15	terpinolene	10.833	1084	1086	0.1
16	linalool	11.260	1094	1095	tr
17	pentyl cyclohexa-1,3-diene	13.661	1153	1156	0.3
18	terpinyl acetate	22.101	1343	1346	0.1
19	γ-muurolene	27.806	1480	1478	0.1
20	germacrene B	30.939	1555	1559	0.1
21	cis-ligustilide	37.831	1734	1734	0.3

R.t. – Retention time, RI – Retention index experimental, RI lit. – Retention index literature reported, tr – detected in traces (less than 0.1%)

Application of dynamic headspace (DHS) extraction method on volatile compounds from lovage fruits originating from Lithuania, determined that the content of β-phellandrene was 79.00%, p-cymene (2.29%), α-pinene (2.09%), α-phellandrene (1.73%), β-ocimene (1.47%), myrcene (1.45%), and limonene (1.02%), while other 20 compounds were present with less than 1% [83]. In this research cis-ligustilide was not detected.

Analysis of lovage fruit essential oil originating from Germany, obtained by hydrodistillation, showed that cis-ligustilide was present in a much higher concentration, between 4.0 and 7.6% [84]. Other components from phthalide class were also recorded: 3-n-butilidenfthalid (7.8–7.9%) and dihydroizobutilidenfthalide (0.3%). In this investigation, β-phellandrene was the dominant compound with 58.5–62.5%.

Lovage root is a potent diuretic; therefore, it is a common ingredient in tea infusions for urine secretion. It is also used as a tincture as well as a pharmaceutical drug [85]. In folk medicine, lovage is used against indigestion and respiratory problems as it is a mild expectorant [86]. However, recent investigations show that lovage have anti-inflammatory activity in carrageenan-induced oedema in experimental animals [87]. Clinical studies show that lovage leaf essential oil inhibits squamous cell carcinoma growth [88]. Moreover, lovage can be a source of active compounds as antiprolifferative therapeutic and antiangiogenic – since it reduces the growth and invasiveness of tumors [89]. However, there is clinical evidence that lovage essential oil causes contact dermatitis in sensitive people [90].

Lovage could be used for its biologically active compounds with antioxidative and antimicrobial properties to prevent lipid oxidation and for ensuring meat quality during cold storage [91]. Antioxidative and antimicrobial activity of lovage was reported by other authors, too [92 - 95]. On the other hand, another study showed that lovage, presumably due to its strong antioxidant properties, could protect against the destructive effects of paraquat on rat hepatocytes [96]. The results of biochemical tests, in which rats were treated with lovage essential oil, lovage fruit infusion and lovage herb infusion after intoxication with acrylamide (hepatic cytolysis and proteosynthesis indicators) highlighted the high antitoxic potential of the lovage volatile oil [97].

Angelica (*Angelica archangelica* L.)

Angelica is a biennial or perennial plant with a hollow, glabrous stem up to 1.5 m tall. It has large leaves (60-90 cm), 2-3 pinnate with wide lobes, and plane stalks which are violet at the base. Inflorescence is convex with a lot of small flowers gathered with white, greenish or pink petals, sometimes even yellowish. Fruit is compressed which gives an oblong to ovate or oblong shape with four wings and visual ridges. All the plant parts have resin canals containing essential oils. Roots contain 0.5-1.0%, fruits 0.6-1.5% and leaves contain 0.2-0.3% of essential oils [98]. Angelica root is used in herbal liqueurs and bitter spirits, in flavoring meat and canned vegetables, while the herb is used for decorating cakes and pastries, and to flavor jams and jellies, confectionaries and liqueurs. Chopped angelica leaf is added to fruit salads, fish dishes, and cottage cheese, while the seeds are used in alcoholic distillates [3].

It is found that *A. archangelica* seed consists of 29 components (Table **8**) among which the most dominant is β-phellandrene (84.7%), followed by α-phellandrene (3.4%), α-pinene (2.5%), myrcene (2.1%) and α-copaene (1.3%). All other compounds were present in less than 1% [99].

According to a study dealing with *A. archangelica* grown wild in Lithuania, the seed essential oil obtained by hydrodistillation contains 67 compounds, in which β-phellandrene was the dominant constituent (33.6-63.4%), followed by α-pinene (4.2-12.8%), α-phellandrene (2.6-7.4%), sabinene (2.5-4.6%) and *d*-germacrene (0.4-3.0%) [100]. The seed of *A. archangelica* from Siberian Region contains 21 compounds in the essential oil, with β-phellandrene (49.3%), α-pinene (15.1%), β-pinene (7.79%) and δ-3-carene (7.5%) as the main compounds [101]. The essential oil of Angelica seed from Iceland contains β-phellandrene (up to 55.2%) and α-pinene (14.4-41.4%) in different relative amounts depending on chemotype [102]. However, our investigations show that angelica roots from plants originating from Serbia contain 0.10% of essential oil with α-pinene (29.7%), δ-3-carene (14.2%), and a mixture of β-phellandrene and limonene (13.2%) as the main compounds [103].

Table 8. Volatile compounds from *Angelica archangelica* fruits.

No	Compound name	R.t. [min]	RI exp.	RI lit.	[%]
1	α-pinene	5.789	932	932	2.5
2	camphene	6.189	941	946	0.1
3	sabinene	6.861	966	969	0.4
4	β-pinene	6.970	974	974	0.4
5	myrcene	7.344	988	988	2.1
6	α-phellandrene	7.800	1003	1002	3.4
7	Δ3-carene	7.999	1008	1008	0.1
8	p-cymene	8.482	1022	1020	0.8
9	β-phellandrene	8.745	1025	1025	84.7
10	cis-β-ocymene	8.917	1030	1032	0.2
11	trans-β-ocymene	9.294	1040	1044	0.2
12	terpinolene	10.833	1084	1086	0.1
13	NI	14.416	1177	/	0.1
14	cryptone	14.896	1182	1183	0.1
15	NI	20.440	1313	/	0.1
16	NI	21.378	1321	/	0.1
17	NI	22.893	1366	/	0.1
18	α-copaene	23.277	1371	1374	1.3
19	β-bourbonene	23.672	1383	1387	0.1
20	trans-caryophyllene	25.160	1419	1417	0.1
21	β-copaene	25.583	1427	1430	0.2
22	α-humulene	26.624	1452	1452	0.6
23	γ-muurolene	27.806	1480	1478	0.4
24	NI	27.860	1481	/	0.1
25	α-zingiberene	28.421	1495	1493	0.4

(Table 8) cont.....

No	Compound name	R.t. [min]	RI exp.	RI lit.	[%]
26	α-muurolene	28.622	1500	1500	0.1
27	β-bisabolene	28.977	1508	1505	0.2
28	NI	29.603	1523	/	0.1
29	β-germacrene	30.939	1555	1559	0.3

R.t. – Retention time, RI – Retention index experimental, RI lit. – Retention index literature reported, NI – not identified, tr – detected in traces (less than 0.1%), / – not detected

Angelica is traditionally used for the treatment of leukoderma, nervous headaches, fever, skin rashes, wounds, rheumatism, toothaches, gastric ulcers, anorexia, migraine, bronchitis, chronic fatigue, menstrual and obstetric complaints and for dental preparation [104]. Investigations show that angelica expresses antiproliferative effect on human pancreas cancer cell line [105], as well as against breast cancer cells [102]. Apart from this, angelica possesses anxiolytic [106] and imunostimulant [107] activities. It is also found that angelica inhibits the malondialdehyde formation in mouse liver homogenates. This plant is a cytoprotective agent effective against chronic ethanol-induced hepatotoxicity, possibly through inhibition of the production of oxygen free radicals that cause lipid peroxidation, and hence indirectly protects the liver from oxidative stress [108].

This plant possesses antioxidative properties [109]. Administration of angelica as natural chelator is significantly recommended as a compensating method against oxidative stress caused by lead poisoning especially after removal of the lead poisoning source [110]. Angelica may be recommended as a safe preservative against molds, aflatoxin contamination and oxidative deterioration of agri-food items during storage and processing [111]. Angelica shows good antimicrobial activity against *Clostridium difficile*, *Clostridium perfringens*, *Enterococcus faecalis*, *Eubacterium limosum*, *Peptostreptococcus anaerobius* and *Candida albicans* [112]. Antibacterial activity was proven in other studies [103, 113]. All this indicated that angelica, apart from the pharmacological importance, has a high potential for being applied as a natural preservative as well as a natural biocontrol agent in agro-food industry [103].

CONCLUSION

Compounds such as β-phellandrene, limonene, α-pinene and β-pinene, sabinene, octyl ester of butanoic acid and 1-octanol are widely distributed flavor ingredients in the Apiaceae family plants. Apart from this, these compounds play a role in metabolic processes and have significant biological activity in humans. Because of this, fruit of Apiaceae species are used as food additives, flavoring agents, functional food and nutraceuticals.

CONSENT FOR PUBLICATION

Not applicable.

CONFLICT OF INTEREST

The author (editor) declares no conflict of interest, financial or otherwise.

ACKNOWLEDGEMENT

This work has been supported by the Serbian Ministry of Education, Science and Technological Development for financial support (Grant number 173032).

REFERENCES

[1]   Wyllie SG, Alves S, Filsoof M, Jennings WG. Headspace sampling: use and abuse.Analysis of Foods and Beverages: Headspace Techniques. New York: Academic Press 1978; pp. 1-15.

[2]   Lazić B, Đurovka M, Marković V, Ilin Ž. Vegetables. University of Novi Sad, Faculty of Agriculture 2001; pp. 1-472.

[3]   Aćimović M, Milić N. Perspectives of the Apiaceae hepatoprotective effects – a review. Nat Prod Commun 2017; 12(2): 309-17.

[4]   Stanković M, Nikolić M, Stanojević Lj, Petrović S, Cakić M. Hydrodistillation kinetics and essential oil composition from fermented parsley seeds. Chem Ind Chem Eng Q 2005; 11(1): 25-9.

[5]   Gorunović M, Lukić P. Pharmacognosy. Belgrade University, Faculty of Pharmacy 2001; pp. 1-785.

[6]   Kišgeci J. Medicinal and aromatic plants. Belgrade: Partenon 2008; pp. 1-402.

[7]   Aćimović M, Todosijević M, Stanković J, *et al.* Headspace analysis of volatile compounds from leafy forms of parsley and celery fruits. Ann Agron 2017; 41(2): 38-44.

[8]   Kurowska A, Gałązka I. Essential oil composition of the parsley seed of cultivars marketed in Poland. Flavour Fragrance J 2006; 21(1): 143-7.

[9]   Mahmoodi L, Valizadegan O, Mahdavi V. Fumigant toxicity of *Petroselinum crispum* L. (Apiaceae) essential oil on *Trialeurodes vaporariorum* (Westwood) (Hemiptera: Aleyrodidae) adults under greenhouse conditions. J Plant Prot Res 2014; 54(3): 294-9.

[10]   Shalaby EA, Nasr NF, El Sherief SM. An *in vitro* study of the antimicrobial and antioxidant efficacy of some nature essential oils. J Med Plants Res 2011; 5(6): 922-31.

[11]   Farzaei MH, Abbasabadi Z, Ardekani MRS, Rahimi R, Farzaei F. Parsley: a review of ethnopharmacology, phytochemistry and biological activities. J Tradit Chin Med 2013; 33(6): 815-26. [PMID: 24660617]

[12]   Linde GA, Gazim ZC, Cardoso BK, *et al.* Antifungal and antibacterial activities of *Petroselinum crispum* essential oil. Genet Mol Res 2016; 15(3) [http://dx.doi.org/10.4238/gmr.15038538] [PMID: 27525894]

[13]   Tang ELH, Rajarajeswaran J, Fung S, Kanthimathi MS. *Petroselinum crispum* has antioxidant properties, protects against DNA damage and inhibits proliferation and migration of cancer cells. J Sci Food Agric 2015; 95(13): 2763-71. [PMID: 25582089]

[14]   Tang ELH, Rajarajeswaran J, Fung S, Kanthimathi MS. *Petroselinum crispum* has antioxidant properties, protects against DNA damage and inhibits proliferation and migration of cancer cells. J Sci Food Agric 2015; 95(13): 2763-71.

[PMID: 25582089]

[15] Petrolini FVB, Lucarini R, de Souza MGM, Pires RH, Cunha WR, Martins CHG. Evaluation of the antibacterial potential of *Petroselinum crispum* and *Rosmarinus officinalis* against bacteria that cause urinary tract infections. Braz J Microbiol 2013; 44(3): 829-34.
[PMID: 24516424]

[16] Alyami FA, Rabah DM. Effect of drinking parsley leaf tea on urinary composition and urinary stones' risk factors. Saudi J Kidney Dis Transpl 2011; 22(3): 511-4.
[PMID: 21566309]

[17] Rezazad M, Farokhi F. Protective effect of Petroselinum crispum extract in abortion using prostadin-induced renal dysfunction in female rats. Avicenna J Phytomed 2014; 4(5): 312-9.
[PMID: 25386393]

[18] Al-Howiriny TA, Al-Sohaibani MO, El-Tahir KH, Rafatullah S. Preliminary evaluation of the anti-inflammatory and anti-hepatotoxic activities of 'parsley' *Petroselinum crispum* in rats. J Nat Rem 2003; 3: 54-62.

[19] Kamal T, Abd-Elhady E, Sadek K, Shukry M. Effect of parsley (*Petroselium crispum*) on carbon tetrachloride-induced acute hepatotoxicity in rats. Res J Pharm Biol Chem Sci 2014; 5: 1524-34.

[20] Troncoso L, Guija E. *Petroselinum sativum* (perejil) antioxidant and hepatoprotective effects in rats with paracetamol-induced hepatic intoxication. An Fac Med, Univ Nac Mayor San Marcos 2007; 68: 333-43.

[21] Jassim AM. Protective effect of *Petroselinum crispum* (parsley) extract on histopathological changes in liver, kidney and pancreas induced by sodium valproate in male rats. Kufa J Vet Med Sci 2013; 4: 20-7.

[22] Bolkent S, Yanardag R, Ozsoy-Sacan O, Karabulut-Bulan O. Effects of parsley (*Petroselinum crispum*) on the liver of diabetic rats: a morphological and biochemical study. Phytother Res 2004; 18(12): 996-9.
[PMID: 15742348]

[23] Fejes S, Kéry A, Blázovics A, *et al.* [Investigation of the *in vitro* antioxidant effect of *Petroselinum crispum* (Mill.) Nym. ex A. W. Hill]. Acta Pharm Hung 1998; 68(3): 150-6.
[PMID: 9703701]

[24] Branković S, Kitić D, Radenković M, Ivetić V, Veljković S, Nesić M. [Relaxant activity of aqueous and ethanol extracts of parsley (Petroselinum crispum (Mill) Nym. ex A. W Hill, Apiaceae) on isolated ileum of rat]. Med Pregl 2010; 63(7-8): 475-8.
[PMID: 21446133]

[25] Akıncı A, Eşrefoğlu M, Taşlıdere E, Ateş B. *Petroselinum crispum* is effective in reducing stress-induced gastric oxidative damage. Balkan Med J 2017; 34(1): 53-9.
[PMID: 28251024]

[26] Aćimović M, Cvetković M. Celery (*Apim graveolens* L.) potential in organic agriculture. Plant Doctor 2016; 44(1): 7-16.

[27] Grubben GJH, Denton OA, Messiaen CM, Schippers RR, Lemmens RHMJ, Oyen LPA. Plant resources of tropical Africa, 2: Vegetables. Wageningen: PROTA Foundation/Backhuys Publishers/CTA 2004; pp. 1-668.

[28] Aćimović M, Cvetković M, Stanković J, Malenčić Đ, Kostadinović Lj. Compound analysis of essential oils from lovageand celery fruits obtained by headspace extracton. Ann Sci Work 2015; 39(1): 44-51.

[29] Olle M, Bender I. The content of oils in umbelliferous crops and its formation. Agron Res (Tartu) 2010; 8: 687-96.

[30] Milutinović M. Effect of celery essential oil (*Apium graveolens* var. dulce (Mill.) Pers) on the growth

of Bacillus cereus in commercial chicken soup as a food model Master Thesis, University of Niń, Faculty of Sciences and Mathematics, Department of Biology and Ecology, 2014; 1-56.

[31]   Bjeldanes LF, Kim IS. Phthalide components of celery essential oil. J Org Chem 1977; 42(3): 2333-5.

[32]   Hassanen NHM, Eissa AMF, Hafez SAM, Mosa EAM. Antioxidant and antimicrobial activity of celery (*Apium graveolens*) and coriander (*Coriandrum sativum*) herb and seed essential oils. Int J Curr Microbiol Appl Sci 2015; 4(3): 284-96.

[33]   Hassanen NHM, Eissa AMF, Hafez SAM, Mosa EAM. Antioxidant and antimicrobial activity of celery (*Apium graveolens*) and coriander (*Coriandrum sativum*) herb and seed essential oils. Int J Curr Microbiol Appl Sci 2015; 4(3): 284-96.

[34]   Sowbhagya HB. Chemistry, technology, and nutraceutical functions of celery (*Apium graveolens* L.): an overview. Crit Rev Food Sci Nutr 2014; 54(3): 389-98.
        [PMID: 24188309]

[35]   Aydemir T, Becerik S. Phenolic content and antioxidant activity of different extracts from *Ocimum basilicum, Apium graveolens* and *Lepidium sativum* seeds. J Food Biochem 2011; 35(1): 62-79.

[36]   Kooti W, Daraei N. A review of the antioxidant activity of celery (*Apium graveolens* L) J Evid-Based Compl Alt 2017; 22(4): 1029-34.

[37]   Momin RA, Nair MG. Antioxidant, cyclooxygenase and topoisomerase inhibitory compounds from *Apium graveolens* Linn. seeds. Phytomedicine 2002; 9(4): 312-8.
        [PMID: 12120812]

[38]   Ugochi NU, Elijah JP, Leonard OI, Charles OO. Antioxidant Properties of *Apium graveolens*. Res J Pharmacogn Phytochem 2011; 3(5): 201-5.

[39]   Umamaheswari M, Ajith MP, Asokkumar K, Sivashanmugam T, Subhadradevi V, Jagannath P. *In vitro* angiotensin converting enzyme inhibitory and antioxidant activities of seed extract of *Apium graveolens* Linn. Ann Biol Res 2012; 3(3): 1274-82.

[40]   Uddin Z, Shad AA, Bakht J, Ullah I, Jan S. *In vitro* antimicrobial, antioxidant activity and phytochemical screening of *Apium graveolens*. Pak J Pharm Sci 2015; 28(5): 1699-704.
        [PMID: 26408890]

[41]   Zor ŞD, Bat M, Peksel A, Alpdoğan G. Optimization of ultrasound-assisted extraction of antioxidants from *Apium graveolens* L. seeds using response surface methodology. J Turk Chem Soc 2017; 4(3): 915-30.

[42]   Al-Sa'aidi JAA, Alrodhan MNA. Ismael, Ahmed K. Antioxidant activity of n-butanol extract of celery (*Apium graveolens*) seed in streptozotocin-induced diabetic male rats. Res Pharm Biotech 2012; 4(2): 24-9.

[43]   Choupankareh S, Beheshti F, Karimi S, *et al.* The effects of aqueous extract of *Apium graveolens* on brain tissues oxidative damage in pentylenetetrazole-induced seizures model in rat. Curr Nutr Food Sci 2018; 14(1): 47-53.

[44]   Shanmugapriya R, Ushadevi T. *In vitro* antibacterial and antioxidant activities of *Apium graveolens* l. seed extracts. Int J Drug Dev Res 2014; 6(3): 165-70.

[45]   Mansi K, Abushoffa AM, Disi A, Aburjai T. Hypolipidemic effects of seed extract of celery (*Apium graveolens*) in rats. Pharmacogn Mag 2009; 5: 301-5.

[46]   Tashakori-Sabzevar F, Ramezani M, Hosseinzadeh H, *et al.* Protective and hypoglycemic effects of celery seed on streptozotocin-induced diabetic rats: experimental and histopathological evaluation. Acta Diabetol 2016; 53(4): 609-19.
        [PMID: 26940333]

[47]   Powanda MC, Rainsford KD. A toxicological investigation of a celery seed extract having anti-inflammatory activity. Inflammopharmacology 2011; 19(4): 227-33.
        [PMID: 20568016]

[48] Teng CM, Lee LG. KFN, Huang TF. Inhibition of platelet aggregation by apigenin from *Apium graveolens*. Asia Pac J Pharm 1988; 3(2): 85-90.

[49] Branković S, Gočmanac-Ignjatović M, Kostić M, *et al.* Spasmolytic activity of the aqueous and ethanol celery leaves (*Apium graveolens* L.) extracts on the contraction of isolated rat ileum. Acta Medica Medianae 2015; 54(2): 11-6.

[50] Köken T, Koca B, Özkurt M, Erkasap N, Kuş G, Karalar M. *Apium graveolens* extract inhibits cell proliferation and expression of vascular endothelial growth factor and induces apoptosis in the human prostatic carcinoma cell line LNCaP. J Med Food 2016; 19(12): 1166-71.
[PMID: 27982754]

[51] Subhadradevi V, Khairunissa K, Asokkumar K, Umamaheswari M, Sivashanmugam A, Jagannath P. Induction of apoptosis and cytotoxic activities of *Apium graveolens* Linn. using *in vitro* models. Middle East J Sci Res 2011; 9(1): 90-4.

[52] Chemical Constituents, Traditional and Modern Medicinal Uses. Ross IA. Medicinal Plants of the WorldTotowa, New Jersey: Humana Press 2005; 3: pp. 1-648.

[53] Aćimović M, Stanković J, Cvetković M, Ignjatov M, Nikolić Lj. Chemical characterization of essential oil from seeds of wild and cultivated carrots from Serbia. Bot Serb 2016; 40(1): 55-60.

[54] Özcan MM, Chalchat JC. Chemical composition of carrot seeds (*Daucus carota* L.) cultivated in Turkey: characterization of the seed oil and essential oil. Grasas Aceites 2007; 58: 359-65.

[55] Kula J, Izydorczyk K, Czajkowska A, Bonikowski R. Chemical composition of carrot umbel oils from *Daucus carota* L. ssp. *sativus* cultivated in Poland. Flavour Fragrance J 2006; 21: 667-9.

[56] Imamu X, Yili A, Aisa HA, Maksimov VV, Veshkurova ON, Salikhov ShI. Chemical composition and antimicrobial activity of essential oil from *Daucus carota sativa* seeds. Chem Nat Compd 2007; 43(4): 495-6.

[57] Siddiqui AA, Wani SM, Rajesh R, Alagarsamy V. Isolation and hyotensive activity of three new phytoconstituents from seeds of *Daucus carota*. Indian J Pharm Sci 2005; 67(6): 716-20.

[58] Gilani AH, Shaheen E, Saeed SA, *et al.* Hypotensive action of coumarin glycosides from Daucus carota. Phytomedicine 2000; 7(5): 423-6.
[PMID: 11081994]

[59] Patil MVK, Kandhare AD, Bhise SD. Anti-inflammatory effect of *Daucus carota* root on experimental colitis in rats. Int J Pharm Pharm Sci 2012; 4(1): 337-43.

[60] Vasudevan M, Gunnam KK, Parle M. Antinociceptive and anti-inflammatory properties of *Daucus carota* seeds extract. J Health Sci 2006; 52(5): 598-606.

[61] Wehbe K, Mroueh M, Daher CF. The potential role of *Daucus carota* aqueous and methanolic extracts on inflammation and gastric ulcers in rats. J Complement Integr Med 2009; 6(1)
[http://dx.doi.org/10.2202/1553-3840.1159]

[62] Ravindra PV, Narayan MS. Antioxidant activity of the anthocyanin from carrot (*Daucus carota*) callus culture. Int J Food Sci Nutr 2003; 54(5): 349-55.
[PMID: 12907406]

[63] Sun T, Simon PW, Tanumihardjo SA. Antioxidant phytochemicals and antioxidant capacity of biofortified carrots (*Daucus carota* L.) of various colors. J Agric Food Chem 2009; 57(10): 4142-7.
[PMID: 19358535]

[64] Shebaby WN, El-Sibai M, Smith KB, Karam MC, Mroueh M, Daher CF. The antioxidant and anticancer effects of wild carrot oil extract. Phytother Res 2013; 27(5): 737-44.
[PMID: 22815230]

[65] Clement YN, Mahase V, Jagroop A, *et al.* Herbal remedies and functional foods used by cancer patients attending specialty oncology clinics in Trinidad. BMC Complement Altern Med 2016; 16(1):

399.
[http://dx.doi.org/10.1186/s12906-016-1380-x] [PMID: 27769229]

[66]   Diab-Assaf M, El-Sharif S, Mroueh M. Evaluation of anti-cancer effect of *Daucus carota* on the human promyelocytic leukemia HL-60 cells. AACR International Conference: Molecular Diagnostics in Cancer Therapeutic Development. Atlanta, Georgia. 2007.

[67]   Najm PI. The Anti-Cancer Activity of 2 Himachalene-6-ol Extracted from Daucus carota ssp. carota Lebanese American University, Master Thesis 2014.

[68]   Singh K, Dhongade H, Singh N, Kashyap P. Hypolipidemic activity of ethanolic extract of *Daucus carota* seeds in normal rats. Int J Biol Adv Res 2010; 1(3): 73-80.

[69]   Afify AMR, Romeilah RRM, Osfor MMH, Elbahnasawy ASM. Evaluation of carrot pomace (*Daucus carota* L.) as hypocholesterolemic and hypolipidemic agent on albino rats. Not Sci Biol 2013; 5(1): 7-14.

[70]   Jasicka-Misiak I, Lipok J, Nowakowska EM, Wieczorek PP, Młynarz P, Kafarski P. Antifungal activity of the carrot seed oil and its major sesquiterpene compounds. Z Natforsch C J Biosci 2004; 59(11-12): 791-6.
[PMID: 15666536]

[71]   Pavlek P, Todorić V, Grubić S. Neke karakteristike pańtrnjaka *Pastinaca sativa* (syn. *Peucedanum sativum*) s posebnim osvrtom na iskustva u proizvodnji (sjetva, gnojidba i suzbijanje korova) na povrńinama OOUR-a —Poljoprivreda| —Podravka| – Koprivnica Agronomski glasnik: Glasilo Hrvatskog agronomskog druńtv 1980; 42(3): 323-32.

[72]   Kurkcuoglu M, Baser KHC, Vural M. Composition of the essential oil of *Pastinaca sativa* L. subsp. *urens* (Req. ex Godron) Celak. Chem Nat Compd 2006; 42(1): 114-5.

[73]   Jovanović SČ, Jovanović OP, Petrović GM, Stojanović GS. Endemic Balkan parsnip *Pastinaca hirsuta*: the chemical profile of essential oils, headspace volatiles and extracts. Nat Prod Commun 2015; 10(4): 661-4.
[PMID: 25973504]

[74]   Kapetanos C, Karioti A, Bojović S, Marin P, Veljić M, Skaltsa H. Chemical and principal-component analyses of the essential oils of Apioideae taxa (Apiaceae) from central Balkan. Chem Biodivers 2008; 5(1): 101-19.
[PMID: 18205131]

[75]   Matejić J, Džamić A, Mihajlov-Krstev T, Ranđelović V, Krivošej Z, Marin P. Antimicrobial potential of essential oil from *Pastinaca sativa* L. Biol Nyssana 2014; 5(1): 31-5.

[76]   Karin A, Karlson B, Valterova IL, Nilsson A. Volatile compounds from flowers of six species in the family Apiaceae: bouquets for different pollinators? Phytochemistry 1993; 35(1): 111-9.

[77]   Tucakov J. Healing with plants. Rad, Belgrade Phytotherapy 2006.

[78]   Skalicka-Woźniak K, Zagaja M, Głowniak K, Łuszczki JJ. Purification and anticonvulsant activity of xanthotoxin (8-methoxypsoralen). Cent Eur J Biol 2014; 9(4): 431-6.

[79]   Nikolić M, Marković T, Ćirić A, Glamočlija J, Marković D, Soković M. Susceptibility of oral *Candida* spp. reference strains and clinical isolates to selected essential oils of Apiaceae species. Lekovite Sirovine 2015; 35(35): 151-62.

[80]   Akbarmivehie M, Baghaei H. The effect of addition parsnip herb and its extract on Momtaze hamburger shelf life. Eur Online J Nat Soc Sci 2016; 5(1): 132-46.

[81]   Aćimović M, Stanković J. Lovage (*Levisticum officinale* Koch.): planth with great potential for use in organic agriculture. Plant Doctor 2015; 43(5): 434-42.

[82]   Blank I, Scbieberle P. Analysis of the seasoning-like flavour substances of a commercial lovage extract *(Levisticum officinale* Koch.). Flavour Fragrance J 1993; 8: 191-5.

[83]   Bylaitė E, Roozen JP, Legger A, Venskutonis RP, Posthumus MA. Dynamic deadspace-gas chromatography-olfactometry analysis of different anatomical parts of lovage (*Levisticum officinale* Koch.) at eight growing stages. J Agric Food Chem 2000; 48(12): 6183-90.
       [PMID: 11312790]

[84]   Fehr D. On the essential oil of *Levisticum officinale*. I. Investigation on the oil from fruit, leaves, stems and roots. Planta Med 1980; 40: 34-40.

[85]   Naber KG. Efficacy and safety of the phytotherapeutic drug Canephron® N in prevention and treatment of urogenital and gestational disease: review of clinical experience in Eastern Europe and Central Asia. Res Rep Urol 2013; 5: 39-46.
       [PMID: 24400233]

[86]   Ovchinnikova SY, Orlovskaya TV, Malikova MK. Carbohydrates from *Levisticum officinale*. Chem Nat Compd 2013; 49(5): 918-9.

[87]   El-Hamid SRA, Abeer YI, Hendawy SF. Anti-inflammatory, antioxidant, anti-tumor and physiological studies on *Levisticum officinale*-Koch plant. Planta Med 2009; 75: PE62.

[88]   Sertel S, Eichhorn T, Plinkert PK, Efferth T. Chemical Composition and antiproliferative activity of essential oil from the leaves of a medicinal herb, *Levisticum officinale*, against UMSCC1 head and neck squamous carcinoma cells. Anticancer Res 2011; 31(1): 185-91.
       [PMID: 21273597]

[89]   Danciu C, Avram Ş, Gaje P, *et al.* An evaluation of three nutraceutical species in the Apiaceae family from the Western part of Romania: antiproliferative and antiangiogenic potential. J Agroaliment Proc Technol 2013; 19(2): 173-9.

[90]   Lapeere H, Boone B, Verhaeghe E, Ongenae K, Lambert J. Contact dermatitis caused by lovage (*Levisticum officinalis*) essential oil. Contact Dermat 2013; 69(3): 181-2.
       [PMID: 23948036]

[91]   Grāmatiņa I, Sazonova S, Krūma Z, Skudra L, Prieciņa L. Herbal extracts for ensuring pork meat quality during cold storage. Proceedings of the Latvian Academy of Sciences. 453-60.

[92]   Mirjalili MH, Salehi P, Sonboli A, Hadian J, Ebrahimi SN, Yousefzadi M. The composition and antibacterial activity of the essential oil of *Levisticum officinale* Koch flowers and fruits at different developmental stages. J Serb Chem Soc 2010; 75(12): 1661-9.

[93]   Popovici C, Capcanari T, Deseatnicova O, Sturza R. Does application of *Petroselinum crispum* and *Levisticum officinale* Koch. extracts improve the thermal stability of vegetable oils? Papers of the Sibiu Alma Mater University Conference. 313-8.

[94]   Mohamadi N, Rajaei P, Moradalizadeh M, Moradalizadeh M, Amiri M. Essential oil composition and antioxidant activity of *Levisticum officinale* Koch. at various phenological stages. Faslnamah-i Giyahan-i Daruyi 2017; 1(61): 45-55.

[95]   Ebrahimi A, Eshraghi A, Mahzoonieh MR, Lotfalian S. Antibacterial and antibiotic-potentiation activities of *Levisticum officinale* L. extracts on pathogenic bacteria. Int J Infect 2017; 4(2): e38768.
       [http://dx.doi.org/10.5812/iji.38768]

[96]   Afarnegan H, Shahraki A, Shahraki J. The hepatoprotective effects of aquatic extract of *Levisticum officinale* against paraquat hepatocyte toxicity. Pak J Pharm Sci 2017; 30(6(Supplementary)): 2363-8.
       [PMID: 29188770]

[97]   Prisăcaru C. Research study on the assessment of the antitoxic action of various phytopreparates derived from the vegetal products of *Levisticum officinale*. Lucrăriştienţifice. Seria Horticultură 2014; 57: 11-6.

[98]   Kylin M. BSc Thesis, Swedish University of Agricultural Sciences, Uppsala.

[99]   Aćimović M, Cvetković M, Stanković J, Filipović V, Nikolić Lj, Dojčinović N. Analysis of volatile compounds from Angelica seeds obtained by headspace method. Arab J Med Arom Plants 2017; 3(1):

10-7.

[100] Nivinskiene O, Butkiene R, Mockute D. Chemical composition of seed (fruit) essential oils of *Angelica archangelica* L. grown wild in Lithuania. Chemija 2005; 16(3-4): 51-4.

[101] Shchipitsyna OS, Efremov AA. Composition of ethereal oil isolated from various vegetative parts of angelica from the Siberian Region. Russ J Bioorganic Chem 2011; 37(7): 888-92.

[102] Sigurdsson S, Ogmundsdottir HM, Gudbjarnason S. The cytotoxic effect of two chemotypes of essential oils from the fruits of *Angelica archangelica* L. Anticancer Res 2005; 25(3B): 1877-80. [PMID: 16158920]

[103] Aćimović M, Pavlović S, Varga A, *et al.* Chemical composition and antibacterial activity of *Angelica archangelica* root essential oil. Nat Prod Commun 2017; 12(2): 205-6.

[104] Kumar D, Bhat ZA, Kumar V, Chashoo IA, Khan NA, Shah MY. Pharmacognostical and phytochemical evaluation of *Angelica archangelica* Linn. Int J Drug Dev Res 2011; 3(3): 173-88.

[105] Sigurdsson S, Ogmundsdottir HM, Gudbjarnason S. Antiproliferative effect of *Angelica archangelica* fruits. Z Natforsch C J Biosci 2004; 59(7-8): 523-7. [PMID: 15813373]

[106] Kumar D, Bhat ZA. Anti-anxiety activity of methanolic extracts of different parts of *Angelica archangelica* Linn. J Tradit Complement Med 2012; 2(3): 235-41. [PMID: 24716138]

[107] Modaresi M. Effect of *Angelica archangelica* root hydro-alcoholic extract on the blood cells of small laboratory mice. Q Hor Med Sci 2013; 18(4): 149-53.

[108] Yeh ML, Liu CF, Huang CL, Huang TC. Hepatoprotective effect of *Angelica archangelica* in chronically ethanol-treated mice. Pharmacology 2003; 68(2): 70-3. [PMID: 12711833]

[109] Wei A, Shibamoto T. Antioxidant activities and volatile constituents of various essential oils. J Agric Food Chem 2007; 55(5): 1737-42. [PMID: 17295511]

[110] Raafat BM, Zahrany SM, Al-Zahrani AS, Tawifiek E, Al-Omery AM. *Angelica archangelica* roots water extraction as a natural antioxidant tolerating ROS production in lead poisoning. Res J Pharm Biol Chem Sci 2012; 3(2): 795-806.

[111] Prakash B, Singh P, Goni R, Raina AK, Dubey NK, Dubey NK. Efficacy of *Angelica archangelica* essential oil, phenyl ethyl alcohol and α- terpineol against isolated molds from walnut and their antiaflatoxigenic and antioxidant activity. J Food Sci Technol 2015; 52(4): 2220-8. [PMID: 25829603]

[112] Fraternale D, Flamini G, Ricci D. Essential oil composition and antimicrobial activity of *Angelica archangelica* L. (Apiaceae) roots. J Med Food 2014; 17(9): 1043-7. [PMID: 24788027]

[113] Rather RA, Rehman SU, Naseer S, Lone SH, Bhat KA, Chouhan A. Flash chromatography guided fractionation and antibacterial activity studies of *Angelica archangelica* root extracts. J Appl Chem 2013; 4(3): 34-8.

CHAPTER 7

Anticancer Properties of Apiaceae

Milica G. Aćimović[1,*], Milica M. Rat[2], Vele V. Tešević[3] and Nevena S. Dojčinović[2]

[1] *Institute of Field and Vegetable Crops Novi Sad, Maksima Gorkog 30, 21000 Novi Sad, Serbia*

[2] *Faculty of Sciences, University of Novi Sad, Trg Dositeja Obradovića 3, 21000 Novi Sad, Serbia*

[3] *Faculty of Chemistry, University of Belgrade, Studentski trg 12-16, 11000 Belgrade, Serbia*

Abstract: The aim of this book chapter was to highlight the great importance of plants from *Apiaceae* family as functional food products, focusing on its anticancer properties. The plants that will be discussed for their anticancer properties include: caraway (*Carum carvi* L.), dill (*Anethum graveolens* L.), anise (*Pimpinella anisum* L.), fennel (*Foeniculum vulgare* Mill.), coriander (*Coriandrum sativum* L.), celery and celeriac (*Apium graveolens* L.), lovage (*Levisticum officinale* Koch.), carrot (*Daucus carota* L.), parsley (*Petroselinum crispum* L.), parsnip (*Pastinaca sativa* L.), angelica (*Angelica archangelica* L.), cumin (*Cuminum cyminum* L.), chervil (*Anthriscus* sp.) and eryngo (*Eryngium campestre* L.). Leaves, roots and seeds of these plants are widely used as spices, flavoring agents and dietary supplements in the folk medicine and pharmaceutical industry. Furthermore, roots and leaves of these plants are valuable sources of phytochemicals used on a daily basis as food with nutraceutical potential. Their essential oils have characteristic aroma and have potent antioxidant and antimicrobial properties, which contribute to their ability to serve as natural food conservatives. Additionally, due to their anticancer, hypoglycemic, hypolipidemic, hepatoprotective and other activities these plants are widely used as alternative and healthy food for the prevention and treatment of many disorders.

Keywords: Angelica, Anise, Caraway, Carrot, Chervil, Celery, Coriander, Cumin, Dill, Eryngo, Fennel, Lovage, Parsley, Parsnip.

INTRODUCTION

Cancer remains one of the leading causes of morbidity and mortality globally. The disease is characterized by cells in the human body continuous multiplying without the immune system having the ability to control or stop its growth. Consequently, these cells lead to the formation of tumors of malignant cells with the potential to be metastatic. Anticancer agents must be able to destroy cancer

---

[*] **Corresponding author Milica G. Aćimović:** Institute of Field and Vegetable Crops, Novi Sad, Serbia; Tel: +021780-365111; E-mail: acimovicbabicmilica@gmail.com

Spyridon A. Petropoulos, Isabel C.F.R. Ferreira and Lillian Barros (Eds.)

cells while limiting side effects on healthy cells and prevent their apoptosis. Current treatments include chemotherapy (routinely used for cancer treatment), radiotherapy and chemically derived drugs. However, chemotherapy can put patients under a lot of strain and further damage their health. Therefore, there is a focus on using alternative treatments and therapies against cancer [1 - 3].

Cancer has been under a constant global battle which yielded and developed new cures and preventive therapies. Cancer chemoprevention with natural phytochemical compounds is an emerging strategy to prevent, impede, delay, or cure cancer. Many plant species are already being used to treat or prevent the development of cancer. The anticancer properties of plants have been recognized since centuries. The anticancer characteristics of a number of plants are still being actively researched and some have shown promising results. Multiple researchers have identified species of plants that have demonstrated anticancer properties with a lot of focus on those that have been used in herbal medicine in developing countries [1, 2, 4].

Increased health awareness has led many consumers to become more vigilant in maintaining good health. Many consumers have incorporated natural health products and functional foods into their daily nutrition, usually as spices and herbs which have been used for food purposes since the ancient times, not only as flavoring agents, but also as food preservatives and medicine ingredients. Functional foods have demonstrated physiological benefits and ability to reduce the risk of chronic diseases, providing the body with the required amounts of nutrients which include minerals, proteins, fibers, carbohydrates, fats, *etc*. Nutraceuticals represent phytochemicals which are non-nutritive plant chemicals that also possess biological activity, such as: polyphenolic compounds, polyacetylenes and terpenoids [5 - 7]. Plant products supplementation has become the standard of care for management of severely ill patients, to which nutritional support was intended to replete substrate deficiencies secondary to stress-induced catabolism. Positive results in the treatment of these patients, have prompted the development of more sophisticated nutritional strategies and concepts [8].

The aim of this book chapter was to highlight the importance of plants from the *Apiaceae* family as functional foods, focusing on its anticancer properties. The plants that will be discussed for their anticancer properties include: caraway (*Carum carvi* L.), dill (*Anethum graveolens* L.), anise (*Pimpinella anisum* L.), fennel (*Foeniculum vulgare* Mill.), coriander (*Coriandrum sativum* L.), celery and celeriac (*Apium graveolens* L.), lovage (*Levisticum officinale* Koch.), carrot (*Daucus carota* L.), parsley (*Petroselinum crispum* L.), parsnip (*Pastinaca sativa* L.), angelica (*Angelica archangelica* L.), cumin (*Cuminum cyminum* L.), chervil (*Anthriscus* sp.) and eryngo (*Eryngium campestre* L.). Leaves, roots and seeds of these plants are widely used as spices, flavoring agents and dietary supplements in the folk medicine and pharmaceutical industry. Furthermore, roots and leaves of these plants are valuable sources of phytochemicals used on a daily basis as food with nutraceutical potential. Their essential oils have characteristic aroma and have potent antioxidant and antimicrobial properties, which contribute to their ability to serve as natural food conservatives. Additionally, due to their anticancer, hypoglycemic, hypolipidemic, hepatoprotective and other activities these plants are widely used as alternative and healthy food for the prevention and treatment of many disorders.

CARAWAY

Caraway (*Carum carvi* L.) is an annual or biannual plant, which is usually grown for its fruit and seeds (*Carvi fructus*). It is used in cakes, cheeses, confections, fresh cabbage, meat dishes, rye bread and salads [9]. Medicinal use of caraway fruit has been used widespread as carminative, for spasmodic gastrointestinal complaints, flatulence, and bloating. Also, it can be used for dyspeptic problems such as mild, spastic condition of the gastrointestinal tract, bloating and fullness [10].

Caraway is an aromatic drug that contains essential oil (*Carvi aetheroleum*), with carvone and limonene forming a mixture that constitutes 97.69 to 98.62% of total oil composition [11]. Apart from essential oil, it also contains fatty oils, proteins, carbohydrates and phenolic acids (Table **1**).

Table 1. Caraway seed constituents (according to EMA [10]).

Essential oil	3-7% v/m, with: d-carvone (50-65%), and (+)- limonene (up to 45%), with less than 1.5% of carveol and dihydrocarveol.
Fatty oil	10-18% of fixed oil, with: petroselinic (30-43%), linoleic (34-37%), oleic (15-25%) and palmitic (4-5%) acids.
Protein	about 20%
Carbohydrates	about 15%
Phenolic compounds	mainly caffeic acid, and traces of flavonoids such as quercetin, kaempferol and their glycosides

Carvone is a monocyclic monoterpene found in the essential oil of caraway seeds and according to various reports it has antioxidant, antimicrobial, anticonvulsant, and antitumor activities. Exploring the potential anticancer activity of different concentrations of carvone (10, 25, 50, 100, 200 and 400 mg/L) revealed that treatment with small concentrations of carvone (25 mg/L) led to an increase in total antioxidant capacity in *in vitro* cultivated primary rat neuron cells, compared with control cells. Also, higher concentrations of carvone (>100 mg/L) increased oxidative stress in both cell types. Treatment with carvone (>100 mg/L) and 3-(4,5-dimethylthiazol-2-yl)-2,5 diphenyltetrazolium bromide showed a reduction in the cell viability rates for both cell types 24 h after the application of carvone. All these results suggest the use of carvone as potential anticancer agent in brain tumors therapy [12].

Further exploration of carvone and its derivatives (obtained by condensation reaction), as candidates for anticancer activity drugs, suggested their anticancer activity against breast (MCF7), cervix (HeLa) and ovary (SK-OV3) cell lines [13]. Furthermore, there are also reported the genomic and proteomic effects of caraway seed extracts. Rich in flavonoids and steroid-like substances, these extracts potentially reversed the rat hepatoma cells co-treated with a 2,3,7,8-tetrachlorodibenzo-p-dioxin dependent induction in cytochrome P450 1A1, which showed their activity at mRNA level [14].

DILL

Dill (*Anethum graveolens* L.) is an annual plant, usually grown for its leaves (*Anethi folium*) or seeds (*Anethi fructus*). Leaves are used in pickles, for garnishing or to flavor salads, vegetable dishes, sea food, soups, yogurt and mayonnaise. Seeds are used in pickled cucumbers, bread, processed meats, sausages, cheeses and condiments [9]. Medicinal use of dill seed is due to different biological properties including antimicrobial, antioxidative and anti-inflammatory [15]. The latter activity is probably due to inhibition of inflammatory mediators [16].

Dill seed contains essential and fatty oils, proteins, carbochydrates and phenolic compounds (Table **2**), while leaves apart from essential oil they also contain vitamin C, carotenoids, chlorophylls, flavonoids, coumarins [17]. Essential oils (*Anethi aetheroleum*) can be obtained from seeds or from the aboveground parts of the plant during the fruit maturation phase. Carvone and limonene are the dominant compounds in dill essential oil which constitute 94.6 to 95.1% of total oil composition [18].

Table 2. Dill seed constituents.

Essential oil	3.3-3.7%, with: d-carvone (51.7-54.5%), and (+)- limonene (40.6-43.1%) [18].
Fatty oil	0.58-1.89% [19, 20] with lauric (1.29%), stearic acids (0.9-3.86%), capric (5.97%), myristic (0.08- 0.25%), palmitic (2.31-4.66%), oleaic (36.38-53.87%), linolenic (0.26-0.4%), linoleic (5.8-45.13%), palmitoleic (0.2%), eicosenoic (0.04%) and arachidoic acids (0.1-1.32%) [21].
Protein	15.68% [22]

(Table 2) cont.....

Carbohydrates	36% [22]
Phenolic compounds	Mainly phenolic acids: vanillic, caffeic, protocathechuic, pcoumaric, ferulic, chlorogenic, syringic, rosmarinic, ocoumaric and trans-cinnamic acid [21]

Essential oil from dill seed showed high anticancer activity against KB-Oral cavity and MCF7-breast cancer cells, having non-cytotoxicity to normal cell [23]. Furthermore, dill application was significantly ameliorative in all the parameters of subchronic model of tramadol hydrochloride toxicity on testicular functions, including tumor necrosis factor-α and nitric oxide in the tested tissues [24].

ANISE

Anise (*Pimpinella anisum* L.) is an annual plant usually grown for its seeds (*Anisi fructus*) which is used in beverages, baked goods, condiments, relishes, oils and fats, frozen dairy products, gravies, meat products and soft candies [9]. Aniseed has been used as a popular medicine to treat dyspeptic complaints, a broad range of adverse symptoms including spasmodic ailments involving altered functional motility of local smooth muscles induced by anomalous hormonal secretions, *Helicobacter* infections, stress and psychological disturbances and other idiopathic causes. Apart from this, anise has been used for catarrh of the respiratory tract and as a mild expectorant [25]. Anise seed contains essential and fatty oils, protein, carbohydrates and phenolic compounds (Table 3). The dominant compound of anise essential oil (*Anisi aetheroleum*) is *trans*-anethole (in amounts of 94.78%) [26].

Table 3. Anise seed constituents.

Essential oil	Minimum 2%, with: trans-anethole (87-94%), estragole (0.5-5.0%), linalool (maximum 1.5%), α-terpineol (maximum 1.2%), cis-anethole (0.1-0.4%), anisaldehyde (0.1-1.4%), pseudoisoeugenyl 2-methylbutirate (0.3-2.0%) [27].
Fatty oil	8-11% of lipids rich in fatty acids, such as palmitic and oleic acids [28]
Protein	18% [28]
Carbohydrates	4% [28]
Phenolic compounds	mainly flavonoids (apigenin and luteolin) followed by phenolic acids (chlorogenic and p-coumaric acid) [29]

As it is reported in a study with gastric cancer cell line (AGS) and angiogenesis on human umbilical vein endothelial cells (HUVEC), alcoholic extracts and essential oil of anise showed antiprolifertive properties on gastric cancer cells and could be used as a plant-based cure for gastric cancer [30]. Other authors, also proposed the ethanolic extract of anise seed as a valuable source of anticancer compounds with antiproliferative and/or apoptotic properties, since they confirmed its cytotoxic activity against human prostate cancer cell line (PC-3) at concentrations found safe to normal cells [31]. Treatment with anise seeds extract caused antiproliferative and apoptotic effects to cancer cells with IC50 values of 400 µg/mL. All the above-mentioned suggests anise as a potential functional food that could contribute to cancer prevention and treatment.

FENNEL

Fennel (*Foeniculum vulgare* Mill.) is an annual (var. *dulce*) or perennial plant (var. *vulgare*). Sweet fennel (var. *dulce*) is grown for its leaves, which fresh and chopped can be used as garnishment for fish dishes, sauces, salads, stews and curries. Leaf stalks (pseudobulbs) are used raw in salads and stuffing, blanched in soups and sauces and baked. Both varieties, sweet and bitter (var. *vulgare*) are grown for its seed (*Foeniculi fructus*) which is used in meat dishes, curries, spice blends, soups, vegetables and breads [9]. The medicinal use of fennel is largely due to antispasmodic, secretolytic, secretomotor and antibacterial effects of its essential oil [32].

Fennel seed contains essential oil, fatty oil, phenolic compounds (Table **4**), while leaves apart from phenolics and flavonoids contain vitamin C, minerals (K, Na, Ca, Mg) and nitrates [33, 34]. Essential oil (*Foeniculi aetheroleum*) is obtained from seeds and herbs during the fruit maturation phase. The main compound in essential oil of both varieties is *trans*-anethole which is present in amounts of 80.0% in sweet and 67.1% in bitter fennel of total oil content [35].

Table 4. Fennel seed constituents.

Essential oil	From bitter fennel contain 55.0-75.0% trans-anethole, fenchone 12.0-25.0%, estragole maximum 6.0%. From sweet fennel contain minimum 80.0% trans-anethole, fenchone maximum 7.5%. estragole maximum 10.0% [27]
Fatty oil	1.28% whereas about 21 fatty acids were identified and quantified, such as caproic acid, caprylic acid, capric acid, undecanoic acid, lauric acid, myristic acid, myristoleic acid, pentadecanoic acid, palmitic acid, heptadecanoic acid, stearic acid, oleic acid, linoleic acid, α-linolenic acid, arachidic acid, eicosanoic acid, cis-11,1--eicosadienoic acid, cis-11,14,17-eicosatrienoic acid + heneicosanoic acid, behenic acid, tricosanoic acid, and lignoceric acid [36]
Protein	1.08-1.37% [36]
Carbohydrates	18.44-22.82% [36]
Phenolic compounds	hydroxyl cinnamic acid derivatives, flavonoid glycosides, and flavonoid aglycones [36]

The fennel ethanolic extract, out of which syringin and 4-methoxycinnamyl alcohol are the most active compounds, exhibited significant toxicity against

MCF-7, HeLa and DU145 cancer cell lines [37]. Other authors reported that methanolic extract of fennel seeds had potent antioxidant activity since it reduced oxidative stress, prevented lipid peroxidation and inhibited the damage of mouse cells caused by reactive oxygen species [38]. The same authors proposed the application of this methanolic extract as a source of natural antioxidants to improve the oxidative stability of fatty foods during storage. Another similar experiment confirmed the 3-fold higher cytoprotective activity of 70% methanolic extract of fennel seed towards normal human blood lymphocytes (micronucleus assay) compared with the standard drug doxorubicin [39]. Also, the same study confirmed antitumor activities against B16F10 melanoma cell line (trypan blue exclusion assay for cell viability) at the concentration of 200µg/mL. Bearing in mind these results, authors suggested fennel as a natural resource of antitumor agents, as well as cytoprotective agents to normal cells. However, fennel essential oil showed low cytotoxic activity against human cervical cancer (HeLa), human colorectal adenocarcinoma (Caco-2), human breast adenocarcinoma (MCF-7), human T lymphoblast leukaemia (CCRF-CEM) and adriamycin resistant leukaemia (CEM/ADR5000) cancer cell lines [40].

CORIANDER

Coriander (*Coriandrum sativum* L.), also known as cilantro, is an annual plant which is grown for its seeds (*Coriandri fructus*) or leaves. Coriander leaves have a strong oriental scent, sometimes described as an unpleasant odour called a "stink bug smell" which is due to the presence of *trans*-tridecen in the essential oil. Leaves are usually used to make chutneys and sauces, green salsas, dips, snacks and soups [9]. Coriander seeds are used as an aromatic, carminative, stomachic, antispasmodic and against gastrointestinal complains such as dyspepsia, flatulance and gastralgia. It is often recommended for insomnia and anxiety, as well as analgetic and antirheumatic agent [41].

However, coriander seed aroma is completely different, with mild, sweet, warm and aromatic flavour, and linalool as its major constituent. Coriander seed is used in couscous, stews and salads, as well as for essential oil distillation (*Coriandri aetheroleum*) [42]. Apart from essential oil, coriander seed contains fatty oil, proteins, carbohydrates and phenolics (Table **5**), while green herb apart from essential oil contains vitamins (A and C), chlorophyll, and carotenoids [43].

Table 5. Coriander seed constituents.

Essential oil	0.3-1.2% [44]
Fatty oil	19-21%, with petroselinic acid as main (up to 80%) [44]
Protein	15.78% [45]

(Table 5) cont.....

Carbohydrates	20% [46]
Phenolic compounds	Rutin, quercetin, chlorogenic acid and caffeic acid were identified and quantified in the multi-component [47]

Aqueous extract of coriander leaves caused significant toxicity toward lymphoma cells (L5178Y-R) at 31.2 µg/mL (Minimum Inhibitory Concentration - MICs), whereas the methanol extract of coriander seed and leaves, caused cytotoxicity at 7.8 and 62.5 µg/mL (MICs), respectively. In addition, coriander leaves aqueous extract stimulated significant splenic cells lymphoproliferation (14-45%), whereas the methanol extracts caused stronger lymphoproliferation (43-59%) at the concentrations tested. Furthermore, aqueous extracts reduce nitric oxide production by LPS-stimulated macrophages [48]. Ethanolic extract of *C.sativum* leaves exhibited reduction of cancer cells viability (HT-29 cell lines) and the death rate of cancer cells was positively correlated with extract concentration. Hence, ethanolic extract of *C. sativum* exhibited high cytotoxicity [49]. Also, coriander herb extract showed effective anticancer activity against oral cancer cell lines [50].

Coriander root is not considered a useful plant part. However, there is a single report about the antioxidant and anticancer properties of coriander root [51]. The ethyl acetate extract of coriander roots showed the highest antiproliferative activity against the breast cancer cell line (MCF-7 cells) and had the highest phenolic content, FRAP and DPPH scavenging activities among the extracts tested. The extract showed anticancer activity against MCF-7 cells by affecting antioxidant enzymes possibly leading to H_2O_2 accumulation, cell cycle arrest at the G_2/M phase and apoptotic cell death. The herb shows potential in preventing oxidative stress-related diseases and would be useful as a supplement used in combination with conventional drugs to improve the treatment of diseases.

CELERY

Celery (*Apium graveolens* L.) is a biannual plant, however when it is grown for its root, leaves and leaf stalks is cultivated as an annual plant. Root, leaves and leaf stalks are used as a vegetable, fresh for salads, for the preparation of juices, in meat dishes, snacks, gravies, sauces. During the second year of cultivation it forms its seed (*Apii fructus*), which are used in pickling of vegetables, salad dressings, breads and biscuits [9]. Various studies have shown that celery plays a role in the prevention of cardiovascular disease, the lowering of blood glucose and serum lipid, the decrease of blood pressure and also has beneficial effects on the cardiovascular system. This herb has also antibacterial, antifungal and antiinflammatory properties [52].

Celery seed contains essential oil (*Apii aetheroleum*), with limonene being the main compound [53, 54]. Also, it contains fatty oil, proteins, carbohydrates, phenolic compounds (Table **6**), as well as minerals (Na, P, K, Ca, Mg, Fe, Cu, Mn, Zn), *etc.* [55]. However, celery leaves contain less essential oil (0.34-0.53%) [56], while essential oil content of roots is even smaller – from 0.01 to 0.08% [57].

Table 6. Celery seed constituents.

Essential oil	2% with limonene (60%) and selinene (20%) [58]
Fatty oil	15% with the fatty acids: petroselenic (64.3%), oleic (8.1%), linoleic (18%), linolenic (0.6%), and palmitic acids [58].
Protein	6.21% [55]
Carbohydrates	41.4% [59]
Phenolic compounds	Furocoumarins included celerin, bergapten, apiumoside, apiumetin, apigravrin, osthenol, isopimpinellin, isoimperatorin, celereoside, and 5 and 8-hydroxy methoxypsoralen. Phenols included graveobioside A and B, apiin, apigenin, isoquercitrin, tannins and phytic acid [60]

Celery seed extracts (aqueous, ethanolic and hexane) showed a concentration-dependent cytotoxic activity on growth of rhabdomysarcom (RD) cancer cell line [61]. The most powerful effect was observed for the hexane extract, especially at the concentrations of 100 and 200 µg/mL. The same authors reported that only the extract at the concentration of 200 µg/mL exhibited strong cytotoxic effect on the murine (L20B) transformed cells.

Another study confirmed that ethanolic extracts had strong antiproliferative effect on the human prostatic carcinoma cell line (LNCaP) and induced apoptosis in cancer cells [62].

Also, a concentration-dependent antiproliferative effect on Dalton's lymphoma ascites (DLA) and mouse lung fibroblast (L929) has been reported for methanolic extract of celery seeds [63]. This extract was cytotoxic toward L-929 cells (72 h, IC_{50}=3.85 µg/mL) and provoked DNA fragmentation, which indicates the induction of apoptosis.

LOVAGE

Lovage (*Levisticum officinale* Koch.) is a perennial plant, which is grown for its leaves, seeds (*Levistici fructus*) and roots (*Levistici radix*). Leaves are used for seasoning soups, sauces, meat dishes, while roots are used for soup seasonings, finishing flavorings in liqueurs and tobacco, and finally seeds are used as spice, cake flavoring, in soups, salads, for pickling of vegetables (especially cabbage and cucumbers) [9]. For centuries lovage is known as carminative and spasmolytic, but it is also approved in the treatment of inflammatory conditions of the urinary tract and against kidney stones [64].

Lovage contains essential oil (*Levistici aetheroleum*), in concentrations which depend on the plant part: roots 0.5-1.0%, leaves 0.08-0.24% and fruits 0.8-1.5% [65], with β-phellandrene as the main compound [54]. Also, the plant has flavonoids, coumarins, phenolic acids, saponins, alkaloids, as well as polyacetylenes [64].

Lovage essential oil extract had an IC_{50} value of 292.6 µg/mL on head and neck squamous carcinoma cells (HNSCC) [66]. Genes involved in apoptosis, cancer, cellular growth and cell cycle regulation were the most prominently affected in microarray analyses. The three pathways to be most significantly regulated were extracellular signal-regulated kinase 5 (ERK5) signaling, integrin-linked kinase (ILK) signaling, virus entry *via* endocytic pathways and p53 signaling. The authors who performed the analyses concluded that lovage essential oil inhibits human HNSCC cell growth.

CARROT

Carrot (*Daucus carota* L.) includes two subspecies – wild carrot (ssp. *carota*) and cultivated carrot (ssp. *sativus*). Wild carrot is the ancestor of the cultivated subspecies. The wild form has been used as a medicinal plant since ancient times [67]. However, the cultivated carrot is mainly used as a root vegetable; raw for juices, salads, cakes, for pickling and cooked in casseroles, soups and stews. Carrot seeds (*Dauci fructus*) are mainly used for essential oil distillation (*Dauci aetheroleum*) which is used in beverages, baked goods, condiments, relishes, meat products [9]. The carrot root contains carotenoids (β-carotene, α-carotene, γ-carotene, lycopene, cryptoxanthin, leutein, violaxanthin) [68].

Acetone-methanol extract from dried wild carrot umbels possesses both antioxidant and promising anticancer activities against human colon (HT-29, Caco-2) and breast (MCF-7, MDA-MB-231) cancer cell lines [69, 70].

PARSLEY

Parsley (*Petroselinum crispum* L.) is a biannual plant which is usually grown as an annual for its leaves and roots. Leaves are used as a garnish (for salads, soups, boiled potatoes and egg dishes), blended in dips, cooked sauces and stews, while roots are used as a vegetable to enhance soups flavor, in stews and condiments. In the second year of vegetation, parsley develops seeds (*Petroselini fructus*), which are used for essential oil distillation (*Petroselini aetheroleum*) [9]. The main compounds in essential oil are α-pinene and β-pinene [53]. Apart from essential oil, parsley contains fatty oil, proteins, carbohydrates and phenolic compounds (Table **7**), as well as vitamin C, tocopherol, carotenoids, flavonoids, coumarins, sterols, triterpenes, *etc.* [71].

Table 7. Parsley seed constituents.

Essential oil	2.52% [72]
Fatty oil	8.85-33.7% [71, 72]
Protein	14% [73]
Carbohydrates	6.3% [74]
Phenolic compounds	Phenolic compounds and flavonoids particularly apigenin, apiin and 6"-Acetylapiin [75].

Alcoholic extracts (10 to 1000 µg/mL, 24h exposure) and essential oil of parsley seed showed cytotoxic activities against human breast cancer cells (MCF-7). The results obtained by 3-(4, 5-dimethylthiazol-2yl)-2, 5-biphenyl tetrazolium bromide (MTT) and neutral red uptake (NRU) assays and cellular morphology by phase contrast inverted microscopy showed that extract and oil significantly reduced cell viability and altered the cellular morphology of MCF-7 cells in a concentration dependent manner [76].

Antioxidant properties (high total polyphenol content, ferric reducing and DPPH radical scavenging activity), prevention of DNA damage in normal 3T3-L1 cells, and the inhibition of proliferation and migration of the MCF-7 cells have been also confirmed for parsley leaves and stems dichloromethane extract. Pre-treatment with parsley extract protected mouse fibroblasts (3T3-L1) against H_2O_2-induced DNA damage and inhibited H_2O_2-induced MCF-7 cell migration. The authors who performed the analyses pointed out the potential of this extract in cancer prevention, as well as in protection against metastasis, since cell migration is necessary for the metastasis of cancer cells [77]. Furthermore, parsley seed extracts significantly reduced the cell viability of human hepatocellular carcinoma cells (HepG2) in a concentration dependent manner [78].

ANGELICA

Angelica (*Angelica archangelica* L.) is a biannual plant which is grown for its herb and leaves as an annual, while for the production of roots and seeds is grown as a biannual plant. Herb is used for decorating cakes and pastry, and to flavor jams and jellies, confectionaries and liqueurs, while chopped leaves are added to fruit salads, fish dishes, and cottage cheese. Roots (*Angelicae radix*) are used in herbal liqueurs and bitter spirits, in flavoring meat and canned vegetables. Seeds

(*Angelicae fructus*) are used in alcoholic distillates [9]. Both, roots and seeds are used for essential oil distillation (*Angelica aetheroleum*). The main component in angelica seed essential oil is β-phellandrene (84.7%), while roots mostly contain α-pinene (29.7%), δ-3-carene (14.2%), and a mixture of β-phellandrene and limonene (13.2%) [79, 80]. Apart from essential oil, angelica contains carbohydrates, protein, fatty acids, and phenolic compounds [81].

It is reported that the essential oil of angelica seeds (10-400 µg/ml) had cytotoxic activity against human pancreas cancer cells (PANC-1) and mouse breast cancer (Crl) cells [82]. *In vivo* (by 3H-thymidine uptake in the Crl cells) and *in vitro* (Crl mouse breast cancer cells) antiproliferative activity were confirmed for angelica leaf extract, however the latter showed antitumour activity, as well [83].

CUMIN

Cumin (*Cuminum cyminum* L.) is an annual plant which is grown for its seeds (*Cumini fructus*), usually used as a flavor component in beverages, confectioneries, baked goods, meat and meat products, condiments and relishes, gravies, snack foods, gelatins and puddings [9]. In medicine, cumin is used to treat a variety of diseases, including hypolipidemia, cancer and diabetes [84], but also possesses anti-nociceptive, antiinflammatory, antimicrobial and antioxidative properties [85].

Cumin contains essential oil (2.0-4.0%), with cumin aldehyde being the most abundant compound, followed by β-pinene, γ-terpinene, γ-terpinene-7-al and p-cymene [85]. Further, cumin contains fatty oil, proteins, carbohydrates, and phenolic compounds (Table **8**).

In vitro anticancer properties of cumin seed ethanolic extract was determined using SRB assay [90], which showed a various degree of activity against SF-295 (25%), Colon 502713 (61%), Colo-205 (40%), Hep-2 and A-549(31%), OVCAR-5 (28%), PC-5 (27%) human cancer cell lines. The maximum anticancer activity of 61% cumin ethanolic extract was detected against Colon 502713 cell line.

Table 8. Cumin seed constituents.

Essential oil	2.3-5.7% [86]
Fatty oil	15.4-17.8%, with palmitic, petroselenic and linoleic acids as predominant [87]
Protein	18-22% [88]
Carbohydrates	55.58-60.05% [86]
Phenolic compounds	phenolic acid, ascorbic acid, tocopherol and pigments [89]

Before these findings, other authors confirmed cytotoxic activity of benzene extracts of cumin seeds against six types of tumour cell lines (HEPG2, HELA, HCT116, MCF7, HEP2, CACO2), with no cytotoxicity on the normal fibroblast cell line (BHK). The active component of this extract was 1-(2-ethyl, 6-heptyl) phenol (EHP), which also possess antifungal and antitumor activities [91].

CHERVIL

There are two species of chervil, cultivated (*Anthriscus cerefolium*), also called garden chervil, and wild (*A. sylvestris*). However, garden chervil is an annual plant usually cultivated for its leaves which are used as fresh and chopped added to soups, salads and fish dishes. Wild chervil is a biannual plant, whose roots have been traditionally used in medicinal formulations.

The dominant compound of the essential oil form aboveground parts of *A. cerefolium* is methyl chavicol (83.10%), followed by 1-allyl-2,4-dimethoxybenzene (15.15%) [92], while the beta-phellandrene (39-45%) was the main component in the leaf essential oil from *A. sylvestris*, and other compounds included beta-myrcene (17%), sabinene (6.2%), cis-beta-ocimene (5.4%) and benzene acetaldehyde (4.1%) [93].

The main biologically active compound from wild chervil roots is deoxypodophyllotoxin (DPPT) which has antitumor and antiproliferative effects [94, 95]. Additionally, excellent inhibitory activity (MTT method) against human hepatocellular carcinoma (HepG2), human osteosarcoma cells (MG-63), melanoma cells (B16) and human cervical carcinoma cells (HeLa) lines was reported for the petroleum ether fraction of wild chervil extract [96].

ERYNGO

Eryngo (*Eryngium campestre*) is a perennial plant spontaneously growing in stony pastures and dry meadows. In the Mediterranean region, these plants have been used as a food or in traditional remedies to treat various ailments. The major compound in essential oil is germacrene D [97]. Apart from essential oil, eryngo roots contain saponins [98].

Its potent cytotoxicity was confirmed by the MTT assay against the following tumour cells: human malignant melanoma (A375), human breast adenocarcinoma cells (MDA-MB 231) and human colon carcinoma cells (HCT116) [97].

CONCLUSION

Plants from Apiaceae family have significant health-promoting properties with the potential to prevent oxidative stress-related diseases, among them cancer, and can be further developed into functional food, *i.e.*. food with potential alternative natural antioxidants. Apart from essential oil which gives this plants its characteristic aroma and biological activity, the phenolics content may contribute directly to the antioxidant activity.

CONSENT FOR PUBLICATION

Not applicable.

CONFLICT OF INTEREST

The author (editor) declares no conflict of interest, financial or otherwise.

ACKNOWLEDGMENT

Declare None.

REFERENCES

[1]   Desai AG, Qazi GN, Ganju RK, *et al.* Medicinal plants and cancer chemoprevention. Curr Drug Metab 2008; 9(7): 581-91.
[http://dx.doi.org/10.2174/138920008785821657] [PMID: 18781909]

[2]   Greenwell M, Rahman PKSM. Medicinal plants: their use in anticancer treatment. Int J Pharm Sci Res 2015; 6(10): 4103-12.
[PMID: 26594645]

[3]   Shabani A. A review of anticancer properties of herbal medicines. J Pharma Care Health Sys 2016; 3: 2.
[http://dx.doi.org/10.4172/2376-0419.1000160]

[4]   Wang H, Khor TO, Shu L, *et al.* Plants *vs.* cancer: a review on natural phytochemicals in preventing and treating cancers and their druggability. Anticancer Agents Med Chem 2012; 12(10): 1281-305.
[http://dx.doi.org/10.2174/187152012803833026] [PMID: 22583408]

[5]   Cencic A, Chingwaru W. The role of functional foods, nutraceuticals, and food supplements in intestinal health. Nutrients 2010; 2(6): 611-25.
[http://dx.doi.org/10.3390/nu2060611] [PMID: 22254045]

[6]   Aćimović M, Kostadinović Lj, Popović S, Dojčinović N. Apiaceae seeds as functional food. Journal of Agricultural Sciences (Belgrade) 2015; 60(3): 237-46.
[http://dx.doi.org/10.2298/JAS1503237A]

[7]   Aćimović M, Kostadinović Lj, Puvača N, Popović S, Urošević M. Phytochemical constituents of selected plants from Apiaceae family and their biological effects in poultry. Food Feed Res 2016; 43(1): 35-41.
[http://dx.doi.org/10.5937/FFR1601035A]

[8]   Santora R, Kozar RA. Molecular mechanisms of pharmaconutrients. J Surg Res 2010; 161(2): 288-94.
[http://dx.doi.org/10.1016/j.jss.2009.06.024] [PMID: 20080249]

[9]   Aćimović M, Milić N. Perspectives of the Apiaceae hepatoprotective effects – a review. Nat Prod Commun 2017; 12(2): 309-17.

[10]   Assessment report on *Carum carvi* L., fructus and *Carum carvi* L., aetheroleum. Committee on Herbal Medicinal Products (HMPC) EMA/HMPC/715093/2013 2015.

[11] Aćimović M, Oljača S, Tešević V, Todosijević M, Đisalov J. Evaluation of caraway essential oil from different production areas of Serbia. Hortic Sci (Prague) 2014; 41(3): 122-30.
[http://dx.doi.org/10.17221/248/2013-HORTSCI]

[12] Aydın E, Türkez H, Keleş MS. Potential anticancer activity of carvone in N2a neuroblastoma cell line. Toxicol Ind Health 2015; 31(8): 764-72.
[http://dx.doi.org/10.1177/0748233713484660] [PMID: 23552268]

[13] Deepak GY, Shashikant YA. Extraction and pharmacological screening of carvone and it's derivatives. International Journal of Pharmaceutical Archive 2014; 3(1): 273-84.

[14] Naderi-Kalali B, Allameh A, Rasaee MJ, *et al.* Suppressive effects of caraway (*Carum carvi*) extracts on 2, 3, 7, 8-tetrachloro-dibenzo-p-dioxin-dependent gene expression of cytochrome P450 1A1 in the rat H4IIE cells. Toxicol In Vitro 2005; 19(3): 373-7.
[http://dx.doi.org/10.1016/j.tiv.2004.11.003] [PMID: 15713544]

[15] Aćimović M, Milić N. Dill in traditional medicine and modern phytotherapy Lekovite sirovine 35; 23-35.
[http://dx.doi.org/10.5937/leksir1535023A]

[16] Rezaee-Asl M, Bakhtiarian A, Nikoui V, *et al.* Antinociceptive properties of hydro alcoholic extracts of *Anethum graveolens* L. (dill) seed and aerial parts in mice. Journal of Clinical and Experimental Pharmacology 2013; 2013(3): 2.
[http://dx.doi.org/10.4172/2161-1459.1000122]

[17] Lisiewska Z, Kmiecik W, Korus A. Content of vitamin C, carotenoids, chlorophylls and polyphenols in green parts of dill (*Anethum graveolens* L.) depending on plant height. J Food Compos Anal 2006; 19(2-3): 134-40.
[http://dx.doi.org/10.1016/j.jfca.2005.04.009]

[18] Aćimović M, Stanković J, Cvetković M, Jaćimović G, Dojčinović N. The study of morphological traits of dill and quality of it essential fruit oil. Annales of Scientific Work, Faculty of Agriculture. University of Novi Sad 2014; 38(1): 69-79.

[19] Badar N, Arshad M, Farooq U. Characteristics of *Anethum graveolens* (Umbelliferae) seed oil: extraction, composition and antimicrobial activity. Int J Agric Biol 2008; 10: 329-32.

[20] Aćimović M, Popović S, Kostadinović Lj, Đuragić O, Lević J. Investigation of fatty acid profile of anise, dill and caraway fruits. Uljarstvo 2016; 47(1): 9-14.

[21] Chahal KK. Chemistry and biological activities of *Anethum graveolens* L. (dill) essential oil: A review. Journal of Pharmacognosy and Phytochemistry 2017; 6(2): 295-306.

[22] Al-Snafi AE. The pharmacological importance of *Anethum graveolens*. A review. Int J Pharm Pharm Sci 2014; 6(4): 11-3.

[23] Peerakam N, Wattanathorn J, Punjaisee S, Buamongkol S, Sirisa-ard P, Chansakaow S. Chemical profiling of essential oil composition and biological evaluation of *Anethum graveolens* L. (seed) grown in Thailand. Journal of Natural Sciences Research 2014; 4(16): 34-41.

[24] Ashin FM, laila IMA (2018): The possible ameliorative role of Anethum graveolens against tramadol-induced testicular damage in male albino rat IOSR Journal of Pharmacy and Biological Sciences 13(1): 82-90.

[25] European Medical Agency (EMA). 2013.Assessment report on *Pimpinella anisum* L., fructus and *Pimpinella anisum* L., aetheroleum Committee on Herbal Medicinal Products (HMPC) EMEA/HMPC/321181/2012

[26] Aćimović M, Korać J, Jaćimović G, Oljača S, Đukanović L, Vuga-Janjatov V. Influence of ecological conditions on seeds traits and essential oil contents in anise (*Pimpinella anisum* L.). Not Bot Horti Agrobot Cluj-Napoca 2014; 42(1): 232-8.
[http://dx.doi.org/10.15835/nbha4219492]

[27]   Council of Europe 2011; pp. Cedex1043-272.

[28]   Besharati-Seidani A, Jabbari A, Yamini Y. Headspace solvent microextraction: a very rapid method for identification of volatile components of Iranian *Pimpinella anisum* seed. Anal Chim Acta 2005; 530(1): 155-61.
[http://dx.doi.org/10.1016/j.aca.2004.09.006]

[29]   Martins N, Barros L, Santos-Buelga C, Ferreira ICFR. Antioxidant potential of two Apiaceae plant extracts: a comparative study focused on the phenolic composition. Ind Crops Prod 2016; 79: 188-94.
[http://dx.doi.org/10.1016/j.indcrop.2015.11.018]

[30]   Rahamooz-Haghighi S, Asadi MH. Anti-proliferative effect of the extracts and essential oil of *Pimpinella anisum* on gastric cancer cells. J HerbMedPharmacol 2016; 5(4): 157-61.

[31]   Kadan S, Rayan M, Rayan A. Anticancer activity of anise (*Pimpinella anisum* L.) seed extract. Open Nutraceuticals J 2013; 6: 1-5.
[http://dx.doi.org/10.2174/1876396001306010001]

[32]   2008.Assessment report on Foeniculum vulgare Miller Committee on Herbal Medicinal Products (HMPC) EMEA/HMPC/137426/2006

[33]   Salama ZA, El Baz FK, Gaafar AA, Zaki MF. Antioxidant activities of phenolics, flavonoids and vitamin C in two cultivars of fennel (*Foeniculum vulgare* Mill.) in responses to organic and bio-organic fertilizers. J Saudi Soc Agric Sci 2015; 14(1): 91-9.
[http://dx.doi.org/10.1016/j.jssas.2013.10.004]

[34]   Koudela M, Petříková K. Nutritional compositions and yield of sweet fennel cultivars–*Foeniculum vulgare* Mill. ssp vulgare var azoricum (Mill) Thell Hort Sci (Prague) 2008; 35(1): 1-6.

[35]   Aćimović M, Popović S, Kostadinović Lj, Stanković J, Cvetković M. Characteristics of fatty acids and essential oil from sweet and bitter fennel fruits grown in Serbia. Sixth International Scientific Symposium "Agrosym 2015", October 15-18, 2015; 2015; pp. Jahorina, Bosnia and Herzegovina949-53. Book of Proceedings

[36]   Badgujar SB, Patel VV, Bandivdekar AH. *Foeniculum vulgare* Mill: a review of its botany, phytochemistry, pharmacology, contemporary application, and toxicology. BioMed Res Int 2014; 2014: 842674.
[http://dx.doi.org/10.1155/2014/842674] [PMID: 25162032]

[37]   Lall N, Kishore N, Binneman B, *et al.* Cytotoxicity of syringin and 4-methoxycinnamyl alcohol isolated from *Foeniculum vulgare* on selected human cell lines. Nat Prod Res 2015; 29(18): 1752-6.
[http://dx.doi.org/10.1080/14786419.2014.999058] [PMID: 25588942]

[38]   Mohamad RH, El-Bastawesy AM, Abdel-Monem MG, *et al.* Antioxidant and anticarcinogenic effects of methanolic extract and volatile oil of fennel seeds (*Foeniculum vulgare*). J Med Food 2011; 14(9): 986-1001.
[http://dx.doi.org/10.1089/jmf.2008.0255] [PMID: 21812646]

[39]   Pradhan M, Sribhuwaneswari S, Karthikeyan D, *et al. In vitro* cytoprotection activity of *Foeniculum vulgare* and *Helicteresisora* in cultured human blood lymphocytes and antitumour activity against B16F10 melanoma cell line. Research Journal of Pharmacology and Technology 2008; 1(4): 450-2.

[40]   Sharopov F, Valiev A, Satyal P, *et al.* Cytotoxicity of the Essential Oil of Fennel (*Foeniculum vulgare*) from Tajikistan. Foods 2017; 6(9): E73.
[http://dx.doi.org/10.3390/foods6090073] [PMID: 28846628]

[41]   Aćimović M, Oljača S, Dražić S. Uses of coriander (*Coriandrum sativum* L.). Lekovite Sirovine 2012; 31(31): 67-82.

[42]   Mandal S, Mandal M. Coriander (*Coriandrum sativum* L.) essential oil: Chemistry and biological activity. Asian Pac J Trop Biomed 2015; 5(6): 421-8.
[http://dx.doi.org/10.1016/j.apjtb.2015.04.001]

[43] Mahamane KA, Ahire PP, Nikam YD. Biochemical investigation of "*Coriandrum sativum*" L. (coriander). International Journal of Applied and Pure Science and Agriculture 2016; 2(8): 154-7.

[44] Kiralan M, Calikoglu E, Ipek A, Bayrak A, Gurbuz B. Fatty acid and volatile oil composition of different coriander (*Coriandrum sativum*) registered varieties cultivated in Turkey. Chem Nat Compd 2009; 45(1): 100-2.
[http://dx.doi.org/10.1007/s10600-009-9240-2]

[45] Abu-Hammour KA, Wittmann D. Seed contents of *Coriandrum sativum* in Jordan Valley. Adv Hortic Sci 2011; 25(4): 207-11.

[46] Mahendra P, Bisht S. *Coriandrum sativum*: A daily use spice with great medicinal effect. Pharmacogn J 2011; 3(21): 84-8.
[http://dx.doi.org/10.5530/pj.2011.21.16]

[47] Rajeshwari CU, Andallu B. Isolation and simultaneous detection of flavonoids in the methanolic and ethanolic extracts of *Coriandrum sativum* L. seeds by RP-HPLC. Pak J Food Sci 2011; 21(1-4): 13-21.

[48] Gomez-Flores R, Hernández-Martínez H, Tamez-Guerra P, *et al.* Antitumor and immunomodulating potential of *Coriandrum sativum, Piper nigrum* and *Cinnamomum zeylanicum*. Journal of Natural Products 2010; 3: 54-63.

[49] Nithya TG, Sumalatha D. Evaluation of *in vitro* anti-oxidant and anticancer activity of *Coriandrum sativum* against human colon cancer HT- 29 cell lines. Int J Pharm Pharm Sci 2014; 6(2): 421-4.

[50] Chouhan S, Priya VV, Gayathri R. Genotoxicity analysis of *Coriandrum sativum* on oral cancer cell line by DNA fragmentation. Int J Pharm Sci Rev Res 2017; 45(1): 18-20.

[51] Tang EL, Rajarajeswaran J, Fung SY, Kanthimathi MS. Antioxidant activity of *Coriandrum sativum* and protection against DNA damage and cancer cell migration. BMC Complement Altern Med 2013; 13: 347.
[http://dx.doi.org/10.1186/1472-6882-13-347] [PMID: 24517259]

[52] Kooti W, Ali-Akbari S, Asadi-Samani M, Ghadery H, Ashtary-Larky D. A review on medicinal plant of *Apium graveolens*. Advanced Herbal Medicine 2014; 1(1): 48-59.

[53] Aćimović M, Todosijević M, Stanković J, *et al.* Headspace analisys of volatile compounds from leafy forms of parsley and celery fruits. Ann Agron 2017; 41(2): 38-44.

[54] Aćimović M, Cvetković M, Stanković J, Malenčić Đ, Kostadinović Lj. Compound analysis of essential oils from lovage and celery fruits obtained by headspace extracton. Annales of Scientific Work 2014; 39(1): 44-51.

[55] Qureshi K, Tabassum F. Neelam, Amin M, Zain Akram M, Zafar M. Investigation of mineral constituents of *Apium graveolens* L available in Khyber Pakhtunkhwa Pakistan. Journal of Pharmacognosy and Phytochemistry 2014; 3(4): 234-9.

[56] Rożek E, Nurzyńska-Wierdak R, Sałata A, Gumiela P. The chemical composition of the essential oil of leaf celery (*Apium graveolens* l. var. *secalinum* Alef.) under the plants' irrigation and harvesting method. Acta Sci Pol Hortorum Cultus 2016; 15(1): 147-57.

[57] Sipailiene A, Venskutonis PR, Sarkinas A, Cypiene V. Composition and antimicrobial activity of celery (*Apium graveolens*) leaf and root extracts obtained with liquid carbon dioxide. Acta Hortic 2005; (677): 71-7.
[http://dx.doi.org/10.17660/ActaHortic.2005.677.9]

[58] Sowbhagya HB. Chemistry, technology, and nutraceutical functions of celery (*Apium graveolens* L.): an overview. Crit Rev Food Sci Nutr 2014; 54(3): 389-98.
[http://dx.doi.org/10.1080/10408398.2011.586740] [PMID: 24188309]

[59] Fazal SS, Singla RK. Review on the Pharmacognostical & Pharmacological Characterization of *Apium Graveolens* Linn. Indo Global Journal of Pharmaceutical Sciences 2012; 2(1): 36-42.

[60] Al-Snafi AE. The Pharmacology of *Apium graveolens* - A Review. International Journal for Pharmaceutical Research Scholars 2014; 3(1): 671-7.

[61] Al-Jumaily RMK. Evaluation of anticancer activities of crude extracts of *Apium graveolens* L. seeds in two cell lines, RD and L20B *in vitro.* Iraqi Journal of Cancer and Medical Genetics 2010; 3(2): 18-23.

[62] Köken T, Koca B, Özkurt M, Erkasap N, Kuş G, Karalar M. *Apium graveolens* extract inhibits cell proliferation and expression of vascular endothelial growth factor and induces apoptosis in the human prostatic carcinoma cell line LNCaP. J Med Food 2016; 19(12): 1166-71.
[http://dx.doi.org/10.1089/jmf.2016.0061] [PMID: 27982754]

[63] Subhadradevi V, Khairunissa K, Asokkumar K, Umamaheswari M, Sivashanmugam A, Jagannath P. Induction of apoptosis and cytotoxic activities of *Apium graveolens* Linn. using *in vitro* models. Middle East J Sci Res 2011; 9(1): 90-4.

[64] 2012.Assessment report on Levisticum officinale Koch, radix Committee on Herbal Medicinal Products (HMPC) EMA/HMPC/524623/2011

[65] Aćimović M, Stanković J. Lovage (*Levisticum officinale* Koch.): plant with great pofor use in organic agriculture. Plant Doctor 2015; 43(5): 434-42.

[66] Sertel S, Eichhorn T, Plinkert PK, Efferth T. Chemical Composition and antiproliferative activity of essential oil from the leaves of a medicinal herb, *Levisticum officinale*, against UMSCC1 head and neck squamous carcinoma cells. Anticancer Res 2011; 31(1): 185-91.
[PMID: 21273597]

[67] Aćimović M, Stanković J, Cvetković M, Ignjatov M, Nikolić Lj. Chemical characterization of essential oil from seeds of wild and cultivated carrots from Serbia. Bot Serb 2016; 40(1): 55-60.

[68] Simpson K, Cerda A, Stange C. Carotenoid Biosynthesis in *Daucus carota.* Subcell Biochem 2016; 79: 199-217.
[http://dx.doi.org/10.1007/978-3-319-39126-7_7] [PMID: 27485223]

[69] Shebaby WN, El-Sibai M, Smith KB, Karam MC, Mroueh M, Daher CF. The antioxidant and anticancer effects of wild carrot oil extract. Phytother Res 2013; 27(5): 737-44.
[http://dx.doi.org/10.1002/ptr.4776] [PMID: 22815230]

[70] Mroueh MA, Shebaby W, Smith K, *et al.* 2013; The anti-cancer effect of the pentane fraction of *Daucus carota* oil extract is mediated through cell cycle arrest and an increase in apoptosis PlantaMedica 79

[71] Stanković M, Stojanović N, Nikolić N, Novković V. The extraction of total lipids from parsley (*Petroselinum crispum* (Mill.) Nym. ex. A.W. Hill) seeds. Chem Ind 2004; 58(12): 563-6.
[http://dx.doi.org/10.2298/HEMIND0412563S]

[72] Mert A, Timur M. Essential oil and fatty acid composition and antioxidant capacity and total phenolic content of parsley seeds (*Petroselinum crispum*) grown in Hatay Region Indian Journal of Pharmaceutical Education and Research ; 51(3): S437-440.

[73] Stanković M, Stanojević Lj, Nikolić N, Cakić M. The effect of parsley (*Petroselinum crispum* (Mill.) Nym. ex. A.W. Hill) seeds milling and fermentation conditions on essential oil yield and composition. CI & CEQ 2005; 11(4): 177-82.
[http://dx.doi.org/10.2298/CICEQ0504177S]

[74] Rezazad M, Farokhi F. Protective effect of Petroselinum crispum extract in abortion using prostadin-induced renal dysfunction in female rats. Avicenna J Phytomed 2014; 4(5): 312-9.
[PMID: 25386393]

[75] Farzaei MH, Abbasabadi Z, Ardekani MRS, Rahimi R, Farzaei F. Parsley: A review of ethnopharmacology, phytochemistry and biological activities J Tradit Chin Med 2013; 15((33)6): 815-26.

[76] Farshori NN, Al-Sheddi ES, Al-Oqail MM, Musarrat J, Al-Khedhairy AA, Siddiqui MA. Anticancer

activity of *Petroselinum sativum* seed extracts on MCF-7 human breast cancer cells. Asian Pac J Cancer Prev 2013; 14(10): 5719-23.
[http://dx.doi.org/10.7314/APJCP.2013.14.10.5719] [PMID: 24289568]

[77] Tang EL, Rajarajeswaran J, Fung S, Kanthimathi MS. *Petroselinum crispum* has antioxidant properties, protects against DNA damage and inhibits proliferation and migration of cancer cells. J Sci Food Agric 2015; 95(13): 2763-71.
[http://dx.doi.org/10.1002/jsfa.7078] [PMID: 25582089]

[78] Farshori NN, Al-Sheddi ES, Al-Oqail MM, Musarrat J, Al-Khedhairy AA, Siddiqui MA. Cytotoxicity assessments of *Portulaca oleracea* and *Petroselinum sativum* seed extracts on human hepatocellular carcinoma cells (HepG2). Asian Pac J Cancer Prev 2014; 15(16): 6633-8.
[http://dx.doi.org/10.7314/APJCP.2014.15.16.6633] [PMID: 25169500]

[79] Aćimović M, Cvetković M, Stanković J, Filipović V, Nikolić Lj, Dojčinović N. Analysis of volatile compounds from Angelica seeds obtained by headspace method. Arabian Journal of Medicinal and Aromatic Plants 2017; 3(1): 10-7.

[80] Aćimović M, Pavlović S, Varga A, *et al*. Chemical composition and antibacterial activity of *Angelica archangelica* root essential oil. Nat Prod Commun 2017; 12(2): 205-6.

[81] Kumar D, Bhat ZA, Kumar V, Chashoo IA, Khan NA, Shah MY. Pharmacognostical and phytochemical evaluation of *Angelica archangelica* Linn. International Journal of Drug Development and Research 2011; 3(3): 173-88.

[82] Sigurdsson S, Ogmundsdottir HM, Gudbjarnason S. The cytotoxic effect of two chemotypes of essential oils from the fruits of *Angelica archangelica* L. Anticancer Res 2005; 25(3B): 1877-80. a
[PMID: 16158920]

[83] Sigurdsson S, Ogmundsdottir HM, Hallgrimsson J, Gudbjarnason S. Antitumour activity of *Angelica archangelica* leaf extract. *In Vivo* 2005; 19(1): 191-4. b
[PMID: 15796173]

[84] Mnif S, Aifa S. Cumin (*Cuminum cyminum* L.) from traditional uses to potential biomedical applications. Chem Biodivers 2015; 12(5): 733-42.
[http://dx.doi.org/10.1002/cbdv.201400305] [PMID: 26010662]

[85] Aćimović M, Tešević V, Mara D, Cvetković M, Stanković J, Filipović V. The analysis of cumin seeds essential oil and total polyphenols from postdestillation waste material. Advanced Technologies 2016; 5(1): 23-30.
[http://dx.doi.org/10.5937/savteh1601023A]

[86] Al-Snafi AE. The pharmacological activities of *Cuminum cyminum* - A review. IOSR Journal of Pharmacy 2016; 6(6): 46-65.

[87] Bettaieb I, Bourgou S, Sriti J, Msaada K, Limam F, Marzouk B. Essential oils and fatty acids composition of Tunisian and Indian cumin (*Cuminum cyminum* L.) seeds: a comparative study. J Sci Food Agric 2011; 91(11): 2100-7.
[http://dx.doi.org/10.1002/jsfa.4513] [PMID: 21681765]

[88] Badr FH, Georgiev EV. Amino acid composition of cumin seed (*Cuminum cyminum*L.). Food Chem 1990; 38(4): 273-8.
[http://dx.doi.org/10.1016/0308-8146(90)90184-6]

[89] Agrawal D, Sharma LK, Rathore SS, Zachariah TZ, Saxena SN. Analysis of total phenolics and antioxidant activity in seed and leaf extracts of cumin genotypes International J. Seed Spices 2016; 6(1): 43-9.

[90] Prakash E, Kumar Gupta D. Cytotoxic activity of ethanolic extract of *Cuminum cyminum* Linn against seven human cancer cell line. Universal Journal of Agricultural Research 2014; 2(1): 27-30.

[91] Mekawey AAI, Mokhtar MM, Farrag RM. Antitumor and antibacterial activities of [1-(2-Ethyl, 6-Heptyl) Phenol] from *Cuminum cyminum* seeds. J Appl Sci Res 2009; 5(11): 1881-8.

[92] Baser KHC, Ermin N, Demirçakmak B. The essential oil of *Anthriscus cerefolium* (L.) Hoffm. (Chervil) growing wild in Turkey. J Essent Oil Res 1998; 10: 463-4.
[http://dx.doi.org/10.1080/10412905.1998.9700944]

[93] Bos R, Koulman A, Woerdenbag HJ, Quax WJ, Pras N. Volatile components from *Anthriscus sylvestris* (L.) Hoffm. J Chromatogr A 2002; 966(1-2): 233-8.
[http://dx.doi.org/10.1016/S0021-9673(02)00704-5] [PMID: 12214699]

[94] Yong Y, Shin SY, Lee YH, Lim Y. Antitumor activity of deoxypodophyllotoxin isolated from *Anthriscus sylvestris*: Induction of G2/M cell cycle arrest and caspase-dependent apoptosis. Bioorg Med Chem Lett 2009; 19(15): 4367-71.
[http://dx.doi.org/10.1016/j.bmcl.2009.05.093] [PMID: 19501508]

[95] Olaru OT, Niţulescu GM, Orţan A, Dinu-Pîrvu CE. Ethnomedicinal, Phytochemical and Pharmacological Profile of Anthriscus sylvestris as an Alternative Source for Anticancer Lignans. Molecules 2015; 20(8): 15003-22.
[http://dx.doi.org/10.3390/molecules200815003] [PMID: 26287153]

[96] Chen H, Jiang HZ, Li YC, Wei GQ, Geng Y, Ma CY. Antitumor constituents from *Anthriscus sylvestris* (L.) Hoffm. Asian Pac J Cancer Prev 2014; 15(6): 2803-7.
[http://dx.doi.org/10.7314/APJCP.2014.15.6.2803] [PMID: 24761904]

[97] Cianfaglione K, Blomme EE, Quassinti L, *et al.* Cytotoxic essential oils from *Eryngium campestre* and *Eryngium amethystinum* (Apiaceae) growing in Central Italy. Chem Biodivers 2017; 14(7)
[http://dx.doi.org/10.1002/cbdv.201700096] [PMID: 28332760]

[98] Kartal M, Mitaine-Offer AC, Paululat T, *et al.* Triterpene saponins from *Eryngium campestre.* J Nat Prod 2006; 69(7): 1105-8.
[http://dx.doi.org/10.1021/np060101w] [PMID: 16872157]

CHAPTER 8

Phytochemicals, Functionality and Breeding for Enrichment of Cole Vegetables (*Brassica oleracea* L.)

Saurabh Singh[1,*], Rajender Singh[2], Prerna Thakur[3] and Raj Kumar[4]

[1] *Division of Vegetable Science, ICAR-Indian Agricultural Research Institute (IARI), New Delhi, India-110012*

[2] *ICAR-Indian Agricultural Research Institute (IARI), Regional Station, Katrain, Kullu Valley, Himachal Pradesh, India-175129*

[3] *Department of Vegetable Science, Punjab Agricultural University (PAU), Ludhiana, Punjab, India-141004*

[4] *Division of Life Sciences, Plant Molecular Biology and Biotechnology Research Center, Research Institute of Natural Science, Gyeongsang National University, Jinju-52828, Republic of Korea*

Abstract: The increase in per capita consumption of vegetables rich in phytochemicals can improve the human nutrition status. Among the different kinds of vegetables, the cole group provides beneficial health effects attributed to the presence of diverse antioxidant compounds such as beta carotene, anthocyanins, ascorbic acid, phenolics, folic acid and organosulphur compounds like glucosinolates and so forth. In all the *Brassica oleracea* vegetables group, there are two different kinds of sulfur-containing phytochemicals vis. glucosinolates and S-methyl cysteine sulfoxide. In the 21st century, efforts are undergoing to improve the quality not only of grains but also of vegetable crops. The different spontaneous mutations representing regulatory genes conferring carotenoid and anthocyanin accumulation in cauliflower have been reported and offer a genetic resource for the development of new varieties with enhanced health-promoting properties and visual appeal. The enhancement of nutritional quality and improvement of glucosinolates for imparting human health benefits or processing fitness require not only the pursuit of breeding efficiency by marker-assisted selection or new analytical methods and modern biotechnological tools but also careful consideration of the organoleptic features of cole vegetables. This review presents a summary of recent advances related with phytochemical value in cole vegetables and future perspectives.

Keywords: Anthocyanin, Ascorbic Acid, Brassica vegetables, Breeding, Carotenoids, Flavonoids, Glucosinolates, Health Benefits, Nutritional Quality, Phytochemicals, Phenols.

---

* **Corresponding author Saurabh Singh:** Division of Vegetable Science, ICAR-Indian Agricultural Research Institute (IARI), New Delhi, India-110012; Tel: +91 9736253953; E-mail: horticulturesaurabh@gmail.com

Spyridon A. Petropoulos, Isabel C.F.R. Ferreira and Lillian Barros (Eds.)

INTRODUCTION

The *Brassicaceae* family comprises numerous economically important species used for fodder, oilseed and food purposes. About 3709 species of 338 genera comes under this large plant family *Brassicaceae* [1], and the genus Brassica belonging to this family is one of the most important genus comprising different ornamental, condiment, fodder, oilseed and vegetable crops [2]. A high intake of cole vegetables (cabbage, cauliflower, broccoli, Brussels sprout, kale, knol khol) is associated with reduction in chronic disease risk, reduction of neurodegenerative diseases, prevention of different types of cancer, diabetes, decrease of cardiovascular diseases and other health promoting benefits [3 - 11].The cole vegetables have been attributed to provide these health promoting benefits as they are rich sources of phytochemicals like phenolics, flavonoids, glucosinolates, carotenoids, anthocyanins, ascorbic acid, vitamin B_6, lutein, vitamin K, folate and vitamin E [3 - 14]. Due to their wide economic and health promoting benefits, Brassica vegetables have been the focus of keen research depending upon the content of different secondary metabolites [4, 15], since besides their significance for plant defense, some of these secondary metabolites are also important for human health.

Phytochemicals are non-nutrient bioactive natural compounds derived from plants having health promoting properties and are naturally present in fruits, vegetables and cereals [16]. The 'phyto' word of 'phytochemicals' is derived from Greek word *phyto*, meaning plant [16], hence phytochemicals are known as plant chemicals. The major classes of phytochemicals are phenolics, carotenoids, alkaloids, nitrogen containing compounds and organosulphur compounds. In the current decade, the interest in finding phytochemicals with useful biological activity in food has been increased owing to their beneficial health improving activities. Within this context, among the Brassicaceae family, the cole vegetables are eye catching Brassica crops as they are considered as an abundant source of phytochemicals as stated above. The anticancer properties of these vegetables have been attributed mainly to decomposition products of secondary metabolites such as glucosinolates, indoles and isothiocyanates, and phytoalexins, other antioxidants like carotenoids, anthocyanins and so forth [15 - 18]. Indole-3-carbinol, a natural component of cole vegetables and particularly of cabbage (*Brassica oleracea* var. *capitata* L.), has potential anticarcinogenic activity *via* different metabolic and hormonal pathways, and helps to reduction of tumors in reproductive organs [19 - 22]. It is also have been proved to be effective in reducing the growth of human breast cancer cells [22, 23]. Indole-3-carbinol is derived by the hydrolysis of glycobrassicin in cruciferous vegetables such as cabbage, broccoli and brussels sprout [23]. Similarly, broccoli (*Brasssica oleracea* L. var. *italica*) is known to contain a large quantity of aliphatic glucosinolate 'Glucoraphanin', which is metabolized to isothiocyanate 'sulforaphane' having superior anticancer activity [24 - 28]. In addition, broccoli, when grown in Se containing soils, is also known to accumulate a good amount of Selenium-methylselenocysteine (SeMSCys) [24, 29, 30] which has been attributed with enhanced anticancer properties [31]. Furthermore, water soluble antioxidants like phenolic compounds are major antioxidants of Brassica vegetables [32, 33]. Phenolics have reactive oxygen species (ROS) scavenging activity due to their electron-donating characteristics [33]. The most widespread polyphenols in Brassica vegetables are flavonoids (flavonols, anthocyanins) and hydroxycinnamic acids. The flavonols like quercetin, kaempferol, and anthocyanidins have higher antioxidant efficacy than vitamin E, vitamin C and carotenoids [33, 34]. Anthocyanins are also considered as important antioxidants in Brasssica vegetables (red cabbage, purple cauliflower, purple broccoli *etc.*). The major anthocyanins found in Brassica vegetables are cyanidin derivatives [32, 35, 36]. Numerous studies have reported that diphenolic compounds such as lignans are also present in Brassica vegetables particularly in broccoli, kale and Brussels sprouts [33, 37], with lariciresinol and pinoresinol being the major lingnans present in these vegetables [38]. Raw broccoli, cabbage and cauliflower also contain a rare vitamin, folic acid, which also acts as a coenzyme in the synthesis of nucleic acids (DNA, RNA) and proteins [33, 39]. Among the carotenoids, lutein and β-carotene are the most prevalent carotenoids in cole vegetables [33, 40]. Hence, it is well established that Brassica vegetables are rich sources of dietary phytochemicals. In the present chapter, we provide a brief account of the role of major phytochemicals present in cole vegetables (*B. oleracea*) and the current status of breeding efforts towards the enrichment of Brassica vegetables.

CAROTENOIDS

Carotenoids (carotenes and xanthophylls) are lipid soluble antioxidants imparting yellow, orange and red pigments in fruit and vegetables, while some of the carotenoids are precursors of vitamin A (β-carotene, β-cryptoxanthin and γ-carotene) [33, 39].There are more than 750 different kinds of carotenoids reported in nature [16, 44]. They contain a 40-carbon (C_{40}) skeleton of isoprene units (Fig. **1**). All of the carotenoids are biosynthesized from the hydrocarbon phytoene or 4,4'-diapophytoene [44]. The anticancer activity of carotenoids is suggested to be due to different mechanisms such as reduction in cell proliferation, antioxidant activity on mutagenesis and genotoxicity, immunomodulatory effects, enhancement of gap-junction cell communication, and increasing apoptosis of cancerous cells [64 - 66].

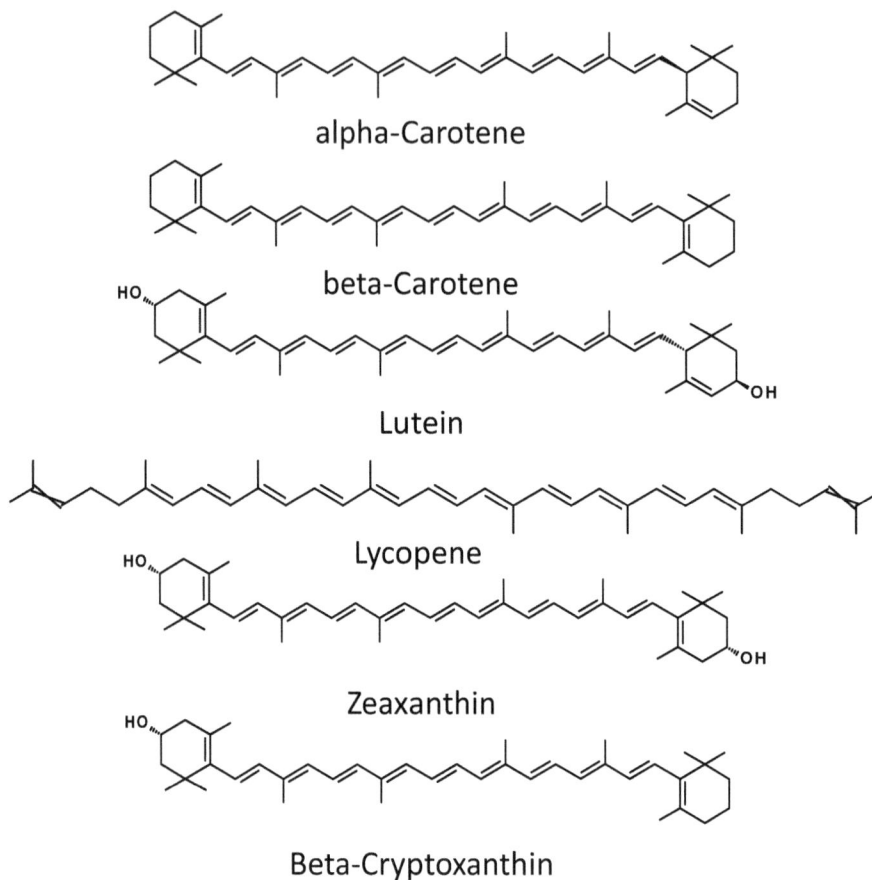

Fig. (1). Chemical structure of common carotenoids [42, 43].

The most abundant forms of carotenoids in Brassica vegetables are lutein and β-carotene [40]. Among the cole vegetables (*Brassica oleracea*), kale (*B. oleracea* var. *acephala*) has the highest carotenoid content (> 10 mg/100 g of edible portion) [41], Brussels sprouts have an intermediate amount of carotenoids (6.1 mg/100 g), whereas broccoli (1.6 mg/100 g), red cabbage (0.43 mg/100 g) and white cabbage 0.26 mg /100 g) contain low amounts of carotenoids [41]. The highest content of lutein + zeaxanthin has been recorded in kale (3.04-39.55 mg/100 g), while moderate amounts have been reported for broccoli and Brussels sprouts (0.78-3.50 mg/100 g) [40]. Dey *et al.* [45] reported an elevated level of carotenoids in two cytoplasmic male sterility (CMS) lines, Ogu13-85-2A and Ogu125-1A, of cauliflower *via* introgression of Ogura hybrid cytoplasm, and suggested their utilization as genetic resources for the development of carotene rich hybrids in Brassica vegetables. In the orange cauliflower curd (line 1227), β-carotene content was reported to range from 5 to 8 µg/g FW, while negligible amounts of lutein, neoxanthin and violaxanthin were recorded [46]. The orange colour of this genotype was due to *Or* gene mutation [46]. On the other hand, the 'Cheddar' cultivar of cauliflower accumulated three different carotenoids *viz.* neoxanthin (40.2 µg/g DW), violaxanthin (19.9 µg/g DW) and β-carotene (86.8 µg/g DW), respectively [47].

Carotenoid Biosynthetic Genes in Brassica Vegetables

Carotenoids are natural pigments derived from general isoprenoid pathway. The enzymes/genes regulating the biosynthetic pathway of carotenoids have been well studied in Arabidopsis [48]. Isopentenyl diphosphate (IPP) and dimethylallyl diphosphate produced by the 2-C-methyl-D-erythritol-4-phosphate (MEP) pathway are main pathways involved in the synthesis of carotenoids [48, 49]. The different enzymes involved in carotenoid biosynthesis from the upstream MEP pathway include 1-deoxy-D-xylulose-5-phosphate synthase (DXS), 1-deoxy-D-xylulose-5-phosphate reductoisomerase (DXR), 2-C-methyl-D-erythritol 4-phosphate cytidylyltransferase, 4-diphosphocytidyl-2-C-methyl-D-erythritol kinase (CMK), 2-C-methyl-D-erythritol 2,4-cyclodiphosphate synthase (MDS), 4-hydroxy-3-methylbut-2-enyl diphosphate synthase (HDS), and 4-hydroxy-3-methylbut-2-enyl diphosphate reductase (HDR) [49].

In case of orange cauliflower, the single semi-dominant '*Or*' gene mutation confers β-carotene accumulation in the curd tissue of cauliflower [46]. In the photosynthetic tissues of plant, carotenoids accumulate in the membranes of chloroplasts [46, 50], on the other hand, in non-photosynthetic tissues carotenoids accumulate in chromoplasts [51]. The orange colour in cauliflower exhibiting a novel '*Or*' gene mutation is due to massive and highly ordered sheets responsible for carotenoid (β-carotene) sequestration in plastids, which are classified as membranous chromoplasts [46, 52]. Thus, this novel gene mutation conferring accumulation of high amounts of β-carotene in L1227 inbred of orange cauliflower represents a useful genetic resource for development of carotenoid rich varieties/hybrids in vegetables including Brassica crops. It has been reported that the *Or* proteins are responsible for posttranscriptional regulation of phytoene synthase (PSY), depicting a major regulatory mechanism underlying carotenoid biosynthesis in plants [53].

Carotenoid Rich Breeding in Brassica Oleracea

Due to antioxidant potential of carotenoids, the breeding for the development of carotenoid rich vegetables is of prime importance in the present era. The advent of molecular markers has given the impetus to molecular breeding techniques and overcomes the limitations of conventional breeding, although the conventional breeding could be an important part of breeding techniques. Through the conventional breeding approach, a β-carotene rich orange cauliflower hybrid 'Jinyu 60' was developed involving CMS line as female parent and homozygous orange inbred 93-4 as donor parent [54]. The interaction of biochemical, physiological and genetic features of a plant species and environmental factors affect the carotenoid accumulation in plant tissues [39]. Farnham and Kopsell reported lutein to be the major carotenoid in broccoli ranging from 65.3 to 139.6 µg/g DW, among the nine inbreds of broccoli tested in three different environments [55]. During the recent years, significant progress has been made in the molecular mapping and identification of carotenoid biosynthetic genes in plant species. In this context, Li and Garvin [56] reported the molecular mapping of cauliflower *Or* gene by using AFLP and RFLP markers in conjunction with bulk segregation analysis (BSA) and these RFLP markers were converted into SCAR markers. Two of the RFLP markers flanked the *Or* locus at 0.5cM distance at one side and another marker flanked at 1.5cM distance [56]. Further, with the help of map based positional cloning strategy with bacterial artificial chromosome (BAC) vectors, the high resolution genetic and physical mapping of *Or* gene was carried out and suggested *Or* gene to be a 50kb DNA fragment within a single BAC clone, which is corresponding to a genetic interval of 0.3cM [57]. The successful introgression of *Or* gene in Indian cauliflower variety through the marker assisted backcross breeding (MABB) approach led to the development of β-carotene rich biofortified cauliflower hybrid 'Pusa Betakesari' with β-carotene content ranging from 800 to 1000 µg/100 g [58].

TOCOPHEROL (VITAMIN E)

Vitamin E is also an important antioxidant belonging to the group of lipid soluble antioxidants. Its antioxidant property is mainly due to free radical scavenging activity of α-tocopherol Fig. (**2**), with α-tocopheroxyl radical being an intermediate [33, 59, 60, 62]. α-tocopherol is the predominant tocopherol in almost all the Brassica vegetables with the exception of cauliflower, in which γ-tocopherol is the predominant tocopherol [33, 40, 63].

Among the Brassica vegetables broccoli (0.82 mg/100 g) contains the highest amount of total tocopherols followed by Brussels sprouts (0.40 mg/100 g), cauliflower (0.35 mg/100 g), red cabbage (0.05 mg/100 g) and white cabbage (0.04 mg/100 g) [40, 63].

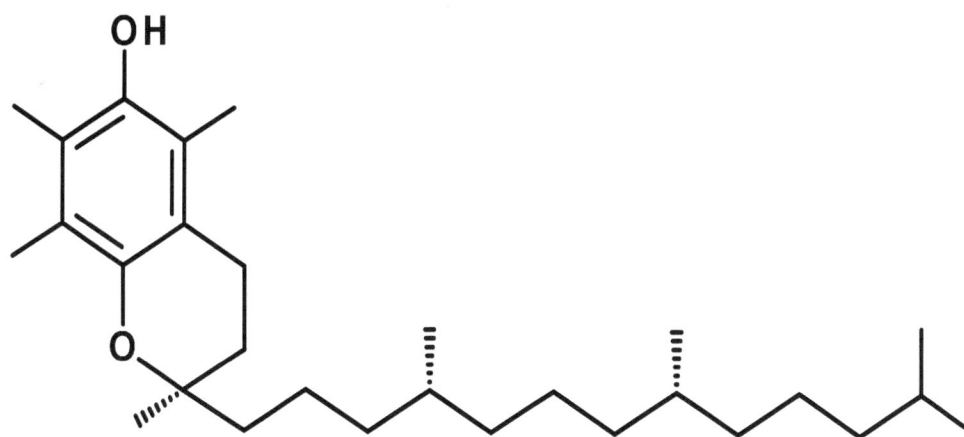

alpha-Tocopherol

Fig. (2). Structure of α-tocopherol [61].

ANTHOCYANINS

Anthocyanins are water soluble flavonoid antioxidants imparting red, blue, purple colour in many flowers, fruits and vegetables. They perform significant biological functions in plant defense mechanism against various biotic and abiotic stresses [35]. They have been proved to have antioxidant properties and impart significant human health benefits [67], like reduction of risks of cancers [68, 69], prevention of cardiovascular diseases [70], night vision improvement [71] and protection from neurological disorders [72]. Anthocyanins provide intense colour under acidic conditions. Anthocyanins possess a characteristic C6-C3-C6 carbon structure [73]. They show the maximum absorption at the visible wavelength range of 465 to 550 nm, but also exhibit significant absorption in the UV range between 270-280 nm [73, 74]. Till now, more than 635 types of anthocyanins have been identified in nature [74]. The word anthocyanin is derived from the Greek words 'anthos' and 'kyanos' meaning flower and dark blue, respectively [74, 75]. Anthocyanins are also used as food colorants mainly in the beverage industry. They significantly quench free radicals by terminating the chain reaction responsible for oxidative damage [74].There are six prevalent anthocyanidins found in edible plants *viz.* pelargonidin, peonidin, cyanidin, malvidin, petunidin, and delphinidin [73, 76].

Anthocyanins regulatory genes/enzymes and enrichment breeding in Brassica

The anthocyanins biosynthetic pathway and regulatory enzymes have been well studied in various plant species [77]. The anthocyanin synthesis is caused by two different classes of genes, the structural genes encoding the enzymes directly involved in the production of flavonoids and anthocyanins, and the dictatorial genes governing the transcription of structural genes [78, 79]. Numerous regulatory genes including R2R3 MYB transcription factors (TF), basic helix-loop-helix (bHLH) transcription factors, and WD40 proteins have been cloned from different plant species, such as Arabidopsis, maize, petunia, *B. rapa,* cauliflower, red cabbage, purple Chinese cabbage, purple kale and other *Brasssica* species, and other species [Table **1**]. Regulatory complexes are formed by them in activating the expression of anthocyanin biosynthetic genes [35, 80 - 82]. In Arabidopsis, different MYB transcription factors, including PAP1 (Production of anthocyanin pigment 1), PAP2, MYB113, MYB114, and MYBL2, three bHLH proteins of TT8 (Transparent Testa 8), GL3 (Glabra 3), and EGL3 (Enhancer of Glabra 3), and a WD40 repeat protein of TTG1 are involved in anthocyanin biosynthesis [35, 90].

Table 1. Transcription factors/genes/enzymes regulating anthocyanin pigmentation in *Brassica* species.

Brassica species	**Transcription factor/genes/enzymes**	**Reference (s)**
Brassica oleracea var. *botrytis*	*Pr* gene mutation encoding R2R3 MYB transcription factor	[35]
Red cabbage (*B. oleracea* var. *capitata*)	*BoMYB2* with *bHLH* transcription factor*BoTT8*	[81]
Purple heading chinese cabbage (*B. rapa* L. ssp. *pekinensis*)	*BrPur*	[83]
Ornamental kale (*B. oleracea* L. var. *acephala*)	*BoPr*	[84]
B. rapa L. ssp. *chinensis* var. *purpurea*	*BrEGL3.1* and *BrEGL3.2* encoding *bHLH* transcription factor	[82]
Purple kale (*B. oleracea* var. *acephala* f. *tricolor*)	*BoPAP 1*	[85]
Brassica rapa	*Anp*	[86]
Brassica napus	*BnaA.PL1*	[87]
Brassica rapa var. *chinensis*	*BrTT8*	[88]
Brassica oleracea var. *gongylodes*	*BoPAP2* and *BoTT8*	[89]

In purple cauliflower a spontaneous mutant, single semi-dominant *Pr* gene encoding MYB transcription factor is responsible for giving tissue specific expression of anthocyanin accumulation in different plant parts Fig. (**3**) [35]. The major anthocyanin compound responsible for anthocyanin pigmentation in purple cauliflower is cyanidin, 3, 5-diglucoside [35]. The anthocyanin accumulation in different tissues of purple cauliflower is due to the up- regulation of three late pathway genes *BoF3'H, BoDFR, and BoLDOX* [35]. The candidate gene analysis and high resolution mapping approach showed that *BoMYB2* is most likely the *Pr* gene [35]. The increased expression of structural genes, CHS (chalcone synthase), F3H (flavonone 3-hydroxylase), F3'H (flavonone 3' hydroxylase), DFR (dihydro flavonol reductase), LDOX (anthocyanidin synthase), GST (glutathione S-trans-ferase) and *bHLH* transcription factor, *BoTT8*and MYB transcription factor, *BoMYB2,* is associated with anthocyanin accumulation in red cabbage [81].

Fig. (3). A) Anthocyanin pigmentation in young leaves. **B)** Anthocyanin expression in curd.

The R2R3MYB transcription factor, *BoPAP1* gene and transcriptional activation of *BoPAP2* and *BoTT8* accounting for up-regulation of anthocyanin biosynthetic structural genes, are responsible for anthocyanin accumulation in purple kale and purple kohlrabi, respectively [85, 89]. Thus due to health promoting activity of anthocyanins these Brassica vegetables can be used as a potential genetic resource for anthocyanin rich breeding in other Brassica species including other plants and for production of antioxidant rich hybrids/varieties. Further, with the advent of advances in molecular breeding technologies it is possible to understand the downstream regulatory mechanism of anthocyanin biosynthesis in various plant species which will enhance the development of phytochemicals rich vegetables.

Glucosinolates

Glucosinolates (GS) are nitrogen- and sulfur-rich plant amino acid-derived secondary metabolites found in *Arabidopsis thaliana* and the *Brassicaceae* family, such as broccoli, cabbage, cauliflower, kale, and turnip [108, 112]. Glucosinolates (*S*-glucopyranosyl thiohydroximates), are naturally occurring S-linked glucosides and enzymatically hydrolyzed to produce sulfate ions, D-glucose, and characteristic degradation products such as isothiocyanate. Glucosinolates are *Brassicaceae*-specific secondary metabolites that act as crop protectants, flavor precursors, and cancer-prevention agents, which show strong evidences of anticarcinogenic, antioxidant, and antimicrobial activities in cruciferous and other vegetables. The chemical structure and total content of glucosinolates vary between species, and between varieties within a species. Within any particular variety of vegetable, the glucosinolate content is influenced greatly by cultivation conditions. Others synonyms of Glucosinolatesare Sinigrin or 2-Propenyl glucosinolate, Glucoside of allyl isothiocyanate or Potassium 1 (β-D-glucopyranosylthio) but-3-enylideneaminooxysulphonate or Glucopyranose, 1-thio-,1-(3-butenohydroximate) NO-(hydrogensulfate), monopotassium salt, β-D- Glucopyranose, 1-thio-, 1-(N-(sulfooxy)-3-butenimidate), monopotassium salt (9Cl) Glucopyranose, 1-thio-, 1-(3-butenohydroximate) NO-(hydrogen sulfate), monopotassium salt, β-D- (8Cl) [91] Fig. (**4**). Among the various varieties of vegetables, *Brassicaceae* vegetables have received the most attention because their unique constituents, glucosinolates, are abundant in edible parts and are regarded as most likely to maintain human health through continuous consumption. In near future, the breeding of *Brassicaceae* vegetables especially cole crops, addressing with beneficial glucosinolates is expected to grow from a breeding point of view.

Despite the variation of glucosinolate side-chains, only seven structures correspond directly to a protein amino acid (alanine, valine, leucine, isoleucine, phenylalanine, tyrosine and tryptophan). The remaining glucosinolates have side-chain structures which arise in three ways. Firstly, many glucosinolates are derived from chain-elongated forms of protein amino acids, notably from methionine and phenylalanine. Secondly, the structure of the side chain may be modified after amino acid elongation and glucosinolate biosynthesis, for example, the oxidation of methionine sulphur to sulphinyl, sulphonyl and by the subsequent loss of the o-methylsulphinyl group to produce a terminal double bond.

Subsequent modifications may also involve hydroxylation and methoxylation of the side chain. Thirdly, some other glucosinolates occur which contain relatively complex side-chains such as o-(α-L-rhamnopylransoyloxy)- benzyl glucosinolate in *Reseda odorata* [92] and glucosinolates containing a sinapoyl moiety in *Raphanus sativus* [93].

A

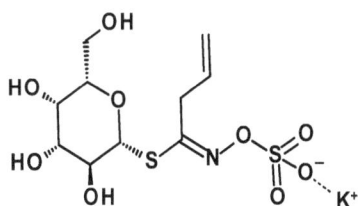

B

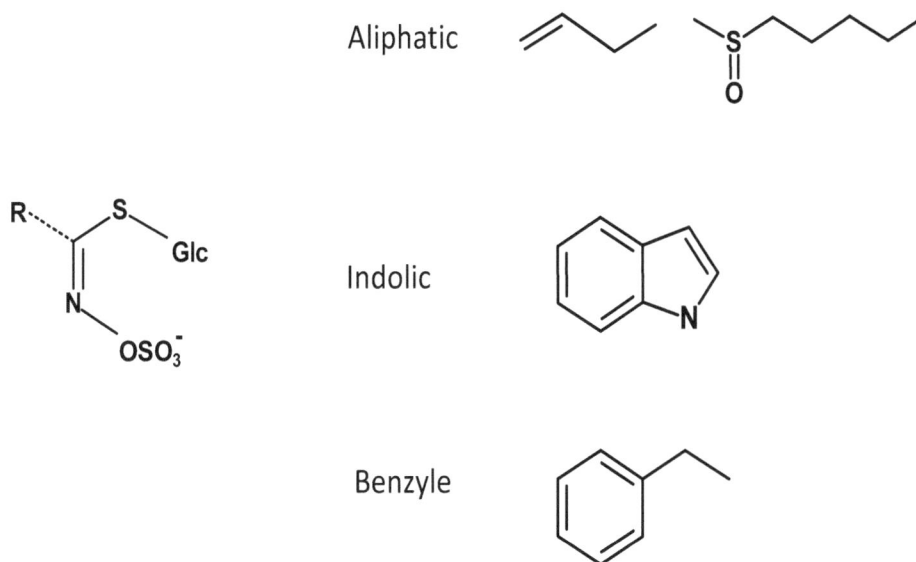

Fig. (4). A-Chemical structure of Glucosinolates; B-Basic chemical structure of glucosinolate. R variable side chain, S sulfur, N nitrogen and Glc glucose moiety. Right panel shows examples of variable side chain (R).

The Biosynthetic Pathway of Glucosinolates

The biosynthetic pathways of glucosinolates and the break-down of these biomolecules present aspecies variation, especially since these pathways are quite different in *Brassica oleracea* compared to *Arabidopsis* and *Brassica rapa*. Previously, several studies have been performed in order to increase the understanding of the glucosinolate biosynthetic pathway, specifically the genes involved in their biosynthetic pathway and their regulation. Glucosinolates are classified in three groups depending on the amino acid from which they are biosynthesized. These compounds are synthesized through three different steps: i) side chain elongation of amino acids; ii) formation of the core glucosinolate structure; iii) secondary modifications of side-chain Figs. **(4)**, **(5a)**, **(5b)** and **(6)** [94, 95]. During the chain elongation, aliphatic and aromatic amino acids are elongated by introducing methylene group but this does not occur in the formation of indole glucosinolates [96].

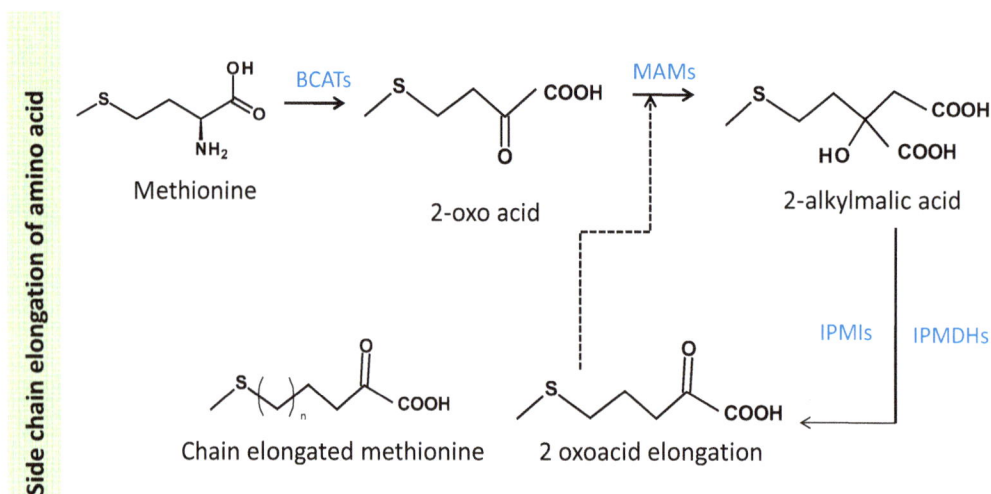

Fig. (5a). The chain elongation pathway of aliphatic glucosinolates from methionine. BCATs branched –chain amino acid aminotransferase, MAMs (methylthioalkylmalate synthase), IPMIs (Isopropylmalate isomerase), IPMDHs (Isopropylmalate dehydrogenase).

In this process, branched-chain amino acid aminotransferase (BCAT), bile acid transporter 5 (BAT5), methylthioalkylmalate synthase (MAM), isopropylmalate isomerase (IPMI), and isopropylmalate dehydrogenase (IPMDH) are involved. For this, aliphatic and aromatic amino acids are deaminated to form the corresponding 2-oxo acid by the action of branched-chain amino acid aminotransferase (BCATs). Then, the 2-oxo acid undergoes three successive reactions, in which 2-oxo acid is condensed with acetyl-CoA by the action of methylthioalkylmalate synthase (MAMs). Then, an isomerization process is catalyzed by isopropylmalate isomerase (IPMIs), and finally an oxidative decarboxylation is produced in a reaction catalyzed by isopropylmalate dehydrogenases (IPMDHs) [96 - 98] (Fig. **5a**).

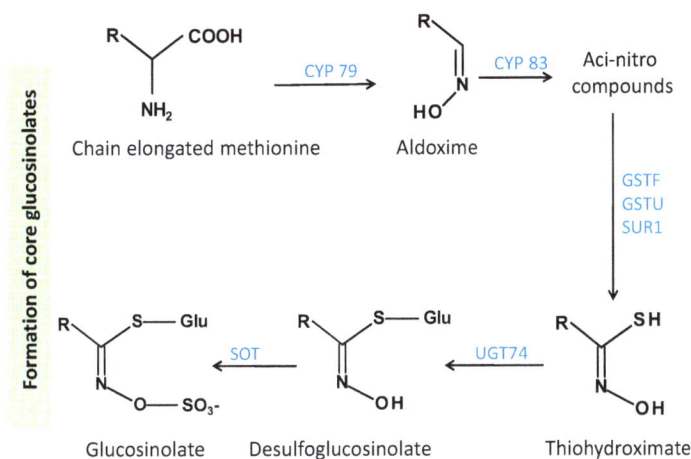

Fig. (5b). Biosynthesis of the glucosinolate core structure. GSTF (phi glutathione S-transferase), GSTU (tau glutathione S-transfease), SUR1 (Carbon-sulfur lyase), UGT74 (Uridine diphosphate glycosyltransferase 74), SOT (Sulfotransferase).

The formation of the core glucosinolate structure is initiated through the conversion of elongated amino acids to aldoximes by the action of cytochrome P450 monooxygenases (CYP79, (Fig. **5b**)). Aldoximes are oxidized by CYP83 to form aci-nitro compounds, and subsequently they are transformed into S-alky--thiohydroximate and thiohydroximate by the action of phi and tau glutathione S-transferases (GSTF and GSTU) and carbon–sulfur lyase (SUR1). This thiohydroximate undergoes two subsequent enzymatic reactions in order to form glucosinolate structure by uridine diphosphate glycosyltransferase 74 (UGT74) and sulfotransferases (SOT) (Fig. **5b**). In the case of indole glucosinolates, tryptophan is not elongated by introducing methylene group but the subsequent steps for the formation of glucosinolate structure are similar to that described for the aliphatic glucosinolates [97]. The last step involves the modifications of side chains by oxidation, hydroxylation, methoxylation, alkenylation, and benzoylation [98]. Thus, in the formation of aliphatic glucosinolates, the S-oxygenation is carried out by a flavinmonooxygenase (FMOGS-OXs, (Fig. **6**)). In the case of indolic glucosinolates, CYP81F has been identified as the gene that encodes the enzymes involved in the oxidation of indolyl-3-methyl glucosinolate (gluco-brassicin) to form 4-hydroxyindol-3-ylmethyl glucosinolate (hydroxygluco-brassicin), 4-methoxyindol-3-ylmethyl glucosinolate (methoxyglucobrassicin), and 1-methoxyindol-3-ylmethyl glucosinolate (neoglucobrassicin). These three different methoxylations occur by non-identified O-methyltransferases to form these three types of glucosinolates [96 - 98].

As regards the regulation process, several MYB transcription factors, belonging to subgroup 12 of the R2R3 MYB family, have been identified in *A. thaliana* [96, 98]. In fact, MYB34, MYB51, and MYB122 transcription factors are involved in the transcriptional regulation of indole glucosinolates [99] while MYB28, MYB29, and MYB76 transcription factors are involved in the transcriptional regulation of aliphatic glucosinolates [98, 100, 101]. Several studies have shown that dominant mutants or lines overexpressing MYB transcription factors were able to increase both the production of indole glucosinolates, as well as the expression of genes involved in their biosynthetic pathway [99, 100, 102]. Moreover, Frerigmann and Gigolashvili [103] showed that MYB34 and MYB51 transcription factors act through controlling the biosynthetic pathway of indolyl-3-methyl glucosinolate in shoots and roots of *A. thaliana*, respectively. These authors also showed that MYB51 is the key factor of indole glucosinolates when triple knockout mutants (myb34/51/122) plants were treated with salicylic acid andethylene while MYB34 is the key factor in *Arabidopsis* plants treated with abscisic acid and jasmonic acid. In addition, MYB122 is the key factor in the glucosinolate biosynthetic pathway induced by jasmonate and ethylene.

MYB28 and MYB29 transcription factors have been identified as key regulators of the biosynthesis of aliphatic glucosinolates. In fact, analysis of a knockout mutant and ectopic expression of MYB28 showed that this gene is a positive regulator for the biosynthesis of aliphatic glucosinolates. In addition, MYB29 plays a key role in methyl jasmonate-mediated responses inducing a set of genes involved in aliphatic glucosinolate biosynthetic pathway [105]. Similarly, Gigolashvili *et al.* [104] used activation tag approach to demonstrate that MYB28 is also a key regulator of the biosynthesis of aliphatic glucosinolates. On the other hand, other transcription factors, such as IQD1 and Dof1.1 (a member of the DNA-binding-with one- finger transcription factors), are also able to activate both aliphatic and indole glucosinolate biosynthesis [106 - 109]. In fact, Skirycz *et al.* [110] showed that constitutive and inducible over-expression of Dof1.1 in *A. thaliana* plants induced the expression of CYP83B1 gene which is involved in the biosynthesis of indole glucosinolates, indicating that Dof1.1 plays a regulatory role controlling biosynthesis of glucosinolate in *Arabidopsis*. In the same way, IQD1 gene, a calmodulin binding nuclear protein, was also able to alter the expression of several genes involved in glucosinolate metabolism. In fact, loss-of-function IQD1 alleles provoked a decrease in glucosinolate levels in *A. thaliana* plants [111].

Elongation of Side Chain

Only Methionine and phenylalanine as precursor amino acids undergo side-chain elongation. This phase requires five reactions: an initial and final transamination, an acetyl-CoA condensation, an isomerisation, and an oxidative decarboxylation [112, 113]. Initially, the parent amino acid is deaminated by cytosolic branched-chain amino acid aminotransferase 4 (BCAT4) to form the corresponding 2-oxo acid [114]. As the subsequent enzymes are localized in chloroplast, 2-oxo acid is transported into chloroplast by plastidic transporter bile acid transporter 5 (BAT5) [115]. Next, in chloroplast, there is a three-step cycle in which 2-oxo acid reacts with acetyl-CoA in a condensation reaction catalysed by the methylthioalkylmalate synthases MAM1 to MAM3 to form a 2-alkylmalic acid [116, 117]. The 2 alkylmalic acid isomerises *via* a 1,2-hydroxyl shift to a 3-alkylmalic acid by an isopropylmalate isomerase (IPMI) [118]. The 3-alkylmalic acid undergoes oxidation- decarboxylation to yield a 2-oxo acid with one more methylene group than the starting compound in the side chain. This reaction is assumed to be catalysed by isopropylmalate dehydrogenase (IPMDH) [119]. The resulting 2-keto acid then can either be transaminated to form the corresponding chain-elongated amino acids by the chloroplastic BCAT3 [120] and then be exported from the chloroplast into the cytosol by so far unknown transporter(s) to be further converted into glucosinolates, or enter a new condensation cycle that insert up to nine methylene unit [121].

Phenolic Compounds

Phenolic compounds are known as the ubiquitous class of secondary metabolites which exhibit wide structural and functional diversity in plants [137, 199]. Among different groups, *Brassica oleracea* group is considered as significant source of secondary plant metabolites particularly flavonoids and polyphenols [128, 160] and high intraspecific variability in *Brassica oleracea* group extends to this aspect as well [171]. Phenolic compounds have demanded attention in the recent past due to increasing importance for human health, crop development, defense and adaptation [149]. Besides, phenolic compounds have functionality in plants as signaling pathway, physiological active compounds, attractants, feeding deterrents, antifungal activity, antiviral and antibacterial properties, offer plant resistance, regulate gene expression and modulate enzymatic pathway [153, 212, 190]. Phenolic compounds are constituents of pollen grain walls [195] and play an important role in the regulation of growth factors like auxin. Flavonoids are known to play various crucial functions in the plants; involving photo protection (Ultraviolet), reduction of reactive oxygen species (ROS), prevention of oxidative damage [122], antimicrobial effects [138]. The main function of antioxidants is to delay oxidation by inhibiting initiation or propagation of oxidizing reactions by free radicals. These free radicals cause oxidative damage that acts as causative factor for various chronic diseases [175, 182, 186]. It was also reported they are important in early plant development, crucial for cell wall biosynthesis and antioxidant activity [205]. Flavonoids enable plants to cope with adverse environmental stresses *viz.* heat, frost, drought [134, 210], and metal toxicity [125]. The flavonoids have been identified to act as chemical messengers or internal physiological regulators within the plant system [129]. These are localized in the nucleus of mesophyll cells and within the core of ROS generation sites. To perform function as antioxidants and UV protectants, flavonoids are accumulated in epidermal cell layers or cuticle of leaves and fruits [123, 166, 180, 188, 198, 208, 210]. Antioxidant activity of *Brassica* group has been ascribed mainly to the phenolic compounds and vitamin C that occurs through polyphenol propanoid pathway [218].

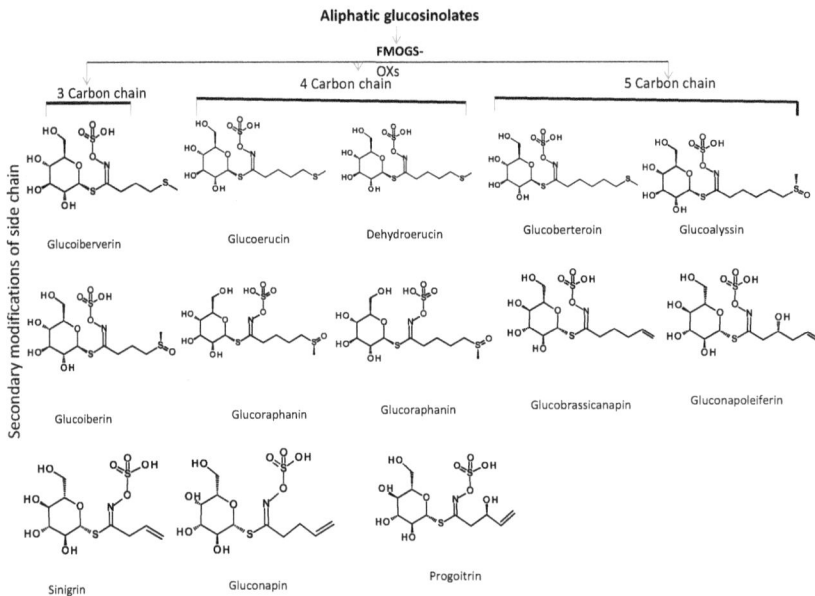

Fig. 6 cont.....

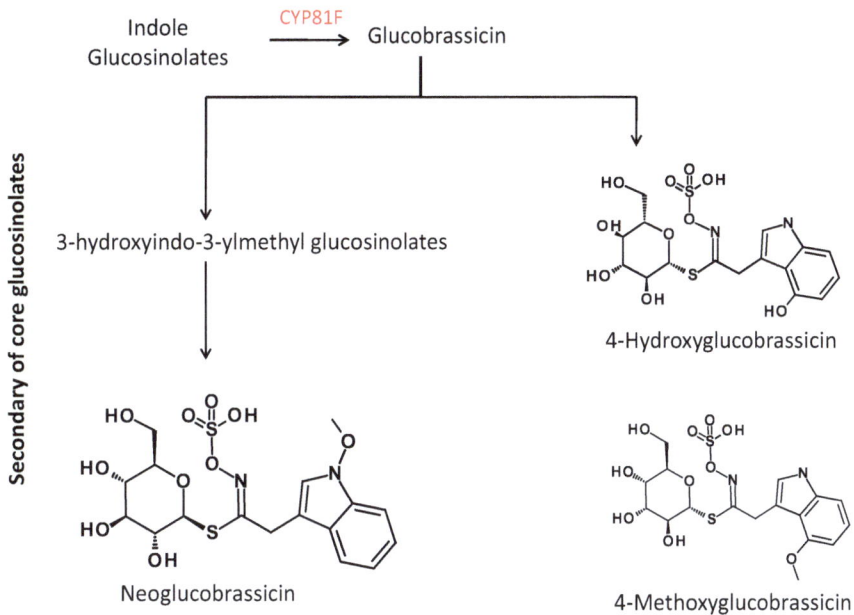

Fig. (6). Secondary modifications of side-chain and common structure; **a)** Aliphatic glucosinolates; **b)** Indole glucosinolates. FMOGS-Oxs flavinmonooxygenase.

Vitamin C includes ascorbic acid and dehydroascorbic acid (oxidation product), has important role in human body [189]. Among all, L-ascorbic acid is not formed in the human body and therefore needs to be administered in diet on a regular basis [144]. It is a labile, water soluble dietary antioxidant that has been long known to avert the adverse effects of free radicals that create oxidative stress and damage DNA, lipids and proteins and consequently lead to various ailments including cancer, cardiovascular, neurodegenerative and cataractogenesis diseases [156]. Ascorbate plays key role in collagen tissue synthesis, metal ion metabolism, enhancement of immune system and anti-histamine reactions [142].

Structure

Phenolic compounds constitute the most important group studied for its phytonutrient activities. These compounds are produced *via* shikimic acid pathway [137, 199]. Over 8,000 compounds were identified under these generic form. These can be classified as simple single aromatic ringed low molecular weight compounds, as well as large complex tannins and polyphenols [143, 187] phenols, phenolic acids, derivatives of hydroxycinnamic acid and flavonoids [161, 169, 214]. The phenylpropanoid metabolism involve many steps catalyzed by three main enzymes: PAL (phenylalanine ammonia-lyase), C4H (cinnamate 4-hydroxylase), 4CL (4-Coumarate: CoA ligase) [155]. The flavonoids are basically polyphenolic compounds with fifteen carbons and two aromatic rings [133]. These can also be further classified on the basis of chemical structure as flavonols, flavanones, flavones, flavanols, isoflavones and anthocyanins [173]. Hydroxycinnamic acids are aromatic compounds with C_6-C_3 structure. The flavanones don not contribute towards direct accumulation but act as precursors of other flavonoids [127]. Flavonols constitute of glycosylated forms of quercetin and kaempferol. Other constituents present in Brassica in high quantities include caffeic, p-coumaric, ferulic and sinapic acids which are found in conjugation with sugar moieties or other hydroxycinnamic acids [127, 158, 189].All the flavonoids that occur naturally have three hydroxyl groups, two on ring A at fifth and seventh position and a third one on ring B at third position (Fig. **7**). These aromatic rings are connected by 3 carbon bridges (C_6-C_3-C_6 carbon skeleton). The biochemical activity of flavonoids depends on the presence and position of various substituent groups that in turn influence the metabolism of compounds. The configuration, substitution and number of hydroxyl groups tend to influence antioxidant activities like radical scavenging and metal chelation [158, 178, 179, 185]. Oslen *et al.* [184] reported that flavonols in the plants occur as complex conjugates, one to five sugar moieties are bound to the aglycone and often acylated with hydroxycinnamic acids.

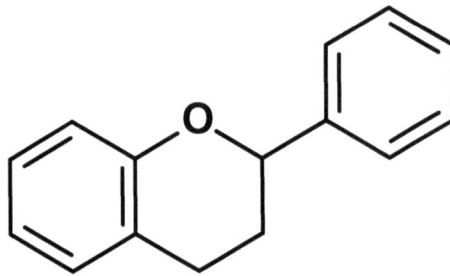

Fig. (7). Basic structure of flavonoid [136].

The uptake, bioavailability, metabolism and antioxidant activity of flavonoids depends on chemical structure and food matrix [207].

Vitamin C consists of ascorbic acid and its oxidation products, dehydroascrobic acid and ascorbigens. *Brassica* vegetables also include ascorbigens, formed as a result of myrosinase-catalysed degradation reaction between ascorbic acid and degradation products of indol-3-ylmethylglucosinolates. About 30 to 60% of indol-3- ylmethylglucosinolates are transformed into ascorbigens [130, 162]. The flavonoids in kale mostly consist of flavonol derivatives that include mono to penta glycosylation products of the aglycones quercetin, kaempferol [184, 202] or isorhamnetin with occasional acylation with hydroxycinnamic acids in broccoli florets [191, 192]. Owing to high variation, quercetin/kaempferol (Q/K) ratio has emerged as potential alternative for categorization of kale cultivars [198, 208]. Ayaz *et al.* [124] identified five derivatives of hydroxybenzoic acid and four hydroxycinnamic acids in kale leaves with 1.24 mg/100 g of fresh weight as the total content. Later, Fiol *et al.* [148] stated that quercetine derivatives have high radical scavenging behavior in kale and differences in antioxidant activity depend on the employed assays and structural pattern. In cabbage, Nielsen *et al.* [183] found a mixture of over 20 compounds out of which seven have been identified as kaempferol and quercetin derivatives with or without further acylation with hydroxycinnamic acids. High antioxidant acidity in swamp cabbage was attributed to high content of carotenoids, tocopherols and ascorbic acid [211] while the antioxidant activity of phenolic compounds is due to their redox properties that enable them to act as hydrogen donors, reducing agents, metal chelators and singlet oxygen quenchers [197].

Content

The main dietary source of flavonoids comprises of vegetables, some cereals, fruits, seeds, certain spices and wine. However, the flavonoids content in vegetables depends on the variety, the growing environment, the cultivation method, and the plant tissue/organ consumed [157]. The quantification of total flavonoid content is impossible due to unavailability of standards, immoderate number of flavonoid compounds and the complexity in derivative forms [167]. Sikora *et al.* [204] reported two fold higher total polyphenol content in kale than cauliflower, broccoli and Brussels sprouts. Mageney *et al.* [177] and Schmidt *et al.* [20] identified a wide variation in Q/K (quercetin and kaempferol) ratio that ranged from 0.11 – 2.31 and 0.17-1.02 respectively among kale cultivars. Pyke [194] reported a vitamin C content between 30 and 65 mg/100 g and Kurilich *et al.* [39] found that ascorbic acid in cabbage accessions varied from 22.6-32.9 mg/100 g, while Singh *et al.* [219] reported values between 5.66 mg/100 g (Sprint Ball) and 23.50 mg/100 g (Kirch-11).

Variable content of phenolics has been estimated by various researchers *viz.* broccoli (80.76 mg/100 g) and cabbage (36.66 mg/100 g) [139, 170]. Singh *et al.* [219] found that phenolics in cabbage varieties ranged from 12.58 mg/100 g (Pusa Mukta) to 34.41 mg/100 g (Green Cornell). Cao *et al.* [131] and Kurilich *et al.* [39] found that kale possessed higher antioxidant activity than cabbage. Similarly, Singh *et al.* [219] reported maximum vitamin C andphenolic contents were present in red cabbage. Heimler *et al.* [159] stated a high correlation between total phenolics and flavonoids content ($R^2 = 0.974$) and between antioxidant activity and total phenolic content (R^2 = 0.9388) and flavonoid content (R^2 = 0.9131) in various cole vegetables (white cabbage, broccoli, Italian kale, savoy cabbage, green cauliflower, cauliflower and Brussels sprouts). Chun *et al.* [141] observed a high correlation (R^2 = 0.95) between antioxidant activity and total flavonoid and phenolic content in raw green cabbage, red cabbage and savoy cabbage. Proteggente *et al.* [193] found a good correlation between TEAC (trolox equivalent antioxidant capacity) and total phenolic and vitamin C content. Contrarily, Kurilich *et al.* [174] reported no correlation between antioxidant acitivity and flavonoid or vitamin C content in investigated broccoli genotypes. Kurilich *et al.* [39] observed a decrease in vitamin C content from 27.32 to 74.71 mg/100 g fresh weight in broccoli, cabbage and cauliflower for total phenolics and flavonoids. A high correlation (R^2 = 0.974) was estimated between total phenolics and flavonoids content. Similarly, a high correlation was estimated for antioxidant activity with total phenolic content (R^2 = 0.9388) and flavonoid content (R^2 = 0.9131). The highest total phenolic content (13.8 mg/g dry weight) was observed in Italian kale and the lowest in Savoy cabbage (4.30 mg/g dry weight), whereas the maximum flavonoids content was recorded for Broccoli and the minimum for Savoy cabbage. Antioxidant behavior varied among these vegetables, wherein broccoli and Italian kale were 4 to 23 times less efficient than the other cole vegetables. The contribution of vitamin C should be taken into consideration while estimating correlation between antioxidant activity and flavonoids and phenolic content. Chu *et al.* [140] investigated *Brassica* vegetables and revealed higher levels of flavones than flavonol with apigenin as the predominant flavones aglycone except in Chinese cabbage. Oslen *et al.* [184] evaluated flavonoid profile content in kale leaves and reported that among the identified aglycones, quercetin content varied from 34.2 to 56.2 mg/100 g of fresh weight and kaempferol from 52.2 to 61.4 mg/100 g of fresh weight. Justesen *et al.* [168], Huang *et al.* [163] and Zhang *et al.* [216] reported comparatively lower quercetin level between 7.7 – 31.8 mg/100 g of fresh weight and kaempferol content between 23.5 – 90.3 mg/100 g of fresh weight. Total phenolic content in kale was higher than the other Brassica species on a dry weight basis [184].

Ismail *et al.* [165] identified kale, broccoli and cabbage with high antioxidant activity and as a rich source of ascorbic acid, tocopherols, carotenoids and polyphenol components. Kurilich *et al.* [39] recorded ascorbic acid content among *Brassica oleracea* subspecies and the mean values for broccoli, cauliflower, and cabbage were 74.71, 41.98, 27.32 mg/100 g, respectively. As reviewed by Podsedek [189], ascorbic acid varied over 4 fold in broccoli and cauliflower, 2.5 fold in Brussels sprouts and cabbage and approximately twice in kale [39, 213]. These variations can be attributed to genotypic differences, climatic conditions, maturity at harvest, cultural practices, storage conditions and sample preparation method.

Improvement strategies

The economic and health importance of cole vegetables necessitates target-oriented breeding strategy and availability of elaborative quantification data for flavonoid accumulation [177]. Various factors have been elucidated to impact the phenolic content and antioxidant activity in plants [217] (Fig. **8**). The environmental factors have substantial effect on variation in flavonoid content with respect to climatic conditions, year, plant tissue, maturity at harvest, postharvest treatments and inter and intra species variability. Other eco-physiological and edaphic factors like UV radiations, temperature, hydric stress, nutrient supply, ozone and soil and water state promote flavonoid synthesis [135, 147, 196, 201, 202, 217, 220]. The complexity of various biotic, abiotic, preharvest and postharvest parameters affecting flavonoids in plants has been depicted in Fig. (**8**) [177].Gardener *et al.* [151] analyzed mapping population of broccoli and concluded both genetic and environmental factors are responsible for extent of phenolic compound accumulation, former being highly significant.

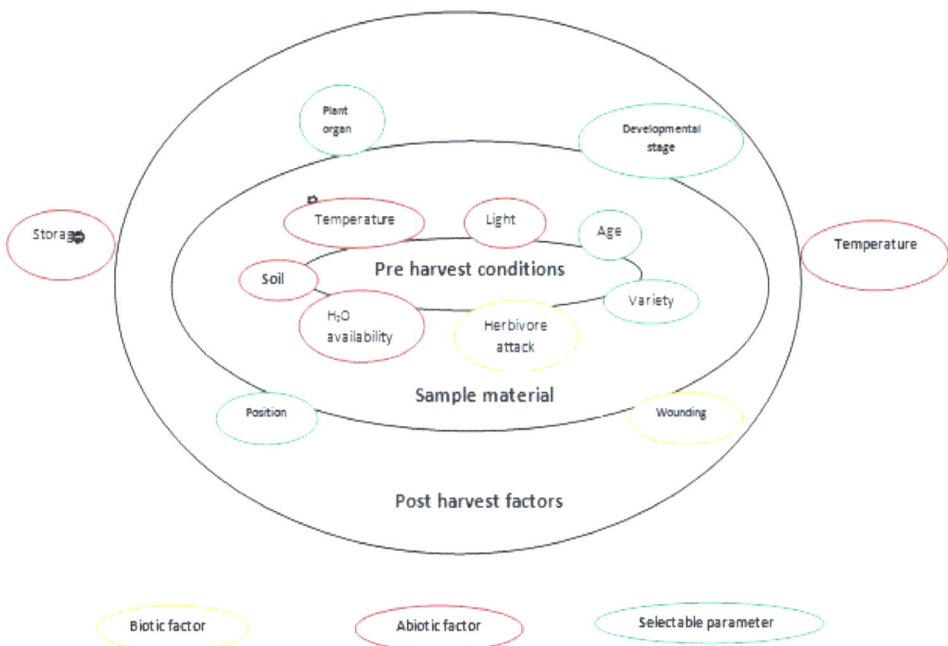

Fig. (8). Biotic, abiotic and selectable factors that influence content and composition of flavonoid in plants.

Oslen *et al.* [184] reported significant variations (16-23%) for phenolic content, total flavonols content and total hydroxycinnamic acids in curry kale leaves harvested from a single cultivar at the same time from the same field. The variations have been reported from other studies as well and can be attributed to biosynthesis of the phenolic compounds in plants, sampling of leaves and difficult comparison of data from different varieties and studies [177].

Ontogenetic Development of Flavonoids

The remarkable influence of ontogenetic development on the metabolic profile of plants has been well studied in plants including cole crops, although the extent of influence remains unclear. Soengas *et al.* [13] suggested the use of ontogenetic differences in flavonoid production as criteria to differentiate and characterize varieties of *B. oleracea* group. In *B. oleracea* var. *costata* the phenol content decreased by 85% in 10 days [205]. Contrarily, a significant increase in flavonoid content was observed in cabbage (*B. oleracea* var. *capitata*) and Chinese cabbage (*B. rapa* var. *pekinensis*) from four to twelve weeks after germination followed by a gradual decrease [200].

The accumulation of flavonoids in variable concentration in different tissues or plant organs has been documented by various researchers [146, 150, 206]. Flavonoids variation in immature primary florets, mature primary florets and secondary florets of three broccoli cultivars was quantified without significant difference in total flavonoid content among varieties by Erdman *et al.* [146]. In some species, injury or wounding by pathogens resulted in enhanced accumulation of catechin and proanthocyanidin [212].

Effect of Nitrogen

Groenbaek *et al.* [154] studied the effect of plant age, nitrogen split dose and frost on phytochemical content of curly kale and stated that split dose failed to alter composition and concentration of phytochemicals but low (50% or 33%) nitrogen dose resulted in higher concentration of kaempferol than 100% dose. The reason for similar occurrence in *Labisia pumila* was attributed to reduced PAL activity in the biosynthetic pathway of flavonols [164]. During the early and late period, the concentration of hydroxycinnamic acid derivatives, kaempferol and total flavonoid glycosides were significantly lower than 13 week old plants. This could be due to strong susceptibility and efficacy during early and late period respectively. Kurilich *et al.* [39] observed variability among *Brassica oleracea* subspecies for various vitamins including ascorbic acid. Significant variation was observed for vitamin content in broccoli subspecies which indicated that health benefits directly depend on the genotype consumed. The variation can arise as a result of various factors like genetic constitution, environmental factors, type of plant tissue analyzed or differences in harvesting time. In broccoli, 55% variation was due to genetic differences among the accessions. The authors construed that the variability of vitamins between or within subspecies is of relevance as it would enable the estimation of concentration that can be achieved through genetic improvement approaches. The extent of association of variability with genetic control and interaction of these compounds in biosynthetic pathways is a prerequisite for improving vitamin content in cole vegetables through plant breeding.

Thermal Treatments

Thermal treatments are known to affect the antioxidant activity of various vegetables [152] whereas cooking methods influence the bioavailability and improve the extractability of phytochemicals from vegetables. Dos Reis *et al.* [145] studied the antioxidant activity, phenolic and flavonoids content in raw and cooked broccoli and cauliflower. Fresh and microwaved cauliflower had high phenolic compounds and quercetin while boiling improved vitamin content. In broccoli, boiling and microwave cooking increased by 37% and 46% kaempferolquercetin content, respectively whereas phenolic compounds were by 49% and 17% higher for the respective methods. Lisiewska and Kmiecik [176] reported that blanching of broccoli and cauliflower reduced vitamin C content by 41-42% and 28-32%,respectively,whileVolden *et al.* [215] studied effect of blanching and freezer storage on cauliflower varieties and noted that blanching before freezing reduced vitamin C and total phenols by 19 and 15%, respectively. Freezer storage decreased ascorbic acid content by 24% in white and green cauliflower varieties. Cooking methods have been found to affect the availability of these nutrients. The antioxidant activity of fresh swamp cabbage, cabbage and kale was 60.3%, 59.3% and 50.2% respectively, whereas after 1 minute of thermal treatment/boiling it decreased to 56.5%, 53.4% and 45.9%. A similar negative impact was observed for total phenolic content in swamp cabbage, kale and cabbage wherein values decreased from 4175 mg/100 g, 3689 mg/100 g and 1107 mg/100 g to 3095 mg/100 g, 3251 mg/100 g, and 886 mg/100 g, respectively.

Identification of Genes

The regulation of phenol production has been found to occur majorly through alteration in the transcription rate of the biosynthetic genes [132, 172, 187, 209]. Among Brassica vegetables, rape has received maximum attention as the molecular regulation of sinapine metabolism has been well investigated [181, 126]. Sotelo *et al.* [218] developed a doubled haploids line and attempted the identification of genes related to phenolic, carotenoid and anthocyanin traits in flower buds and leaves. In total, 19 QTLs were detected for all the traits; ten of which were positioned in only one of the organs indicating an organ-dependent regulation of genes. Antioxidant activity of *Brassica* is primarily related to phenolic compounds, while for the content of metabolites 2 QTLs in leaves and 4

QTLs in flower buds were identified. The candidate genes proposed for playing a key role in phenylpropanoid biosynthesis with *in-silico* mapping, revealed location of 4CL-1 gene in AA-C7, HCT gene in AA-C3 AND C3′H gene between pX105Ce to pW 120Cx on chromosome. Similarly, Gardener *et al.* [151] reported 60 QTLs from variation of twelve traits-year assay in broccoli. Maximum stability and consistency was exhibited by QTL no. 7 across all assays with 9.6-18.6% phenotypic variation for traits, followed by 9.9-17.6% variation from other QTL. Among the candidate genes, larger share was attributed to genes involved during early stages of phenylpropanoid biosynthesis and MYB transcription factors. PAL (first step in phenylpropanoid biosynthesis), 4CL and CHI were identified as the most stable core genes across environments with potential to contribute in breeding progress. The study suggested that the overexpression or repression of MYB activity utilizing transgenes and miRNA approach has the potential to change the phenolic profile of *Brassica* vegetables.

Metabolite profiling was attempted by Francisco *et al.* [149] in three organs (leaves, flower buds and seeds) in a double haploid of *B. leraceae BolTBDH* mapping population to investigate the genetic basis of phenol accumulation in relation to plant ontogeny. The identified flavonoids were O-glycosides with a substituent at 3- and/or 7- positions of kaempferol, isorhamnetin and quercetin and/or in conjugation with various hydroxycinnamic acids. Plant part analyzed affected the phenolic content and profile of the mapping population. Average content of total phenolics in various organs included 35.61 µmol/g dry weight in leaves, 53.49 µmol/g dw in flower buds and 75.09 µmol/g dw in seeds. Flavonoids in flower buds (19.02 µmol/g dw) were detected in higher concentration than the leaves (9.28 µmol/g dw). The organ-dependent accumulation of some phenolics might depend on *de novo* biosynthesis and catabolism or re-allocation of compounds. This feature allows plants to develop protective adaptation for constantly changing environmental conditions during life cycle. The variation in quality and quantity of flavonoids might also have a role in organ differentiation, development and defense mechanisms.

Furthermore, seventy nine QTLs were identified for twenty nine analyzed traits in all the organs and were found unevenly spread all over the nine linkage groups as revealed by the presence of hotspots for genetic control of metabolite content. Transgressive distribution was observed for phenolic content and the reason for that can be attributed to epistatic interactions among loci or new combination of additive alleles for phenolics accumulation. The common QTLs (QTL-7.3, QTL-9.5) were present in flower buds, leaves and seeds and found related with seven traits that included flavonoids, hydroxycinnamic acids and total phenolic compounds. This indicated that the genes from these QTLs could play a role in the core pathway of phenolic synthesis. For phenotypic variation of QTL-7.3,

*Bol43270*was identified as a candidate gene which suggested the presence of a regulatory loci in the QTL region or a tight linkage among the genes involved in the accumulation of metabolites.

CONCLUSION

The improvement in society's quality of life has resulted in anincrease in the demand for food of high nutritional quality. *Brassica* vegetables have been related to the reduction of the risk of chronic diseases including cardiovascular diseases and cancer as *Brassica* foods are very nutritive, providing nutrients and health-promoting phytochemicals such as vitamins, carotenoids, fiber, soluble sugars, minerals, glucosinolates and phenolic compounds. Systematic biological technologies in genomics, transcriptomics, proteomics, and metabolomics have been utilized in the elucidation of the biosynthesis and degradation of nutrients. Recent integrated-omics approaches applied to the phytonutrients in Brassica vegetables have paved the way for the elucidation of the genetic mechanism underlying their biosynthesis and provided new insights for genetic engineering and quality breeding of Brassica crops. Compared with cereal crops such as rice, maize, and barley, the study of Brassica is relatively less efficient, but the advances in genetic engineering and breeding are happening rapidly due to the efforts of Brassica community. Genes discovery and regulatory studies open the possibility for genetic modification of nutritional metabolites and obtain desirable vegetable Brassica resources. Despite recent progress in QTL/ gene mapping and transcriptome studies of different metabolites in vegetable Brassicas, our understanding of the molecular mechanisms underlying their accumulation and regulation in both *B. rapa* and *B. oleracea* vegetables is yet limited and has to be increased in the near future.

CONSENT FOR PUBLICATION

Not applicable.

CONFLICT OF INTEREST

The author (editor) declares no conflict of interest, financial or otherwise.

ACKNOWLEDGMENT

Declare None.

REFERENCES

[1]   Warwick SL, Francis A, Shehbaz LA. Brassicaceae: species checklist and database on CD- rom. Plant
      Syst Evol 2006; 259: 249-58.
      [http://dx.doi.org/10.1007/s00606-006-0422-0]

[2]   Gomez-Campo C. Morphology and morpho-taxonomy of the tribe Brassiceae.Brassica crops and wild allies, biology and breeding. Tokyo: Japan Scientific Societies Press 1980; pp. 3-31.

[3]   Cartea ME, Velasco P. Glucosinolates in *Brassica* foods: bioavailability in food and significance for human health. Phytochem Rev 2008; 7: 213-29.
[http://dx.doi.org/10.1007/s11101-007-9072-2]

[4]   Traka M, Mithen R. Glucosinolates, isothiocynates and human health. Phytochem Rev 2009; 8: 269-82.
[http://dx.doi.org/10.1007/s11101-008-9103-7]

[5]   Velasco P, Francisco M, Moreno DA, Ferreres F, García-Viguera C, Cartea ME. Phytochemical fingerprinting of vegetable *Brassica oleracea* and *Brassica napus* by simultaneous identification of glucosinolates and phenolics. Phytochem Anal 2011; 22(2): 144-52.
[http://dx.doi.org/10.1002/pca.1259] [PMID: 21259374]

[6]   Balasubashini MS, Rukkumani R, Viswanathan P, Menon VP. Ferulic acid alleviates lipid peroxidation in diabetic rats. Phytother Res 2004; 18(4): 310-4.
[http://dx.doi.org/10.1002/ptr.1440] [PMID: 15162367]

[7]   Rodríguez-Cantú LN, Gutiérrez-Uribe JA, Arriola-Vucovich J, Díaz-De La Garza RI, Fahey JW, Serna-Saldivar SO. Broccoli (Brassica oleracea var. italica) sprouts and extracts rich in glucosinolates and isothiocyanates affect cholesterol metabolism and genes involved in lipid homeostasis in hamsters. J Agric Food Chem 2011; 59(4): 1095-103.
[http://dx.doi.org/10.1021/jf103513w] [PMID: 21254774]

[8]   Villarreal-Garcia D, Jacobo-Velazquez DA. Glucosinolates from broccoli: nutraceutical properties and their purification. J Nutraceuticals Food Sci 2016; 1: 5.

[9]   Wagner AE, Terschluesen AM, Rimbach G. Health promoting effects of brassica-derived phytochemicals: from chemopreventive and anti-inflammatory activities to epigenetic regulation. Oxid Med Cell Longev 2013; 2013: 964539.
[http://dx.doi.org/10.1155/2013/964539] [PMID: 24454992]

[10]   Björkman M, Klingen I, Birch ANE, *et al.* Phytochemicals of *Brassicaceae* in plant protection and human health--influences of climate, environment and agronomic practice. Phytochemistry 2011; 72(7): 538-56.
[http://dx.doi.org/10.1016/j.phytochem.2011.01.014] [PMID: 21315385]

[11]   Kristal AR, Lampe JW. *Brassica* vegetables and prostate cancer risk: a review of the epidemiological evidence. Nutr Cancer 2002; 42(1): 1-9.
[http://dx.doi.org/10.1207/S15327914NC421_1] [PMID: 12235639]

[12]   Sotelo T, Cartea ME, Velasco P, Soengas P. Identification of antioxidant capacity -related QTLs in *Brassica oleracea.* PLoS One 2014; 9(9): e107290.
[http://dx.doi.org/10.1371/journal.pone.0107290] [PMID: 25198771]

[13]   Soengas P, Cartea ME, Francisco M, Sotelo T, Velasco P. New insights into antioxidant activity of Brassica crops. Food Chem 2012; 134(2): 725-33.
[http://dx.doi.org/10.1016/j.foodchem.2012.02.169] [PMID: 23107684]

[14]   Hanschen FS, Schreiner M. Isothiocynates, nitriles, and epithionitriles from glucosinolates are affected by genotype and developmental stage in *Brassica oleracea* varieties. Front Plant Sci 2017; 8: 1095.
[http://dx.doi.org/10.3389/fpls.2017.01095] [PMID: 28690627]

[15]   Verkerk R, Schreiner M, Krumbein A, *et al.* Glucosinolates in Brassica vegetables: the influence of the food supply chain on intake, bioavailability and human health. Mol Nutr Food Res 2009; 53 (Suppl. 2): S219.
[http://dx.doi.org/10.1002/mnfr.200800065] [PMID: 19035553]

[16]   Liu RH. Potential synergy of phytochemicals in cancer prevention: mechanism of action. J Nutr 2004; 134(12) (Suppl.): 3479S-85S.

[http://dx.doi.org/10.1093/jn/134.12.3479S] [PMID: 15570057]

[17]   Zukalova H, Vasak J. The role and effect of glucosinolates of *Brassica* species-a review. Rostlinna Vyroba 2002; 48: 175-80.

[18]   Samaila D, Ezekwudo DE, Yimam KK, *et al.* Bioactive plant compounds inhibited the proliferation and induced apoptosis in human cancer cell lines, *in vitro.* Trans Int Biomed Inform Enabling Tech Symp J 2004; 1: 34-42.

[19]   Staub RE, Feng C, Onisko B, Bailey GS, Firestone GL, Bjeldanes LF. Fate of indole-3-carbinol in cultured human breast tumor cells. Chem Res Toxicol 2002; 15(2): 101-9.
[http://dx.doi.org/10.1021/tx010056m] [PMID: 11849035]

[20]   Hanf V, Gonder U. Nutrition and primary prevention of breast cancer: foods, nutrients and breast cancer risk. Eur J Obstet Gynecol Reprod Biol 2005; 123(2): 139-49.
[http://dx.doi.org/10.1016/j.ejogrb.2005.05.011] [PMID: 16316809]

[21]   Singh J, Upadhyay AK, Bahadur A, *et al.* Antioxidant phytochemicals in cabbage (*Brassica oleraceae* L. var. *capitata*). Sci Hortic (Amsterdam) 2006; 108: 233-7.
[http://dx.doi.org/10.1016/j.scienta.2006.01.017]

[22]   Aggarwal BB, Ichikawa H. Molecular targets and anticancer potential of indole-3-carbinol and its derivatives. Cell Cycle 2005; 4(9): 1201-15.
[http://dx.doi.org/10.4161/cc.4.9.1993] [PMID: 16082211]

[23]   Cover CM, Hsieh SJ, Tran SH, *et al.* Indole-3-carbinol inhibits the expression of cyclin-dependent kinase-6 and induces a G1 cell cycle arrest of human breast cancer cells independent of estrogen receptor signaling. J Biol Chem 1998; 273(7): 3838-47.
[http://dx.doi.org/10.1074/jbc.273.7.3838] [PMID: 9461564]

[24]   Ávila FW, Faquin V, Yang Y, *et al.* Assessment of the anticancer compounds *Se*-methylselenocysteine and glucosinolates in Se-biofortified broccoli (*Brassica oleracea* L. var. *italica*) sprouts and florets. J Agric Food Chem 2013; 61(26): 6216-23.
[http://dx.doi.org/10.1021/jf4016834] [PMID: 23763668]

[25]   Li Y, Zhang T, Korkaya H, *et al.* Sulforaphane, a dietary component of broccoli/broccoli sprouts, inhibits breast cancer stem cells. Clin Cancer Res 2010; 16(9): 2580-90.
[http://dx.doi.org/10.1158/1078-0432.CCR-09-2937] [PMID: 20388854]

[26]   Abbaoui B, Riedl KM, Ralston RA, *et al.* Inhibition of bladder cancer by broccoli isothiocyanates sulforaphane and erucin: characterization, metabolism, and interconversion. Mol Nutr Food Res 2012; 56(11): 1675-87.
[http://dx.doi.org/10.1002/mnfr.201200276] [PMID: 23038615]

[27]   Aires A, Rosa E, Carvalho R. Effect of nitrogen and sulfur fertilization on glucosinolates in the leaves and roots of broccoli sprouts (*Brassica oleracea* var. *italica*). J Sci Food Agric 2006; 86: 1512-6.
[http://dx.doi.org/10.1002/jsfa.2535]

[28]   Gu ZX, Guo QH, Gu YJ. Factors influencing glucoraphanin and sulforaphane formation in Brassica plants: a review. J Integr Agric 2012; 11: 1804-16.
[http://dx.doi.org/10.1016/S2095-3119(12)60185-3]

[29]   Roberge MT, Borgerding AJ, Finley JW. Speciation of selenium compounds from high selenium broccoli is affected by the extracting solution. J Agric Food Chem 2003; 51(15): 4191-7.
[http://dx.doi.org/10.1021/jf021247d] [PMID: 12848483]

[30]   Ramos SJ, Yuan Y, Faquin V, Guilherme LR, Li L. Evaluation of genotypic variation of broccoli (Brassica oleracea var. italic) in response to selenium treatment. J Agric Food Chem 2011; 59(8): 3657-65.
[http://dx.doi.org/10.1021/jf104731f] [PMID: 21417275]

[31]   Irion CW. Growing alliums and brassicas in selenium-enriched soils increases their anticarcinogenic potentials. Med Hypotheses 1999; 53(3): 232-5.

[http://dx.doi.org/10.1054/mehy.1998.0751] [PMID: 10580529]

[32]   Jahangir M, Kim HK, Choi YH, *et al.* Health affecting compounds in *Brassicaceae*. Compr Rev Food Sci Food Saf 2009; 8: 31-43.
[http://dx.doi.org/10.1111/j.1541-4337.2008.00065.x]

[33]   Soengas P, Sotelo T, Velasco P, *et al.* Antioxidant properties of *Brassica* vegetables. Funct Plant Sci Biotechnol 2011; 5: 43-55.

[34]   Vinson JA, Dabbagh YA, Serry MM, Jang J. Plant flavonoids, especially tea flavonoids, are powerful antioxidants using an *in vitro* oxidation model for heart disease. J Agric Food Chem 1995; 43: 2800-2.
[http://dx.doi.org/10.1021/jf00059a005]

[35]   Chiu LW, Zhou X, Burke S, Wu X, Prior RL, Li L. The purple cauliflower arises from activation of a MYB transcription factor. Plant Physiol 2010; 154(3): 1470-80.
[http://dx.doi.org/10.1104/pp.110.164160] [PMID: 20855520]

[36]   Dyrby M, Westergaard N, Stapelfeldt H. Light and heat sensitivity of red cabbage extraxt in soft drink model systems. Food Chem 2001; 72: 431-7.
[http://dx.doi.org/10.1016/S0308-8146(00)00251-X]

[37]   Heinonen S, Nurmi T, Liukkonen K, *et al. In vitro* metabolism of plant lignans: new precursors of mammalian lignans enterolactone and enterodiol. J Agric Food Chem 2001; 49(7): 3178-86.
[http://dx.doi.org/10.1021/jf010038a] [PMID: 11453749]

[38]   Milder IEJ, Arts ICW, van de Putte B, Venema DP, Hollman PC. Lignan contents of Dutch plant foods: a database including lariciresinol, pinoresinol, secoisolariciresinol and matairesinol. Br J Nutr 2005; 93(3): 393-402.
[http://dx.doi.org/10.1079/BJN20051371] [PMID: 15877880]

[39]   Kurilich AC, Tsau GJ, Brown A, *et al.* Carotene, tocopherol, and ascorbate contents in subspecies of *Brassica oleracea.* J Agric Food Chem 1999; 47(4): 1576-81.
[http://dx.doi.org/10.1021/jf9810158] [PMID: 10564019]

[40]   Podsedek A. Natural antioxidants and antioxidant capacity of *Brassica* vegetables: a review. Lebensm Wiss Technol 2007; 40: 1-11.
[http://dx.doi.org/10.1016/j.lwt.2005.07.023]

[41]   Muller H. Determination of the carotenoid content in selected vegetablesand fruit by HPLC and photodiode array detection. Z Lebensm Unters F A 1997; 204: 88-94.
[http://dx.doi.org/10.1007/s002170050042]

[42]   Kiokias S, Oreopoulou V. Antioxidant properties of natural carotenoid extracts against the AAPH-initiated oxidation of food emulsions. Innov Food Sci Emerg Technol 2006; 7: 132-9.
[http://dx.doi.org/10.1016/j.ifset.2005.12.004]

[43]   Kiokias S, Proestos C, Varzakas T. A review of the structure, biosynthesis, absorption of carotenoids-analysis and properties of their common natural extracts. Curr Res Nutr Food Sci 2016; 4: 25-37.
[http://dx.doi.org/10.12944/CRNFSJ.4.Special-Issue1.03]

[44]   Britton G, Liaaen-Jensen S, Pfander H, Eds. Carotenoid Handbook. Basel: Birkhauser Verlag 2004.
[http://dx.doi.org/10.1007/978-3-0348-7836-4]

[45]   Dey SS, Bhatia R, Parkash C, *et al.* Alteration in important quality traits and antioxidant activities in *Brassica oleracea* with *Ogura* cybrid cytoplasm. Plant Breed 2017; 136: 400-9.
[http://dx.doi.org/10.1111/pbr.12478]

[46]   Li L, Paolillo DJ, Parthasarathy MV, Dimuzio EM, Garvin DF. A novel gene mutation that confers abnormal patterns of β-carotene accumulation in cauliflower (*Brassica oleracea* var. *botrytis*). Plant J 2001; 26(1): 59-67.
[http://dx.doi.org/10.1046/j.1365-313x.2001.01008.x] [PMID: 11359610]

[47]   Guzman I, Yousef GG, Brown AF. Simultaneous extraction and quantitation of carotenoids,

chlorophylls, and tocopherols in *Brassica* vegetables. J Agric Food Chem 2012; 60(29): 7238-44.
[http://dx.doi.org/10.1021/jf302475d] [PMID: 22734504]

[48] Ruiz-Sola MÁ, Rodríguez-Concepción M. Carotenoid biosynthesis in Arabidopsis: a colorful pathway. Arabidopsis Book 2012; 10: e0158.
[http://dx.doi.org/10.1199/tab.0158] [PMID: 22582030]

[49] Li P, Zhang S, Zhang S, *et al.* Carotenoid biosynthetic genes in *Brassica rapa*: comparative genomic analysis, phylogenetic analysis, and expression profiling. BMC Genomics 2015; 16: 492.
[http://dx.doi.org/10.1186/s12864-015-1655-5] [PMID: 26138916]

[50] Siefermann-Harms D. the light-harvesting and protective functions of carotenoids in photosynthetic membranes. Physiol Plant 1987; 69: 561-8.
[http://dx.doi.org/10.1111/j.1399-3054.1987.tb09240.x]

[51] Camara B, Hugueney P, Bouvier F, Kuntz M, Monéger R. Biochemistry and molecular biology of chromoplast development. Int Rev Cytol 1995; 163: 175-247.
[http://dx.doi.org/10.1016/S0074-7696(08)62211-1] [PMID: 8522420]

[52] Paolillo DJ Jr, Garvin DF, Parthasarathy MV. The chromoplasts of *Or* mutants of cauliflower (*Brassica oleracea* L. var. *botrytis*). Protoplasma 2004; 224(3-4): 245-53.
[http://dx.doi.org/10.1007/s00709-004-0059-1] [PMID: 15614485]

[53] Zhou X, Welsch R, Yang Y, *et al. Arabidopsis OR* proteins are the major posttranscriptional regulators of phytoene synthase in controlling carotenoid biosynthesis. Proc Natl Acad Sci USA 2015; 112(11): 3558-63.
[http://dx.doi.org/10.1073/pnas.1420831112] [PMID: 25675505]

[54] Ding Y, Jian Y. An orange cauliflower hybrid 'Jinyu 60' with high β-carotene. Acta Hortic 2010; (856):
[http://dx.doi.org/10.17660/ActaHortic.2010.856.37]

[55] Farnham MW, Kopsell DA. Importance of genotype on carotenoid and chlorophyll levels in broccoli heads. HortScience 2009; 44: 1248-53.

[56] Li L, Garvin DF. Molecular mapping of *Or*, a gene inducing beta-carotene accumulation in cauliflower (*Brassica oleracea* L. var. *botrytis*). Genome 2003; 46(4): 588-94.
[http://dx.doi.org/10.1139/g03-043] [PMID: 12897867]

[57] Li L, Lu S, O'Halloran DM, Garvin DF, Vrebalov J. High-resolution genetic and physical mapping of the cauliflower high-β-carotene gene Or (Orange). Mol Genet Genomics 2003; 270(2): 132-8.
[http://dx.doi.org/10.1007/s00438-003-0904-5] [PMID: 12908106]

[58] Li L, Lu S, O'Halloran DM, Garvin DF, Vrebalov J. High-resolution genetic and physical mapping of the cauliflower high-β-carotene gene Or (Orange). Mol Genet Genomics 2003; 270(2): 132-8.
[http://dx.doi.org/10.1007/s00438-003-0904-5] [PMID: 12908106]

[59] Traber MG, Atkinson J. Vitamin E, antioxidant and nothing more. Free Radic Biol Med 2007; 43(1): 4-15.
[http://dx.doi.org/10.1016/j.freeradbiomed.2007.03.024] [PMID: 17561088]

[60] Engin KN. Alpha-tocopherol: looking beyond an antioxidant. Mol Vis 2009; 15: 855-60.
[PMID: 19390643]

[61] Alpha tocopherol https://www.ncbi.nlm.nih.gov/pccompound?term=alpha-tocopherol

[62] Yamauchi R. Vitamin E: mechanism of its antioxidant activity. Food Sci Technol Int Tokyo 1997; 3: 301-9.
[http://dx.doi.org/10.3136/fsti9596t9798.3.301]

[63] Piironen V, Syvaoja EL, Varo P, *et al.* Tocopherols and tocotrienols in finnish foods: vegetables, fruits, and berries. J Agric Food Chem 1986; 34: 742-6.
[http://dx.doi.org/10.1021/jf00070a038]

[64]   van Poppel G. Carotenoids and cancer: an update with emphasis on human intervention studies. Eur J Cancer 1993; 29A(9): 1335-44.
[http://dx.doi.org/10.1016/0959-8049(93)90087-V] [PMID: 8343282]

[65]   Palozza P, Serini S, Di Nicuolo F, Calviello G. Modulation of apoptotic signalling by carotenoids in cancer cells. Arch Biochem Biophys 2004; 430(1): 104-9.
[http://dx.doi.org/10.1016/j.abb.2004.02.038] [PMID: 15325917]

[66]   Ibrahim KE, Juvik JA. Feasibility for improving phytonutrient content in vegetable crops using conventional breeding strategies: case study with carotenoids and tocopherols in sweet corn and broccoli. J Agric Food Chem 2009; 57(11): 4636-44.
[http://dx.doi.org/10.1021/jf900260d] [PMID: 19489619]

[67]   Lila MA. Interactions between flavonoids that benefit human health.New York: Springer Anthocyanins: biosynthesis, functions, and applications 2009; pp. 305-23.

[68]   Thomasset S, Teller N, Cai H, *et al.* Do anthocyanins and anthocyanidins, cancer chemopreventive pigments in the diet, merit development as potential drugs? Cancer Chemother Pharmacol 2009; 64(1): 201-11.
[http://dx.doi.org/10.1007/s00280-009-0976-y] [PMID: 19294386]

[69]   Wang LS, Carmella S, Keyes R, *et al.* Anthocyanins and cancer prevention.Nutraceuticals and cancer. Netherlands: Springer 2012; pp. 201-29.
[http://dx.doi.org/10.1007/978-94-007-2630-7_11]

[70]   Lazzè MC, Pizzala R, Perucca P, *et al.* Anthocyanidins decrease endothelin-1 production and increase endothelial nitric oxide synthase in human endothelial cells. Mol Nutr Food Res 2006; 50(1): 44-51.
[http://dx.doi.org/10.1002/mnfr.200500134] [PMID: 16288501]

[71]   Lee J, Lee HK, Kim CY, *et al.* Purified high-dose anthocyanoside oligomer administration improves nocturnal vision and clinical symptoms in myopia subjects. Br J Nutr 2005; 93(6): 895-9.
[http://dx.doi.org/10.1079/BJN20051438] [PMID: 16022759]

[72]   Miller MG, Shukitt-Hale B. Berry fruit enhances beneficial signaling in the brain. J Agric Food Chem 2012; 60(23): 5709-15.
[http://dx.doi.org/10.1021/jf2036033] [PMID: 22264107]

[73]   Wallace TC. Anthocyanins in cardiovascular disease. Adv Nutr 2011; 2(1): 1-7.
[http://dx.doi.org/10.3945/an.110.000042] [PMID: 22211184]

[74]   He J, Giusti MM. Anthocyanins: natural colorants with health-promoting properties. Annu Rev Food Sci Technol 2010; 1: 163-87.
[http://dx.doi.org/10.1146/annurev.food.080708.100754] [PMID: 22129334]

[75]   Delgado-Vargas F, Paredes-Lopez O. Anthocyanins and betalins.Natural colorants for food and nutraceutical uses. Boca Raton, FL: CRC Press 2003; pp. 167-219.

[76]   Welch CR, Wu Q, Simon JE. Recent advances in anthocyanin analysis and characterization. Curr Anal Chem 2008; 4(2): 75-101.
[http://dx.doi.org/10.2174/157341108784587795] [PMID: 19946465]

[77]   Jaakola L. New insights into the regulation of anthocyanin biosynthesis in fruits. Trends Plant Sci 2013; 18(9): 477-83.
[http://dx.doi.org/10.1016/j.tplants.2013.06.003] [PMID: 23870661]

[78]   Pervaiz T, Songtao J, Faghihi F, *et al.* Naturally occuring anthocyanin, structure, functions and biosynthetic pathway in fruit plants. J Plant Biochem Physiol 2017; 5: 2.
[http://dx.doi.org/10.4172/2329-9029.1000187]

[79]   Jaakola L, Maatta K, Pirttila AM, *et al.* Expression of genes involved in relation to anthocyanin, proanthocyanidin, and flavonol levels during bilberry fruit development. Plant Physiol 2002; 130: 729-39.

[http://dx.doi.org/10.1104/pp.006957] [PMID: 12376640]

[80] Dixon RA, Xie DY, Sharma SB. Proanthocyanidins--a final frontier in flavonoid research? New Phytol 2005; 165(1): 9-28.
[http://dx.doi.org/10.1111/j.1469-8137.2004.01217.x] [PMID: 15720617]

[81] Yuan Y, Chiu L-W, Li L. Transcriptional regulation of anthocyanin biosynthesis in red cabbage. Planta 2009; 230(6): 1141-53.
[http://dx.doi.org/10.1007/s00425-009-1013-4] [PMID: 19756724]

[82] Guo N, Wu J, Zheng S, *et al.* Anthocyanin profile characterization and quantitative trait locus mapping in zicaitai (*Brassica rapa* L. ssp. *chinensis* var. *purpurea*). Mol Breed 2015; 35: 113.
[http://dx.doi.org/10.1007/s11032-015-0237-1]

[83] Junquing W, Jing Z, Meiling Q, *et al.* Genetic analysis and mapping of the purple gene in purple heading chinese cabbage. Hortic Plant J 2016; 2: 351-6.
[http://dx.doi.org/10.1016/j.hpj.2016.11.007]

[84] Liu X-P, Gao B-Z, Han FQ, *et al.* Genetics and fine mapping of a purple leaf gene, *BoPr,* in ornamental kale (*Brassica oleracea* L. var. *acephala*). BMC Genomics 2017; 18(1): 230.
[http://dx.doi.org/10.1186/s12864-017-3613-x] [PMID: 28288583]

[85] Zhang B, Hu Z, Zhang Y, Li Y, Zhou S, Chen G. A putative functional MYB transcription factor induced by low temperature regulates anthocyanin biosynthesis in purple kale (Brassica Oleracea var. acephala f. tricolor). Plant Cell Rep 2012; 31(2): 281-9.
[http://dx.doi.org/10.1007/s00299-011-1162-3] [PMID: 21987119]

[86] Hayashi K, Matsumoto S, Tsukazaki H, *et al.* Mapping of a novel locus regulating anthocyanin pigmentation in *Brassica rapa.* Breed Sci 2010; 60: 76-80.
[http://dx.doi.org/10.1270/jsbbs.60.76]

[87] Li H, Zhu L, Yuan G, *et al.* Fine mapping and candidate gene analysis of an anthocyanin-rich gene, *BnaA.PL1,* conferring purple leaves in *Brassica napus* L. Mol Genet Genomics 2016; 291(4): 1523-34.
[http://dx.doi.org/10.1007/s00438-016-1199-7] [PMID: 27003438]

[88] Zhang Y, Chen G, Dong T, *et al.* Anthocyanin accumulation and transcriptional regulation of anthocyanin biosynthesis in purple bok choy (*Brassica rapa* var. *chinensis*). J Agric Food Chem 2014; 62(51): 12366-76.
[http://dx.doi.org/10.1021/jf503453e] [PMID: 25419600]

[89] Zhang Y, Hu Z, Zhu M, *et al.* Anthocyanin accumulation and molecular analysis of correlated genes in purple kohlrabi (*Brassica oleracea* var. *gongylodes* L.). J Agric Food Chem 2015; 63(16): 4160-9.
[http://dx.doi.org/10.1021/acs.jafc.5b00473] [PMID: 25853486]

[90] Shi MZ, Xie D-Y. Biosynthesis and metabolic engineering of anthocyanins in *Arabidopsis thaliana.* Recent Pat Biotechnol 2014; 8(1): 47-60.
[http://dx.doi.org/10.2174/1872208307666131218123538] [PMID: 24354533]

[91] Sinigrin
https://pubchem.ncbi.nlm.nih.gov/compound/23684362#section=Synonymsncbi.nlm.nih.gov2018. Pubchem.ncbi.nlm.nih.gov, Data retrieved on dated 7th Jan

[92] Olsen O, Sùrensen H. Isolation of glucosinolates and the identification of o-(α-L-Rhamnopylransoyloxy) benzylglucosinolate from *Reseda odorata.* Phytochemistry 1979; 18: 1547-53.
[http://dx.doi.org/10.1016/S0031-9422(00)98494-2]

[93] Linscheid M, Wendisch D, Strack D. The structure ofsinapic acid esters and their metabolism in cotyledons of *Raphanus sativus.* Z Pflanzenphysiol 1980; 35c: 907-12.

[94] Halkier BA, Gershenzon J. Biology and biochemistry of glucosinolates. Annu Rev Plant Biol 2006; 57: 303-33.

[http://dx.doi.org/10.1146/annurev.arplant.57.032905.105228] [PMID: 16669764]

[95]   Sotelo T, Velasco P, Soengas P, Rodríguez VM, Cartea ME. Modification of leaf glucosinolate contents in *Brassica oleracea* by divergent selection and effect on expression of genes controlling glucosinolate pathway. Front Plant Sci 2016; 7: 1012.
[http://dx.doi.org/10.3389/fpls.2016.01012] [PMID: 27471510]

[96]   Ishida M, Hara M, Fukino N, Kakizaki T, Morimitsu Y. Glucosinolate metabolism, functionality and breeding for the improvement of *Brassicaceae* vegetables. Breed Sci 2014; 64(1): 48-59.
[http://dx.doi.org/10.1270/jsbbs.64.48] [PMID: 24987290]

[97]   Robin AHK, Yi GE, Laila R, *et al.* Expression profiling of glucosinolate biosynthetic genes in *Brassica oleracea* L. var. *capitata* inbred lines reveals their association with glucosinolate content. Molecules 2016; 21(6): 787.
[http://dx.doi.org/10.3390/molecules21060787] [PMID: 27322230]

[98]   Sønderby IE, Geu-Flores F, Halkier BA. Biosynthesis of glucosinolates--gene discovery and beyond. Trends Plant Sci 2010; 15(5): 283-90.
[http://dx.doi.org/10.1016/j.tplants.2010.02.005] [PMID: 20303821]

[99]   Gigolashvili T, Berger B, Flugge UI. Specific and coordinated control of indolic and aliphatic glucosinolate biosynthesis by R2R3-MYB transcription factors in *Arabidopsis thaliana.* Phytochem Rev 2009; 8(1): 3-13.
[http://dx.doi.org/10.1007/s11101-008-9112-6]

[100] Malitsky S, Blum E, Less H, *et al.* The transcript and metabolite networks affected by the two clades of *Arabidopsis* glucosinolate biosynthesis regulators. Plant Physiol 2008; 148(4): 2021-49.
[http://dx.doi.org/10.1104/pp.108.124784] [PMID: 18829985]

[101] Sønderby IE, Hansen BG, Bjarnholt N, Ticconi C, Halkier BA, Kliebenstein DJ. A systems biology approach identifies a R2R3 MYB gene subfamily with distinct and overlapping functions in regulation of aliphatic glucosinolates. PLoS One 2007; 2(12): e1322.
[http://dx.doi.org/10.1371/journal.pone.0001322] [PMID: 18094747]

[102] Celenza JL, Quiel JA, Smolen GA, *et al.* The *Arabidopsis* ATR1 Myb transcription factor controls indolic glucosinolate homeostasis. Plant Physiol 2005; 137(1): 253-62.
[http://dx.doi.org/10.1104/pp.104.054395] [PMID: 15579661]

[103] Frerigmann H, Gigolashvili T. MYB34, MYB51, and MYB122 distinctly regulate indolic glucosinolate biosynthesis in *Arabidopsis thaliana.* Mol Plant 2014; 7(5): 814-28.
[http://dx.doi.org/10.1093/mp/ssu004] [PMID: 24431192]

[104] Gigolashvili T, Berger B, Mock HP, Müller C, Weisshaar B, Flügge UI. The transcription factor HIG1/MYB51 regulates indolic glucosinolate biosynthesis in *Arabidopsis thaliana.* Plant J 2007; 50(5): 886-901.
[http://dx.doi.org/10.1111/j.1365-313X.2007.03099.x] [PMID: 17461791]

[105] Hirai MY, Sugiyama K, Sawada Y, *et al.* Omics-based identification of *Arabidopsis* Myb transcription factors regulating aliphatic glucosinolate biosynthesis. Proc Natl Acad Sci USA 2007; 104(15): 6478-83.
[http://dx.doi.org/10.1073/pnas.0611629104] [PMID: 17420480]

[106] Mewis I, Appel HM, Hom A, Raina R, Schultz JC. Major signaling pathways modulate *Arabidopsis* glucosinolate accumulation and response to both phloem-feeding and chewing insects. Plant Physiol 2005; 138(2): 1149-62.
[http://dx.doi.org/10.1104/pp.104.053389] [PMID: 15923339]

[107] Mikkelsen MD, Petersen BL, Glawischnig E, Jensen AB, Andreasson E, Halkier BA. Modulation of CYP79 genes and glucosinolate profiles in *Arabidopsis* by defense signaling pathways. Plant Physiol 2003; 131(1): 298-308.
[http://dx.doi.org/10.1104/pp.011015] [PMID: 12529537]

[108] Kliebenstein D, Pedersen D, Barker B, Mitchell-Olds T. Comparative analysis of quantitative trait loci controlling glucosinolates, myrosinase and insect resistance in *Arabidopsis thaliana*. Genetics 2002; 161(1): 325-32. [PMID: 12019246]

[109] Brader G, Tas E , Palva ET. Jasmonate-dependent induction of indole glucosinolates in *Arabidopsis* by culture filtrates of the nonspecific *pathogen Erwinia carotovora*. Plant Physiol 2001; 126(2): 849-60. [http://dx.doi.org/10.1104/pp.126.2.849] [PMID: 11402212]

[110] Skirycz A, Reichelt M, Burow M, *et al.* DOF transcription factor AtDof1.1 (OBP2) is part of a regulatory network controlling glucosinolate biosynthesis in *Arabidopsis*. Plant J 2006; 47(1): 10-24. [http://dx.doi.org/10.1111/j.1365-313X.2006.02767.x] [PMID: 16740150]

[111] Levy M, Wang Q, Kaspi R, Parrella MP, Abel S. *Arabidopsis* IQD1, a novel calmodulin-binding nuclear protein, stimulates glucosinolate accumulation and plant defense. Plant J 2005; 43(1): 79-96. [http://dx.doi.org/10.1111/j.1365-313X.2005.02435.x] [PMID: 15960618]

[112] Wittstock U, Halkier BA. Glucosinolate research in the *Arabidopsis* era. Trends Plant Sci 2002; 7(6): 263-70. [http://dx.doi.org/10.1016/S1360-1385(02)02273-2] [PMID: 12049923]

[113] Mikkelsen MD, Petersen BL, Olsen CE, Halkier BA. Biosynthesis and metabolic engineering of glucosinolates. Amino Acids 2002; 22(3): 279-95. [http://dx.doi.org/10.1007/s007260200014] [PMID: 12083070]

[114] Schuster J, Knill T, Reichelt M, Gershenzon J, Binder S. Branched-chain aminotransferase4 is part of the chain elongation pathway in the biosynthesis of methionine-derived glucosinolates in Arabidopsis. Plant Cell 2006; 18(10): 2664-79. [http://dx.doi.org/10.1105/tpc.105.039339] [PMID: 17056707]

[115] Gigolashvili T, Yatusevich R, Rollwitz I, Humphry M, Gershenzon J, Flügge UI. The plastidic bile acid transporter 5 is required for the biosynthesis of methionine-derived glucosinolates in *Arabidopsis thaliana*. Plant Cell 2009; 21(6): 1813-29. [http://dx.doi.org/10.1105/tpc.109.066399] [PMID: 19542295]

[116] Kroymann J, Textor S, Tokuhisa JG, *et al.* A gene controlling variation in *Arabidopsis* glucosinolate composition is part of the methionine chain elongation pathway. Plant Physiol 2001; 127(3): 1077-88. [http://dx.doi.org/10.1104/pp.010416] [PMID: 11706188]

[117] Textor S, de Kraker JW, Hause B, Gershenzon J, Tokuhisa JG. MAM3 catalyzes the formation of all aliphatic glucosinolate chain lengths in *Arabidopsis*. Plant Physiol 2007; 144(1): 60-71. [http://dx.doi.org/10.1104/pp.106.091579] [PMID: 17369439]

[118] Knill T, Reichelt M, Paetz C, Gershenzon J, Binder S. *Arabidopsis thaliana* encodes a bacterial-type heterodimeric isopropylmalate isomerase involved in both Leu biosynthesis and the Met chain elongation pathway of glucosinolate formation. Plant Mol Biol 2009; 71(3): 227-39. [http://dx.doi.org/10.1007/s11103-009-9519-5] [PMID: 19597944]

[119] He Y, Mawhinney TP, Preuss ML, *et al.* A redox-active isopropylmalate dehydrogenase functions in the biosynthesis of glucosinolates and leucine in *Arabidopsis*. Plant J 2009; 60(4): 679-90. [http://dx.doi.org/10.1111/j.1365-313X.2009.03990.x] [PMID: 19674406]

[120] Knill T, Schuster J, Reichelt M, Gershenzon J, Binder S. *Arabidopsis* branched-chain aminotransferase 3 functions in both amino acid and glucosinolate biosynthesis. Plant Physiol 2008; 146(3): 1028-39. [http://dx.doi.org/10.1104/pp.107.111609] [PMID: 18162591]

[121] Falk KL, Tokuhisa JG, Gershenzon J. The effect of sulfur nutrition on plant glucosinolate content: physiology and molecular mechanisms. Plant Biol (Stuttg) 2007; 9(5): 573-81. [http://dx.doi.org/10.1055/s-2007-965431] [PMID: 17853357]

[122] Agati G, Azzarello E, Pollastri S, Tattini M. Flavonoids as antioxidants in plants: location and functional significance. Plant Sci 2012; 196: 67-76.

[http://dx.doi.org/10.1016/j.plantsci.2012.07.014] [PMID: 23017900]

[123] Alcerito T, Barbo FE, Negri G, *et al.* Foliar epicuticular wax of Arrabidaea brachypoda: flavonoids and antifungal activity. Biochem Syst Ecol 2002; 30(7): 677-83.
[http://dx.doi.org/10.1016/S0305-1978(01)00149-1]

[124] Ayaz FA, Hayırlıoglu-Ayaz S, Alpay-Karaoglu S, *et al.* Phenolic acid contents of kale (Brassica oleraceae L. var. acephala DC.) extracts and their antioxidant and antibacterial activities. Food Chem 2008; 107(1): 19-25.
[http://dx.doi.org/10.1016/j.foodchem.2007.07.003]

[125] Barcelo J, Poschenrieder C. Fast root growth responses, root exudates, and internal detoxification as clues to the mechanisms of aluminium toxicity and resistance: a review. Environ Exp Bot 2002; 48(1): 75-92.
[http://dx.doi.org/10.1016/S0098-8472(02)00013-8]

[126] Baumert A, Milkowski C, Schmidt J, Nimtz M, Wray V, Strack D. Formation of a complex pattern of sinapate esters in Brassica napus seeds, catalyzed by enzymes of a serine carboxypeptidase-like acyltransferase family? Phytochemistry 2005; 66(11): 1334-45.
[http://dx.doi.org/10.1016/j.phytochem.2005.02.031] [PMID: 15907956]

[127] Beecher GR. Overview of dietary flavonoids: nomenclature, occurrence and intake. J Nutr 2003; 133(10): 3248S-54S.
[http://dx.doi.org/10.1093/jn/133.10.3248S] [PMID: 14519822]

[128] Bilyk A, Sapers GM. Distribution of quercetin and kaempferol in lettuce, kale, chive, garlic chive, leek, horseradish, red radish, and red cabbage tissues. J Agric Food Chem 1985; 33(2): 226-68.
[http://dx.doi.org/10.1021/jf00062a017]

[129] Buer CS, Imin N, Djordjevic MA. Flavonoids: new roles for old molecules. J Integr Plant Biol 2010; 52(1): 98-111.
[http://dx.doi.org/10.1111/j.1744-7909.2010.00905.x] [PMID: 20074144]

[130] Buskov S, Hansen LB, Olsen CE, Sørensen JC, Sørensen H, Sørensen S. Determination of ascorbigens in autolysates of various Brassica species using supercritical fluid chromatography. J Agric Food Chem 2000; 48(7): 2693-701.
[http://dx.doi.org/10.1021/jf000165r] [PMID: 10898607]

[131] Cao G, Sofic E, Prior RL. Antioxidant capacity of tea and common vegetables. J Agric Food Chem 1996; 44(11): 3426-31.
[http://dx.doi.org/10.1021/jf9602535]

[132] Carbonell-Bejerano P, Diago MP, Martínez-Abaigar J, Martínez-Zapater JM, Tardáguila J, Núñez-Olivera E. Solar ultraviolet radiation is necessary to enhance grapevine fruit ripening transcriptional and phenolic responses. BMC Plant Biol 2014; 14(1): 183.
[http://dx.doi.org/10.1186/1471-2229-14-183] [PMID: 25012688]

[133] Cartea ME, Francisco M, Soengas P, Velasco P. Phenolic compounds in Brassica vegetables. Molecules 2010; 16(1): 251-80.
[http://dx.doi.org/10.3390/molecules16010251] [PMID: 21193847]

[134] Chalker-Scott L, Krahmer RL. Microscopic studies of tannin formation and distribution in plant tissues.Chemistry and significance of condensed tannins. Boston, MA: Springer 1989; pp. 345-68.
[http://dx.doi.org/10.1007/978-1-4684-7511-1_22]

[135] Chaves N, Escudero JC. Variation of flavonoid synthesis induced by ecological factors.Boca Raton: CRC PressPrinciples and practices in plant ecology 1999; pp. 267-85.

[136] Kumar S, Pandey AK. Chemistry and biological activities of flavonoids: an overview. Sci World J 2013; 2013: 162750.
[http://dx.doi.org/10.1155/2013/162750] [PMID: 24470791]

[137] Cheynier V, Comte G, Davies KM, Lattanzio V, Martens S. Plant phenolics: recent advances on their

biosynthesis, genetics, and ecophysiology. Plant Physiol Biochem 2013; 72: 1-20.
[http://dx.doi.org/10.1016/j.plaphy.2013.05.009] [PMID: 23774057]

[138] Chou CH. Roles of allelopathy in plant biodiversity and sustainable agriculture. Crit Rev Plant Sci 1999; 18(5): 609-36.
[http://dx.doi.org/10.1080/07352689991309414]

[139] Chu YF, Sun J, Wu X, Liu RH. Antioxidant and antiproliferative activities of common vegetables. J Agric Food Chem 2002; 50(23): 6910-6.
[http://dx.doi.org/10.1021/jf020665f] [PMID: 12405796]

[140] Chu YH, Chang CL, Hsu HF. Flavonoid content of several vegetables and their antioxidant activity. J Sci Food Agric 2000; 80(5): 561-6.
[http://dx.doi.org/10.1002/(SICI)1097-0010(200004)80:5<561::AID-JSFA574>3.0.CO;2-#]

[141] Chun OK, Smith N, Sakagawa A, Lee CY. Antioxidant properties of raw and processed cabbages. Int J Food Sci Nutr 2004; 55(3): 191-9.
[http://dx.doi.org/10.1080/09637480410001725148] [PMID: 15223595]

[142] Combs GF. The Vitamins. San Diego, CA: Academic Press 1992.

[143] Crozier A, Jaganath IB, Clifford MN. Phenols, polyphenols and tannins: An overview.Plant secondary metabolites: Occurrence, structure and role in the human diet. Oxford, UK: Blackwell 2006; pp. 1-24.
[http://dx.doi.org/10.1002/9780470988558.ch1]

[144] Davey MW, Montagu MV, Inze D, *et al.* Plant L-ascorbic acid: chemistry, function, metabolism, bioavailability and effects of processing. J Sci Food Agric 2000; 80(7): 825-60.
[http://dx.doi.org/10.1002/(SICI)1097-0010(20000515)80:7<825::AID-JSFA598>3.0.CO;2-6]

[145] Dos Reis LC, de Oliveira VR, Hagen ME, Jablonski A, Flôres SH, de Oliveira Rios A. Carotenoids, flavonoids, chlorophylls, phenolic compounds and antioxidant activity in fresh and cooked broccoli (*Brassica oleracea* var. Avenger) and cauliflower (*Brassica oleracea* var. Alphina F$_1$). Food Sci Technol 2015; 63(1): 177-83.

[146] Erdman JW Jr, Balentine D, Arab L, *et al.* Flavonoids and heart health: proceedings of the ILSI North America flavonoids workshop, May 31–June 1, 2005, Washington, DC. J Nutr 2007; 137(3) (Suppl. 1): 718S-37S.
[http://dx.doi.org/10.1093/jn/137.3.718S] [PMID: 17311968]

[147] Fallovo C, Schreiner M, Schwarz D, Colla G, Krumbein A. Phytochemical changes induced by different nitrogen supply forms and radiation levels in two leafy Brassica species. J Agric Food Chem 2011; 59(8): 4198-207.
[http://dx.doi.org/10.1021/jf1048904] [PMID: 21395334]

[148] Fiol M, Adermann S, Neugart S, *et al.* Highly glycosylated and acylated flavonols isolated from kale (*Brassica oleracea* var. *sabellica*)-Structure-antioxidant activity relationship. Food Res Int 2012; 47(1): 80-9.
[http://dx.doi.org/10.1016/j.foodres.2012.01.014]

[149] Francisco M, Ali M, Ferreres F, Moreno DA, Velasco P, Soengas P. Organ-specific quantitative genetics and candidate genes of phenylpropanoid metabolism in Brassica oleracea. Front Plant Sci 2016; 6: 1240.
[http://dx.doi.org/10.3389/fpls.2015.01240] [PMID: 26858727]

[150] Francisco M, Moreno DA, Cartea ME, Ferreres F, García-Viguera C, Velasco P. Simultaneous identification of glucosinolates and phenolic compounds in a representative collection of vegetable *Brassica rapa.* J Chromatogr A 2009; 1216(38): 6611-9.
[http://dx.doi.org/10.1016/j.chroma.2009.07.055] [PMID: 19683241]

[151] Gardner AM, Brown AF, Juvik JA. QTL analysis for the identification of candidate genes controlling phenolic compound accumulation in broccoli (*Brassica oleracea* l. var. *italica*). Mol Breed 2016; 36(6): 81.

[http://dx.doi.org/10.1007/s11032-016-0497-4]

[152] Gazzani G, Papetti A, Massolini G, Daglia M. Anti-and prooxidant activity of water soluble components of some common diet vegetables and the effect of thermal treatment. J Agric Food Chem 1998; 46(10): 4118-22.
[http://dx.doi.org/10.1021/jf980300o]

[153] Grayer RJ, Harborne JB. A survey of antifungal compounds from higher plants, 1982–1993. Phytochemistry 1994; 37(1): 19-42.
[http://dx.doi.org/10.1016/0031-9422(94)85005-4]

[154] Groenbaek M, Jensen S, Neugart S, Schreiner M, Kidmose U, Kristensen HL. Nitrogen split dose fertilization, plant age and frost effects on phytochemical content and sensory properties of curly kale (*Brassica oleracea* L. var. *sabellica*). Food Chem 2016; 197(Pt A): 530-8.
[http://dx.doi.org/10.1016/j.foodchem.2015.10.108] [PMID: 26616985]

[155] Hahlbrock K, Scheel D. Physiology and molecular biology of phenylpropanoid metabolism. Annu Rev Plant Biol 1989; 40(1): 347-69.
[http://dx.doi.org/10.1146/annurev.pp.40.060189.002023]

[156] Halliwell B, Gutteridge JM. Free radicals in biology and medicine. USA: Oxford University Press 1999.

[157] Hallmann E, Rembiałowska E, Szafirowska A, Grudzień K. Importance of fruits and vegetables from organicproduction in preventive medicine at the example ofpeppers from organic farming. Rocz Panstw Zakl Hig 2007; 58(1): 77-82. [in Polish].
[PMID: 17711094]

[158] Heim KE, Tagliaferro AR, Bobilya DJ. Flavonoid antioxidants: chemistry, metabolism and structure-activity relationships. J Nutr Biochem 2002; 13(10): 572-84.
[http://dx.doi.org/10.1016/S0955-2863(02)00208-5] [PMID: 12550068]

[159] Heimler D, Vignolini P, Dini MG, Vincieri FF, Romani A. Antiradical activity and polyphenol composition of local Brassicaceae edible varieties. Food Chem 2006; 99(3): 464-9.
[http://dx.doi.org/10.1016/j.foodchem.2005.07.057]

[160] Hertog MG, Hollman PC, Katan MB. Content of potentially anticarcinogenic flavonoids of 28 vegetables and 9 fruits commonly consumed in the Netherlands. J Agric Food Chem 1992; 40(12): 2379-83.
[http://dx.doi.org/10.1021/jf00024a011]

[161] Ho CT. Phenolic compounds in food: An overview.Phenolic compounds in food and their effects on health II: Antioxidants and cancer prevention. ACS symposium series2-7.
[http://dx.doi.org/10.1021/bk-1992-0507.ch001]

[162] Hrncirik K, Valusek J, Velisek J. Investigation of ascorbigen as a breakdown product of glucobrassicin autolysis in Brassica vegetables. Eur Food Res Technol 2001; 212(5): 576-81.
[http://dx.doi.org/10.1007/s002170100291]

[163] Huang Z, Wang B, Eaves DH, Shikany JM, Pace RD. Phenolic compound profile of selected vegetables frequently consumed by African Americans in the southeast United States. Food Chem 2007; 103(4): 1395-402.
[http://dx.doi.org/10.1016/j.foodchem.2006.10.077]

[164] Ibrahim MH, Jaafar HZ, Rahmat A, Rahman ZA. Effects of nitrogen fertilization on synthesis of primary and secondary metabolites in three varieties of Kacip Fatimah (Labisia pumila Blume). Int J Mol Sci 2011; 12(8): 5238-54.
[http://dx.doi.org/10.3390/ijms12085238] [PMID: 21954355]

[165] Ismail A, Marjan ZM, Foong CW. Total antioxidant activity and phenolic content in selected vegetables. Food Chem 2004; 87(4): 581-6.
[http://dx.doi.org/10.1016/j.foodchem.2004.01.010]

[166] Ju Z, Bramlage WJ. Phenolics and lipid-soluble antioxidants in fruit cuticle of apples and their antioxidant activities in model systems. Postharvest Biol Technol 1999; 16(2): 107-18. [http://dx.doi.org/10.1016/S0925-5214(99)00006-X]

[167] Julkunen-Tiitto R, Nenadis N, Neugart S, *et al.* Assessing the response of plant flavonoids to UV radiation: an overview of appropriate techniques. Phytochem Rev 2015; 14(2): 273-97. [http://dx.doi.org/10.1007/s11101-014-9362-4]

[168] Justesen U, Knuthsen P, Leth T. Quantitative analysis of flavonols, flavones, and flavanones in fruits, vegetables and beverages by high-performance liquid chromatography with photo-diode array and mass spectrometric detection. J Chromatogr A 1998; 799(1-2): 101-10. [http://dx.doi.org/10.1016/S0021-9673(97)01061-3] [PMID: 9550103]

[169] Kähkönen MP, Hopia AI, Vuorela HJ, *et al.* Antioxidant activity of plant extracts containing phenolic compounds. J Agric Food Chem 1999; 47(10): 3954-62. [http://dx.doi.org/10.1021/jf990146l] [PMID: 10552749]

[170] Kaur C, Kapoor HC. Antioxidant activity and total phenolic content of some Asian vegetables. Int J Food Sci Technol 2002; 37(2): 153-61. [http://dx.doi.org/10.1046/j.1365-2621.2002.00552.x]

[171] Koh E, Wimalasiri KM, Chassy AW, Mitchell AE. Content of ascorbic acid, quercetin, kaempferol and total phenolics in commercial broccoli. J Food Compos Anal 2009; 22(7-8): 637-43. [http://dx.doi.org/10.1016/j.jfca.2009.01.019]

[172] Koyama K, Numata M, Nakajima I, Goto-Yamamoto N, Matsumura H, Tanaka N. Functional characterization of a new grapevine MYB transcription factor and regulation of proanthocyanidin biosynthesis in grapes. J Exp Bot 2014; 65(15): 4433-49. [http://dx.doi.org/10.1093/jxb/eru213] [PMID: 24860184]

[173] Kozłowska A, Szostak-Wegierek D. Flavonoids--food sources and health benefits. Rocz Panstw Zakl Hig 2014; 65(2): 79-85. [PMID: 25272572]

[174] Kurilich AC, Jeffery EH, Juvik JA, Wallig MA, Klein BP. Antioxidant capacity of different broccoli (Brassica oleracea) genotypes using the oxygen radical absorbance capacity (ORAC) assay. J Agric Food Chem 2002; 50(18): 5053-7. [http://dx.doi.org/10.1021/jf025535l] [PMID: 12188607]

[175] Lindley MG. The impact of food processing on antioxidants in vegetable oils, fruits and vegetables. Trends Food Sci Technol 1998; 9(8-9): 336-40. [http://dx.doi.org/10.1016/S0924-2244(98)00050-8]

[176] Lisiewska Z, Kmiecik W. Effects of level of nitrogen fertilizer, processing conditions and period of storage of frozen broccoli and cauliflower on vitamin C retention. Food Chem 1996; 57(2): 267-70. [http://dx.doi.org/10.1016/0308-8146(95)00218-9]

[177] Mageney V, Neugart S, Albach DC. A Guide to the Variability of Flavonoids in *Brassica oleracea.* Molecules 2017; 22(2): 252-68. [http://dx.doi.org/10.3390/molecules22020252] [PMID: 28208739]

[178] Majewska M, Czeczot H. Flavonoids in prevention and therapy of diseases. Pol Merkuriusz Lek 2009; 65(5): 369-77. [in Polish].

[179] Małolepsza U, Urbanek H. Plant flavonoids as biochemically active compounds. Wiad Bot 2000; 44(3/4): 27-37. [in Polish].

[180] Merzlyak MN, Solovchenko AE, Chivkunova OB. Patterns of pigment changes in apple fruits during adaptation to high sunlight and sunscald development. Plant Physiol Biochem 2002; 40(6-8): 679-84. [http://dx.doi.org/10.1016/S0981-9428(02)01408-0]

[181] Milkowski C, Baumert A, Schmidt D, Nehlin L, Strack D. Molecular regulation of sinapate ester

metabolism in Brassica napus: expression of genes, properties of the encoded proteins and correlation of enzyme activities with metabolite accumulation. Plant J 2004; 38(1): 80-92.
[http://dx.doi.org/10.1111/j.1365-313X.2004.02036.x] [PMID: 15053762]

[182]   Namiki M. Antioxidants/antimutagens in food. Crit Rev Food Sci Nutr 1990; 29(4): 273-300.
[http://dx.doi.org/10.1080/10408399009527528] [PMID: 2257080]

[183]   Nielsen JK, Olsen CE, Petersen MK. Acylated flavonol glycosides from cabbage leaves. Phytochemistry 1993; 34(2): 539-44.
[http://dx.doi.org/10.1016/0031-9422(93)80042-Q] [PMID: 7764144]

[184]   Olsen H, Aaby K, Borge GI. Characterization and quantification of flavonoids and hydroxycinnamic acids in curly kale (*Brassica oleracea* L. Convar. *acephala* Var. *sabellica*) by HPLC-DAD-ESI-MSn. J Agric Food Chem 2009; 57(7): 2816-25.
[http://dx.doi.org/10.1021/jf803693t] [PMID: 19253943]

[185]   Pandey AK, Mishra AK, Mishra A. Antifungal and antioxidative potential of oil and extracts derived from leaves of Indian spice plant *Cinnamomum tamala.* Cell Mol Biol 2012; 58(1): 142-7.
[PMID: 23273204]

[186]   Papas AM. Diet and antioxidant status. Food Chem Toxicol 1999; 37(9-10): 999-1007.
[http://dx.doi.org/10.1016/S0278-6915(99)00088-5] [PMID: 10541457]

[187]   Pereira DM, Valentão P, Pereira JA, Andrade PB. Phenolics: From chemistry to biology. Molecules 2009; 14: 2202-11.
[http://dx.doi.org/10.3390/molecules14062202]

[188]   Pietrini F, Iannelli MA, Massacci A. Anthocyanin accumulation in the illuminated surface of maize leaves enhances protection from photoinhibitory risks at low temperature, without further limitation to photosynthesis. Plant Cell Environ 2002; 25(10): 1251-9.
[http://dx.doi.org/10.1046/j.1365-3040.2002.00917.x]

[189]   Podsędek A. Natural antioxidants and antioxidant capacity of Brassica vegetables: A review. Lebensm Wiss Technol 2007; 40(1): 1-11.
[http://dx.doi.org/10.1016/j.lwt.2005.07.023]

[190]   Pollastri S, Tattini M. Flavonols: old compounds for old roles. Ann Bot 2011; 108(7): 1225-33.
[http://dx.doi.org/10.1093/aob/mcr234] [PMID: 21880658]

[191]   Price KR, Casuscelli F, Colquhoun IJ, Rhodes MJ. Composition and content of flavonol glycosides in broccoli florets (Brassica olearacea) and their fate during cooking. J Sci Food Agric 1998; 77(4): 468-72.
[http://dx.doi.org/10.1002/(SICI)1097-0010(199808)77:4<468::AID-JSFA66>3.0.CO;2-B]

[192]   Price KR, Casuscelli F, Colquhoun IJ, Rhodes MJ. Hydroxycinnamic acid esters from broccoli florets. Phytochemistry 1997; 45(8): 1683-7.
[http://dx.doi.org/10.1016/S0031-9422(97)00246-X]

[193]   Proteggente AR, Pannala AS, Paganga G, *et al.* The antioxidant activity of regularly consumed fruit and vegetables reflects their phenolic and vitamin C composition. Free Radic Res 2002; 36(2): 217-33.
[http://dx.doi.org/10.1080/10715760290006484] [PMID: 11999391]

[194]   Pyke M. The vitamin content of vegetables. J Chem Technol Biotechnol 1942; 61(10): 149-51.
[http://dx.doi.org/10.1002/jctb.5000611001]

[195]   Quilichini TD, Samuels AL, Douglas CJ. ABCG26-mediated polyketide trafficking and hydroxycinnamoyl spermidines contribute to pollen wall exine formation in Arabidopsis. Plant Cell 2014; 26(11): 4483-98.
[http://dx.doi.org/10.1105/tpc.114.130484] [PMID: 25415974]

[196]   Reilly K, Valverde J, Finn L, *et al.* Potential of cultivar and crop management to affect phytochemical content in winter-grown sprouting broccoli (*Brassica oleracea* L. var. *italica*). J Sci Food Agric 2014; 94(2): 322-30.

[http://dx.doi.org/10.1002/jsfa.6263] [PMID: 23761132]

[197]   Rice-Evans C, Miller N, Paganga G. Antioxidant properties of phenolic compounds. Trends Plant Sci 1997; 2(4): 152-9.
[http://dx.doi.org/10.1016/S1360-1385(97)01018-2]

[198]   Rice-Evans CA, Miller NJ, Paganga G. Structure-antioxidant activity relationships of flavonoids and phenolic acids. Free Radic Biol Med 1996; 20(7): 933-56.
[http://dx.doi.org/10.1016/0891-5849(95)02227-9] [PMID: 8743980]

[199]   Saito K, Yonekura-Sakakibara K, Nakabayashi R, *et al.* The flavonoid biosynthetic pathway in Arabidopsis: structural and genetic diversity. Plant Physiol Biochem 2013; 72: 21-34.
[http://dx.doi.org/10.1016/j.plaphy.2013.02.001] [PMID: 23473981]

[200]   Šamec D, Piljac-Žegarac J, Bogović M, Habjanič K, Grúz J. Antioxidant potency of white (*Brassica oleracea* L. var. *capitata*) and Chinese (*Brassica rapa* L. var. *pekinensis* (Lour.)) cabbage: The influence of development stage, cultivar choice and seed selection. Sci Hortic (Amsterdam) 2011; 128(2): 78-83.
[http://dx.doi.org/10.1016/j.scienta.2011.01.009]

[201]   Sandermann H Jr, Ernst D, Heller W, Langebartels C. Ozone: an abiotic elicitor of plant defence reactions. Trends Plant Sci 1998; 3(2): 47-50.
[http://dx.doi.org/10.1016/S1360-1385(97)01162-X]

[202]   Schmidt S, Zietz M, Schreiner M, Rohn S, Kroh LW, Krumbein A. Genotypic and climatic influences on the concentration and composition of flavonoids in kale (*Brassica oleracea* var. *sabellica*). Food Chem 2010; 119(4): 1293-9.
[http://dx.doi.org/10.1016/j.foodchem.2009.09.004]

[203]   Schmidt S, Zietz M, Schreiner M, Rohn S, Kroh LW, Krumbein A. Identification of complex, naturally occurring flavonoid glycosides in kale (*Brassica oleracea* var. *sabellica*) by high-performance liquid chromatography diode-array detection/electrospray ionization multi-stage mass spectrometry. Rapid Commun Mass Spectrom 2010; 24(14): 2009-22.
[http://dx.doi.org/10.1002/rcm.4605] [PMID: 20552580]

[204]   Sikora E, Cieślik E, Leszczyńska T, Filipiak-Florkiewicz A, Pisulewski PM. The antioxidant activity of selected cruciferous vegetables subjected to aquathermal processing. Food Chem 2008; 107(1): 55-9.
[http://dx.doi.org/10.1016/j.foodchem.2007.07.023]

[205]   Sousa C, Lopes G, Pereira DM, *et al.* Screening of antioxidant compounds during sprouting of *Brassica oleracea* L. var. *costata* DC. Comb Chem High Throughput Screen 2007; 10(5): 377-86.
[http://dx.doi.org/10.2174/138620707781662817] [PMID: 17896933]

[206]   Sousa C, Pereira DM, Pereira JA, *et al.* Multivariate analysis of tronchuda cabbage (*Brassica oleracea* L. var. *costata* DC) phenolics: influence of fertilizers. J Agric Food Chem 2008; 56(6): 2231-9.
[http://dx.doi.org/10.1021/jf073041o] [PMID: 18290619]

[207]   Stevenson DE, Hurst RD. Polyphenolic phytochemicals--just antioxidants or much more? Cell Mol Life Sci 2007; 64(22): 2900-16.
[http://dx.doi.org/10.1007/s00018-007-7237-1] [PMID: 17726576]

[208]   Tabart J, Kevers C, Pincemail J, Defraigne JO, Dommes J. Comparative antioxidant capacities of phenolic compounds measured by various tests. Food Chem 2009; 113(4): 1226-33.
[http://dx.doi.org/10.1016/j.foodchem.2008.08.013]

[209]   Takos AM, Ubi BE, Robinson SP, Walker AR. Condensed tannin biosynthesis genes are regulated separately from other flavonoid biosynthesis genes in apple fruit skin. Plant Sci 2006; 170(3): 487-99.
[http://dx.doi.org/10.1016/j.plantsci.2005.10.001]

[210]   Tattini M, Galardi C, Pinelli P, Massai R, Remorini D, Agati G. Differential accumulation of flavonoids and hydroxycinnamates in leaves of *Ligustrum vulgare* under excess light and drought

stress. New Phytol 2004; 163(3): 547-61.
[http://dx.doi.org/10.1111/j.1469-8137.2004.01126.x]

[211] Tee ES, Lim CL, Chong YH, Khor SC. A study of the biological utilization of carotenoids of carrot and swamp cabbage in rats. Food Chem 1996; 56(1): 21-32.
[http://dx.doi.org/10.1016/0308-8146(95)00145-X]

[212] Treutter D. Significance of flavonoids in plant resistance: a review. Environ Chem Lett 2006; 4(3): 147.
[http://dx.doi.org/10.1007/s10311-006-0068-8]

[213] Vallejo F, Tomás-Barberán FA, García-Viguera C. Potential bioactive compounds in health promotion from broccoli cultivars grown in Spain. J Sci Food Agric 2002; 82(11): 1293-7.
[http://dx.doi.org/10.1002/jsfa.1183]

[214] Velioglu YS, Mazza G, Gao L, Oomah BD. Antioxidant activity and total phenolics in selected fruits, vegetables, and grain products. J Agric Food Chem 1998; 46(10): 4113-7.
[http://dx.doi.org/10.1021/jf9801973]

[215] Volden J, Bengtsson GB, Wicklund T. Glucosinolates, L-ascorbic acid, total phenols, anthocyanins, antioxidant capacities and colour in cauliflower (*Brassica oleracea* L. ssp. *botrytis*); effects of long-term freezer storage. Food Chem 2009; 112(4): 967-76.
[http://dx.doi.org/10.1016/j.foodchem.2008.07.018]

[216] Zhang J, Satterfield MB, Brodbelt JS, Britz SJ, Clevidence B, Novotny JA. Structural characterization and detection of kale flavonoids by electrospray ionization mass spectrometry. Anal Chem 2003; 75(23): 6401-7.
[http://dx.doi.org/10.1021/ac034795e] [PMID: 14640707]

[217] Zietz M, Weckmüller A, Schmidt S, *et al.* Genotypic and climatic influence on the antioxidant activity of flavonoids in Kale (*Brassica oleracea* var. *sabellica*). J Agric Food Chem 2010; 58(4): 2123-30.
[http://dx.doi.org/10.1021/jf9033909] [PMID: 20095605]

[218] Sotelo T, Cartea ME, Velasco P, Soengas P. Identification of antioxidant capacity -related QTLs in *Brassica oleracea*. PLoS One 2014; 9(9): e107290.
[http://dx.doi.org/10.1371/journal.pone.0107290] [PMID: 25198771]

[219] Singh J, Upadhyay AK, Bahadur A, Singh B, Singh KP, Rai M. Antioxidant phytochemicals in cabbage (*Brassica oleracea* L. var. *capitata*). Sci Hortic (Amsterdam) 2006; •••: 233-7.
[http://dx.doi.org/10.1016/j.scienta.2006.01.017]

[220] Hagen SF, Borge GIA, Bengtsson GB, *et al.* Phenolic contents and other health and sensory related properties of apple fruit (*Malus domestica* Borkh., cv. Aroma): Effect of postharvest UV-B irradiation. Postharvest Biol Technol 2007; 45: 1-10.
[http://dx.doi.org/10.1016/j.postharvbio.2007.02.002]

CHAPTER 9

Solanaceae: A Family Well-known and Still Surprising

Blanka Svobodová and **Vlastimil Kubáň**[*]

Department of Food Technology, Faculty of Technology, Tomas Bata University in Zlin, Vavreckova 275, 76272, Czech Republic

Abstract: The family *Solanaceae* features many popular vegetable crops, mainly belonging to its three major genera *Solanum, Capsicum*, and *Physalis*. They have played an important role in human nutrition and health since ancient times, most of them being native to South and Central America, but nowadays domesticated worldwide. Edible fruit, leaves, tubers as well as non-edible plant parts are a valuable source of often unique compounds with multiple biological activities. Despite numerous studies and intensive research, there are still novel compounds being discovered from both cultivated and wild species, that could bring more benefits into treatment of civilisation diseases not only as part of pharmaceutical products, but also as functional foods important in everyday prevention of health problems, especially those related to oxidative stress. Apart from the edible parts, the organic waste from production of these vegetables is a useful source of bioactive substances and extracts that can be used both in the food and pharmaceutical industry. In this chapter, the most important species from three genera are described in detail with emphasis being given on the research studies published within the last two decades. Major bioactive constituents representing each genus and biological activity of extracts and individual compounds, with special attention to the most interesting findings regarding antioxidant, anti-inflammatory and anticancer activity, are also discussed in the corresponding subsections.

Keywords: Biological activity, Chilli, Capsaicinoids, Capsinoids, Capsanthin, *Capsicum*, Eggplant, Lutein, Lycopene, Nightshade vegetables, *Physalis*, Pepper, Physalins, *Solanum*, Steroidal alkaloids, Solamargine, Solanine, Tomato, Tomatine, Withanolides.

INTRODUCTION

The *Solanaceae*, also known as the nightshade family, is one of the most important families from the Angiosperms group of plants for humans from the

[*] **Corresponding author Vlastimil Kubáň:** Department of Food Technology, Faculty of Technology, Tomas Bata University in Zlin, Vavreckova 275, 76272, Czech Republic; Tel: +420 576 033 018; E-mail: kuban@ft.utb.cz

Spyridon A. Petropoulos, Isabel C.F.R. Ferreira and Lillian Barros (Eds.)

dietary, economical and pharmaceutical point of view. It is the third largest family in the plant kingdom after the *Poaceae* and *Fabaceae* [1]. For such a large family it is not surprising, that it brings complications in systematic nomenclature and identification of the species. According to The Plant List, it includes 115 genera with over 8,400 species names, of which only 32% are accepted and another 31% are detected synonyms. The remaining 37% of unassessed species are yet to be validated by the means of modern methods, such as DNA analysis and molecular techniques [2, 3]. Quite often each plant species has more than 10 synonyms and the orientation in the literature is a bit challenging. A typical example is tomato, for which both synonyms *Solanum lycopersicum* and *Lycopersicon esculentum* are simultaneously used in scientific literature up to date. Therefore, a good attention must be paid and the correct accepted names should be used in order to avoid duplicity in research.

The original region of *Solanaceae* points to South America due to the great genetic diversity, from where it spread to every green continent, mostly in tropical areas. The family is represented by different herbs, shrubs, small trees, or some woody vines; with or without spines. Very often the fruit are edible in raw state, but the rest of the plant body is poisonous to humans and/or animals [4]. From the edible species, many plants are cultivated for the plant part that may be eaten as fruit or as vegetable (usually cooked). Therefore, it is difficult to sort some species in terms of their edible products as they can represent both groups, depending on the traditional use.

The term vegetable may have a botanical, culinary or legal explanation. Usually it refers to the traditional culinary use of the plant part in a main dish, appetizer or salad. Vegetables include edible stems and stalks, roots, tubers, bulbs, leaves, flowers, some fruits. Most of the *Solanaceae* vegetables are in fact fruits from the botanical point of view (tomato, pepper, black nightshade, pepino, ground cherry, tomatillo, *etc.*). On the other hand, potato is an enlarged modified underground stem called tuber and often considered not as a vegetable, but as a starchy field crop together with wheat, maize and rice [5].

The fruit group contains *Physalis* and *Solanum* species, such as cape gooseberry (*P. peruvia*na) and wild gooseberry (*P. angulata*), goji berry (*Lycium barbarum* and *L. chinense*), cocona (*S. sessiliflorum*), naranjilla (*S. quitoense*), pepino (*S. muricatum*), and tamarrilo (*S. betaceum*). Nevertheless, some of them (*P. angulata*) have leaves that are cooked and served as vegetable and therefore might be considered a vegetable as well [6].

The vegetable group includes eggplant (*Solanum melongena*), chilli and bell peppers (*Capsicum annuum, C. baccatum, C. chinense, C. frustescens, C. pube-*

scens), potato (*S. tuberosum*), tomato (*S. lycopersicum*), tomatillo (*Physalis philadelphica*), turkey berry (*S. torvum)*, nipple fruit (*S. mammosum*), popular African indigenous vegetables of *Solanum nigrum* complex (*S. americanum, S. scabrum, S. villosum)*, the eggplant relatives (S. *macrocarpon, S. aethiopicum*) and other minor species [6 - 8].

Some of the species became an everyday part of human diet worldwide and their cultivation brought many new varieties in order to perfect the yield and characteristics of the desired crop [9]. According to the latest Food and Agriculture Organization (FAO) annual statistics for 2016, 38 million hectares of the most important solanaceous food crops (eggplant, tomato, potato, capsicums) were cultivated globally, with a total production of 733 million tonnes [10].

As it has been mentioned, the most valuable vegetable crops from this family are *Solanum tuberosum* (potato), *S. lycopersicum* (tomato), *S. melongena* (eggplant), *Capsicum annuum* (bell pepper) and *C. frutescens* (chilli pepper). The research on these species is very broad and numerous papers were published on their biological activities both *in vitro* and *in vitro*. Some of the species serve as model systems in plant biology studies for research of defence responses, fruit ripening, metabolism and development [11].

Among other species from *Solanaceae* family, *Nicotiana tabacum* (tobacco), *Atropa belladonna* (belladonna), *Datura stramonium* (Jimsonweed), *Mandragora officinarum* (mandrake), *Hyoscyamus niger* (henbane), *Withania somnifera* (ashwagandha) and *Solanum nigrum* (black nightshade) are precious sources of valuable compounds for the pharmaceutical and food industry [12]. From 60 to 70% of the species in *Solanaceae* and particularly some genera like *Solanum*, are poisonous due to the content of self-protection products of secondary metabolism, such as alkaloids [13]. There are also numerous ornamental plant species belonging to genera, such as *Brunfelsia, Cestrum, Lycium, Nierembergia, Petunia, Physalis, Schizanthus* or *Solandra.*

However, not only the edible parts of vegetables from *Solanaceae* are a good source of phytochemicals, the importance of other non-edible plant parts, that are usually considered as a waste or by-product of commercial cultivation, is on the rise as many useful compounds may be obtained from discarded plant tissues by separation methods [14]. Very often such plant parts are toxic for human consumption, nevertheless contain medicinally more valuable compounds with biological activity and therefore have their place in traditional medicine as well as in the pharmaceutical and food industry as starting compounds for medicinal drugs synthesis or functional food ingredients, for example [15].

In addition to the main solanaceous crops that underwent domestication and cultivation, there are many wild species used worldwide in traditional cuisine of rural communities [16]. They are used as leafy or fruit vegetables and for medicinal purposes. Recently, many wild *Solanaceae* species are being studied as a valuable source of bioactive compounds for the modern food industry due to the quantity and quality of active metabolites that may be applied in functional foods, nutraceuticals or other food products. Wild species often tend to be rich in bioactive compounds due to the strongly competitive environment they are grown, especially in tropical regions [17]. People usually collect them from the wild or grow them in the gardens. For example, *Solanum americanum*, *S. scabrum*, *S. villosum*, *S. tarderemotum*, *S. florulentum* from Africa and *S. betaceum* (tree tomato), *S. quitoense* (naranjilla) or *S. muricatum* (pepino) from America belong to the lesser known crops [12, 16].

This chapter intends to contribute to the latest knowledge about the main cultivated Solanaceae species (except for potato, which is covered in a different chapter for tuberous crops), especially covering the last two decades. Despite the research which has been done on the cultivated species, they are still able to surprise with the revealing of biological activity and unknown metabolites which are still being discovered up to date. The mechanisms of action in even well-known compounds already in market, such as capsaicin, still awaits unfolding in detail. In conclusion, this family has surely been important for both human nutrition and health and will always have its place on the table and intrigue the scientists.

CAPSICUM GENUS

The genus *Capsicum* (peppers) consists of about 30 species, broadly cultivated for their edible and often pungent fruit. There are five major cultivated species featuring 26 chromozomes Fig. (**1**): *Capsicum annuum* L., *C. baccatum* L., *C. frutescens* L., *C. chinense* Jacq., and *C. pubescens* Ruiz and Pav. The progenitor of the domesticated *C. annuum* is most probably the wild *C. annuum* var. *glabriusculum* (Dunal) Heiser and Pickersgill. Additionally, *C. frutescens* is the probable ancestor of *C. chinense*. From the phylogenetical point of view, the species are organized into two branches: the white-flowered group (*C. annuum* and *C. baccatum*) and the purple-flowered group (*C. pubescens* and the wild species *C. cardenasii* and *C. eximium*). The domesticated *C. pubescens* has also unique black seeds. During last decades, an intense research on wild species with 26 chromosomes has been done and several unique species, especially from Brazilian forests, have been categorized, such as *C. rhomboideum* Kuntze with yellow flower and dark brown seeds [18].

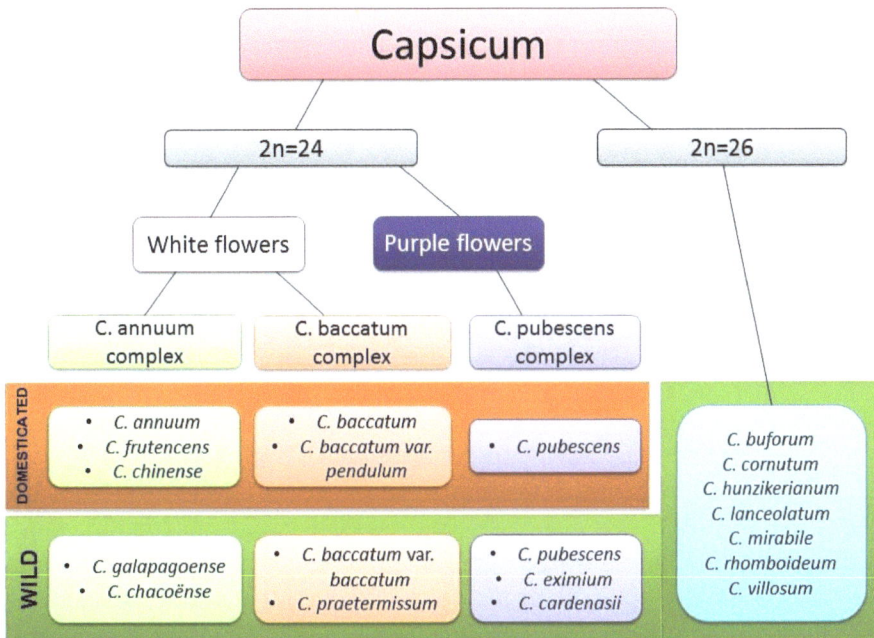

Fig. (1). Biosystematical categorization of genus *Capsicum*.

The global production is growing and reached 52 million tons of green peppers and chilli and 4 million tons of dry peppers and chilli in 2016 [10]. Lately, under the pressure of consumers, the food industry is interested in using peppers as natural food coloring agents in food products instead of synthetic colorants [19].

Peppers are native to tropical America, where they are perennials, however they have been domesticated and grown worldwide as annuals, especially in the temperate climate [6].

Peppers have been used since ancient times for their fruit, consumed both raw and cooked, as condiment or spice, probably more for the antimicrobial effects in food preservation rather than spiciness [20]. They have been valued also medicinally, for example in the treatment of asthma, coughs, digestive problems, sore throats, or toothache [21].

Fig. (2). Different fruit of *Capsicum* spp.: 1. *C. annuum* Jalapeno Mammoth, 2. *C. baccatum* Pimenta Trepadeira do Werner, 3. *C. frutescens* Zimbabwe Bird/ African Devil, 4. *C. chinense* Trinidad Red Scorpion, 5. *C. pubescens* Manzano Rocoto Red. (Photographs reproduced with permission from www.seminka-chilli.cz).

*Capsicum annuum L.*is an important vegetable and one of the most consumed spices spread worldwide from its place of original diversity, Mexico and Latin America. The species covers both pungent and non-pungent varieties and its berry fruit are variously shaped and come in different colours, usually from green to yellow, orange and red, just like other species of *Capsicum* (Fig. **2**). White and purple varieties do not transition to green colour during ripening. The fruits are eaten raw for their fleshy and firm pericarp, cooked, and dried as seasoning and spice, while the seed oil is used for flavouring. The pepper oleoresin is part of many food, drug and cosmetic products. Young leaves and flowers are traditionally added to soups and stews in some parts of Asia [6].

This species is not only valued for its nutritional value but also has been associated with the treatment of various health problems. They are well known its benefits as a digestive, tonic, antiseptic and antimicrobial agent. It may be applied topically or internally for the treatment of asthma, fever, neuralgia, obesity, rheumatism, upset stomach, and scabies. The root is used to treat asthenia and gonorrhoea, while stems are being used for the treatment of rheumatism. Leaves are applied in case of emesis and dysentery [6, 13].

Previously, the fruits of *C. baccatum* exhibited many biological activities in relation to human health – to treat digestive problems, hemorrhoids, rheumatism, sepsis, and being an irritant, sialagogue and tonic. The plant hot fruits have been used both topically and internally to treat problems, such as asthma, cold stage of fevers, digestive problems, neurodegenerative diseases of old age, or varicose veins, while externally it is used for the treatment of pleuritis, sprain, neuralgia, *etc*. [6, 22].

C. frutescens (Tabasco, Piri Piri) originated in Central or South America and spread to other tropical and subtropical regions, the Caribbean, Africa, Asia, Pacific Island and Australia. This annual species is typical to *Capsicum* genus having erect small fruits that ripen *via* yellow to red colour.

The pungent fruit are traditionally part of sauces and curry powder in many national cousins, such as Mexican, Indian or Indonesian. Traditional healers all over the world have been using fruit, leaves and roots of this species for treating various ailments like asthma, arthritis, diabetes, toothache, diarrhoea, wounds, as local anaesthetic and stimulant, and against ring worms [1, 6].

C. chinense Jacq. (Habanero, Scotch bonnet, Naga morich) originated in Amazonas and is popular in the Caribbean and America as well as India. Eaten fresh or dried made into hot sauces and purees that are known worldwide. Flowers and fruit appear on nodes in clusters of 2 or more, unlike other *Capsicum* species that bear only one per node [6].

C. pubescens (rocoto, tree pepper) is only known as cultivated pepper with no wild relatives occurring mainly in higher altitudes of Peru, Bolivia, Ecuador and Argentina and as such it is adapted to low temperatures. The fruit reminds an apple by shape and the species features several typical characteristics – purple flowers, black seeds and hairy leaves, which makes it easy to distinguish from other *Capsicum* species [23].

In the following text, the most important groups of bioactive compounds from this genus will be described and the studied bioactivity of extracts and individual compounds, if known, will be discussed.

Chemical Constituents in Genus *Capsicum*

The fruits of *Capsicums* are a valuable source of bioactive compounds, such as carotenoids, vitamins (A, B, C, E), flavonoid glycosides, steroidal alkaloids, saponins, antifungal proteins and alkaloids called capsaicinoids. These compounds often present synergy in the natural complex. Synergistic effects have been proved between quercetin and vitamin C and between capsaicin and vitamin E. Therefore the consumption of peppers on a daily basis is more efficient for the maintenance of health *via* antioxidants rather than supplements with separate phytochemicals [24]. Other parts of the plant also contain various bioactive compounds. For example, seeds and roots contain steroidal glycosides, capsicosides A – D and proto-degalactotigonin (isolated from the roots and seeds of *C. annuum* var. *conoides* and *C. annuum* var. *fasciculatum* [25].

Capsaicinoids and Capsinoids

Capsaicinoids, the *Capsicum* non-volatile alkaloids, are the acyls of vanillylamine (capsaicin, dihydrocapsaicin, nordihydrocapsaicin, homodihydrocapsaicin, and homocapsaicin). These secondary metabolites are responsible for the typical hot taste - pungency - of *Capsicum* spp. fruit. In fact, *C. annuum* species are distinguished and sorted as pungent chilli peppers and sweet bell peppers with no pungency. The degree of pungency is affected by every variation in the chemical structure of capsaicinoids and an organoleptic test with a scale reflecting the pungency has been developed and defined as one part per million (ppm) of capsaicin having a pungency of 15 SHU - Scoville heat units [26]. High-performance liquid chromatography (HPLC) is more costly yet more objective method analysing not only the overall pungency, but individual capsaicinoids responsible for the heat [27, 28]. Capillary electrophoresis has also been used in analysing capsaicinoids, too [29].

The breeders have been competing in developing the most pungent chilli in the world to reach the Guinness World Record prize, and varieties with over 2 million SHU are on the market up to date. Among them, Trinidad Moruga Scorpion (*C. chinense*) [30]. As such, it is a crop of interest for the future exploitation in both the food and pharmaceutical industry as a rich source of capsaicinoids.

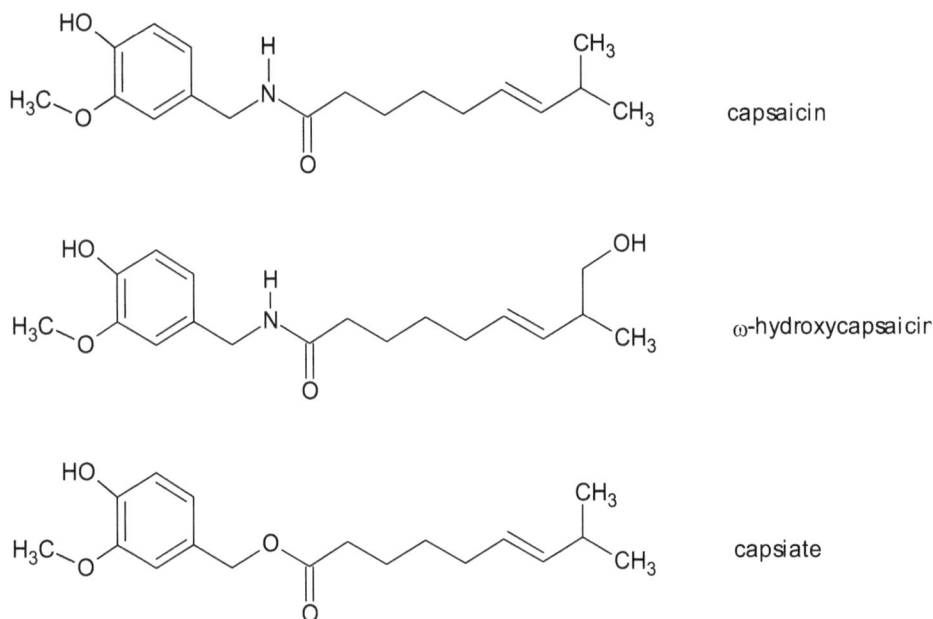

Fig. (3). Pungent (capsaicin) and non-pungent vanilloid representatives.

In fact, 90% of the total amount of capsaicinoids in hot peppers is represented by capsaicin (*trans*-8-methyl-N-vanillylnon-6-enamide) and dihydrocapsaicin accumulated in the vesicles - yellow sacs usually attached to placenta and seed tissue rather than in the pericarp [31], whereas the bell peppers contain low-pungent capsinoids, such as capsiate, dihydrocapsiate, and nordihydrocapsiate [32]. In addition, non-pungent derivatives of capsaicin were also described previously, for example ω-hydroxycapsaicin and 6″,7″-dihydro-5′,5‴-dicapsaicin [33].

Capsinoids, the capsaicinoid-like substances, have also a vanillyl and a fatty acid moiety, however bound through an ester instead of amide bond, as can be seen in Fig. (**3**), which seems to result in a thousand times lower pungency due to the easy bond breakage in aqueous conditions. In contrary, capsaicinoids are stable in polar and non-polar solvents and are able to produce a burning and heat sensation after contact [34]. Capsinoids are less toxic than capsaicinoids and possess some interesting biological properties, including antioxidant, antitumor, and anti-obesity [35]. This makes them good candidates for clinical applications as cancer preventive or weight loss substances [36].

Phytochemicals responsible for pungency and aroma of *C. pubescens* fruit were determined in a study by Kollmannsberger *et al.* [37]. Substituted 2-methoxypyrazines and lipoxygenase cleavage products (*e.g.* 2-nonenals and 2,6-nonadienal) are responsible for the aroma. The ratio of capsaicin:dihydrocapsaicin (15.2:33.0 µg/mL) was different from other *Capsicum* species, since the amount of dihydrocapsaicin dominated. The accession 'Canario' (*C. pubescens*) showed the equal quantity of phenolic compounds in the unripe (green) and ripe (yellow) stage, yet the amounts were much lower than those in *C. annuum* varieties. Only small amount of vitamin C (180 µg/g FW) was determined in ripe fruit, which was 10-times lower than *C. annuum* in the test [38]. In a different study, vitamin C level for *C. pubescens* fruit varied from 2.4 to 4.6 mg/g FW [39]. Carotenoids were studied in six accessions of this species originated from Andean regions. The major compound was capsanthin, followed by β-carotene, *cis*-violaxanthin, antheraxanthin, and lutein in yellow/orange rocotos. The accession Bol57y could be considered as a valuable source of lutein (13 mg/g fresh weight) as it exceeded the *C. annuum* and *C. baccatum* values.

Thirty-two accessions from the Peruvian *Capsicum* germplasm collection were assayed for phytochemicals contents. The total capsaicinoid amount (0.5 to 4.1 mg/g = 8,400 - 60,000 SHU), total polyphenols (18 to 25 mg GAE/g), tocopherols (68 to 184 µg/g) were calculated from the results [40].

Carotenoids

Capsicum genus is one of the richest vegetable crop sources of carotenoids. Carotenoids act as natural pigments in flowers, fruit and vegetables and are responsible for their attractive color. They have an important role in photosynthesis together with chlorophylls. There are more than 30 pigments identified in the *Capsicum* spp. fruit. Carotenoid pigments are the major compounds responsible for the attractive color of green, yellow, orange and red fruit in peppers [41]. In fruit, carotenoids are characterized as C40 isoprenoids with 9 conjugated double bonds in the central polyenic chain, with various end groups that affect their chromophore properties [42]. Over 60% of the total carotenoids are represented by capsanthin and capsorubin. Their content increases along ripening *via* a spectacular synthesis. Capsanthin is a more stable compound than capsorubin [43].

The final colour of the fruit depends on the concentration and type of carotenoids, as well as on the presence of other pigments. For example, the purple and black *C. annuum* fruit derives its colour from anthocyanin delphinidin-3-p-coumaroyl-rutinoside-5-glucoside [44]. Chlorophylls and carotenoids (lutein, violaxanthin, and neoxanthin) are the major compounds in unripe fruit of *Capsicum* species. During ripening, the chlorophylls are degraded and the content of carotenoids increases. Capsanthin and capsorubin are the major carotenoids in red-fruited peppers, and antheraxanthin, β-carotene, β-cryptoxanthin, cucurbitaxanthin zeaxanthin, and violaxanthin dominate in yellow fruit (Fig. **4**). Orange colour of fruit is represented by the same carotenoids, however in a different quantity profile. Brownish fruit are typical of a mixture of chlorophyll band carotenoids in the mature fruit [45, 46].

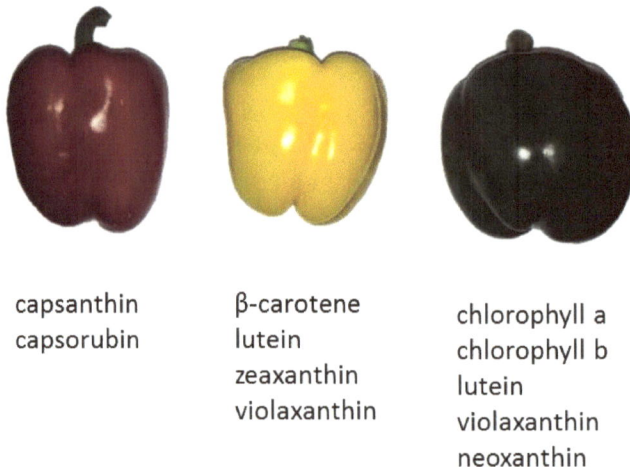

capsanthin	β-carotene	chlorophyll a
capsorubin	lutein	chlorophyll b
	zeaxanthin	lutein
	violaxanthin	violaxanthin
		neoxanthin

Fig. (4). Carotenoids and other compounds responsible for the fruit colour in *Capsicum* spp.

Various cultivars of three species (*C. chinense, C. annuum, C. frutescens*) were investigated and 52 carotenoids were identified with a significant variation in carotenoid composition among them [47]. Red coloured fruit showed high β-carotene content, on the other hand Tabasco and Jalapeno showed high levels of capsanthin but complete absence of β-carotene. Habanero golden and Scotch Bonnet contained dominantly α-carotene, β-carotene and lutein, while orange fruit were rich in antheraxanthin, capsanthin and zeaxanthin. A collection of 63 *C. chinense* accessions was tested for β-carotene level in fruits. The accession PI-355817 from Ecuador contained the highest concentrations of 8 mg/g FW [48].

Vitamins

All pepper species contain significant levels of vitamin C and E in the pericarp. In a study of 32 *Capsicum* spp. accessions of different origin and fruit colour, representing *C. annuum*, *C. baccatum, C. frutescens*, and *C. chinense*, all representatives revealed outstanding levels of vitamin C, up to 2 mg/g fresh weight (FW), actually ten times higher than vitamin C level associated to tomato which is considered as a good vegetable source of this vitamin [49]. The highest level of α-tocopherol (163 μg/g FW) was detected in *C. chinense* AC2212, whereas for the other accessions it varied from 8 μg/g FW in *C. chinense* RU72–194 to 95 μg/g FW in *C. annuum* AC1979, which also contained the highest amount of β-tocopherol (42 μg/g FW). The content of vitamin C (ascorbic acid) varied from 204 μg/g FW in *C. baccatum* Aji Blanco Christal to 2.1 mg/g FW in *C. annuum* Long Sweet. The provitamin A levels (α -carotene, β-carotene, β-cryptoxanthin) ranged from traces up to 185 μg/g FW in *C. annuum* l2 Tit Super [45].

Recently, Perla *et al*. analyzed 123 genotypes of *Capsicum baccatum* L. from 22 countries for vitamin C content in order to consider the impact of origin and genotype [49]. The values ranged between 2.5 and 50 mg/g dry weight (DW) of pericarp and higher levels were detected in mature unripe fruit compared to ripe fruits. Fruit from four genotypes may cover 100 – 500% of the recommended daily allowance (RDA) of vitamin C which is 90 mg per day for adult male, and additionally, 10 genotypes could supply more than 50% of the RDA of vitamin C with only 2 g DW or 22 g FW of fruit pericarp [50].

Antonious *et al*. [48] performed a study on vitamin C content in 63 accessions of *C. chinense* from North and Latin America. The highest concentrations of ascorbic acid of 1.2 and 1.1 mg/g FW were observed in accessions PI-152452 (Brazil) and PI-360726 (Ecuador), respectively. The fruit of *C. chinense* were described to be rich in vitamins, such as vitamin C. Total values of ascorbic acid in 216 varieties and landraces ranged from 0.3 to 14 mg/g FW [51]. Howard *et al*.

[52] determined the concentrations of ascorbic acid in two cultivars of *C. chinense* to be about 1.2 mg/g FW.

Phenolic Compounds

Polyphenols are natural secondary metabolites founds in fruit and vegetables with a major role in plant defensive systems against both abiotic and biotic stress and are responsible also for characteristics, such as color, flavor, odor or bitterness. About 8,000 polyphenols have been identified in various plants and their contribution to human health has been proved in many studies. The regular intake of polyphenols in human diet may prevent civilization diseases, for example various types of cancer, cardiovascular diseases, diabetes, osteoporosis and neurodegenerative diseases. They usually appear in conjugated forms and can be categorized into several main classes including phenolic acids, flavonoids, stilbenes and lignans [53].

Capsicum spp. are rich in polyphenols and attention has been paid especially to flavonoids and phenolic acids that are known bioactive compounds with several activities, such as antioxidant, antimicrobial, anti-inflammatory and anticancer [54].

Certain differences appear among the species of *Capsicum* genus. For example, the cultivars *C. annuum* and *C. frutescens* are appreciably richer in total flavonoids than *C. chinense* cultivars, all at mature stage [52]. Marín *et al.* [55] characterized 23 flavonoids from the pericarp of sweet pepper and found the majority of the phenolic compounds in the peel, mostly *p*-coumaric acid and caffeic acid derivatives, such as *p*-coumaroylglucopyranoside. Patras *et al.* investigated that peppers show lower contents of caffeoyl glucoses in comparison to tomato, with chilli samples giving the highest amounts, whereas conventional bell peppers contained much lower amounts [56].

In comparison to other *Capsicum* species, the contents of phenolic and flavonoid compounds in *C. baccatum* accessions were usually higher, which could result in stronger activity in antioxidant and anti-inflammatory tests, usually correlated with these types of phytochemicals [57 - 59].

Antimicrobial Plant Peptides

Antimicrobial plant peptides (AMPs) represent a novel natural alternative to agrochemicals. Despite they were discovered more than 45 years ago, the unfolding of their overall antimicrobial, insecticidal and antifungal activities has been only of recent interest of the scientists and breeders.

The AMPs are a class of small protein structures built from 12 to 54 amino acids, rich in cysteine units, such as defensins and cyclotides [60]. The expression of these peptides takes place during storage and reproduction. They are associated with self-defence and possess antibacterial and antifungal activities. The AMP defensins, cecropins and magainins are able to kill bacteria. The mechanism involves a detergent-like effect and pore formation in membrane [60, 61]. The bacteria fail in developing resistance because this process does not involve any special receptor, is fast and concentration independent.

Many *Capsicum* spp. produce a group of AMPs called defensins. Plant defensins appear mainly in the family *Brassicaceae*, *Fabaceae*, and *Solanaceae*, but appear through the entire plant kingdom [62]. These small molecules with 12 – 45 amino acids (around 5 kDa) are highly basic and usually consist of 8 – 10 cysteine units. The disulphide bridges stabilize these peptides [63]. For example, the defensin from habanero pepper *C. chinense* named Deffito1 was described previously [60, 64].

Lipid Transfer Proteins (LTPs) are small proteins rich in cysteine with molecular mass not exceeding 10 kDa [65]. LTP structure is more stable and more resistant to heat denaturation due to the disulphide bonds formed in α-helices. Usually, four or five α-helices are present [66]. These compounds have ability to penetrate the membrane thanks to their hydrophobic residues. They are expressed mainly in fruits during ripening as a defence tool [67].

Another group of plant proteins are 2S albumins that are stored in vacuoles of seeds and kernels, as well as in leaves, which as serve as reserves for self-defence. They are synthesised as large precursors polypeptides (18 – 21 kDa) and post-modified into two subunits of 8 – 14 and 3 – 10 kDa. The peptides contain preferably cationic and glutamine residues. Fraction of peptides with sequences of 2S albumins was found in the seeds of chilli pepper (*C. annuum*) [68].

Previously, defensin J1 from the bell pepper (*Capsicum annuum* var. yolo), Ca-LTP1 protein (9461 Da) from seeds of chilli (*C. annuum*), a thionin-like peptide CaThi from *C. annuum* fruit, and a DING peptide (Mr ~7.57 kDa) from *C. chinense* seeds, were isolated and their activity against various pathogens was analyzed [69 - 72].

Biological Activity in Genus *Capsicum*

Many studies proved that *Capsicum* species are a valuable source of bioactive phytochemicals (Table **1**). These compounds are not only helping plant against biotic and abiotic stress, but also have a beneficial impact on human health [20]. Capsaicinoids, carotenoids, vitamins, phytosterols, and policosanols, were identified and isolated from pepper fruit and their pharmacological effects were evaluated. The levels of health-promoting secondary metabolites in the fruit are strongly affected by genotype, maturity, environmental conditions, and processing [45, 73, 74]. The current aim of breeders is to develop new varieties rich in biologically active phytochemicals. For example, a violet pepper cultivar from hybridization of pepper with eggplant is rich in anthocyanins [75] and a pale green pepper developed in Korea revealed a strong α-glucosidase inhibition activity [76].

Table 1. Biological activity of selected active compounds of *capsicum* species

Active compound	Species	Plant part	Biological activity	References
Antimicrobial peptides				
Ca Thi	*C. annuum*	fruit	antifungal, antibacterial	[70, 77, 78]
Ca-LTP1	*C. annuum*	seeds	antifungal	[71]
HyPep	*C. annuum*	seeds	antimicrobial, insecticidal	[79]
defensin HisXarJ1-1	*C. chinense*		antimicrobial	[69]
Capsaicinoids				
capsaicin	*C. annuum, C. chinense, C. chacoense, C. baccatum, C. frutescens, C. pubescens*	fruit, leaves, stem	antioxidant, anti-obesity, antimicrobial,anti-inflammatory, analgesic, anticancer antiviral hypoglycemic, gastroprotective, antiplatelet, insecticidal, anti-allergic, hypocholesterolemic, antidiabetic	[18, 57, 71, 80 - 90]
dihydrocapsaicin	*C. annuum, C. chinense, C. baccatum, C. frutescens, C. pubescens*	fruit stem leaves	analgesic, anticancer, antidiabetic, hypolipidemic,antihypercholesterolemic, antimicrobial, anti-inflammatory, anti-obesity, antioxidant	[36, 80 - 85, 91 - 93]
homocapsaicin	*C. annuum, C. chinense*	fruit	analgesic, anticancer, antidiabetic, hypolipidemic, antihypercholesterolemic, anti-inflammatory, antiobesity, antioxidant,	[36, 80, 82, 94]
homodihydrocapsaicin	*C. annuum, C. chinense*	fruit	analgesic, anticancer, antidiabetic, hypolipidemic, antihypercholesterolemic, anti-inflammatory, anti-obesity, antioxidant,	[36, 80 - 82]
nordihydrocapsaicin	*C. frutescens, C. baccatum, C. pubescens*	fruit	analgesic, anticancer, antidiabetic, hypolipidemic, antihypercholesterolemic, anti-inflammatory, anti-obesity, antioxidant,	[36, 80, 82, 83, 93]
ω-hydroxycapsaicin	*C. annuum*	fruit	anticancer, antioxidant, anti-obesity	[33]
capsaicin-β-D-glucopyranoside	*C. annuum*	fruit	anticancer, antioxidant, anti-obesity	[95]
dihydrocapsaicin-β-D-glucopyranoside	*C. annuum*	fruit	anticancer, antioxidant, anti-obesity	[95]
6' ',7' 'dihydro-5',5' "-dicapsaicin	*C. annuum*	fruit	anticancer, antioxidant, anti-obesity	[33]

(Table 1) cont.....

Active compound	Species	Plant part	Biological activity	References
Capsinoids				
capsiate	*C. annuum* var. *annuum,* *C. annuum* var. *glabriusculum,* *C. baccatum,* *C. chinense,* *C. frutescens*		anticancer, anti-inflammation, antioxidant, anti-obesity, anti-atopic, anti-allergic	[96–98]
E-capsiate	*C. annuum,* *C. chinense,* *C. baccatum,* *C. frutescens*	fruit	anticancer, antioxidant, antiobesity	[99]
nordihydrocapsiate	*C. annuum*	fruit	immunosuppressive	[100, 101]
dihydrocapsiate	*C. annuum*	fruit	anticancer,anti-inflammation, antioxidant	[36, 102]
Capsianosides				
capsianoside VI	*C. annuum* var. *fasciculatum,* *C. annuum* var. *conoide*	stem, leaves	anticancer, antioxidant, anti-obesity	[103]
capsianosides II, A, B, C and D	*C. annuum* var. *fasciculatum,* *C. annuum* var. *conoide*	stem, leaves	anticancer, antioxidant, anti-obesity	[103]
Carotenoids				
β-carotene	*C. annuum* *C. chinense*	fruit	antioxidant, anticancer	[6, 104 - 115]
capsanthin	*C. baccatum,* *C. pubescens,* *C. annuum,* *C. chinense*	fruit	antioxidant, anticancer, anti-inflammatory, hypocholesterolemic	[80, 90, 105, 108, 111, 116 - 120]
capsanthin 3,3'-diester	*C. annuum*	fruit	anticancer	[118]
capsanthin 3'-ester	*C. annuum*	fruit	anticancer	[118]
capsorubin	*C. annuum*	fruit	antioxidant, anti-inflammatory	[120, 121]
lutein	*C. baccatum,* *C. pubescens,* *C. annuum*	fruit	anticancer, antioxidant,	[106, 116, 117]
mutatoxanthin	*C. annuum*	fruit	anticancer, antioxidant, anti-obesity	[122]
neoxanthin	*C. annuum*	fruit	anticancer, antioxidant, anti-obesity	[106, 117]
nigroxanthin	*C. annuum*	fruit	anticancer, antioxidant, anti-obesity	[122]
Flavonoids				

(Table 1) cont.....

Active compound	Species	Plant part	Biological activity	References
luteolin	*C. annuum*	fruit	antioxidant	[123]
apigenin 7-*O*-β-D-apiofuranosyl (1—>2)β-D-glucopyranoside	*C. annuum*	leaves	antioxidant	[124]
catechine	*C. annuum*	fruit	antioxidant	[125]
kaempferol	*C. annuum*	fruit	antioxidant	[110]
quercetin	*C. annuum*	fruit	antioxidant	[108, 110, 126, 127]
quercetin 3-*O*-α-L-rhamnopyranoside-7- *O*-β-D-glucopyranoside	*C. annuum*	fruit	antioxidant	[123]
quercetin-3-*O*-L-rhamnoside	*C. annuum*	fruit	antioxidant	[123]
chrysoeriol	*C. frutescence*	fruit	antimicrobial	[128]
Phenolic acid glycosides				
trans-p-feruloyl-β-D-glucopyranoside	*C. annuum*	fruit	antioxidant	[123]
trans-p-ferulyl alcohol-4-*O*-[6- (2-methyl-3-hydroxypropionyl] glucopyranoside	*C. annuum*	fruit	antioxidant	[123]
trans-p-sinapoyl-β-D-glucopyranosid	*C. annuum*	fruit	antioxidant	[123]
Phenolic acids				
caffeic acid	*C. annuum*	fruit	antimicrobial	[129]
cinnamic acid	*C. annuum*	fruit	antimicrobial	[85]
t-cinnamic acid	*C. annuum*	fruit	antimicrobial	[129]
ferulic acid	*C. annuum*	fruit, stem	antimicrobial	[126, 129, 130]
chlorogenic acid	*C. annuum*	fruit	antioxidant	[110]
gallic acid	*C. annuum*	fruit	antioxidant	[110]
m-coumaric acid	*C. annuum*	fruit	antimicrobial	[85]
o-coumaric acid	*C. annuum*	fruit	antimicrobial	[85]
sinapic acid	*C. annuum*	fruit	antioxidant	[126]
vanillic acid	*C. annuum*	stems, leaves	antioxidant	[124, 130]
Furostanol saponins				
capsicoside G	*C. annuum*	seed	anti-adipogenic	[131]
Vitamins				
vitamin C	*C. annuum, C. frutescens, C. chinense*	fruit	antioxidant, anti-inflammatory	[52, 110, 132]
vitamin E (γ-tocopherol)	*C. annuum*	leaves	anticarcinogenic	[133]

The fruit of *Capsicums* and their active compounds exhibited antioxidant, anti-inflammatory, antimicrobial, antidiabetic, anticancer, hypolipidemic, anti-obesity, gastroprotective and anti-psoriatic effects, mostly attributed to the content of capsaicinoids [6]. The most interesting findings are listed below and categorized by biological activity.

Both capsaicinoids and capsinoids have attracted great attention due to their numerous biological activities and pharmacological application in the treatment of bacterial, viral and fungal infections, cardiovascular and gastrointestinal disorders, cancer, diabetes, inflammation, pain, obesity, neurogenic bladder, respiratory problems and dermatologic conditions. The corresponding biological effects have been evaluated both in *in vitro* and *in vitro* studies, as well as in clinical trials, and capsaicin-containing products have been manufactured and developed [36, 134, 135].

Antioxidant Activity

The inhibition of methyl linolate oxidation initiated by 2,2'-azobis-(2,4-dimethylvaleronitrile) was observed by *Capsicum annum* pepper carotenoids decreasing as follows: capsorubin > capsanthin 3,6-epoxide > capsanthin > cycloviolaxanthin > β-carotene [121].

A strong lipid peroxidation activity in TBARS assay (thiobarbituric acid-reactive substances) was observed for non-pungent derivatives of capsaicin. Moreover, 6'',7''-dihydro-5',5'''-dicapsaicin was similar to capsaicin and about 25 times more potent than α-tocopherol (IC_{50} = 10 μM) [33].

Nine cultivars of *C. annuum* were subjected to antioxidant assays. The highest total phenol content was observed with Portafortuna cultivar (935 mg of chlorogenic acid equivalents/100 g DW), whereas best results in antioxidant assays were achieved with Pellegrino and Idealino against 2,2'-azino-bis (3-ethylbenzothiazoline-6-sulphonic acid radical (IC_{50} = 45.2 and 45.7 μg/mL, respectively) [136].

Regarding the antioxidant activity, it was found that antioxidant properties of the extract decrease during fruit ripening, despite the content of vitamin C rises. A variety of *C. annuum* prevents Fe^{2+}-induced lipid peroxidation in rat brain *in vitro*, thanks to the total phenol content [137].

The role of prevention of atherosclerosis and the increased risk of cardiovascular diseases was discussed in relation to antioxidant and anti-inflammatory activities of red pepper *C. baccatum* var. *pendulum* with emphasis on its carotenoid and capsaicinoid composition [138].

Seeds of *C. baccatum* were previously analyzed for antioxidant activity using a

DPPH radical scavenging assay and EC_{50} values ranging from 229 to 820 µg/mL were observed [22].

Nascimento *et al.* [128] concentrated on the determination of active principles in different plant parts (whole fruits, peel and seeds) of *C. frutescence* and quantified capsaicin, dihydrocapsaicin and chrysoeriol content. A high concentration of chrysoeriol (0.4 - 11.4 mg/g extract) was confirmed. Antioxidant activity of both extracts of *C. frutescence* and compounds was tested. Capsaicin performed the lowest EC_{50} value (23.1 µg/mL) similar to ascorbic acid in DPPH test.

Zhuang *et al.* [139] performed antioxidant scavenging assay with DPPH to find EC_{50} of 135 µg/mL for *C. frutescens* fruits. Antioxidant activity of hot short pepper (*C. frutescens* L. var. *abbreviatum*) flesh (pericarp) and seeds was associated with the prevention of cyclophosphamide-induced oxidative stress in brain. An inhibition of serum glutamate oxaloacetate transaminase (SGOT) and glutamate pyruvate transaminase (SGPT) was also observed in the *in vitro* test with mice [140].

Inhibition of lipid peroxidation and cyclooxygenase (COX-1, COX-2) enzymes was observed with capsaicin and dihydrocapsaicin in Bhut Jolokia (*Capsicum chinense/Capsicum frutescens*), one of the hottest peppers in the world. In comparison, the concentration of these two compounds was 5.7%, which is about 338-times greater than that of Scotch Bonnet and 18-times greater than Jalapeno pepper cultivars [141].

The antioxidant capacity for 32 accessions from the Peruvian *Capsicum* germplasm was determined in Trolox equivalent antioxidant capacity (TEAC) assay and ranged from 24 to 46 µmol Trolox/g DW [40].

Oboh *et al.* [142] found that removal of seeds from fruit pepper results in a 50% decrease of total phenol content and antioxidant activity. The unripe peppers were higher in total phenol content that ripe ones and exhibited lipid peroxidation in a dose-dependent manner [142]. Both ripe and unripe *C. pubescens* extracts inhibit lipid peroxidation in rat brain *in vitro*. High amounts of vitamin C were also determined for ripe fresh fruit with values of 231 µg/g FW [143].

Anti-Inflammatory Activity

Previously, capsanthin and capsorubin from peppers showed anti-inflammatory activity by suppressing the NO production in LPS-stimulated RAW 264.7 cells, which are associated with acute and chronic inflammation [120].

A strong anti-inflammatory activity was presented in a carrageenan-induced pleurisy model in mice pretreated with 200 mg/kg, p.o. (oral administration) of ethanol and butanol extracts in *C. baccatum*. An inhibition of leukocyte migration and reduced exudate formation was described and attributed to total phenolic compounds content and specifically flavonoids [22].

C. baccatum butanol and residual aqueous extracts showed *in vitro* anti-inflammatory activity in mice. The extracts inhibited the neutrophil migration induced by carrageenan and paw edema induced by carrageenan, prostaglandin E2, and histamine. Furthermore, the extracts *in vitro* inhibited the nitric oxide and tumor necrosis factor-alpha production by lipopolysaccharide/interferon gamma (IFN-c)-stimulated macrophages [144].

Juice from *Capsicum baccatum* L. var. *pendulum* (Willd.) Eshbaugh, currently the most consumed species in Brazil, was examined in animal models. Tests of acute inflammation induced by carrageenan and immune inflammation induced by methylated bovine serum albumin were performed in rats and mice previously treated with pepper juice. This treatment led to a decrease of the leucocyte and neutrophil migration, it reduced the vascular permeability on carrageenan-induced peritonitis and lowered the levels of pro-inflammatory cytokines, such as TNF-a and IL-1b in mice. A crude extract of *C. baccatum* presented anti-inflammatory activity in rodents at doses of 2 and 20 g/kg [57].

C. frutescens ethyl acetate extract and capsaicin performed analgesic and anti-inflammatory activity, with results comparable to diclofenac, in an egg albumin-induced oedema on the rat hind paw [145].

Anticancer Activity

The cancer-related activity of capsaicinoids was reported in a large set of studies. Capsaicinoids were found to suppress carcinogenesis by inducing the apoptosis in cancer cells in the breast, colon, lung, pancreas, prostate, bladder, and skin cancer while leaving normal cells unharmed [135, 146, 147]. The mechanism of apoptosis induction of capsaicin is not fully elucidated. Several ways have been suggested, including the activation of nonselective cation channel TRPV-1 (transient receptor potential vanilloid type 1), inhibition of the plasma membrane NADH oxidase, reduction of complex I and III activity leading to reactive oxygen species (ROS) production, activation of AMPK/AMP-dependent protein kinase) or *via* targeting p53 protein through downregulation of sirtuin 1 [148]. The apoptosis can happen *via* direct pathway (control of electron transport and influencing the amount of ROS inducing cell damage) and indirect pathway *via* receptor TRPV-1, (accumulation of Ca^{2+} in cancer cells and late elements of apoptosis) [149].

Oh *et al.* [150] described that dihydrocapsaicin induces the autophagy in human colon cancer cells lines HCT116 and that it was also more toxic to MCF-7 and WI38 than capsaicin. Choi *et al.* [91] reported that dihydrocapsaicin reduced the ROS accumulation and induced autophagy in lung cancer cells.

Nevertheless, there have been some evidence that capsaicinoids and chilli extracts fail in the prevention of cancer acting as co-carcinogen or tumour promoters [58]. The frequent consumers of chilli peppers are at higher risk of gastric cancer and reactive phenoxy radicals (metabolites of capsaicin) are able to attack the DNA and lead up to malignant transformation [151]. On the other hand, capsiate and dihydrocapsiate from sweet red peppers were found to exert similar anticancer and chemopreventive activity to capsaicin without the irritating action [102].

Capsanthin and capsorubin are considered as resistance modifiers in chemotherapy since they can improve the cytotoxic activity [118]. Also, β-caro-tene has shown strong anticancer activity, but interestingly, capsanthin, crocetin lutein, and zeaxanthin seem to be more potent in cancer prevention [152].

The DING peptide with Mr ~7.57 kDa and pI ~5.06 was detected in G10P1.7.57 protein fraction from *C. chinense* Jacq. seeds. The fraction showed anti-proliferative effect on carcinogenic cell lines PC-3, SiHa and Hep-G2, as well as the non-carcinogenic Vero cell line [153].

Antimicrobial Activity

The antimicrobial properties were evaluated to some extent in various *Capsicum* species and compounds present in the extracts and fractions were tested. The results are quite inconsistent, yet capsaicinoids seem not to be the most active principles and phenolic compounds are believed to be responsible for the antimicrobial activity. Crude extracts from different *C. annuum* cultivars inhibited *Bacillus, Clostridium, Pseudomonas, Listeria, Salmonella, Staphylococcus*, and *Streptococcus* strains [87, 154]. Recently, Mokhtar *et al.* [155] tested capsaicinoids (capsaicin, and dihydrocapsaicin), typically found in *C. annuum* peppers, against 13 pathogenic bacterial strains and three probiotic bacilli. The probiotic microorganism were not inhibited, but various strains of *Staphylococcus aureus* and *Bacillus subtilis, Listeria monocytogenes, Pseudomonas aeruginosa, Proteus mirabilis, Escherichia coli, Salmonella typhimurium* were inhibited and minimal inhibition concentrations (MICs) for capsaicinoid extract (2.5 – 10 mg/mL) were stated.

Mokhtar *et al.* [155] tested polyphenols (coumarin, caffeic acid, kaempferol, narangin, quercetin, and rutin) from *C. annuum* peppers, against pathogenic bacterial and probiotic strains. The probiotic microorganism showed no sensitivity to these compounds, nevertheless *S. aureus, B. subtilis, L. monocytogenes, P. aeruginosa, P. mirabilis, E. coli,* and *S. typhimurium* were inhibited and minimal inhibition concentrations (MICs) for polyphenols were determined. Quercetin, kaempferol and caffeic acid were the most efficient. A synergistic effect was also observed among these polyphenols.

Staphylococcus aureus and *Listeria monocytogenes* were inhibited by all extracts of 9 cultivars of *C. annuum*, with best results in Effix and Fantasia peppers (MIC = 12.5 and 25 mg/mL, respectively) [136].

Previously, it was confirmed that *meta*-courmaric acid and *trans*-cinnamic acid in chilli pepper extracts from *C. annuum* bell pepper inhibit *Erwinia carotovora* subsp. *carotovora*, but capsaicin and dihydrocapsaicin had no activity against this pathogen [156].

A lectin with antifungal activity against *Aspergillus flavus* and *Fusarium moniliforme* was identified in *C. frutescens* L. var. *fasciculatum* from Thailand. The lectin (0.27 mM) performed mitogenic activity against spleen cells isolated from BALB/c mice [157]. Antimicrobial activity assay was performed with *C. frutescens* aqueous and methanolic extracts. They were found to be effective against *Vibrio cholerae, Staphylococcus aureus,* and *Salmonella typhimurium,* however the extract from *C. annuum* showed a higher antibacterial activity [158].

Capsaicin, dihydrocapsaicin and chrysoeriol from *C. frutescence* inhibited both Gram positive (*E. faecalis, B. subtilis*, and *S. aureus*) and Gram negative bacteria (*E. coli, P. aeruginosa*, and *K. pneumoniae*), with chrysoeriol giving the lowest MICs (0.06 µg/mL for *E. coli*), yet not able to inhibit *C. albicans* [128].

C. frutescens var. *longum* leaves were found to contain substances with antimicrobial potency as the methanolic extract was active against *S. aureus, K. pneumoniae* and *P. aeruginosa*. A moderate anthelmintic activity was observed as well against Indian earthworm [159].

Interesting results were obtained with two similar steroidal saponins (1081 Da and 919 Da) isolated from cayenne pepper (*C. frutescens*). Despite sharing the same steroidal moiety, the number of glucose moieties resulted in the absence of antifungal activity. Whilst the larger saponin molecule was fungicidal against *Aspergillus flavus, A. niger, A. parasiticus, A. fumigatus, Fusarium oxysporum, F. moniliforme*, and *F. graminearum*, the second saponin (missing one glucose molecule) was inactive against these fungi [160].

Fresh and cooked extracts of *C. chinense* revealed antimicrobial activity inhibiting all microorganism in a test performed by disc diffusion assay, namely the plain and heated extracts were found to exhibit varying degrees of inhibition against *Bacillus cereus, B. subtilis, Candida albicans, Clostridium sporogenes, C. tetani,* and *Streptococcus pyogenes* [87].

Antimicrobial activity against *Streptococcus pyogenes, Clostridium sporogenes, C. tetani, Bacillus subtilis* and *B. cereus* was observed for *C. pubescens* USDA PI 387838 variety extract from fresh fruit in a disk diffusion test [87].

Defensin J1 from the bell pepper (*Capsicum annuum* var. *yolo*) protects the fruit against pathogens as tested in disk diffusion assay against *Botrytis cinerea* and *Fusarium oxysporum* [161]. Further, this defensin at concentration of 1 mg/mL can inhibit the growth of *Colletotrichum gloeosporioides* and is intensively produced upon the infection [162].

Diz *et al.* [163] described the isolation of three fractions from pepper seeds F1 – F3. Three peptides (6 to 10 kDa) with homology to lipid transfer proteins composed the F1 fraction, which showed strong antifungal activity against *Candida albicans, Saccharomyces cerevisiae* and *Schizosaccharomyces pombe.* The same team demonstrated the mechanism of membrane permeabilization in *Candida tropicalis* and *Colletotrichum lindemunthianum* of Ca-LTP1 protein isolated from seeds [71]. Peptide enriched fractions isolated from pepper leaves were associated with antimicrobial activity against the plant pathogenic bacteria *Ralstonia solanacearum* and *Clavibacter michiganensis* sp. *michiganensis* [164].

A peptide from chilli pepper (*C. annuum* L.) seeds called Ca-LTP1 (9461 Da) featured significant antifungal activity against *Candida albicans, C. lindemunthianum, C. tropicalis, F. oxysporum,* and three yeasts *Pichia membranifaciens Saccharomyces cerevisiae,* and *Schizosaccharomyces pombe* [71, 163, 165]. It was able to disrupt the membrane in *C. tropicalis* and to inhibit the media acidification caused by *S. cerevisiae* [163, 165].

The 2S albumin-like proteins from seeds of chilli pepper (*C. annuum*) affected the growth of several yeast strains - *Candida albicans, C. guilliermondii, C. parapsilosis, C. tropicalis, S. cerevisiae, Pichia membranifaciens,* and *Kluyveromyces marxiannus.* Moreover, the fraction caused disorganisation of cell wall in *S. cerevisiae* model [68].

Antimicrobial activity was confirmed also for thionin-like peptide CaThi, isolated from *C. annuum* fruit, against six pathogenic *Candida* species (*C. tropicalis, C. albicans, C. parapsilosis. C. pelliculosa, C. buinensis,* and *C. mogii*). Synergistic effect was observed with fluconazole resulting in candidicidal activity [70].

An inhibition activity against yeast *Saccharomyces cerevisiae* in G10P1.7.57 protein fraction from *C. chinense* Jacq. seeds was confirmed [153]. Previously, several plant and human pathogenic bacteria were also inhibited by G10P1.7.57, namely *Erwinia carotovora, Pseudomonas syringae, P. aeruginosa, Xanthomonas campestris, Shigella flexnerii,* and *Staphylococcus aureus* [166].

Anti-Obesity Activity

Non-pungent capsinoids proved to be good candidates as potential cure for obesity as these compounds suppressed fat accumulation *in vitro* and *in vitro* in mice, promoting the lipid metabolism in liver and adipocytes. Increased levels of HMG-CoA reductase, CPT-1, FAT/CD36 and GLUT4 were achieved in liver tissue [167]. Regular ingestion of capsaicin as well as capsinoids is helpful in obesity prevention. The decrease in body fat is resulting from the increase of the energy expenditure through the activation and recruitment of brown adipose tissue (BAT) in humans [168].

Suppressive effect of pepper carotenoids capsanthin and capsorubin on chronic inflammation in adipocyte cells was studied in experiments with 3T3-L1 adipocyte cells and macrophage cell co-culture. These carotenoids promoted the differentiation of 3T3-L1 adipocyte cells. The effect of capsanthin and capsorubin on the co-culture was similar to troglitazone, a PPARγ ligand medicine as the cells' interleukin-6 (IL-6), monocyte chemotactic protein-1 (MCP-1), tumor necrosis factor-α (TNF-α), and resistin mRNA expression were suppressed. The carotenoids adjusted the adipocytokine secretion, which may be useful in ameliorating the obesity-induced chronic inflammation in adipocytes [169].

Other Biological Activities

Capsaicin has been used topically or orally to reduce inflammation heat and alleviate post-surgical, osteoarthritis, postherpetic neuralgia, diabetic neuropathy, rheumatoid arthritis, or fibromyalgia pain. The concentration of capsaicin in commercial creams is quite low (0.075% or less) and injection preparations and site-specific therapy with high doses of capsaicin are preferable in chronic pain analgesic treatment [88]. The mechanism responsible for its effect is associated to the transient receptor potential vanilloid subfamily member 1 (TRPV1), to which capsaicinoids act surprisingly not as antagonists (blocking the pain provoking substances in activating the receptor) but as agonists. Initially, they excite the sensory neurons and a desensitization follows resulting in the analgesic effect [170].

The gastroprotective effect of capsaicinoids is strongly dose-dependent and time-related, which has been proved in animal models of gastric mucosal injuries

[171]. High doses lead to the damage of capsaicin-sensitive sensory *via* exhaustion of the neurotransmitters [172], whereas low dosage increases the gastric mucus secretion, basal gastric mucosal blood flow and helps with the gastric epithelial restitution [173].

Lee *et al.* [98] reported that capsiate applied topically inhibits DNFB-induced atopic dermatitis in the NC/Nga mice. They observed lower serum IgE levels, suppression of CD4+ T-cells and mast cells and also inhibition of the expression of pro-inflammatory cytokines. Unlike capsaicin, it does not irritate the affected area and therefore is a better candidate for allergies and atopic dermatitis treatment.

Capsaicin used topically was found to reduce cough symptoms and improve non-allergic rhinitis by desensitization of the transient receptor potential vanilloid subfamily member 1 (TRPV1) [89]. Hypocholesterolemic properties in animal assays were confirmed in *C. annuum* [90].

Carotenoids as part of the diet may contribute to the improvement of plasma lipid profile and lower the risk of cardiovascular diseases. For example, capsanthin is able to rise plasma HDL-cholesterol accompanied by an increase in hepatic apoA5 and LCAT mRNA expression in rats [90]. Red chilli peppers of *C. frutescens* were reported to be effective in the psoriasis treatment [174].

PHYSALIS GENUS

This genus counts more than 90 species, annuals or perennials, usually semi-erect, herbaceous or semi-woody plants with fruit typically inside the calyx husks. Primarily centered in the Americas with Mexico being the center of biodiversity, where ca. 70 species can be found, over 40 of them being endemic. Today they are both growing wild and cultivated worldwide. There are three species widely cultivated [6].

P. peruviana L. (uchuva, uvilla del monte) from Andes, nowadays popular in Asia, South Africa, and Australia. *P. philadelphica* Lam. (tomatillo, miltomate) gained popularity in the mountains of Guatemala and Mexico. Its unripe fruit are consumed cooked as vegetable and the husk is traditionally added to dough for stretchiness. Some cultivars produce purple-colored fruits, whereas the majority of them is green. As third, *P. alkekengi* L. (Chinese lantern) is believed to be native to China and it is planted worldwide either as an ornamental plant for its bright orange husks, while some cultivars with edible fruits are used for culinary purposes [175]. *P. angulata* L. is a less known weed used as vegetable in some countries that has been lately studied in detail for its medicinal properties.

P. angulata L. (camapu, popa, cutleaf ground cherry, balloon cherry) is a common erect annual weed originating from sub-tropical or tropical America. It bears edible fruits, which are rich in vitamin C, juicy and may be used raw in salads or cooked as vegetable. The young leaves are eaten as vegetable in Southeast Asia [175] and in tropical Africa. Teas and tinctures are prepared from the roots and traditionally taken as medicines to treat hepatitis, diabetes, while leaf infusions are used to treat inflammations, dermatitis, rheumatism, infections, asthma, malaria, gonorrhea, fever, toothache, influenza, bronchitis, cancer and many others [6].

Physalis philadelphica Lam. and *P. ixocarpa* Brot. ex Hornem. are often categorised as synonyms, however they are different species. Both synonyms are commonly known as tomatillo or husk tomato. *P. philadelphica* is grown commercially for its green edible fruit that are used in salsa verde and enchiladas in some countries of Latin America. In Mexico, it is one of the main vegetables for domestic sale. The raw fruit is generally recognised as non-edible. *P. philadelphica* produces purple fruit, but they are usually harvested green before full ripening to avoid overly seedy and insipid in flavour [176].

Physalis ixocarpa Brot. or Mexican husk tomato, as it is commonly known, was an important crop in the Mayan and Aztec economy. Only about 60 years ago it was introduced to India, Australia and Africa. In contrast to cape gooseberry, it is more used as vegetable, often ripe, but mainly as part of the salsa verde, in soups or in curries [177]. *P. ixocarpa´*s mature fruits are pale yellow.

Another minor species is *P. coztomatl* (tomate agrio), which has been used both in Mexican cuisine and traditional medicine. Its use was described already in the sixteenth century in Florentine codex. It is used as an antidiarrheal, to treat asthma, stomach pain and pulpitis [178].

Chemical Constituents in Genus *Physalis*

Species belonging to the genus *Physalis* produce numerous biologically active compounds. Only in the past two decades, over 200 compounds were isolated and determined in genus *Physalis*. Among them, a group of steroids called withanolides and physalins is typical for these plants. Further, the genus is represented by labdane diterpenes, flavonoids, sucrose esters, ceramides, and other phytochemicals [179].

P. angulata extracts are rich in polyphenol content; however the quantity may vary according to several parameters. Carniel *et al.* determined major polyphenols and observed variations for total polyphenols, gallic acid, ellagic acid, caffeic acid, rutin, and mangiferin [180]. Fruits of *P. philadelphica* Lam. are a valuable source of several bioactive withanolides, including physalin B [181], ixocarpalactone A and B [182], ixocarpanolide [183], and withaphysacarpin [181]. Addingly, α-tomatine and ascorbic acid were found in fruit and phygrine alkaloid in roots and aerial parts [184]. Philadelphicalactones A, C and D were found in aerial parts [185]. Nutritionally, the *P. ixocarpa* fruit contains approximately 310 cal/g, 11% protein, 18% fat, and 5% total dietary fiber (DW) [186]. Husk tomato waste is a rich source of high methoxyl pectins [187].

The ceramides isolated from *P philadelphica* stems and leaves were elucidated as (2S,3S,4R)-2-tetracosanoylamino-1,3,4-octadecanetriol,   (2S,3S,4R,9E)-1,3,4-trihydroxy-2-[(2'R)-2'-hydroxytetracosanoylamino]-9-octadecene,   and (2S,3S,4R)-2-[(2'R)-2'-hydroxytetracosanoylamino]-1,3,4-octadecanetriol [188].

The first labdane diterpenes reported from genus *Physalis*, physacoztomatin and labd-13(*E*)-ene-8α,15-diol, were isolated from aerial parts of *P. cozcomatl* (Mociño & Sessé) Ex Dunal together with withanolides physocoztolides A-E [178].

Withasteroids

Withasteroids, mainly withanolides and physalins (Fig. **5**), belong to the interesting compounds with multiple biological activities derived from this genus.

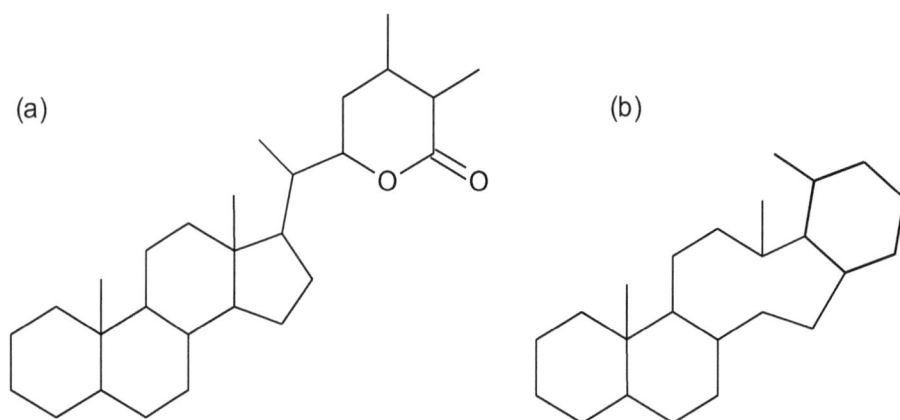

Fig. (5). Basic skeletons of withanolides (a) and physalins (b).

Withanolides are a group of modified and highly-oxygenated ergostane-type steroids with C17 lactone/lactol side-chain substituents [189]. They are the most abundant compounds in this genus. Several subgroups, depending on skeleton modifications, are described, among them 5-ene withanolides, 5β,6β-epoxide withanolides, physalins, neophysalins, withaphysalins, ixocarpolactones, perulactones [189]. Unusual structures may appear, for example that of aminophysalin A, which contain a N-atom in its skeleton [190].

Physalins are secosteroids unique for *Physalis* species. They are highly oxygenated and demonstrated various *in vitro* biological activity, such as antimicrobial, antimycobacterial, antioxidant, hepatoprotective, anti-leukemic, immunomodulatory effects, antiinflammatory effects related to arthritis [191 - 194].

Recently, new physalins were isolated from the leaves and stems of *P. angulata*, namely physalin V to IX, as well as 11 known analogues, such as physalin B, D, F, G, H, I, P, R, physalin D_1 25β-hydroxyphysalin D, and isophysalin B [195].

A rare steroid with a seven-membered ring was isolated and named physanolide A. With it, two other new compounds physalin U and V were found in *P. angulata* [196]. Another rare skeleton containing a nitrogen atom was identified for aminophysalin A [190].

Biological Activity in Genus *Physalis*

Previously, withanolides have revealed immunomodulating, anti-inflammatory, antitumor, hepatoprotective and anti-feedant activity [197]. Antioxidant, antipyretic, analgesic and anti-inflammatory effects were observed for the gallic acid, ellagic acid, caffeic acid, rutin, and mangiferin from *P. angulata* [180].

Biologically active compounds from other *Physalis* species were studied, including *P. alkekengi*, *P. crassifolia*, *P. chenopodifolia*, *P. cinerascens*, *P. coztomatl*, *P. hispida*, *P. longifolia*, *P. minima*, and *P. pubescens* [179]. Biological activity of selected active compounds of *Physalis* species are presented in (Table **2**).

Table 2. Biological activity of selected active compounds of *physalis* species

Active compound or extract	Species	Plant part	Biological activity	Ref.
Steroidal lactones				
physaline B	*P. angulata*	leaves, fruits	anti-inflammatory, antimalarial, anticancer, antileishmanial, antimycobacterial, anti-melanoma, hepatoprotective	[201 - 206]

(Table 2) cont.....

Active compound or extract	Species	Plant part	Biological activity	Ref.
physalin D	*P. angulata*	leaves, fruits	antibacterial, anticancer, antimalarial, antimycobacterial	[202, 205 - 207]
physalin E	*P. angulata*	leaves, fruits	anti-inflammatory	[208]
physalin F	*P. angulata*	leaves, fruits	antimalarial, anticancer, antileishmanial, immunosuppressive, anti-inflammatory	[202, 204, 209 - 212]
physalin G	*P. angulata*	leaves, fruits	antimalarial, antileishmanial, immunomodulatory	[202, 204]
physalin X	*P. angulata*	whole plant	antiinflammatory	[213]
withangulatin A	*P. angulata*	whole plant	anticancer, trypanocidal, anti-inflammatory	[214 - 216]
withangulatin B	*P. angulata*	whole plant	anticancer	[217]
withangulatin I	*P. angulata*	whole plant	anticancer	[218]
withaphysacarpin	*P. philadelphica*	whole plant	anticancer	[184, 185]
24,25-dihydrowithanolide D	*P. philadelphica*	whole plant	anticancer	[184]
2,3-dihydro-3-methoxywithaphysacarpin	*P. philadelphica*	whole plant	anticancer	[184]
physangulidine A	*P. angulata*	whole plant	antiproliferative	[219]
physagulide B	*P. angulata*	calyxes	anticancer	[220]
physagulide P	*P. angulata*	calyxes	anti-proliferative	[221]
aromaphysalin B	*P. angulata*	stems, leaves	anti-inflammatory	[213, 222]
ixocarpalactone A	*P. philadelphica*	fruits	anticancer	[223]
ixocarpalactone B	*P. philadelphica*	fruits	anticancer	[223]
philadelphicalactone B	*P. philadelphica*	fruits	anticancer	[223]
diacetylphiladelphicalactone C	*P. angulata*	aerial parts	anticancer	[185]
physaminimin B and E	*P.minima*	whole plant	anti-inflammatory	[224]

(Table 2) cont.....

Active compound or extract	Species	Plant part	Biological activity	Ref.
Other compounds				
physanguloside A (phenol glycoside)	*P. angulata*	whole plant	anti-inflammatory	[225]
sucrose esters	*P. philadelphica*	fruit coating	anti-inflammatory	[226]
physangulatosides F-J (labdane diterpenoid glycosides)	*P. angulata*	stems, leaves	anti-inflammatory	[227]
myricetin 3-O-neohesperidoside (flavonoid glycoside)	*P. angulata*	leaves	anticancer	[228]
Various extracts				
methanolic extract	*P. angulata*	leaves	anti-histaminic	[229]
water extract	*P. angulata*	stems, leaves	anti-adipogenetic	[230]
methanolic extract	*P.ixocarpa*	leaves, fruit, stem	antifungal, antibacterial	[231]
methanolic extract and chloroform fraction	*P. minima*	whole plant	anti-inflammatory, analgesic, antipyretic	[232]
	P. ixocarpa	fruits	antioxidant	[200]
methanolic extract	*P. ixocarpa*	leaves and fruit	antimicrobial	[231]

Antioxidant Activity

Extracts of roots, stems, and leaves of *P. angulata* showed significant antioxidant capacity [198]. Young leaves of *P. angulata* are high in flavonoids content (38 µg/g FW) and exert reducing power [199].

Interestingly, the antioxidant activity in DPPH test did not correlate with total phenolics in purple anthocyanin-rich fruits of several *P. ixocarpa* genotypes, the samples with the best activity had in fact the lowest level of phenolics [200].

Anti-Inflammatory Activity

The number of evidence related to the biological activities of *P. angulata* L. has been constantly on rise. Yang *et al.* studied the anti-inflammatory effects of physalin E on RAW 264.7 mouse macrophages stimulated by lipopolysaccharide (LPS). Physalin E inhibits the expression and secretion of the tumor necrosis factor-α (TNF-α) and interleukin-6 (IL-6). Interestingly, this feature may not be blocked by miferstone (RU486), such as in the case of dexamethasone. Physalin E also helps to reduce the inflammatory cytokines in the NF-κB signaling pathway [233]. Eight compounds from the leaves and stems of *P. angulata* (physalins V, VII, B, D_1, F, and isophysalin B), exhibited inhibitory activities against nitric oxide (NO) production and may be beneficial in the inflammation-related ailments [195]. Physalin F revealed immunosuppressive activity in collagen-induced arthritis model with different mechanism from that of glucocorticoids [210]. Soares *et al.* also proved that seco-steroids physalin B, F or G (but not D) from *P. angulata* are potent immunomodulatory compounds with a mechanism different from that of dexamethasone [211].

Recently, two novel seco-steroids physalin X and aromaphysalin B were isolated from *P. angulata* L. They exhibited anti-inflammatory activity on NO production (IC_{50} = 68.5 and 29.7 µM, respectively) [213].

Regarding the anti-inflammatory effect, some withanolides also showed inhibitory effects on NO production induced by LPS in macrophages [234]. The new withanolide, aromaphysalin A, isolated from stems and leaves of *P. angulata* by the same research team, exhibited inhibitory activity on NO production (IC_{50} = 51.6 µM) [222].

Among other interesting compounds, twelve new labdane-type diterpenoid glycosides, physangulatosides A-L were isolated from stems and leaves of *P. angulata* and tested for inhibitory activities against LPS-induced NO production in RAW 264.7 macrophages. Four compounds inhibited NO production with IC_{50} ranging from 15.9 to 60.7 µM [227]. The whole plant extract of *P. minima* showed analgesic and anti-inflammatory activity in a study conducted in rats and mice [232].

Not only withanolides have the anti-inflammatory effect. Sucrose esters isolated from sticky layer of *P. philadelphica* fruits showed activity comparable to aspirin or ibuprofen in the test with cyclooxygenase 1 and 2 in a recent study [226].

Anticancer Activity

Antiproliferative activity against several human cancer cell lines (C4-2B, 22Rv1, 786-O, A-498, ACHN, and A375-S2) was confirmed for physalin B and F from the leaves and stems of *P. angulata* [195].

In 2017, another novel physalin, 25-hydroxylisophysalin B was identified together with 3 megastigmane glucosides and 8 already known physalins, all of them active against MCF-7 human breast cancer and HepG2 human hepatoma cell lines [235].

The compound physalin F showed cytotoxicity against colon (Colo-205), lung (Calu-1), hepatoma (HA225), cervix (HeLa), and nasopharynx (KB) human cancer cell lines and displayed antitumor activity in the murine P388 lymphocytic leukemia *in vitro* test system [236]. Another compound, withangulatin A promotes Type II DNA topoisomerase-mediated DNA damage *in vitro* [237, 238].

In 2016, sixteen novel withanolides (physangulatins A-N) and withaphysalins Y and Z, together with 12 known analogues, were isolated from stems and leaves and several of them exhibited antiproliferative effects against human cancer lines. More precisely, human melanoma cells (A375-S2), human renal carcinoma cells (786-O, A-498, and ACHN), and prostate cancer cells (C4-2B and 22Rvl) [222].

From a recent study, a new withanolide physagulide I (Fig. **6**) was isolated from aerial parts and cytotoxicity against the human osteosarcoma MG-63, HepG2 hepatoma and MDA-MB-231 breast cancer cell lines was detected for this compound with IC_{50} 3.5, 4.4 and 15.7 µM, respectively [239]. Physagulide P showed anti-proliferative effect in triple-negative breast cancer cells tested in MDA-MB-231 and MDA-MB-468 cells *via* G2/M arrest leading to apoptosis [221]. This compound seems to be very promising for new anticancer drug development.

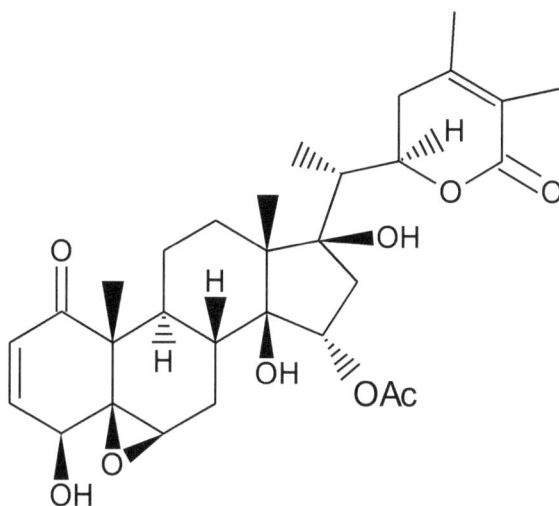

Fig. (6). Chemical structure of physagulide P.

The water and ethanol extracts of leaves from *P. angulata* extracts act as immunomodulators as they cause an increase in lymphocyte cell proliferation [240].

A flavonol glycoside, myricetin-3-O-neohesperidoside, exhibited antitumor activities against several cell lines, such as epidermoid carcinoma of the nasopharynx KB-16 cells, murine leukemia cell line P-388, and lung adenocarcinoma A-549 [241].

Withaphysacarpin and 24,25-dihydrowithanolide D, together with a novel compound 2,3-dihydro-3-methoxywithaphysacarpin induced quinone reductase activity in Hepa1c1c7 murine hepatoma cells [184].

Ixocarpalactone A, ixocarpalactone B, philadelphicalactone B, and withaphysacarpin isolated from fruits of *P. philadelphica* were studied for their cancer chemoprotective activity in SW480 human colon cancer cells. Ixocarpalactone A (Fig. **7**) induces apoptosis and shows cell cycle arrest in the G2/M phase [223]. Its quantity in fresh fruit was determined as 143 ± 4 ppb [242].

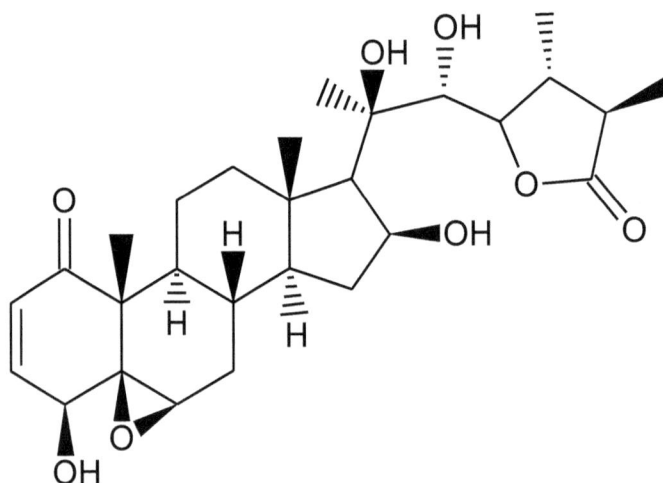

Fig. (7). Chemical structure of ixocarpalactone A.

Cytotoxicity assays with six human cancer cell lines, namely colorectal adenocarcinoma (HCT-15), chronic myelogenous leukemia (K-562), glioblastome (U-251), mammary adenocarcinoma (MCF-7), lung adenocarcinoma (SKLU-1), and prostatic adenocarcinoma (PC-3) were conducted by Maldonado *et al.* [185]. Withaphysacarpin, philadelphicalactone, ixocarpalactone A, and diacetylphiladelphicalactone C obtained from aerial parts revealed potent activity. Withaphysacarpin showed the most promising results with IC_{50} range between 0.4 – 0.9 μM in 5 cell lines, followed by diacetylphiladelphicalactone C with values lower than 1.0 μM in U-251 and K-562 lines, which proves that esterification promotes the cytotoxic potency in withanolides [185].

In search for new prostate cancer therapeutics, the extract of wild *P.crassifolia* was the most active among 18,000 samples. The blockage of androgen-induced expression genes was observed and five withanolides were isolated, four of them were new structures: 15α-acetoxyphysachenolide D, 15α-acetoxy-28-hydroxyphysachenolide D. 18-acetoxy-17-*epi*-withanolide K, and 15α,18-diacetoxy-17-*epi*-withanolide K [243]. Physachenolide D was already known from *P. chenopodifolia* [244].

The C5-C6 double bond seems to be responsible for the hepatoprotective activity in physalin B. This compound exhibited *in vitro* potential against leukemia in mice [245] and together with physalin F was active against some human leukemia cells as well [191].

Antimicrobial Activity

Antimicrobial potential of extracts (using several solvents) from *P. ixocarpa* from different plant parts was tested in disc diffusion assay [231]. Leaf and fruit extracts inhibited *S. aureus* and *K. pneumoniae* and stem sample controlled the growth of *E. coli* and *K. pneumoniae*. Antifungal activity was reported for the methanolic extract of leaf against *Aspergillus niger, Penicillium chrysogenum*, and *Rhizopus stolinifer*.

Antimycobacterial activity of *P. angulata* extracts against *Mycobacterium tuberculosis* $H_{37}Rv$ strain was described by Januário *et al.* [206]. From several isolated physalins, physalin B displayed the strongest effect with $IC_{50} = 32$ μg/mL.

Other Biological Activities

The anti-adipogenetic activity was demonstrated in a 3T3-L1 cell line. The stem and leaf extract of *P. angulata* produced over 90% inhibition of adipogenesis [230]. Anti-histaminic activity of *P. angulata* leaves extract was investigated in animal smooth muscle models in order to study its benefits in treatment of asthma [229]. Antimalarial activity was studied and confirmed for physalin B, D, F, and G [202]. A study dedicated to antileishmanial effect of seco-steroids from *P. angulata* revealed a potential of physalins B, F and G on this field both *in vitro* and *in vitro* [204]. According to their study, physalin F contributed to reduction of parasite load and lesion size in mice infected with *Leishmania amazonensis*.

SOLANUM GENUS

The *Solanum* genus is the most numerous genus in family *Solanaceae*. It includes important crops for human diet, while most of the species contain bioactive compounds important in pharmacology.

With about 1,500 species this genus is well represented all around the globe. The species are rich in phytochemicals with interesting biological activities and therefore have been used not only as food, but also as medicines. Among characteristic compounds, toxic alkaloids are present in all plant parts [175, 246].

As a defence against pathogens and predators, the *Solanum* plants synthesize a variety of secondary metabolites including glycoalkaloids, vitamins, carotenoids, phytoalexins, phenolic compounds and antimicrobial peptides. These compounds do not only guarantee the plant resistance to biotic and abiotic stress, but also bring health benefits to their consumers, as proved in numerous epidemiological studies up to date [247].

According to the latest findings in the search of eggplant origin using DNA correlations, Africa may be the home of eggplant progenitors and place from where it spread to Asia [248]. *S. melongena* complex covers morphologically different signs from spineless plants bearing large fruit up to spiny plants with little fruit. The fruit (eggplant, aubergine, brinjal) of this annual or short-living perennial shrub is widely used in lots of cousins all around the world and consumed raw or processed in many ways, for example in curries, baked or cooked [6]. Since ancient times it has been used for its medicinal properties, for treating inflammations, neuralgia, cholera, bronchitis, asthma, hypertension, fever and pain [249].

Typically, the ripe eggplant fruit is variable in size (2-35 cm long and 2-20 cm wide), colour (green, white, yellow, purple or mixed) and shape (globose, ellipsoid, ovoid) as can be seen in Fig. (**8**) [6].

Fig. (8). The diversity of eggplant shapes. (Reproduced with permission from the copyright owner: David Cavagnaro).

The categorisation and common names reflect these properties. Round, oval or egg-shaped fruit belong to botanical variety *S. melongena* var. *esculentum* Dunal (Nees), long slender fruit to *S. melongena* var. *serpentinum* L. and the small, mini fruit, and early types as *S. melongena* var. *depressum* L. Regarding the vernacular names, Japanese, Indian, Thai, oriental eggplants are known to describe a particular cultivar [250].

Tomato (*Solanum lycopersicum* Lam., syn. *Lycopersicon esculentum* Mill.) is the second most important vegetable crop worldwide after potato. Globally, 117 million tons of fresh fruit were produced in 2016 with China, India, Nigeria and USA being the main producers [10]. Originally native to South America (from Ecuador to Chile), it spread worldwide after it was introduced to Europe from the colonists of the New World [6]. Currently it is an important ingredient of many national cuisines and tomato is basically a part of the everyday diet in most countries, eaten raw, cooked, baked or processed into pastes, sauces or purées, or ketchups. Usually, the ripe fruit is consumed; however, the unripe green tomato is also used in some recipes as part of salsa, pickled, breaded or deep fried. Who would not know the famous dishes like spaghetti Bolognese, gazpacho soup, Tom Yum soup or the fast food "must" in form of ketchup that would not be possible without this fruit? Despite being botanically classified as fruit, it is the best known vegetable and categorized as such in market and in law as well. A famous quote from Miles Kington reminds it: *"Knowledge is knowing that a tomato is a fruit but wisdom is not putting it into a fruit salad."*

Another important Solanum species is *S. nigrum*. The utilization of *S. nigrum* as medicine is older than 2,000 years, being used as an analgesic, laxative, expectorant, to treat diabetes, fever, externally for wounds, cuts, burns and skin inflammations [251]. Some pharmaceutical properties of the species include hypotensive, antispasmodic, anti-HIV, insecticidal, hypocholesterolemic, antimicrobial and gastroprotective [251].

African eggplant or Scarlet, as it is commonly known, is an important vegetable in sub-Saharan Africa and its production quantity belongs to the top, together with tomato, onion and okra. In Brazil, *S. aethiopicum* (jiló) from Gilo group also belongs to the top-rated vegetable and consumed on a daily basis. The cultivars of *S. aethiopicum* are divided into 4 groups: Gilo, Kumba, Sham, and Aculeatum, which differ mainly by appearance of leaves and fruit.

The species of Aculeatum group are used as rootstock for eggplants and tomatoes for their agronomic properties, but the fruit is not consumed, neither the leaves because of prickles and stellate hairs. Usually, unripe or fully ripe fruits, leaves and young sprouts are eaten, mostly cooked. In traditional medicine, especially the bitter cultivars are applied in treatment of hypertension and colic, as a sedative (root, fruit), while leaves are used as antiemetic, for enema or as a sedative too [249, 252].

S. erianthum (syn. *S. verbascifolium*) is a weed (commonly known as potato tree, tropillo or salvadora), native from southern North America down to South America, naturalised all around the globe after being spread after colonisation of Americas. The fruit is globose, up to 12 mm in diameter, yellow after ripening. Despite the raw fruit is considered poisonous, it is widely consumed after cooking. In India it is added into curries. Regular consummation may cause nausea, headache and cramps [251]. In traditional medicines, the leaves have multiple applications in treatment of wounds, leprosis, malaria, venereal diseases, vertigo, headache, scrofula, haemorrhoids. The root decoction is diarrhoea, dysentery, stomach pain, fever and arthritis [251]. The species also serves as a rootstock for eggplant [253].

Chemical Constituents in Genus *Solanum*

Solanum species, such as tomato and eggplant, are a rich source of bioactive carotenoids, phenolic compounds, glycoalkaloids, lectins, antimicrobial proteins and vitamins [254] and other compounds that often carry a health benefit for consumers [255]. Cao *et al.* counted eggplant amount top 10 common vegetables with the highest antioxidant activity [256].

Carotenoids

Carotenoids are an important part of human diet because animals are unable to synthesize these compounds that play a key role in antioxidant defence mechanisms [257]. The most important carotenoids of tomato are the well-studied pigments lycopene and β-carotene. Lycopene is a polyunsaturated molecule containing eleven conjugated and two non-conjugated double bonds in its straight chain [258]. Lycopene gives the fruit its red colour and its quantity increases during ripening of the fruit. Lycopene accounts for over 80% of carotenoids in fully ripe red tomato fruit, followed by α-carotene, β-carotene and lutein [259]. Other carotenoids present in tomatoes are phytoene, phytofluene, neurosporene, and γ-carotene [260]. The structures of some carotenoids are shown in Fig. (**9**).

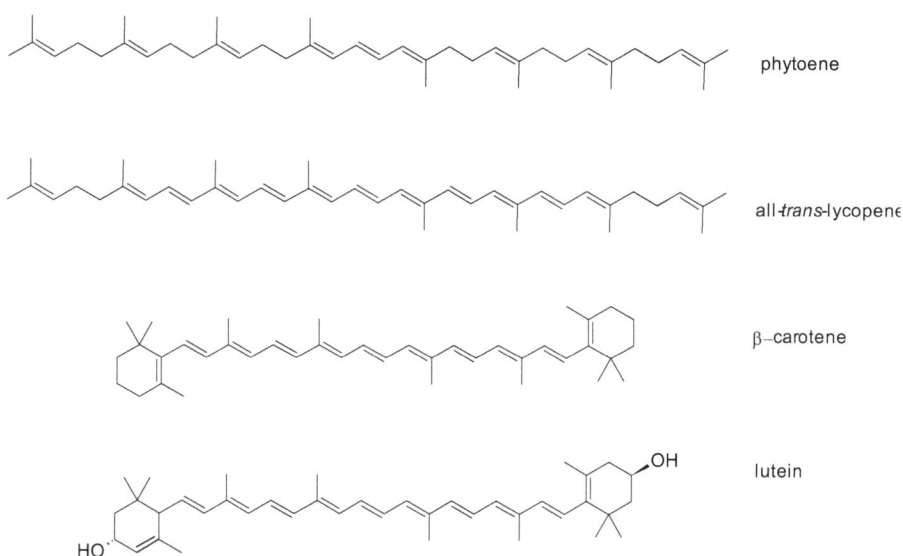

Fig. (9). Chemical structure of the most important carotenoids in tomato (*Solanum lycopersicum*).

Usually, the *trans* isomer of lycopene appears in the highest quantity, however it is the *cis* configuration that exhibits the better *in vitro* bioactivity [258]. The content of lycopene is 2 to 3 times higher in peel than in the pulp [259]. The conjugated double bonds in lycopene molecule, as well as the opening of β-ionone ring stand behind the ability of quenching singlet oxygen radicals if compared to other antioxidants, such as vitamin E or β-carotene [261].

Vitamins

The content of provitamin A and vitamin E is low in *S. melongena*. Nevertheless, high level of vitamins C and B in eggplant have been detected [262 - 264].

Tomato fruit are an important source of hydrophilic vitamin C and lipophilic vitamin E in human diet due to the impossibility of self-production. Vitamin C stands for L-ascorbic acid and L-dehydroascorbic acid, both interchangeable by enzymes *in vitro*. This compound is a vital and powerful antioxidant which can donate an electron [265]. The average content of vitamin C (ascorbic acid) for tomato is reported to be approximately 200 μg/g FW, however it depends on several conditions related to genotype, the environment and plant part. The cherry varieties usually contain higher amounts of ascorbic acid, probably because of the peel fraction in the total weight of the fruit, as the concentrations were found to be higher in peel (90 – 560 μg/g FW) [266]. Vitamin E refers to eight isoforms, four tocopherols and four tocotrienols. All of these forms are able to donate hydrogen thanks to the chromanol ring and its hydroxyl group. From all absorbed isoforms of vitamin E, only α-tocopherol is maintained [267]. They quantity of α-tocopherol in tomatoes varies from 0.01 to about 12 µg/g FW [268].

Glycoalkaloids

These characteristic secondary metabolites of the family *Solanaceae* play a key role in plant resistance and display important pharmacological effects both in humans and animals [254].

Polyhydroxylated nortropane alkaloids were found in fruits and leaves of *S. melongena* [269]. Solamargine and solasonine (Fig. **10**), with their aglycon solasodine, and calystegines A3, B1, B2, and C1 were detected in eggplant [270].

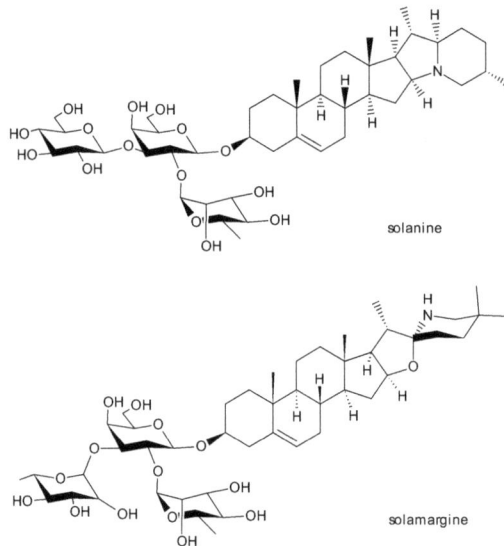

Fig. (10). Chemical structure of the most important glycoalkaloids from *Solanum melongena*.

The major glycoalkaloids synthesized in tomato plant are tomatine and esculenoside A. In fact, tomatine consists of two compounds, α-tomatine and dehydrotomatine in ratios of about 10:1 in fruit and 2:1 - 8:1 in other plant parts [271]. Friedman *et al.* determined their level to be up to 431 and 44 µg/g FW in green tomatoes and 1.7 and 0.16 µg/g FW in red tomatoes, respectively, whereas the cherry varieties gave these highest results, several fold higher than in regular-sized tomato fruits [272]. Similarly, Kozukue *et al.* found that the quantity of dehydrotomatine and α-tomatine content ranged from 42 to 1498 and 521 to 16285 µg/g FW, respectively [271]. Dehydrotomatine has a double bond present in B ring of the steroidal aglycons tomatidenol. The aglycon of α-tomatine is called tomatidine (Fig. **11**). Both compounds consist of the corresponding aglycon and a tetra saccharide side chain, lycotetraose, which seems to be important for the biological activity [273, 274].

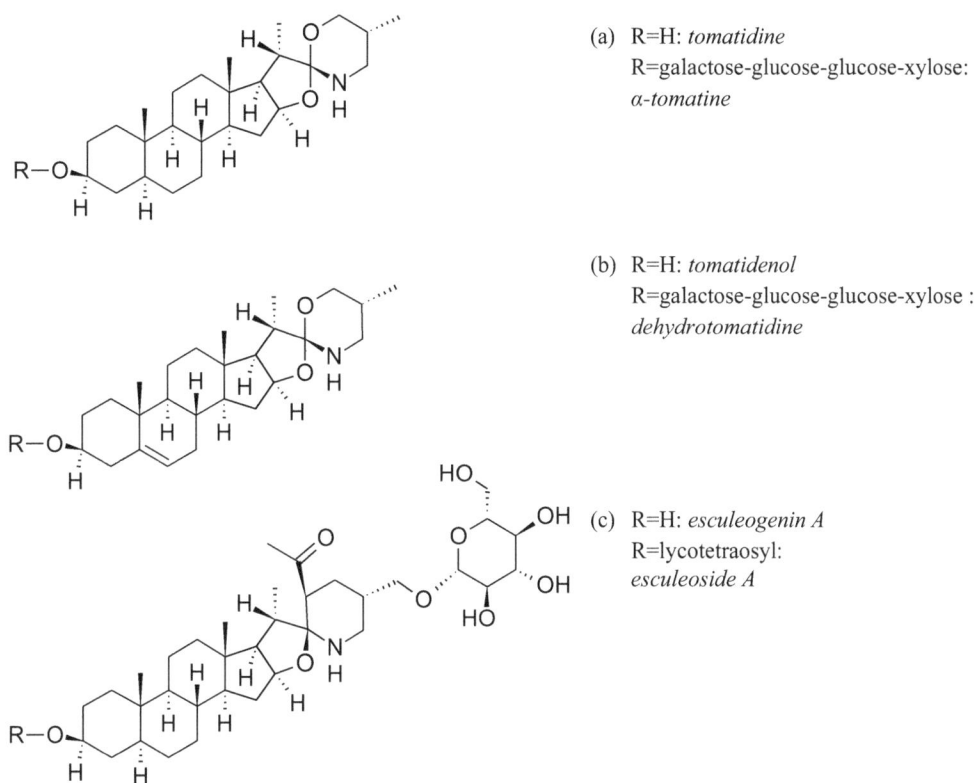

(a) R=H: *tomatidine*
R=galactose-glucose-glucose-xylose:
α-tomatine

(b) R=H: *tomatidenol*
R=galactose-glucose-glucose-xylose :
dehydrotomatidine

(c) R=H: *esculeogenin A*
R=lycotetraosyl:
esculeoside A

Fig. (11). Chemical structures of (a) α-tomatine, (b) dehydrotomatine, (c) esculeoside A and their aglycons.

Tomatine is present in all plant parts, mainly in unripe fruit, flowers, and leaves, usually accompanied by the products of lycotetaose hydrolysis: β_1–tomatine (after loss of xylose), β_2–tomatine (after loss of glucose), and γ–tomatine (after loss of xylose and glucose) [275]. In general, the level of tomatine is high in unripe tomato fruits (up to 500 mg/kg) of fresh weight, but decreases during ripening down to 5 mg/kg FW [276]. The quantity of esculenoside A increases along the maturing process, contrary to tomatine, and is comparable or higher than the quantity of lycopene in ripe fruits [273, 277, 278]. Katsumata *et al.* found levels of esculenoside A in cherry and Momotaro tomatoes to be 21-times and 9-times higher, respectively, than those of lycopene [279]. Cataldi *et al*. identified an isomer of α-tomatine named filotomatine with soladulcidine as its aglycon in the leaves of tomato plant [280]. Further, esculenoside C and D and lycoperoside G and unnamed 3-*O*-β-lycotetraosyl 3β,26-dihydroxycholestan-16, 22-dione 26-*O*-α-L-arabinopyranosyl-(1→6)–β-D-glucopyranoside were found in ripe fruits of cherry tomato *Lycopersicon esculentum* var. *carasiforme*. Other steroidal alkaloid glycosides named lycoperosides A, B and C were identified in tomato leaves and fruits and elucidated as 3-*O-β*-lycotetraosides of (23*R*)-23-acetoxy-tomatidine, (23*S*)-23-acetoxysoladulcidine and (23*S*, 25*S*)-23-acetoxy-5 α, 22 α *N*-spirosola--3 *β* -ol [281]. Later on, lycoperosides D and F-H and a lycoperodine-1 were also isolated from fruit [282]. Tomatoside A, a furostanol glycoside identified belong to the three major triterpenoid glycosides in unripe tomatoes. Its dehydrogenated analogue dehydrotomatoside was also found [283]. Yoshizaki *et al*. isolated isoesculeogenin A and esculeogenin B, two sapogenols deriving from their steroidal alkaloid glycosides lycoperoside F and esculeoside B and characterized them as (5α,22R,23R,25S)-3β,23,27- trihydroxyspirosolane and (5α,22S,23R,25S)-22,26-epimino-16β,23-epoxy-3β,23,27-trihydroxycholestane, respectively [284].

The quantity of α-tomatine in green fruits is high (up to average 500 μg/g FW and decreasing to only approximately 5 mg/kg FW in the fully mature red fruit [273, 276]. The content of bioactive compounds is strongly influenced by several factors, including variety, environmental factors during growth, maturity, origin, processing, *etc*. Additionally, the agricultural approach plays a vital role as organically grown tomatoes showed to have twice the amount of tomatine than conventionally cultivated ones.

Even during the last decade, novel compounds have been identified in tomatoes and despite being of minor quantity, they may hold a more potent biological activity. In such case, these compounds could be synthesized from its precursors. For example, Ohno *et al*. isolated a minor steroidal glycoside of solanocapcine-type named esculeoside B-5, determined as (5*S*,22*R*,23*S*,24*R*,25*S*)-22,26-epimi-o-16*β*,23-epoxy-3*β*,23,24-trihydroxycholestane 3-*O-β*-lycotetraoside from Momotaro cherry tomato [285].

Unripe green fruit of *S. nigrum* were described to contain steroidal alkaloids, mostly solasodine (0.7%), diosgenin (0.2%) and tigogenin (0.15%), whereas leaves contain much higher amount of tigogenin (1.3%) [251]. Due to this high content of alkaloids, fruit and leaves are bitter and considered quite poisonous and regular consumption may cause dizziness, stomachache and vomiting [286].

S. erianthum is a valuable source of spirosolane alkaloids, steroidal saponins and free genins. Solasodine, a nitrogen-containing analogue of diosgenin, isolated from this plant is a starting material for medicinal steroids production in the pharmaceutical industry, such as anabolic steroids, corticosteroids, and contraceptive steroids. The concentration of solasodine is higher that tomatidine (0.26 and 0.05%, respectively). The fruit contain up to 0.7% of solasodine. Total alkaloid content in dry aerial parts rises to 0.4% [287]. Several alpha-linolenic acid analogs, 2 benzofuran-type lactones, and 2 steroidal alkaloids were isolated from leaves, namely alpha-linolenic acid; 13S-hydroxy-9(Z),11(E)-octadecadienoic acid; 9S-hydroxy-10(E),12(Z), 15(Z)-octadectrienoic acid; 9(Z),11(E)-octadecadienoic acid; and octadecanoic acid; loliolide and dihydroactinidiolide; solasonine and solamargine.

Phenolic Compounds

The main polyphenols synthesized in eggplants are phenolic acids (caffeic acid, chlorogenic acid, *p*-coumaric acid). *S. melongena* (eggplant) is a rich source of anthocyanins like nasunin and delphinidin conjugates [288]. Nasunin in its *cis*-form can be seen in Fig. (**12**).

The predominant compound found in the eggplant fruit is chlorogenic acid (5-*O*-caffeoylquinic acid), which was identified together with its 3-*O*-, 4-*O*-, and 5-*O*-*cis* isomers representing over 75% of the total phenolics acids. Other phenolics, like 3,5- and 4,5-dicaffeoylquinic acid isomers, dicaffeoylquinic and 3-*O*-acetyl chlorogenic acids were also extracted [263, 264]. Unlike tomato, purple eggplant usually contains anthocyanins in the fruit peel. Delphinidin 3-(*p*-coumaroylrutinoside)-5-glucoside (nasunin), delphinidin 3-rutinoside, delphinidin 3-glucoside, and petunidin 3-(*p*-coumaroylrutinoside)-5-glucoside (petunidin 3RGc5G) belong to the major anthocyanins [289]. Ichiyanagi *et al.* isolated two anthocyanin isomeric structures of nasunin delphinidin 3-[4-(*cis-p*-coumaroyl)-1-rhamnosyl(1 → 6)glucopyranoside]-5-glucopyranoside and delphinidin 3-[4-(*trans-p*-coumaroyl)-1-rhamnosyl-(1 → 6)glucopyranoside]-5-glucopyranoside, which were interconvertible under room light [217]. Singh *et al.* discovered that organically grown eggplants contained higher amounts of phenolics than conventionally grown plants. The most abundant compounds were N-caffeoylputrescine, 5-caffeoylquinic acid, and 3-acetyl-5-caffeoylquinic acid [290]. García-Salas *et al.* described 14 phenolics compounds for the first time in eggplant fruits, among them bis(dihydrocaffeoyl)spermidine, kaempferol-3-O-rutinoside, homovanillic acid hexose, and caffeoylshikimic acid [291].

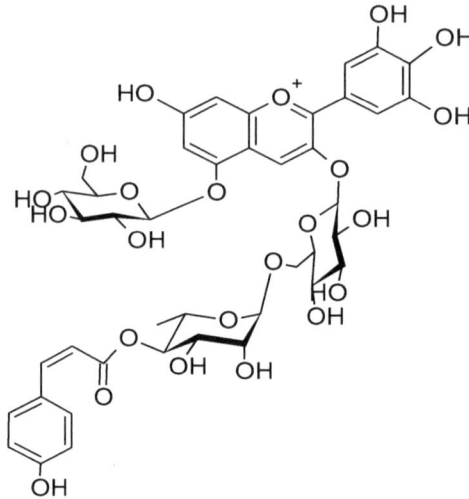

Fig. (12). Chemical structure of *cis*-nasunin.

In tomato, flavonoids are represented by flavonols (rutin, quercetin and kaempferol derivatives) and flavanones (naringenin and chrysin derivatives), usually abundant in the peel [254]. Hydroxycinnamic acids are the main phenolics acids in whole tomatoes, with caffeic, chlorogenic, *p*-coumaric and ferulic acid as major compounds [292]. They comprise about 60% of simple polyphenols, flavonoids being the second group accounting for 30% total polyphenols. Curiously, in commercial tomato by-products, the ratio is opposite and flavonoids predominate with over 73%, with naringenin dominating with over 87% of all flavonoids present [293].

The most abundant phenolic compound in the African *S. aethiopicum* is chlorogenic acid (63 to 96% of total phenolics) [294]. Therefore, the content in different cultivars was screened in several studies and the results varied from 0.2 to 9.9 g/kg [295]. Leaves contain more β-carotene (0.4 and 6.4 mg/100 g FW) and vitamin C (8 and 67 mg/100 g FW) than fruit [249].

Further, a flavonol glycoside and a flavone: camelliaside C and 5-methoxy-(3,4 "-dihydro-3",4"-diacetoxy)-2",2'-dimethylpyrano-(7,8:5",6")-flavone were isolated in *S. erianthum* [296]. Flavonoid glycosides isocytisoside 7-O- β-d-glucoside and embinoidin were identified in leaves, including a new 7,4'-dimethyl-apigenin-6-C-beta-glucopyranosyl-2"-O-alpha-L-arabinopyranoside [297]. Zhoe *et al.* isolated vanillic acid and cinnamide derivatives N (p-hydroxyphenylethyl) p-coumaramide and N-2-hydroxy-2-(p-hydroxyphenylethyl) p-coumaramide from the stems of *S. verbascifolium* (synonym) [298].

Other Compounds

Interesting is the content of nicotine in eggplant of 0.1 µg/g of fruit, however this amount is not comparable to amounts from passive smoking [299]. In contrary, a population-based study revealed that dietary nicotine may reduce the risk of Parkinson disease [300].

The main sterols detected in tomatoes are β-sitosterol and stigmasterol [293]. Among the terpenes, β-amyrin, ursolic acid, and cycloartenol have been reported to be present in tomato fruit surface wax [293].

Recently, new steroidal oligoglycosides nigrumnins I and II and known β-2-solamargine, solamargine and degalactotigonin were isolated from *S. nigrum* leaves [301]. An interesting glycoprotein of 150 kDa has been isolated from *S. nigrum* and its anticancer activity has been studied [302].

An epiminocholestane alkaloid named solaverbascine, assigned as (22S:25R)-22,26-epiminocholest-5-ene-3β, 16β-diol was extracted from leaves of *S. erianthum* [303]. Solanerianones A and B, new norsesquiterpenoids were isolated from the roots of *S. erianthum*, together with sesquiterpenoids (−)-solavetivone, (+)-anhydro-β-rotunol, solafuranone, and lycifuranone A; alkaloid N-trans-feruloyltyramine, as well as β-sitosterol and stigmasterol [304].

Biological Activity in Genus Solanum

Biological activities of selected active compounds from *Solanum* species are summarized in (Table **3**). Bioactive phytochemicals isolated from tomato fruit appear to exhibit antibiotic, immunostimulatory effects, as well as antioxidant and anti-inflammatory activity, which are closely related to the set of civilisation chronic problems, such as cancer, diabetes, neurodegenerative disorders and cardiovascular system diseases [293].

Several compounds of the *S. melongena* have been reported to have antioxidant, anti-inflammatory, anticancer, antiviral, hypocholesterolemic, antimalarial, hepatoprotective effects and other biological activities [255].

In addition, *S. nigrum* has been reported to possess antioxidant [305], antitumor [302, 306], hepatoprotective [307], cytoprotective [308], antimicrobial [309], antiviral [310], antiulcerogenic [311], antihistaminic [312], larvicidal [313], and centrally mediated depressant activities [12].

According to reports, a diet rich in tomatoes is protective against cardiovascular problems, neurodegenerative processes and several types of cancer. The majority of the effects is related to the antioxidant and anti-inflammatory activity of compounds, because these processes are crucial in the development of the mentioned ailments [314]. Not only fruits, but also leaves contain a large number of bioactive substances that usually exert some synergism in the natural mixture and therefore present high protection *in vitro* [315].

As proven by numerous research articles, tomato by-products and production waste, such as leaves, seeds and pulps, are a low cost source of valuable bioactive phytochemicals for the food and pharmaceutical industry.

Table 3. Biological activity of selected active compounds of *solanum* species

Active Compound or Extract	Species	Plant Part	Biological Activity	Ref.
Steroidal alkaloids and their glycosides				
solasodine-3-*O*-β-D-glucopyranoside	*S. nigrum*	fruits	antifungal	[316]
solamargine	*S. nigrum* *S. incanum*	fruits	anti-metastatic, cytotoxic, antiviral, antimalarial	[255, 317 - 321]
solasonine	*S. nigrum*	fruits	antiviral, cytotoxic, antimalarial	[255, 321, 322]
solasodine	*S. lycopersicon*	fruits	antiviral, anticancer,	[323]
β₁-solasonine	*S. nigrum*	whole plant	cytotoxic	[324]
solanine A	*S. nigrum*	fruits	cytotoxic	[322]
esculeogenin A	*S. lycopersicon*	fruits	antiatherosclerotic, anticholesterolemic	[277, 278, 325]
esculeoside A	*S. lycopersicon*	fruits	cytotoxic	[326]
α -tomatine	*S. lycopersicon*	fruit	antimicrobial, anti-inflammatory, antioxidative, cardiovascular, antiproliferative, immunostimulatory, hypocholesterolemic, anti-obesity	[271, 273, 327 - 331]
dehydrotomatine	*S. lycopersicon*	fruit	anticancer	[332]
tomatidine	*S. lycopersicon*	fruits	antibacterial	[333, 334]
solanigroside P	*S. nigrum*	whole plant	cytotoxic	[324]
Carotenoids				

(Table 3) cont.....

Active Compound or Extract	Species	Plant Part	Biological Activity	Ref.
β-carotene	S. nigrum S. villosum, S. scabrum, S. americanum	leaves	antioxidant, cardioprotective	[335, 336]
lycopene	S. lycopersicon	fruits	antibacterial	[337]
Flavonoids and flavonoid glycosides				
kaempferol-3-O-(2″,6″-di-O-p-trans-coumaroyl)-β-glucoside	S. melongena	roots	anti-inflammatory	[338]
delphinidin	S. melongena	fruit peel	anticancer	[339]
naringenin			antiviral	[340]
nasunin	S. melongena	fruit	antioxidant, antiangiogenic, anticholesterolemic, anticancer, cardioprotective	[289, 336, 341 - 344]
delphinidin 3-caffeoylrutinoside-5-glucoside	S. melongena	fruit	antioxidant	[289]
petunidin 3-(p-coumaroylrutinoside)-5-glucoside	S. melongena	fruit	antioxidant	[289]
solanoflavone	S. melongena	aerial parts	anti-inflammatory	[345]
7,4′-dimethyl-apigenin-6-C-β-glucopyranosyl-2″-O-α-l-arabinopyranoside	S. erianthum	leaves	anticancer	[297]
isocytisoside 7-O-β-d-glucoside and embinoidin	S. erianthum	leaves	anticancer	[297]
camelliaside C	S. erianthum	leaves	antiviral	[296]
Proteins				
cystine-knot miniproteins	S. lycopersicon	fruit	antiangiogenic	[346]
SN2 snakin peptide	S. lycopersicum		antimicrobial	[347]
Steroidal saponins				
protodioscin, methyl-protodioscin, indioside D	S. incanum	roots	antioxidant, cytotoxic	[348]
Lignanamides				
melongenamides B–D, cannabisin D, grossamide, cannabisin F, cannabisin G	S. melongena	roots	anti-inflammatory	[349]
Other compounds				
ursolic acid	S. lycopersicon	fruit surface wax	antiproliferative	[350]
melongenolide A	S. melongena	roots	anti-inflammatory	[338]

(Table 3) cont.....

Active Compound or Extract	Species	Plant Part	Biological Activity	Ref.
α-tocopherol	*S. nigrum* *S. villosum,* *S. scabrum,* *S. americanum*	leaves	antioxidant	[335]

Beneficial impact of α-tomatine on human health includes antimicrobial, anti-inflammatory, antioxidative, cardiovascular, antiproliferative and immunostimulatory activity and it plays important role in lowering cholesterol and triglycerols [271, 273, 327 - 329].

Lycopene possess anti-proliferative effects against several human cancer cell lines, such as breast, prostate, colon, myeloid leukaemia, lymphoma, and lung *via* cell cycle arrest, apoptosis induction and mainly the involvement in antioxidant system enhancement [293, 351, 352].

Antioxidant Activity

The high levels of phenolic compounds are usually correlated with the antioxidant activity of eggplant. *in vitro* antioxidant capacity was ascribed to 3,5-diglycosylated structures of anthocyanins in acetone extract of purple eggplant peel [75]. Azuma *et al.* referred to delphinidin 3RGcaf5G, nasunin and petunidin 3RGc5G having the highest radical-scavenging potential in DPPH (1,1-diphenyl-2-picrylhydrazyl) assay [289]. Nasunin was found to present a good protection against hydrogen peroxide-induced lipid peroxidation as it is a good O2* scavenger and iron chelator [341, 342]. In an *in vitro* study with rats, 1 mg of flavonoids from eggplant per 100 grams of body weight a day enhanced the catalase activity and reduced concentration of malondialdehyde and hydroperoxides in normal and cholesterol fed animals [353]. Recently, Nisha *et al.* compared several eggplant fruit of different colour and size to find out that small purple fruit demonstrate higher antioxidant activity and associated it with their higher phenolic and anthocyanin content [354]. Interestingly, Lo Scalzo *et al.* observed higher antioxidant activity in cooked eggplants compared to raw samples (effective dilution 1.25 µg/mL *vs* 10 µg/mL). They explained this feature by a higher level of chlorogenic and caffeic acids in the thermally treated samples [355]. Hepatoprotective activity conducted with HepG2 cells was correlated with antioxidant potential and high total phenolic compounds and flavonoids in different eggplant accessions [356].

The hydromethanolic extracts of tomato leaves of several cultivars displayed strong antioxidant activity scavenging $O_2^{\cdot-}$ radicals ($IC_{50} = 0.12 - 0.43$ mg/mL) [327].

Synergism has been observed in mixtures of lycopene with other carotenoids and vitamins and the combination with vitamin E in tests of antioxidant activity in liposome system seems to be the most efficient [357].

Anti-Inflammatory Activity

Anti-inflammatory activity was studied in an *in vitro* test with *S. melongena* water extract at a dose of 10 mg/kg. The extract inhibited PAR2 agonist-induced paw edema and vascular permeability, and the expression of the tumour necrosis factor (TNF)-α [358]. An active biflavonol glycoside solanoflavone was isolated from aerial parts of white fruited cultivar [345]. The fruit extract of *S. nigrum* inhibited carrageenan-induced edema [359].

In evaluation of *S. melongena* roots extract, a promising inhibition of NO production in LPS-induced RAW 264.7 macrophages was detected *in vitro* with a lignin (+)-syringaresinol, flavonoid kaempferol-3-*O*-(2″,6″-di-*O*-*p*-*trans*-coumaroyl)-β-glucoside, triterpenoid saponin called arjunolic acid and a novel γ-alkylated-γ-butyrolactone melongenolide A with IC_{50} 5.6, 11.5, 27.8, and 40 μM/L, respectively [338].

In search for superior candidates suitable for eggplant breeding in order to enhance the content of biologically active and health-promoting phytochemicals, chlorogenic acid levels, antioxidant and anti-inflammatory activity of 48 accessions of *S. aethiopicum* was determined [360]. The Shum group accessions gave the highest average amount of 3 g/kg, but Kemba group accessions exhibited the highest reducing power within the other groups of scarlet eggplant. The LPS-induced NO production in macrophage cells was tested and the best effect was observed in high-chlorogenic acid accessions.

The sesquiterpenoid (−)-solavetivone was tested for anti-inflammatory activity in lipopolysaccharide (LPS)-activated murine macrophages RAW264.7 and showed a significant potential with IC_{50} 65.5 ± 0.2 μM [304].

Significant anti-inflammatory activity was described *in vitro* in rats using carrageenan-induced paw edema test for methanolic extracts of *S. nigrum* fruits [361] and ethanolic extract, where the effect at 500 mg/kg dose was comparable to diclogenac standard (50 mg/kg) [362].

Anticancer Activity

Solamargine and solasonine showed antiproliferative activity against human colon HT29 and liver HepG2 cell lines, whereas their hydrolysis products were less active [270]. Solamargine induced cell death in human hepatoma cells Hep3B

[363], as well as HepG2 and SMMC-7721 [319]. Liu *et al.* described solamargine as a potent inducer of apoptosis in human lung cancer cell lines H441, H520, H661, and H69 (IC_{50} = 3.0, 6.7, 7.2 and 5.8 µM, resp.) [364]. Another research group observed the potential of solamargine against TNFs- and cisplatin-resistant lung cancer as it suppresses the growth of human lung adenocarcinoma A549 cell line [365]. A synergic effect of solamargine and epirubicin caused an accelerated apoptosis in non-small lung cancer cells NSCLC [366]. These steroidal alkaloids augmented the death of H661 and H69 lung cancer cells [365]. Similarly, solamargine was effective against human breast cancer cell lines MCF-7 and SK-BR-3 with results comparable to 5-fluorouracil, epirubicin, cisplatin or cyclophosphamide [367, 368]. Human leukaemia cells K562 were also susceptible to solamargine [369]. Next to solamargine, solasodine, solasonine, β-sitosterol-3-*O*-β-D-glucoside and poriferasterol-3-*O*-β-D-glucoside exhibited promising activity against MCF7 breast, HCT116 colon cancer, HeLa cervix cancer, HepG2 liver, and Hep2 lung cell lines. Additionally, methanol extract from peels showed a dose-dependent anticancer activity *in vitro* against CCl4-induced hepatocellular carcinoma in rats [370].

Delphinidin, the pigment from eggplant peel, was effective against HT-1080 human fibrosarcoma *in vitro* [339]. Its derivative nasunin (delphinidin-3-(*p*-coumaroylrutinoside)-5-glucoside) isolated also from peel inhibited angiogenesis in an *ex vivo* test on rat outgrowth ring [344].

Cytotoxic effects of leaves extracts from several tomato cultivars (Abuela, Anairis, Caramba, Negro, and Valentine) were tested against human gastric adenocarcinoma cells and antioxidant activity against several radicals and lipid peroxidation was evaluated [327]. The methanolic extracts with the main flavonoid quercetin-3-*O*-rutinoside and chlorogenic acid reduced the cell viability. The best results were confirmed in alkaloid extract and ascribed to tomatine. Its aglycon tomatidine is less toxic to cancer cells as the presence of sugar chain has an impact on target cell membrane integrity [371].

The α-tomatine from green tomatoes was inhibited by the growth of human breast (MCF-7), colon (HT-29), gastric (AGS), and liver (HepG2) cancer cell lines. However, dehydrotomatine, and the aglycons tomatidenol and tomatidine did not show any significant activity [371]. In a different study, glycoalkaloids from Solanum vegetables were examined in order to judge the antiproliferative activity against human liver (HepG2) and colon (HT29) cancer cell lines. Tomato-derived α-tomatine, $β_1$-tomatine, γ-tomatine, and δ-tomatine and their aglycon tomatidine, as well as solamargine and solasonine from eggplant, were active in the microculture tetrazolium (MTT) assay. The products of hydrolysis were less active than the starting glycoalkaloid and liver cells were more susceptible. The effect of α-tomatine against the liver carcinoma cells was higher than those determined with the anticancer drugs doxorubicin and camptothecin at concentration of 1 μg/mL. Nevertheless, α-solanine and α-tomatine inhibited also normal human liver HeLa (Chang) cells, and therefore safety of use and exact dosage shall be considered and toxicologically determined [270].

Acetone, alkaloid-rich and hydromethanolic extracts of tomato supressed cell growth in human gastric adenocarcinoma cells (IC$_{50}$ = 9 – 55 μg/mL for alkaloid-rich extract) [327].

S. erianthum flower methanol extract was highly cytotoxic in the MB49 bladder, LM2 mammary adenocarcinoma, B16 melanoma, and A549 lung cancer cells [372]. The leaves extract of *S. erianthum* was cytotoxic towards human gastric adenocarcinoma (AGS) cells and TRAIL-resistant human colon cancer (DLD1/TR). The flavonoid glycosides 7,4′-dimethyl-apigenin-6- C-β-glucopyranosyl-2″-O-α-l-arabinopyranoside, along with isocytisoside 7-O-β-D-glucoside and embinoidin, acted synergistically [297].

The biological activity of vitamin E isomers was studied and its role in cancer was associated to the strong antioxidant potential, as well as other oxidative stress-related ailments (chronic inflammations, cardiovascular disease, and neurological disorders) [373].

Antimicrobial Activity

A SN2 snakin peptide from *S. lycopersicum* perforated the membranes of microconidia cells and hyphae (observed IC$_{50}$ = 8 μM and 2.2 μM, resp.) and possessed strong antimicrobial activity and anti-agglomeration effect in *Agrobacterium tumefaciens, Fusarium solani, P. pastoris, E. coli, Micrococcus luteus,* and *Staphylococcus cohnii* [347]. Snakins are small peptides composed of 12 cysteines with 6 disulfidic bonds. Their highly conserved C-terminal plays a key role in their biological activity [67].

Aqueous extract from *S. erianthum* leaves showed antimicrobial activity against MRSA methicillin-resistant *Staphylococcus aureus* (6.3 mg/mL) [374].

Antidiabetic Activity

The antidiabetic activity in tomato leaves was described for the first time in 2016 as the methanolic extract inhibited key enzymes α-glucosidase and α-amylase involved in diabetes mellitus. Capacity to supress enzymes in Alzheimer´s disease was observed. Acetylcholinesterase (AChE), butyrylcholinesterase (BuChE) and 5-lipoxygenase (LOX) were inhibited by the methanolic extracts rich in phenolics

(3.8 - 9.2 mg/g of dry extract), namely chlorogenic and neochlorogenic acid, quercetin-3-*O*-rutinoside, and quercetin-3-*O*-pentosyl-rutinoside [315].

Ripe tomato fruits showed antiplatelet activity and adenosine was isolated as the bioactive compound responsible for this effect. Platelet aggregation and thrombus formation was inhibited and therefore this compound may have potential in stroke prevention [375].

Inhibition of enzymes important in management of the diabetes type 2 and hypertension was studied *in vitro* by Kwon *et al.* in *S. melongena* [376]. Extracts of White and Graffiti cultivars of *S. melongena* inhibited α-glucosidase and strong inhibition of angiotensin I-converting enzyme.

Antiatherosclerotic Activity

Esculeogenin A prevents the accumulation of cholesterol esters in macrophages *via* acyl-CoA:cholesterol acyl transferase (ACAT) [277]. Fujiwara *et al.* studied *in vitro* effects of this compound and its aglycons esculeogenin A on the foam cell formation and *in vitro* atherogenesis in apoE-deficient mice and stated that purified esculeoside A is able to prevent the expression of ACAT-1 protein which results in the reduction of atherogenesis [261]. The levels of serum cholesterol, triglycerides, and LDL-cholesterol were reduced after oral administration of this compound and the atherosclerotic lesions were also reduced. Moreover, no serious side effects were detected [325].

Other iological Activities

Solamargine and solasonine showed antimalarial activity against *Plasmodium yoelii* [321]. Solasodine exhibited antiviral activity against HSV-1 and the herpes virus type 1 [323]. A hypolipidemic effect was noticed in hypercholesterolemic rabbits by Odetola *et al.* [377]. Total serum cholesterol was reduced, HDL/LDL cholesterol ratio increased [377]. Das *et al.* described cardioprotective effects of compounds from eggplant, such as nasunin, vitamins A and C, or β-carotene, reducing myocardial infarct and apoptosis of cardiomyocytes [336].

Calystegines A3, B1, B2, and C1 (polyhydroxylated nortropane alkaloids) from leaves and fruits of eggplant were found to be strong competitive inhibitors of β-glucosidase, α-galactosidase and β-galactosidase *in vitro* in human liver suggesting the toxicity of large amounts of these compounds in human diet [269].

Aqueous extract of *S. melongena* leaves showed peripheral and central analgesic effect comparable to aspirin in animal study [378]. A study regarding the antipyretic and analgesic potential of leaves was conducted in albino rats and a dose-dependent effect was confirmed in both yeast-induced pyrexia and in acetic acid-induced writhing test [379]. A clinical trial in order to judge the role of eggplant in asthma treatment was evaluated and significant improvement of asthma symptoms in asthmatics was found [380]. Total cholesterol, LDL-cholesterol and apolipoprotein B were reduced significantly in blood of humans in a clinical trial with *S. melongena* [381].

Anti-hepatitis B virus (anti-HCV) activity assay was conducted for 11 isolated compounds from *S. erianthum* leaves extract in methanol. Camelliaside C was the only active compound against HBeAg (IC_{50} 36 µM). Solamargine and solasonine possessed the highest activity against HBsAg (IC_{50} 1.57 µM and IC_{50} 5.89 µM, respectively) [296]. *S. nigrum* seed extract also displayed anti-HCV activity and in combination with interferon it could be beneficial in chronic HCV treatment [310].

Larvicidal properties of the leaf extract were noticed in *S. nigrum* against malaria vector *Anopheles culicifecies*, and further against filariasis vectors *Culex quinquefasciatus* and *Aedes aegyptii* [382]. Aqueous extract of leaves were documented to have anti-inflammatory, antinociceptive, and antipyretic effects [383].

CONSENT FOR PUBLICATION

Not applicable.

CONFLICT OF INTEREST

The author (editor) declares no conflict of interest, financial or otherwise.

ACKNOWLEDGEMENT

Declare None.

REFERENCES

[1]   Shah VV, Shah ND, Patrekar PV. Medicinal Plants from Solanaceae Family. Res J Pharm Technol 2013; 6(2): 143-51.

[2]   The Plant List [Internet] Version 11 2017. [cited 2017 Dec 20] Available from: http://www.theplantlist.org/

[3]   Olmstead RG, Sweere JA, Spangler RE, Bohs L, Palmer JD. Phylogeny and provisional classification of the Solanaceae based on chloroplast DNA. Solanaceae IV 1999; 1(1): 1-137.

[4]   Hunziker AT. South American Solanaceae: A synoptic survey.The biology and taxonomy of the Solanaceae. New York: Academic Press 1979; pp. 49-85.

[5]   Vainio H, Bianchini F. Fruit and vegetables: IARC handbooks of cancer prevention. Lyon, France: IARC Press 2003; Vol. 8.

[6]   Lim TK. Edible Medicinal And Non-Medicinal Plants: Volume 6, Fruits. Springer Netherlands; 2013. 606 p.

[7]   Shackleton CM, Pasquini MW, Drescher AW. African indigenous vegetables in urban agriculture Routledge. UK: Earthscan 2009; p. 336.

[8]   van Rensburg WSJ, Van Averbeke W, Slabbert R, Faber M, Van Jaarsveld P, Van Heerden I, *et al.* African leafy vegetables in South Africa. Water SA 2007; 33(3): 317-26.

[9]   Samuels J. The Solanaceae-novel crops with high potential. Org Grow 2009; 9: 32-4.

[10]   FAOSTAT Crop Production Data for 2016 [Internet] 2017. [cited 2017 Dec 27] Available from: http://faostat.fao.org

[11]   Cárdenas PD, Sonawane PD, Heinig U, Bocobza SE, Burdman S, Aharoni A. The bitter side of the nightshades: Genomics drives discovery in Solanaceae steroidal alkaloid metabolism. Phytochemistry 2015; 113: 24-32.
        [http://dx.doi.org/10.1016/j.phytochem.2014.12.010] [PMID: 25556315]

[12]   Olet EA, Heun M, Lye KA. African crop or poisonous nightshade; the enigma of poisonous or edible black nightshade solved. Afr J Ecol 2005; 43(2): 158-61.
        [http://dx.doi.org/10.1111/j.1365-2028.2005.00556.x]

[13]   Eich E. Solanaceae and Convolvulaceae: Secondary metabolites: Biosynthesis, chemotaxonomy, biological and economic significance (a handbook). Springer Science & Business Media 2008.
        [http://dx.doi.org/10.1007/978-3-540-74541-9]

[14]   Singh J, Kaur L. Advances in potato chemistry and technology. New York: Academic Press 2016.

[15]   Svobodova B, Barros L, Sopik T, *et al.* Non-edible parts of Solanum stramoniifolium Jacq. - a new potent source of bioactive extracts rich in phenolic compounds for functional foods. Food Funct 2017; 8(5): 2013-21.
        [http://dx.doi.org/10.1039/C7FO00297A] [PMID: 28488719]

[16]   Yousaf Z, Akram A, Rehman HA. A monograph on solanum torvum Swartz A Monograph on Solanum torvum Swartz 2013; 1-91.

[17]   Svobodova B, Barros L, Calhelha RC, Heleno S, Alves MJ, Walcott S, *et al.* Bioactive properties and phenolic profile of Momordica charantia L. medicinal plant growing wild in Trinidad and Tobago. Ind Crops Prod 2017; 95: 365-73.
        [http://dx.doi.org/10.1016/j.indcrop.2016.10.046]

[18]   Djian-Caporilano C, Lefebvre V, Sage-Daubeze AM, Palloix A. Genetic resources, chromosome engineering, and crop improvement: vegetable crops. Boca Raton, FL: CRC Press 2006; Vol. 3: p. 552.

[19]   Carocho M, Barreiro MF, Morales P, Ferreira ICFR. Adding Molecules to Food, Pros and Cons: A Review on Synthetic and Natural Food Additives. Compr Rev Food Sci Food Saf 2014; 13(4): 377-99.
        [http://dx.doi.org/10.1111/1541-4337.12065]

[20]   Schulze B, Spiteller D. Capsaicin: tailored chemical defence against unwanted "frugivores". ChemBioChem 2009; 10(3): 428-9.
        [http://dx.doi.org/10.1002/cbic.200800755] [PMID: 19130454]

[21]   Zachariah TJ, Gobinath P. Paprika and Chilli.Chemistry of Spices. London, UK: CABI 2008; p. 260.
        [http://dx.doi.org/10.1079/9781845934057.0260]

[22]   Zimmer AR, Leonardi B, Miron D, Schapoval E, Oliveira JR, Gosmann G. Antioxidant and anti-inflammatory properties of Capsicum baccatum: from traditional use to scientific approach. J Ethnopharmacol 2012; 139(1): 228-33.
        [http://dx.doi.org/10.1016/j.jep.2011.11.005] [PMID: 22100562]

[23]   DeWitt D, Bosland PW. The Complete Chile Pepper Book: A Gardener's Guide to Choosing,

Growing, Preserving, and Cooking. Portland: Timber Press 2009; p. 336.

[24]   Maksimova V, Gudeva LK, Gulaboski R, Nieber K. Co-extracted bioactive compounds in Capsicum fruit extracts prevent the cytotoxic effects of capsaicin on B104 neuroblastoma cells. Rev Bras Farmacogn 2016; 26(6): 744-50.
[http://dx.doi.org/10.1016/j.bjp.2016.06.009]

[25]   Yahara S, Ura T, Sakamoto C, Nohara T. Steroidal glycosides from *Capsicum annuum*. Phytochemistry 1994; 37(3): 831-5.
[http://dx.doi.org/10.1016/S0031-9422(00)90366-2] [PMID: 7765694]

[26]   Wang J, Peng Z, Zhou S, Zhang J, Zhang S, Zhou X, *et al.* A study of pungency of capsaicinoid as affected by their molecular structure alteration. Pharmacol Pharm 2011; 2(3): 109-15.
[http://dx.doi.org/10.4236/pp.2011.23014]

[27]   Collins MD, Wasmund LM, Bosland PW. Improved method for quantifying capsaicinoids in Capsicum using high-performance liquid chromatography. HortScience 1995; 30(1): 137-9.

[28]   Barbero GF, Liazid A, Palma M, Barroso CG. Fast determination of capsaicinoids from peppers by high-performance liquid chromatography using a reversed phase monolithic column. Food Chem 2008; 107(3): 1276-82.
[http://dx.doi.org/10.1016/j.foodchem.2007.06.065]

[29]   Liu L, Chen X, Liu J, Deng X, Duan W, Tan S. Determination of capsaicin and dihydrocapsaicin in Capsicum anuum and related products by capillary electrophoresis with a mixed surfactant system. Food Chem 2010; 119(3): 1228-32.
[http://dx.doi.org/10.1016/j.foodchem.2009.08.045]

[30]   Bosland PM, Coon D, Reeves G. "Trinidad Moruga Scorpion" Pepper is the World's Hottest Measured Chile Pepper at More Than Two Million Scoville Heat Units. Horttechnology 2012; 22(4): 534-8.

[31]   Howard LR, Wildman REC. Handbook of nutraceuticals and functional foods.Boca Raton, FL: CRC Press 2016; pp. 165-91.

[32]   Kobata K, Todo T, Yazawa S, Kazuo I, Watanabe T. Novel Capsaicinoid-like Substances, Capsiate and Dihydrocapsiate, from the Fruits of a Nonpungent Cultivar, CH-19 Sweet, of Pepper (*Capsicum annuum* L.). J Agric Food Chem 1998; 46(5): 1695-7.
[http://dx.doi.org/10.1021/jf980135c]

[33]   Ochi T, Takaishi Y, Kogure K, Yamauti I. Antioxidant activity of a new capsaicin derivative from *Capsicum annuum*. J Nat Prod 2003; 66(8): 1094-6.
[http://dx.doi.org/10.1021/np020465y] [PMID: 12932131]

[34]   Tanaka Y, Hosokawa M, Otsu K, Watanabe T, Yazawa S. Assessment of capsiconinoid composition, nonpungent capsaicinoid analogues, in capsicum cultivars. J Agric Food Chem 2009; 57(12): 5407-12.
[http://dx.doi.org/10.1021/jf900634s] [PMID: 19489540]

[35]   Watanabe E, Kodama T, Masuyama T, *et al.* Studies of the toxicological potential of capsinoids: VIII. A 13-week toxicity study of commercial-grade dihydrocapsiate in rats. Int J Toxicol 2008; 27(3) (Suppl. 3): 101-18.
[http://dx.doi.org/10.1080/10915810802513619] [PMID: 19037802]

[36]   Luo X-J, Peng J, Li Y-J. Recent advances in the study on capsaicinoids and capsinoids. Eur J Pharmacol 2011; 650(1): 1-7.
[http://dx.doi.org/10.1016/j.ejphar.2010.09.074] [PMID: 20946891]

[37]   Kollmannsberger H, Rodríguez-Burruezo A, Nitz S, Nuez F. Volatile and capsaicinoid composition of ají (Capsicum baccatum) and rocoto (Capsicum pubescens), two Andean species of chile peppers. J Sci Food Agric 2011; 91(9): 1598-611.
[http://dx.doi.org/10.1002/jsfa.4354] [PMID: 21445890]

[38]   Vera-Guzmán AM, Chávez-Servia JL, Carrillo-Rodríguez JC, López MG. Phytochemical evaluation

of wild and cultivated pepper (*Capsicum annuum* L. and C. pubescens Ruiz & Pav.) from Oaxaca, Mexico. Chil J Agric Res 2011; 71(4): 578-85.
[http://dx.doi.org/10.4067/S0718-58392011000400013]

[39]   Cruz Pérez AB, González Hernández VA, Soto Hernández RM, Gutiérrez Espinosa MA, Gardea Béjar AA, Pérez Grajalez M. Capsaicinoides, vitamina C y heterosis durante el desarrollo del fruto de chile manzano. Agrociencia 2007; 41(6): 627-35.

[40]   Meckelmann SW, Jansen C, Riegel DW, van Zonneveld M, Ríos L, Peña K, *et al.* Phytochemicals in native Peruvian Capsicum pubescens (Rocoto). Eur Food Res Technol 2015; 241(6): 817-25.
[http://dx.doi.org/10.1007/s00217-015-2506-y]

[41]   Bosland PW, Votava EJ, Votava EM. Peppers: vegetable and spice capsicums. Cabi 2012; Vol. 22.
[http://dx.doi.org/10.1079/9781845938253.0000]

[42]   Kumar R, Dwivedi N, Singh RK, Kumar S, Rai VP, Singh M. A Review on Molecular Characterization of Pepper for Capsaicin and Oleoresin. Int J Plant Breed Genet 2011; 5(2): 99-110.
[http://dx.doi.org/10.3923/ijpbg.2011.99.110]

[43]   Bosland PW. Capsicums: Innovative uses of an ancient crop.Progress in new crops Arlington. VA: ASHS Press 1996; pp. 479-87.

[44]   Lightbourn GJ, Griesbach RJ, Novotny JA, Clevidence BA, Rao DD, Stommel JR. Effects of Anthocyanin and Carotenoid Combinations on Foliage and Immature Fruit Color of *Capsicum annuum* L J Hered 2008; 99(2): 105-1.

[45]   Wahyuni Y, Ballester AR, Sudarmonowati E, Bino RJ, Bovy AG. Metabolite biodiversity in pepper (Capsicum) fruits of thirty-two diverse accessions: variation in health-related compounds and implications for breeding. Phytochemistry 2011; 72(11-12): 1358-70.
[http://dx.doi.org/10.1016/j.phytochem.2011.03.016] [PMID: 21514607]

[46]   Guzman I, Hamby S, Romero J, Bosland PW, O'Connell MA. Variability of Carotenoid Biosynthesis in Orange Colored Capsicum spp. Plant Sci 2010; 179(1-2): 49-59.
[http://dx.doi.org/10.1016/j.plantsci.2010.04.014] [PMID: 20582146]

[47]   Giuffrida D, Dugo P, Torre G, *et al.* Characterization of 12 Capsicum varieties by evaluation of their carotenoid profile and pungency determination. Food Chem 2013; 140(4): 794-802.
[http://dx.doi.org/10.1016/j.foodchem.2012.09.060] [PMID: 23692768]

[48]   Antonious GF, Lobel L, Kochhar T, Berke T, Jarret RL. Antioxidants in *Capsicum chinense*: variation among countries of origin. J Environ Sci Health B 2009; 44(6): 621-6.
[http://dx.doi.org/10.1080/03601230903000727] [PMID: 20183071]

[49]   Padayatty SJ, Katz A, Wang Y, *et al.* Vitamin C as an antioxidant: evaluation of its role in disease prevention. J Am Coll Nutr 2003; 22(1): 18-35.
[http://dx.doi.org/10.1080/07315724.2003.10719272] [PMID: 12569111]

[50]   Perla V, Nimmakayala P, Nadimi M, *et al.* Vitamin C and reducing sugars in the world collection of Capsicum baccatum L. genotypes. Food Chem 2016; 202: 189-98.
[http://dx.doi.org/10.1016/j.foodchem.2016.01.135] [PMID: 26920284]

[51]   Jarret RL, Berke T, Baldwin EA, Antonious GF. Variability for free sugars and organic acids in Capsicum chinense. Chem Biodivers 2009; 6(2): 138-45.
[http://dx.doi.org/10.1002/cbdv.200800046] [PMID: 19235156]

[52]   Howard LR, Talcott ST, Brenes CH, Villalon B. Changes in phytochemical and antioxidant activity of selected pepper cultivars (Capsicum species) as influenced by maturity. J Agric Food Chem 2000; 48(5): 1713-20.
[http://dx.doi.org/10.1021/jf990916t] [PMID: 10820084]

[53]   Pandey KB, Rizvi SI. Plant polyphenols as dietary antioxidants in human health and disease. Oxid Med Cell Longev 2009; 2(5): 270-8.
[http://dx.doi.org/10.4161/oxim.2.5.9498] [PMID: 20716914]

[54] Heleno SA, Martins A, Queiroz MJRP, Ferreira ICFR. Bioactivity of phenolic acids: metabolites *versus* parent compounds: a review. Food Chem 2015; 173: 501-13.
[http://dx.doi.org/10.1016/j.foodchem.2014.10.057] [PMID: 25466052]

[55] Marín A, Ferreres F, Tomás-Barberán FA, Gil MI. Characterization and quantitation of antioxidant constituents of sweet pepper (*Capsicum annuum* L.). J Agric Food Chem 2004; 52(12): 3861-9.
[http://dx.doi.org/10.1021/jf0497915] [PMID: 15186108]

[56] Patras MA, Jaiswal R, Kuhnert N. Profiling and quantification of regioisomeric caffeoyl glucoses in Solanaceae vegetables. Food Chem 2017; 237: 659-66.
[http://dx.doi.org/10.1016/j.foodchem.2017.05.150] [PMID: 28764050]

[57] Spiller F, Alves MK, Vieira SM, *et al.* Anti-inflammatory effects of red pepper (*Capsicum baccatum*) on carrageenan- and antigen-induced inflammation. J Pharm Pharmacol 2008; 60(4): 473-8.
[http://dx.doi.org/10.1211/jpp.60.4.0010] [PMID: 18380920]

[58] Surh Y-J, Lee SS. Capsaicin in hot chili pepper: carcinogen, co-carcinogen or anticarcinogen? Food Chem Toxicol 1996; 34(3): 313-6.
[http://dx.doi.org/10.1016/0278-6915(95)00108-5] [PMID: 8621114]

[59] Mueller LA, Solow TH, Taylor N, *et al.* The SOL Genomics Network: a comparative resource for Solanaceae biology and beyond. Plant Physiol 2005; 138(3): 1310-7.
[http://dx.doi.org/10.1104/pp.105.060707] [PMID: 16010005]

[60] Barbosa Pelegrini P, Del Sarto RP, Silva ON, Franco OL, Grossi-de-Sa MF. Antibacterial peptides from plants: what they are and how they probably work. Biochem Res Int 2011; 2011: 250349.
[http://dx.doi.org/10.1155/2011/250349] [PMID: 21403856]

[61] Hancock REW, Rozek A. Role of membranes in the activities of antimicrobial cationic peptides. FEMS Microbiol Lett 2002; 206(2): 143-9.
[http://dx.doi.org/10.1111/j.1574-6968.2002.tb11000.x] [PMID: 11814654]

[62] Carvalho A de O, Gomes VM. Plant defensins and defensin-like peptides - biological activities and biotechnological applications. Curr Pharm Des 2011; 17(38): 4270-93.
[http://dx.doi.org/10.2174/138161211798999447] [PMID: 22204427]

[63] Corrêa RCG, de Souza AH, Calhelha RC, *et al.* Bioactive formulations prepared from fruiting bodies and submerged culture mycelia of the Brazilian edible mushroom Pleurotus ostreatoroseus Singer. Food Funct 2015; 6(7): 2155-64.
[http://dx.doi.org/10.1039/C5FO00465A] [PMID: 26065398]

[64] De Lucca AJ, Cleveland TE, Wedge DE. Plant-derived antifungal proteins and peptides. Can J Microbiol 2005; 51(12): 1001-14.
[http://dx.doi.org/10.1139/w05-063] [PMID: 16462858]

[65] Kader J-C. Lipid-transfer proteins in plants. Annu Rev Plant Physiol Plant Mol Biol 1996; 47(1): 627-54.
[http://dx.doi.org/10.1146/annurev.arplant.47.1.627] [PMID: 15012303]

[66] Edstam MM, Edqvist J. Involvement of GPI-anchored lipid transfer proteins in the development of seed coats and pollen in Arabidopsis thaliana. Physiol Plant 2014; 152(1): 32-42.
[http://dx.doi.org/10.1111/ppl.12156] [PMID: 24460633]

[67] Meneguetti BT, Machado LD, Oshiro KGN, Nogueira ML, Carvalho CME, Franco OL. Antimicrobial Peptides from Fruits and Their Potential Use as Biotechnological Tools-A Review and Outlook. Front Microbiol 2017; 7: 2136.
[http://dx.doi.org/10.3389/fmicb.2016.02136] [PMID: 28119671]

[68] Ribeiro SFF, Carvalho AO, Da Cunha M, *et al.* Isolation and characterization of novel peptides from chilli pepper seeds: antimicrobial activities against pathogenic yeasts. Toxicon 2007; 50(5): 600-11.
[http://dx.doi.org/10.1016/j.toxicon.2007.05.005] [PMID: 17572465]

[69]   Guillén-Chable F, Arenas-Sosa I, Islas-Flores I, Corzo G, Martinez-Liu C, Estrada G. Antibacterial activity and phospholipid recognition of the recombinant defensin J1-1 from Capsicum genus. Protein Expr Purif 2017; 136: 45-51.
[http://dx.doi.org/10.1016/j.pep.2017.06.007] [PMID: 28624494]

[70]   Taveira GB, Mathias LS, da Motta OV, *et al.* Thionin-like peptides from *Capsicum annuum* fruits with high activity against human pathogenic bacteria and yeasts. Biopolymers 2014; 102(1): 30-9.
[http://dx.doi.org/10.1002/bip.22351] [PMID: 23896704]

[71]   Diz MS, Carvalho AO, Ribeiro SFF, *et al.* Characterisation, immunolocalisation and antifungal activity of a lipid transfer protein from chili pepper (*Capsicum annuum*) seeds with novel α-amylase inhibitory properties. Physiol Plant 2011; 142(3): 233-46.
[http://dx.doi.org/10.1111/j.1399-3054.2011.01464.x] [PMID: 21382036]

[72]   Brito-Argáez L, Tamayo-Sansores JA, Madera-Piña D, *et al.* Biochemical characterization and immunolocalization studies of a Capsicum chinense Jacq. protein fraction containing DING proteins and anti-microbial activity. Plant Physiol Biochem 2016; 109: 502-14.
[http://dx.doi.org/10.1016/j.plaphy.2016.10.031] [PMID: 27835848]

[73]   Gouni-Berthold I, Berthold HK. Policosanol: clinical pharmacology and therapeutic significance of a new lipid-lowering agent. Am Heart J 2002; 143(2): 356-65.
[http://dx.doi.org/10.1067/mhj.2002.119997] [PMID: 11835043]

[74]   Wahyuni Y, Ballester A-R, Sudarmonowati E, Bino RJ, Bovy AG. Secondary metabolites of Capsicum species and their importance in the human diet. J Nat Prod 2013; 76(4): 783-93.
[http://dx.doi.org/10.1021/np300898z] [PMID: 23477482]

[75]   Sadilova E, Stintzing FC, Carle R. Anthocyanins, colour and antioxidant properties of eggplant (*Solanum melongena* L.) and violet pepper (*Capsicum annuum* L.) peel extracts. Z Natforsch C J Biosci 2006; 61(7-8): 527-35.
[http://dx.doi.org/10.1515/znc-2006-7-810] [PMID: 16989312]

[76]   Cho MC, Park DB, Yang EY, Pae DH, Won SR, Yu WK. Selection and Horticultural Characteristics Evalution of High α-Glucosidase Inhibitor in Pepper. J Bio-Environment Control 2007; 16: 233-9.

[77]   Taveira GB, Mello EO, Carvalho AO, *et al.* Antimicrobial activity and mechanism of action of a thionin-like peptide from *Capsicum annuum* fruits and combinatorial treatment with fluconazole against Fusarium solani. Biopolymers 2017; 108(3): 1-14.
[http://dx.doi.org/10.1002/bip.23008] [PMID: 28073158]

[78]   Taveira GB, Carvalho AO, Rodrigues R, Trindade FG, Da Cunha M, Gomes VM. Thionin-like peptide from *Capsicum annuum* fruits: mechanism of action and synergism with fluconazole against Candida species. BMC Microbiol 2016; 16(1): 12.
[http://dx.doi.org/10.1186/s12866-016-0626-6] [PMID: 26819228]

[79]   Vieira Bard GC, Nascimento VV, Ribeiro SFF, *et al.* Characterization of peptides from *Capsicum annuum* hybrid seeds with inhibitory activity against α-Amylase, serine proteinases and fungi. Protein J 2015; 34(2): 122-9.
[http://dx.doi.org/10.1007/s10930-015-9604-3] [PMID: 25750185]

[80]   Reilly CA, Crouch DJ, Yost GS. Quantitative analysis of capsaicinoids in fresh peppers, oleoresin capsicum and pepper spray products. J Forensic Sci 2001; 46(3): 502-9.
[http://dx.doi.org/10.1520/JFS14999J] [PMID: 11372985]

[81]   Barbero GF, Palma M, Barroso CG. Pressurized liquid extraction of capsaicinoids from peppers. J Agric Food Chem 2006; 54(9): 3231-6.
[http://dx.doi.org/10.1021/jf060021y] [PMID: 16637678]

[82]   Kozukue N, Han J-S, Kozukue E, *et al.* Analysis of eight capsaicinoids in peppers and pepper-containing foods by high-performance liquid chromatography and liquid chromatography-mass spectrometry. J Agric Food Chem 2005; 53(23): 9172-81.

[http://dx.doi.org/10.1021/jf050469j] [PMID: 16277419]

[83] Antonious GF, Jarret RL. Screening Capsicum accessions for capsaicinoids content. J Environ Sci Health B 2006; 41(5): 717-29.
[http://dx.doi.org/10.1080/03601230600701908] [PMID: 16785178]

[84] Estrada B, Bernal MA, Díaz J, Pomar F, Merino F. Capsaicinoids in vegetative organs of *Capsicum annuum* L. in relation to fruiting. J Agric Food Chem 2002; 50(5): 1188-91.
[http://dx.doi.org/10.1021/jf011270j] [PMID: 11853502]

[85] Dorantes L, Colmenero R, Hernandez H, Mota L, Jaramillo ME, Fernandez E, *et al.* Inhibition of growth of some foodborne pathogenic bacteria by Capsicum annum extracts. Int J Food Microbiol 2000; 57(1): 125-8.
[http://dx.doi.org/10.1016/S0168-1605(00)00216-6]

[86] Ohnuki K, Haramizu S, Oki K, Watanabe T, Yazawa S, Fushiki T. Administration of capsiate, a non-pungent capsaicin analog, promotes energy metabolism and suppresses body fat accumulation in mice. Biosci Biotechnol Biochem 2001; 65(12): 2735-40.
[http://dx.doi.org/10.1271/bbb.65.2735] [PMID: 11826971]

[87] Cichewicz RH, Thorpe PA. The antimicrobial properties of chile peppers (Capsicum species) and their uses in Mayan medicine. J Ethnopharmacol 1996; 52(2): 61-70.
[http://dx.doi.org/10.1016/0378-8741(96)01384-0] [PMID: 8735449]

[88] Knotkova H, Pappagallo M, Szallasi A. Capsaicin (TRPV1 Agonist) therapy for pain relief: farewell or revival? Clin J Pain 2008; 24(2): 142-54.
[http://dx.doi.org/10.1097/AJP.0b013e318158ed9e] [PMID: 18209521]

[89] Ternesten-Hasséus E, Johansson E-L, Millqvist E. Cough reduction using capsaicin. Respir Med 2015; 109(1): 27-37.
[http://dx.doi.org/10.1016/j.rmed.2014.11.001] [PMID: 25468411]

[90] Aizawa K, Inakuma T. Dietary capsanthin, the main carotenoid in paprika (*Capsicum annuum*), alters plasma high-density lipoprotein-cholesterol levels and hepatic gene expression in rats. Br J Nutr 2009; 102(12): 1760-6.
[http://dx.doi.org/10.1017/S0007114509991309] [PMID: 19646292]

[91] Choi C-H, Jung Y-K, Oh S-H. Selective induction of catalase-mediated autophagy by dihydrocapsaicin in lung cell lines. Free Radic Biol Med 2010; 49(2): 245-57.
[http://dx.doi.org/10.1016/j.freeradbiomed.2010.04.014] [PMID: 20417273]

[92] Ramírez-Romero R, Gallup JM, Sonea IM, Ackermann MR. Dihydrocapsaicin treatment depletes peptidergic nerve fibers of substance P and alters mast cell density in the respiratory tract of neonatal sheep. Regul Pept 2000; 91(1-3): 97-106.
[http://dx.doi.org/10.1016/S0167-0115(00)00124-5] [PMID: 10967206]

[93] Manirakiza P, Covaci A, Schepens P. Solid-phase extraction and gas chromatography with mass spectrometric determination of capsaicin and some of its analogues from chili peppers (Capsicum spp.). J AOAC Int 1999; 82: 1399-405.

[94] Thompson RQ. Homocapsaicin: nomenclature, indexing and identification. Flavour Fragrance J 2007; 22(4): 243-8.
[http://dx.doi.org/10.1002/ffj.1814]

[95] Higashiguchi F, Nakamura H, Hayashi H, Kometani T. Purification and structure determination of glucosides of capsaicin and dihydrocapsaicin from various Capsicum fruits. J Agric Food Chem 2006; 54(16): 5948-53.
[http://dx.doi.org/10.1021/jf0607720] [PMID: 16881699]

[96] Moon D-G, Cho K-M, Kim C-H, Seong K-C, Son D, Cho M-W, *et al.* Content of Vitamin C and Physiological Properties of Bitter Gourd Cultivars in Plastic Greenhouse. International Symposium of New Technologies for Environment Control, Energy-saving and Crop Production; 2014; pp. Plant

Factory- Greensys 20131037: 4072-12.

[97] Faraut B, Giannesini B, Matarazzo V, *et al.* Capsiate administration results in an uncoupling protein-3 downregulation, an enhanced muscle oxidative capacity and a decreased abdominal fat content *in vivo*. Int J Obes 2009; 33(12): 1348-55.
[http://dx.doi.org/10.1038/ijo.2009.182] [PMID: 19773740]

[98] Lee JH, Lee YS, Lee E-J, Lee JH, Kim T-Y. Capsiate Inhibits DNFB-Induced Atopic Dermatitis in NC/Nga Mice through Mast Cell and CD4+ T-Cell Inactivation. J Invest Dermatol 2015; 135(8): 1977-85.
[http://dx.doi.org/10.1038/jid.2015.117] [PMID: 25806854]

[99] Singh S, Jarret R, Russo V, *et al.* Determination of capsinoids by HPLC-DAD in capsicum species. J Agric Food Chem 2009; 57(9): 3452-7.
[http://dx.doi.org/10.1021/jf8040287] [PMID: 19415923]

[100] Sancho R, Lucena C, Macho A, *et al.* Immunosuppressive activity of capsaicinoids: capsiate derived from sweet peppers inhibits NF-kappaB activation and is a potent antiinflammatory compound *in vivo*. Eur J Immunol 2002; 32(6): 1753-63.
[http://dx.doi.org/10.1002/1521-4141(200206)32:6<1753::AID-IMMU1753>3.0.CO;2-2] [PMID: 12115659]

[101] Kobata K, Sutoh K, Todo T, Yazawa S, Iwai K, Watanabe T. Nordihydrocapsiate, a new capsinoid from the fruits of a nonpungent pepper, *Capsicum annuum*. J Nat Prod 1999; 62(2): 335-6.
[http://dx.doi.org/10.1021/np9803373] [PMID: 10075779]

[102] Pyun B-J, Choi S, Lee Y, *et al.* Capsiate, a nonpungent capsaicin-like compound, inhibits angiogenesis and vascular permeability *via* a direct inhibition of Src kinase activity. Cancer Res 2008; 68(1): 227-35.
[http://dx.doi.org/10.1158/0008-5472.CAN-07-2799] [PMID: 18172315]

[103] Yahara S, Kobayashi N, Izumitani Y, Nohara T. Studies on the Solanaceous Plants. Part XXIII. New Acyclic Diterpene Glycosides, Capsianosides VI, G and H from the Leaves and Stems of *Capsicum annuum* L. Chem Pharm Bull (Tokyo) 1991; 39: 3258-60.
[http://dx.doi.org/10.1248/cpb.39.3258]

[104] Hornero-Méndez D, Gómez-Ladrón De Guevara R, Mínguez-Mosquera MI. Carotenoid biosynthesis changes in five red pepper (*Capsicum annuum* L.) cultivars during ripening. Cultivar selection for breeding. J Agric Food Chem 2000; 48(9): 3857-64.
[http://dx.doi.org/10.1021/jf991020r] [PMID: 10995282]

[105] Deli J, Molnár P, Matus Z, Tóth G. Carotenoid composition in the fruits of red paprika (*Capsicum annuum* var. lycopersiciforme rubrum) during ripening; biosynthesis of carotenoids in red paprika. J Agric Food Chem 2001; 49(3): 1517-23.
[http://dx.doi.org/10.1021/jf000958d] [PMID: 11312889]

[106] Azevedo-Meleiro CH, Rodriguez-Amaya DB. Qualitative and quantitative differences in carotenoid composition among Cucurbita moschata, Cucurbita maxima, and Cucurbita pepo. J Agric Food Chem 2007; 55(10): 4027-33.
[http://dx.doi.org/10.1021/jf063413d] [PMID: 17444652]

[107] Jarén-Galán M, Nienaber U, Schwartz SJ. Paprika (*Capsicum annuum*) oleoresin extraction with supercritical carbon dioxide. J Agric Food Chem 1999; 47(9): 3558-64.
[http://dx.doi.org/10.1021/jf9900985] [PMID: 10552685]

[108] Sun T, Xu Z, Wu C-T, Janes M, Prinyawiwatkul W, No HK. Antioxidant activities of different colored sweet bell peppers (*Capsicum annuum* L.). J Food Sci 2007; 72(2): S98-S102.
[http://dx.doi.org/10.1111/j.1750-3841.2006.00245.x] [PMID: 17995862]

[109] Márkus F, Daood HG, Kapitány J, Biacs PA. Change in the carotenoid and antioxidant content of spice red pepper (paprika) as a function of ripening and some technological factors. J Agric Food Chem 1999; 47(1): 100-7.

[http://dx.doi.org/10.1021/jf980485z] [PMID: 10563856]

[110] Hallmann E, Rembiałkowska E. Characterisation of antioxidant compounds in sweet bell pepper (*Capsicum annuum* L.) under organic and conventional growing systems. J Sci Food Agric 2012; 92(12): 2409-15.
[http://dx.doi.org/10.1002/jsfa.5624] [PMID: 22368104]

[111] Matsufuji H, Nakamura H, Chino M, Takeda M. Antioxidant activity of capsanthin and the fatty acid esters in paprika (*Capsicum annuum*). J Agric Food Chem 1998; 46(9): 3468-72.
[http://dx.doi.org/10.1021/jf980200i]

[112] Benson AM, Hunkeler MJ, Talalay P. Increase of NAD(P)H:quinone reductase by dietary antioxidants: possible role in protection against carcinogenesis and toxicity. Proc Natl Acad Sci USA 1980; 77(9): 5216-20.
[http://dx.doi.org/10.1073/pnas.77.9.5216] [PMID: 6933553]

[113] Talalay P, De Long MJ, Prochaska HJ. Identification of a common chemical signal regulating the induction of enzymes that protect against chemical carcinogenesis. Proc Natl Acad Sci USA 1988; 85(21): 8261-5.
[http://dx.doi.org/10.1073/pnas.85.21.8261] [PMID: 3141925]

[114] Talalay P, Fahey JW, Holtzclaw WD, Prestera T, Zhang Y. Chemoprotection against cancer by phase 2 enzyme induction. Toxicol Lett 1995; 82-83: 173-9.
[http://dx.doi.org/10.1016/0378-4274(95)03553-2] [PMID: 8597048]

[115] Ku KM, Kang YH. Quinone reductase inductive activity of *Capsicum annuum* leaves and isolation of the active compounds. J Korean Soc Appl Biol Chem 2010; 53(6): 709-15.
[http://dx.doi.org/10.3839/jksabc.2010.107]

[116] Rodríguez-Burruezo A, González-Mas M del C, Nuez F. Carotenoid composition and vitamin A value in ají (Capsicum baccatum L.) and rocoto (C. pubescens R. & P.), 2 pepper species from the Andean region. J Food Sci 2010; 75(8): S446-53.
[http://dx.doi.org/10.1111/j.1750-3841.2010.01795.x] [PMID: 21535519]

[117] Hornero-Méndez D, Mínguez-Mosquera MI. Xanthophyll esterification accompanying carotenoid overaccumulation in chromoplast of *Capsicum annuum* ripening fruits is a constitutive process and useful for ripeness index. J Agric Food Chem 2000; 48(5): 1617-22.
[http://dx.doi.org/10.1021/jf9912046] [PMID: 10820068]

[118] Maoka T, Mochida K, Kozuka M, *et al.* Cancer chemopreventive activity of carotenoids in the fruits of red paprika *Capsicum annuum* L. Cancer Lett 2001; 172(2): 103-9.
[http://dx.doi.org/10.1016/S0304-3835(01)00635-8] [PMID: 11566483]

[119] Maoka T, Fujiwara Y, Hashimoto K, Akimoto N. Isolation of a series of apocarotenoids from the fruits of the red paprika *Capsicum annuum* L. J Agric Food Chem 2001; 49(3): 1601-6.
[http://dx.doi.org/10.1021/jf0013149] [PMID: 11312902]

[120] Murakami A, Nakashima M, Koshiba T, *et al.* Modifying effects of carotenoids on superoxide and nitric oxide generation from stimulated leukocytes. Cancer Lett 2000; 149(1-2): 115-23.
[http://dx.doi.org/10.1016/S0304-3835(99)00351-1] [PMID: 10737715]

[121] Maoka T, Goto Y, Isobe K, Fujiwara Y, Hashimoto K, Mochida K. Antioxidative activity of capsorubin and related compounds from paprika (*Capsicum annuum*). J Oleo Sci 2001; 50(8): 663-5.
[http://dx.doi.org/10.5650/jos.50.663]

[122] Deli J, Tóth G. Carotenoid composition of the fruits of *Capsicum annuum* Cv. Bovet 4 during ripening. Z Lebensm Unters F A 1997; 205(5): 388-91.
[http://dx.doi.org/10.1007/s002170050186]

[123] Materska M, Perucka I. Antioxidant activity of the main phenolic compounds isolated from hot pepper fruit (*Capsicum annuum* L). J Agric Food Chem 2005; 53(5): 1750-6.
[http://dx.doi.org/10.1021/jf035331k] [PMID: 15740069]

[124] Choi J-G, Hur J-M, Cho H-W, Park J-C. Phenolic compounds from *Capsicum annuum* leaves showing radical scavenging effect. Korean J Pharmacogn 2007; 38: 258-62.

[125] Belza A, Frandsen E, Kondrup J. Body fat loss achieved by stimulation of thermogenesis by a combination of bioactive food ingredients: a placebo-controlled, double-blind 8-week intervention in obese subjects. Int J Obes 2007; 31(1): 121-30.
[http://dx.doi.org/10.1038/sj.ijo.0803351] [PMID: 16652130]

[126] Materska M, Perucka I, Stochmal A, Piacente S, Oleszek W. Quantitative and qualitative determination of flavonoids and phenolic acid derivatives from pericarp of hot pepper fruit cv Bronowicka Ostra Pol J Food Nutr Sci 2003; 12/53(SI2): 72-6.

[127] Lee Y, Howard L, Villalón B. Flavonoids and Antioxidant Activity of Fresh Pepper (*Capsicum annuum*) Cultivars. J Food Sci 2006; 60: 473-6.
[http://dx.doi.org/10.1111/j.1365-2621.1995.tb09806.x]

[128] Nascimento PL, Nascimento TC, Ramos NS, *et al.* Quantification, antioxidant and antimicrobial activity of phenolics isolated from different extracts of Capsicum frutescens (Pimenta Malagueta). Molecules 2014; 19(4): 5434-47.
[http://dx.doi.org/10.3390/molecules19045434] [PMID: 24879587]

[129] Acero-Ortega C, Dorantes L, Hernandez H, Gutiérrez-López G, Aparicio G, Jaramillo-Flores M. Evaluation of Phenylpropanoids in Ten *Capsicum annuum* L. Varieties and Their Inhibitory Effects on Listeria monocytogenes Murray, Webb and Swann Scott A. Food Sci Technol Int 2005; 11(1): 5-10.
[http://dx.doi.org/10.1177/1082013205050902]

[130] Chen C-Y, Yeh Y-T, Yang W-L. Amides from the stem of *Capsicum annuum*. Nat Prod Commun 2011; 6(2): 227-9.
[PMID: 21425680]

[131] Sung J, Lee J. Capsicoside G, a furostanol saponin from pepper (*Capsicum annuum* L.) seeds, suppresses adipogenesis through activation of AMP-activated protein kinase in 3T3-L1 cells. J Funct Foods 2016; 20: 148-58.
[http://dx.doi.org/10.1016/j.jff.2015.10.024]

[132] Antonious GF, Kochhar TS, Jarret RL, Snyder JC. Antioxidants in hot pepper: variation among accessions. J Environ Sci Health B 2006; 41(7): 1237-43.
[http://dx.doi.org/10.1080/03601230600857114] [PMID: 16923603]

[133] Lee DY, Lee DG, Cho JG, *et al.* Lignans from the fruits of the red pepper (*Capsicum annuum* L.) and their antioxidant effects. Arch Pharm Res 2009; 32(10): 1345-9.
[http://dx.doi.org/10.1007/s12272-009-2001-8] [PMID: 19898795]

[134] Patowary P, Pathak MP, Zaman K, Raju PS, Chattopadhyay P. Research progress of capsaicin responses to various pharmacological challenges. Biomed Pharmacother 2017; 96: 1501-12.
[http://dx.doi.org/10.1016/j.biopha.2017.11.124] [PMID: 29198921]

[135] Lu M, Ho C-T, Huang Q. Extraction, bioavailability, and bioefficacy of capsaicinoids. Yao Wu Shi Pin Fen Xi 2017; 25(1): 27-36.
[http://dx.doi.org/10.1016/j.jfda.2016.10.023] [PMID: 28911540]

[136] Loizzo MR, Bonesi M, Serio A, Chaves-López C, Falco T, Paparella A, *et al.* Application of nine air-dried *Capsicum annum* cultivars as food preservative: Micronutrient content, antioxidant activity, and foodborne pathogens inhibitory effects. Int J Food Prop 2017; 20(4): 899-910.
[http://dx.doi.org/10.1080/10942912.2016.1188310]

[137] Oboh G, Puntel RL, Rocha JBT. Hot pepper (*Capsicum annuum*, Tepin and Capsicum chinese, Habanero) prevents Fe^{2+}-induced lipid peroxidation in brain – *in vitro*. Food Chem 2007; 102(1): 178-85.
[http://dx.doi.org/10.1016/j.foodchem.2006.05.048]

[138] Kappel VD, Costa GM, Scola G, *et al.* Phenolic content and antioxidant and antimicrobial properties

of fruits of Capsicum baccatum L. var. pendulum at different maturity stages. J Med Food 2008; 11(2): 267-74.
[http://dx.doi.org/10.1089/jmf.2007.626] [PMID: 18598168]

[139] Zhuang Y, Chen L, Sun L, Cao J. Bioactive characteristics and antioxidant activities of nine peppers. J Funct Foods 2012; 4(1): 331-8.
[http://dx.doi.org/10.1016/j.jff.2012.01.001]

[140] Oboh G, Ogunruku OO. Cyclophosphamide-induced oxidative stress in brain: protective effect of hot short pepper (Capsicum frutescens L. var. abbreviatum). Exp Toxicol Pathol 2010; 62(3): 227-33.
[http://dx.doi.org/10.1016/j.etp.2009.03.011] [PMID: 19447589]

[141] Liu Y, Nair MG. Capsaicinoids in the hottest pepper Bhut Jolokia and its antioxidant and antiinflammatory activities. Nat Prod Commun 2010; 5(1): 91-4.
[PMID: 20184029]

[142] Oboh G, Rocha JBT. Distribution and antioxidant activity of polyphenols in ripe and unripe tree pepper (Capsicum pubescens). J Food Biochem 2007; 31(4): 456-73.

[143] Oboh G, Rocha JBT. Water extractable phytochemicals from Capsicum pubescens (tree pepper) inhibit lipid peroxidation induced by different pro-oxidant agents in brain: *in vitro*. Eur Food Res Technol 2008; 226(4): 707-13.
[http://dx.doi.org/10.1007/s00217-007-0580-5]

[144] Allemand A, Leonardi BF, Zimmer AR, Moreno S, Romão PRT, Gosmann G. Red Pepper (*Capsicum baccatum*) Extracts Present Anti-Inflammatory Effects *in vitro* and Inhibit the Production of TNF-α and NO *in vitro*. J Med Food 2016; 19(8): 759-67.
[http://dx.doi.org/10.1089/jmf.2015.0101] [PMID: 27533650]

[145] Jolayemi AT, Ojewole JA. Comparative anti-inflammatory properties of Capsaicin and ethyl-aAcetate extract of *Capsicum frutescens* linn [Solanaceae] in rats. Afr Health Sci 2013; 13(2): 357-61.
[PMID: 24235936]

[146] Gilardini Montani MS, D'Eliseo D, Cirone M, *et al.* Capsaicin-mediated apoptosis of human bladder cancer cells activates dendritic cells *via* CD91. Nutrition 2015; 31(4): 578-81.
[http://dx.doi.org/10.1016/j.nut.2014.05.005] [PMID: 25220876]

[147] Mori A, Lehmann S, O'Kelly J, *et al.* Capsaicin, a component of red peppers, inhibits the growth of androgen-independent, p53 mutant prostate cancer cells. Cancer Res 2006; 66(6): 3222-9.
[http://dx.doi.org/10.1158/0008-5472.CAN-05-0087] [PMID: 16540674]

[148] Chapa-Oliver AM, Mejía-Teniente L. Capsaicin: From Plants to a Cancer-Suppressing Agent. Molecules 2016; 21(8): 931.
[http://dx.doi.org/10.3390/molecules21080931] [PMID: 27472308]

[149] Ziglioli F, Frattini A, Maestroni U, Dinale F, Ciufifeda M, Cortellini P. Vanilloid-mediated apoptosis in prostate cancer cells through a TRPV-1 dependent and a TRPV-1-independent mechanism. Acta Biomed 2009; 80(1): 13-20.
[PMID: 19705615]

[150] Oh SH, Kim YS, Lim SC, Hou YF, Chang IY, You HJ. Dihydrocapsaicin (DHC), a saturated structural analog of capsaicin, induces autophagy in human cancer cells in a catalase-regulated manner. Autophagy 2008; 4(8): 1009-19.
[http://dx.doi.org/10.4161/auto.6886] [PMID: 18818525]

[151] Báez S, Tsuchiya Y, Calvo A, *et al.* Genetic variants involved in gallstone formation and capsaicin metabolism, and the risk of gallbladder cancer in Chilean women. World J Gastroenterol 2010; 16(3): 372-8.
[http://dx.doi.org/10.3748/wjg.v16.i3.372] [PMID: 20082485]

[152] Nishino A, Yasui H, Maoka T. Reaction of Paprika Carotenoids, Capsanthin and Capsorubin, with Reactive Oxygen Species. J Agric Food Chem 2016; 64(23): 4786-92.

[http://dx.doi.org/10.1021/acs.jafc.6b01706] [PMID: 27229653]

[153] Brito-Argáez L, Tamayo-Sansores JA, Madera-Piña D, *et al.* Biochemical characterization and immunolocalization studies of a Capsicum chinense Jacq. protein fraction containing DING proteins and anti-microbial activity. Plant Physiol Biochem 2016; 109: 502-14.
[http://dx.doi.org/10.1016/j.plaphy.2016.10.031] [PMID: 27835848]

[154] Bacon K, Boyer R, Denbow C, O'Keefe S, Neilson A, Williams R. Evaluation of different solvents to extract antibacterial compounds from jalapeño peppers. Food Sci Nutr 2016; 5(3): 497-503.
[http://dx.doi.org/10.1002/fsn3.423] [PMID: 28572934]

[155] Mokhtar M, Ginestra G, Youcefi F, Filocamo A, Bisignano C, Riazi A. Antimicrobial Activity of Selected Polyphenols and Capsaicinoids Identified in Pepper (*Capsicum annuum* L.) and Their Possible Mode of Interaction. Curr Microbiol 2017; 74(11): 1253-60.
[http://dx.doi.org/10.1007/s00284-017-1310-2] [PMID: 28721659]

[156] Acero Ortega C, Dorantes Alvarez L, Jaramillo Flores ME, Hernández Sánchez H, López Malo A. Effect of Chili (*Capsicum annuum* L.) extracts and derived compounds on growth of Erwinia carotovora subsp. carotovora (Jones) Bergey, Harrison, Breed, Hammer and Huntoon. Rev Mex Fitopatol 2003; 21(2): 233-7.

[157] Ngai PHK, Ng TB. A lectin with antifungal and mitogenic activities from red cluster pepper (Capsicum frutescens) seeds. Appl Microbiol Biotechnol 2007; 74(2): 366-71.
[http://dx.doi.org/10.1007/s00253-006-0685-y] [PMID: 17082928]

[158] Koffi-Nevry R, Kouassi KC, Nanga ZY, Koussémon M, Loukou GY. Antibacterial Activity of Two Bell Pepper Extracts: *Capsicum annuum* L. and *Capsicum frutescens*. Int J Food Prop 2012; 15(5): 961-71.
[http://dx.doi.org/10.1080/10942912.2010.509896]

[159] Vinayaka K, Nandini K, Rakshitha M, Ramya M, Shruthi J. Proximate Composition, Antibacterial and Anthelmintic Activity of Capsicum frutescens (L.) Var. Longa (Solanaceae) Leaves. Pharmacogn J 2010; 2(12): 486-91.
[http://dx.doi.org/10.1016/S0975-3575(10)80036-7]

[160] De Lucca AJ, Boue S, Palmgren MS, Maskos K, Cleveland TE. Fungicidal properties of two saponins from *Capsicum frutescens* and the relationship of structure and fungicidal activity. Can J Microbiol 2006; 52(4): 336-42.
[http://dx.doi.org/10.1139/w05-137] [PMID: 16699584]

[161] Meyer B, Houlné G, Pozueta-Romero J, Schantz M-L, Schantz R. Fruit-specific expression of a defensin-type gene family in bell pepper. Upregulation during ripening and upon wounding. Plant Physiol 1996; 112(2): 615-22.
[http://dx.doi.org/10.1104/pp.112.2.615] [PMID: 8883377]

[162] Seo H-H, Park S, Park S, *et al.* Overexpression of a defensin enhances resistance to a fruit-specific anthracnose fungus in pepper. PLoS One 2014; 9(5): e97936.
[http://dx.doi.org/10.1371/journal.pone.0097936] [PMID: 24848280]

[163] Diz MSS, Carvalho AO, Rodrigues R, *et al.* Antimicrobial peptides from chili pepper seeds causes yeast plasma membrane permeabilization and inhibits the acidification of the medium by yeast cells. Biochim Biophys Acta 2006; 1760(9): 1323-32.
[http://dx.doi.org/10.1016/j.bbagen.2006.04.010] [PMID: 16784815]

[164] Teixeira FR, Lima M, Almeida HO, Romeiro RS, Silva DJH, Pereira PRG, *et al.* Bioprospection of cationic and anionic antimicrobial peptides from bell pepper leaves for inhibition of Ralstonia solanacearum and Clavibacter michiganensis ssp. michiganensis growth. J Phytopathol 2006; 154(7-8): 418-21.
[http://dx.doi.org/10.1111/j.1439-0434.2006.01119.x]

[165] Cruz LP, Ribeiro SFF, Carvalho AO, *et al.* Isolation and partial characterization of a novel lipid transfer protein (LTP) and antifungal activity of peptides from chilli pepper seeds. Protein Pept Lett

2010; 17(3): 311-8.
[http://dx.doi.org/10.2174/092986610790780305] [PMID: 19508213]

[166] Fernando M-S, Ligia B-A, Mayra D-B, Ignacio I-F. A review of a promising therapeutic and agronomical alternative: Antimicrobial peptides from Capsicum sp. Afr J Biotechnol 2011; 10(86): 19918-28.

[167] Hong Q, Xia C, Xiangying H, Quan Y. Capsinoids suppress fat accumulation *via* lipid metabolism. Mol Med Rep 2015; 11(3): 1669-74.
[http://dx.doi.org/10.3892/mmr.2014.2996] [PMID: 25421144]

[168] Saito M, Yoneshiro T. Capsinoids and related food ingredients activating brown fat thermogenesis and reducing body fat in humans. Curr Opin Lipidol 2013; 24(1): 71-7.
[http://dx.doi.org/10.1097/MOL.0b013e32835a4f40] [PMID: 23298960]

[169] Maeda H, Saito S, Nakamura N, Maoka T. Paprika Pigments Attenuate Obesity-Induced Inflammation in 3T3-L1 Adipocytes. ISRN Inflamm 2013; Article ID 763758, 9 pages.

[170] Gerner P, Binshtok AM, Wang C-F, *et al.* Capsaicin combined with local anesthetics preferentially prolongs sensory/nociceptive block in rat sciatic nerve. Anesthesiology 2008; 109(5): 872-8.
[http://dx.doi.org/10.1097/ALN.0b013e31818958f7] [PMID: 18946300]

[171] Mózsik G, Szolcsányi J, Dömötör A. Capsaicin research as a new tool to approach of the human gastrointestinal physiology, pathology and pharmacology. Inflammopharmacology 2007; 15(6): 232-45.
[http://dx.doi.org/10.1007/s10787-007-1584-2] [PMID: 18236014]

[172] Wang L, Hu C-P, Deng P-Y, *et al.* The protective effects of rutaecarpine on gastric mucosa injury in rats. Planta Med 2005; 71(5): 416-9.
[http://dx.doi.org/10.1055/s-2005-864135] [PMID: 15931578]

[173] Nishihara K, Nozawa Y, Nakano M, Ajioka H, Matsuura N. Sensitizing effects of lafutidine on CGRP-containing afferent nerves in the rat stomach. Br J Pharmacol 2002; 135(6): 1487-94.
[http://dx.doi.org/10.1038/sj.bjp.0704596] [PMID: 11906962]

[174] Reuter J, Wölfle U, Weckesser S, Schempp C. Which plant for which skin disease? Part 1: Atopic dermatitis, psoriasis, acne, condyloma and herpes simplex. J Dtsch Dermatol Ges 2010; 8(10): 788-96.
[PMID: 20707875]

[175] Samuels J. John. Biodiversity of Food Species of the Solanaceae Family: A Preliminary Taxonomic Inventory of Subfamily Solanoideae. Resources 2015; 4(2): 277-322.
[http://dx.doi.org/10.3390/resources4020277]

[176] Zamora-Tavares P, Vargas-Ponce O, Sánchez-Martínez J, Cabrera-Toledo D. Diversity and genetic structure of the husk tomato (Physalis philadelphica Lam.) in Western Mexico. Genet Resour Crop Evol 2015; 62(1): 141-53.
[http://dx.doi.org/10.1007/s10722-014-0163-9]

[177] Morton JF. Fruits of Warm Climates. Miami, FL, USA: Michigan Universtity 1987; p. 505.

[178] Pérez-Castorena A-L, Oropeza RF, Vázquez AR, Martínez M, Maldonado E. Labdanes and withanolides from Physalis coztomatl. J Nat Prod 2006; 69(7): 1029-33.
[http://dx.doi.org/10.1021/np0601354] [PMID: 16872139]

[179] Zhang WN, Tong WY. Chemical constituents and biological activities of plants from the genus physalis. Chem Biodivers 2016; 13(1): 48-65.
[http://dx.doi.org/10.1002/cbdv.201400435] [PMID: 26765352]

[180] Carniel N, Dallago RM, Dariva C, Bender JP, Nunes AL, Zanella O, *et al.* Microwave-Assisted Extraction of Phenolic Acids and Flavonoids from Physalis angulata. J Food Ing 2017; 40(e12433): 1-11.

[181] Subramanian SS, Sethi PD. Steroidal lactones of Physalis ixocarpa leaves. Indian J Pharm 1973.

[182] Kirson I, Cohen A, Greenberg M, Gottlieb HE, Glotter E, Varenne P, *et al.* Isocarpalactones A and B, two unusual naturally occuring steroids of the ergostane type. ChemInform 1979; 10(29)

[183] Abdullaev ND, Vasina OE, Maslennikova VA, Abubakirov NK. Withasteroids of Physalis VI. 1 H and 13 C NMR spectra of withasteroids ixocarpalactone A and ixocarpanolide. Chem Nat Compd 1986; 22(3): 300-5.
[http://dx.doi.org/10.1007/BF00598301]

[184] Kennelly EJ, Gerhäuser C, Song LL, Graham JG, Beecher CWW, Pezzuto JM, *et al.* Induction of quinone reductase by withanolides isolated from Physalis philadelphica (tomatillos). J Agric Food Chem 1997; 45(10): 3771-7.
[http://dx.doi.org/10.1021/jf970246w]

[185] Maldonado E, Pérez-Castorena AL, Garcés C, Martínez M. Philadelphicalactones C and D and other cytotoxic compounds from Physalis philadelphica. Steroids 2011; 76(7): 724-8.
[http://dx.doi.org/10.1016/j.steroids.2011.03.018] [PMID: 21497618]

[186] Bock MA, Sanchez-Pilcher J, McKee LJ, Ortiz M. Selected nutritional and quality analyses of tomatillos (Physalis ixocarpa). Plant Foods Hum Nutr 1995; 48(2): 127-33.
[http://dx.doi.org/10.1007/BF01088308] [PMID: 8837871]

[187] Morales-Contreras BE, Rosas-Flores W, Contreras-Esquivel JC, Wicker L, Morales-Castro J. Pectin from Husk Tomato (Physalis ixocarpa Brot.): Rheological behavior at different extraction conditions. Carbohydr Polym 2018; 179: 282-9.
[http://dx.doi.org/10.1016/j.carbpol.2017.09.097] [PMID: 29111053]

[188] Su B-N, Misico R, Park EJ, Santarsiero BD, Mesecar AD, Fong HHS, *et al.* Isolation and characterization of bioactive principles of the leaves and stems of Physalis philadelphica. Tetrahedron 2002; 58(17): 3453-66.
[http://dx.doi.org/10.1016/S0040-4020(02)00277-6]

[189] Chen L-X, He H, Qiu F. Natural withanolides: an overview. Nat Prod Rep 2011; 28(4): 705-40.
[http://dx.doi.org/10.1039/c0np00045k] [PMID: 21344104]

[190] Men R-Z, Li N, Ding W-J, Hu Z-J, Ma Z-J, Cheng L. Unprecedent aminophysalin from Physalis angulata. Steroids 2014; 88: 60-5.
[http://dx.doi.org/10.1016/j.steroids.2014.06.016] [PMID: 24973634]

[191] Chiang H-C, Jaw SM, Chen PM. Inhibitory effects of physalin B and physalin F on various human leukemia cells *in vitro*. Anticancer Res 1992; 12(4): 1155-62.
[PMID: 1503404]

[192] Lin Y-S, Chiang H-C, Kan W-S, Hone E, Shih S-J, Won M-H. Immunomodulatory activity of various fractions derived from Physalis angulata L extract Am J Chin Med 1992; 20(03n04): 233-43.
[http://dx.doi.org/10.1142/S0192415X92000242]

[193] Olivares-Tenorio M-L, Dekker M, Verkerk R, van Boekel MAJS. Health-promoting compounds in cape gooseberry (Physalis peruviana L.): Review from a supply chain perspective. Trends Food Sci Technol 2016; 57(Part A): 83–92.

[194] Januário AH, Filho ER, Pietro RCLR, Kashima S, Sato DN, França SC. Antimycobacterial physalins from *Physalis angulata* L. (Solanaceae). Phytother Res 2002; 16(5): 445-8.
[http://dx.doi.org/10.1002/ptr.939] [PMID: 12203265]

[195] Sun C-P, Qiu C-Y, Zhao F, Kang N, Chen L-X, Qiu F. Physalins V-IX, 16,24-cyclo-13,14-seco withanolides from Physalis angulata and their antiproliferative and anti-inflammatory activities. Sci Rep 2017; 7(1): 4057.
[http://dx.doi.org/10.1038/s41598-017-03849-9] [PMID: 28642618]

[196] Kuo P-C, Kuo T-H, Damu AG, *et al.* Physanolide A, a novel skeleton steroid, and other cytotoxic principles from Physalis angulata. Org Lett 2006; 8(14): 2953-6.
[http://dx.doi.org/10.1021/ol060801s] [PMID: 16805525]

[197] Glotter E. Withanolides and related ergostane-type steroids. Nat Prod Rep 1991; 8(4): 415-40.
[http://dx.doi.org/10.1039/np9910800415] [PMID: 1787922]

[198] Cobaleda-Velasco M, Almaraz-Abarca N, Elizabeth Alanis-Banuelos R, Natividad Uribe-Soto J, Silvia Gonzalez-Valdez L, Munoz-Hernandez G, *et al.* Rapid Determination of Phenolics, Flavonoids, and Antioxidant Properties of Physalis ixocarpa Brot. ex Hornem. and Physalis angulata L. by Infrared Spectroscopy and Partial Least Squares. Anal Lett 2018; 51(4): 523-36.
[http://dx.doi.org/10.1080/00032719.2017.1331238]

[199] Cobaleda-Velasco M, Alanis-Banuelos RE, Almaraz-Abarca N, Rojas-Lopez M, Gonzalez-Valdez LS, Avila-Reyes JA, *et al.* Phenolic profiles and antioxidant properties of Physalis angulata L. as quality indicators. J Pharm Pharmacogn Res 2017; 5(2): 114-28.

[200] González-Mendoza D, Grimaldo-Juárez O, Soto-Ortiz R, Escoboza-Garcia F, Hernández JFS. Evaluation of total phenolics, anthocyanins and antioxidant capacity in purple tomatillo (Physalis ixocarpa) genotypes. Afr J Biotechnol 2010; 9(32): 5173-6.

[201] Rengifo-Salgado E, Vargas-Arana G. Physalis angulata L.(Bolsa Mullaca): a review of its traditional uses, chemistry and pharmacology. B Latinoam Caribe 2013; 12(5): 431-45.

[202] Sá MS, de Menezes MN, Krettli AU, *et al.* Antimalarial activity of physalins B, D, F, and G. J Nat Prod 2011; 74(10): 2269-72.
[http://dx.doi.org/10.1021/np200260f] [PMID: 21954931]

[203] Vieira AT, Pinho V, Lepsch LB, *et al.* Mechanisms of the anti-inflammatory effects of the natural secosteroids physalins in a model of intestinal ischaemia and reperfusion injury. Br J Pharmacol 2005; 146(2): 244-51.
[http://dx.doi.org/10.1038/sj.bjp.0706321] [PMID: 16025143]

[204] Guimarães ET, Lima MS, Santos LA, Ribeiro IM, Tomassini TBC, dos Santos RR, *et al.* Effects of seco-steroids purified from Physalis angulata L., Solanaceae, on the viability of Leishmania sp. Rev Bras Farmacogn 2010; 20(6): 945-9.
[http://dx.doi.org/10.1590/S0102-695X2010005000036]

[205] Magalhães HIF, Veras ML, Torres MR, *et al. In-vitro* and *in-vivo* antitumour activity of physalins B and D from Physalis angulata. J Pharm Pharmacol 2006; 58(2): 235-41.
[http://dx.doi.org/10.1211/jpp.58.2.0011] [PMID: 16451752]

[206] Januário AH, Filho ER, Pietro RC, Kashima S, Sato DN, França SC. Antimycobacterial physalins from Physalis angulata L. (Solanaceae). Phytother Res 2002; 16(5): 445-8.
[http://dx.doi.org/10.1002/ptr.939] [PMID: 12203265]

[207] Helvaci S, Kökdil G, Kawai M, Duran N, Duran G, Güvenç A. Antimicrobial activity of the extracts and physalin D from Physalis alkekengi and evaluation of antioxidant potential of physalin D. Pharm Biol 2010; 48(2): 142-50.
[http://dx.doi.org/10.3109/13880200903062606] [PMID: 20645830]

[208] Pinto NB, Morais TC, Carvalho KMB, *et al.* Topical anti-inflammatory potential of Physalin E from Physalis angulata on experimental dermatitis in mice. Phytomedicine 2010; 17(10): 740-3.
[http://dx.doi.org/10.1016/j.phymed.2010.01.006] [PMID: 20149612]

[209] Wu SY, Leu YL, Wu TS, Kuo PC, Liao YR, *et al.* Physalin F induces cell apoptosis in human renal carcinoma cells by targeting NF-kappaB and generating reactive oxygen species. PLoS One 2012; 7(7): e40727.

[210] Brustolim D, Vasconcelos JF, Freitas LA, *et al.* Activity of physalin F in a collagen-induced arthritis model. J Nat Prod 2010; 73(8): 1323-6.
[http://dx.doi.org/10.1021/np900691w] [PMID: 20681573]

[211] Soares MBP, Bellintani MC, Ribeiro IM, Tomassini TCB, Ribeiro dos Santos R. Inhibition of macrophage activation and lipopolysaccharide-induced death by seco-steroids purified from Physalis angulata L. Eur J Pharmacol 2003; 459(1): 107-12.

[http://dx.doi.org/10.1016/S0014-2999(02)02829-7] [PMID: 12505539]

[212] Pinto LA, Meira CS, Villarreal CF, *et al.* Physalin F, a seco-steroid from Physalis angulata L., has immunosuppressive activity in peripheral blood mononuclear cells from patients with HTLVl-associated myelopathy. Biomed Pharmacother 2016; 79: 129-34.
[http://dx.doi.org/10.1016/j.biopha.2016.01.041] [PMID: 27044821]

[213] Sun C-P, Oppong MB, Zhao F, Chen L-X, Qiu F. Unprecedented 22, 26-seco physalins from Physalis angulata and their anti-inflammatory potential. Org Biomol Chem 2017; 15(41): 8700-4.

[214] He Q, Ma L, Luo J, He F, Lou L, Hu L. Cytotoxic withanolides from Physalis angulata L. Chem Biodivers 2007; 4(3): 443-9.
[PMID: 17372946]

[215] Nagafuji S, Okabe H, Akahane H, Abe F. Trypanocidal constituents in plants 4. Withanolides from the aerial parts of Physalis angulata. Biol Pharm Bull 2004; 27(2): 193-7.
[http://dx.doi.org/10.1248/bpb.27.193] [PMID: 14758032]

[216] Sun L, Liu J, Cui D, *et al.* Anti-inflammatory function of Withangulatin A by targeted inhibiting COX-2 expression *via* MAPK and NF-kappaB pathways. J Cell Biochem 2010; 109(3): 532-41.
[PMID: 19950196]

[217] Damu AG, Kuo P-C, Su C-R, *et al.* Isolation, structures, and structure - cytotoxic activity relationships of withanolides and physalins from Physalis angulata. J Nat Prod 2007; 70(7): 1146-52.
[http://dx.doi.org/10.1021/np0701374] [PMID: 17580910]

[218] Lee S-W, Pan M-H, Chen C-M, Chen Z-T. Withangulatin I, a new cytotoxic withanolide from Physalis angulata. Chem Pharm Bull (Tokyo) 2008; 56(2): 234-6.
[http://dx.doi.org/10.1248/cpb.56.234] [PMID: 18239318]

[219] Jin Z, Mashuta MS, Stolowich NJ, *et al.* Physangulidines A, B, and C: three new antiproliferative withanolides from Physalis angulata L. Org Lett 2012; 14(5): 1230-3.
[http://dx.doi.org/10.1021/ol203498a] [PMID: 22329497]

[220] Maldonado E, Hurtado NE, Pérez-Castorena AL, Martínez M. Cytotoxic 20,24-epoxywithanolides from Physalis angulata. Steroids 2015; 104: 72-8.
[http://dx.doi.org/10.1016/j.steroids.2015.08.015] [PMID: 26335153]

[221] Yu P, Zhang C, Gao C-Y, *et al.* Anti-proliferation of triple-negative breast cancer cells with physagulide P: ROS/JNK signaling pathway induces apoptosis and autophagic cell death. Oncotarget 2017; 8(38): 64032-49.
[http://dx.doi.org/10.18632/oncotarget.19299] [PMID: 28969050]

[222] Sun C-P, Kutateladze AG, Zhao F, Chen L-X, Qiu F. A novel withanolide with an unprecedented carbon skeleton from Physalis angulata. Org Biomol Chem 2017; 15(5): 1110-4.
[http://dx.doi.org/10.1039/C6OB02656G] [PMID: 28098317]

[223] Choi JK, Murillo G, Su BN, Pezzuto JM, Kinghorn AD, Mehta RG. Ixocarpalactone A isolated from the Mexican tomatillo shows potent antiproliferative and apoptotic activity in colon cancer cells. FEBS J 2006; 273(24): 5714-23.
[http://dx.doi.org/10.1111/j.1742-4658.2006.05560.x] [PMID: 17212786]

[224] Guan Y-Z, Shan S-M, Zhang W, Luo J-G, Kong L-Y. Withanolides from Physalis minima and their inhibitory effects on nitric oxide production. Steroids 2014; 82: 38-43.
[http://dx.doi.org/10.1016/j.steroids.2014.01.004] [PMID: 24480102]

[225] Sun C-P, Nie X-F, Kang N, Zhao F, Chen L-X, Qiu F. A new phenol glycoside from Physalis angulata. Nat Prod Res 2017; 31(9): 1059-65.
[http://dx.doi.org/10.1080/14786419.2016.1269102] [PMID: 28033720]

[226] Zhang C-R, Khan W, Bakht J, Nair MG. New antiinflammatory sucrose esters in the natural sticky coating of tomatillo (Physalis philadelphica), an important culinary fruit. Food Chem 2016; 196: 726-32.

[http://dx.doi.org/10.1016/j.foodchem.2015.10.007] [PMID: 26593547]

[227]   Sun C-P, Yuan T, Wang L, Kang N, Zhao F, Chen L-X, *et al.* Anti-inflammatory labdane-type diterpenoids from Physalis angulata. RSC Advances 2016; 6(80): 76838-47.
[http://dx.doi.org/10.1039/C6RA16424B]

[228]   Ismail N, Alam M. A novel cytotoxic flavonoid glycoside from Physalis angulata. Fitoterapia 2001; 72(6): 676-9.
[http://dx.doi.org/10.1016/S0367-326X(01)00281-7] [PMID: 11543968]

[229]   Rathore C, Dutt KR, Sahu S, Deb L. Antiasthmatic activity of the methanolic extract of Physalis angulata Linn. J Med Plants Res 2011; 5(22): 5351-5.

[230]   Jang YS, Wang Z, Lee J-M, Lee J-Y, Lim SS. Screening of Korean Natural Products for Anti-Adipogenesis Properties and Isolation of Kaempferol-3-O-rutinoside as a Potent Anti-Adipogenetic Compound from Solidago virgaurea. Molecules 2016; 21(2): E226.
[http://dx.doi.org/10.3390/molecules21020226] [PMID: 26901177]

[231]   Khan W, Bakht J, Shafi M. Antimicrobial potentials of different solvent extracted samples from Physalis ixocarpa. Pak J Pharm Sci 2016; 29(2): 467-75.
[PMID: 27087074]

[232]   Khan MA, Khan H, Khan S, Mahmood T, Khan PM, Jabar A. Anti-inflammatory, analgesic and antipyretic activities of Physalis minima Linn. J Enzyme Inhib Med Chem 2009; 24(3): 632-7.
[http://dx.doi.org/10.1080/14756360802321120] [PMID: 18825533]

[233]   Yang Y-J, Yi L, Wang Q, Xie B-B, Dong Y, Sha C-W. Anti-inflammatory effects of physalin E from *Physalis angulata* on lipopolysaccharide-stimulated RAW 264.7 cells through inhibition of NF-κB pathway. Immunopharmacol Immunotoxicol 2017; 39(2): 74-9.
[http://dx.doi.org/10.1080/08923973.2017.1282514] [PMID: 28152630]

[234]   Sun C-P, Qiu C-Y, Yuan T, *et al.* Antiproliferative and Anti-inflammatory Withanolides from Physalis angulata. J Nat Prod 2016; 79(6): 1586-97.
[http://dx.doi.org/10.1021/acs.jnatprod.6b00094] [PMID: 27295506]

[235]   Fan J, Liu X, Zheng X, Zhao H, Xia H, Sun Y. A Novel Cytotoxic Physalin from Physalis angulata. Nat Prod Commun 2017; 12(10): 1589-91.

[236]   Chiang H-C, Jaw SM, Chen CF, Kan WS. Antitumor agent, physalin F from Physalis angulata L. Anticancer Res 1992; 12(3): 837-43.
[PMID: 1622143]

[237]   Juang J-K, Huang HW, Chen C-M, Liu HJ. A new compound, withangulatin A, promotes type II DNA topoisomerase-mediated DNA damage. Biochem Biophys Res Commun 1989; 159(3): 1128-34.
[http://dx.doi.org/10.1016/0006-291X(89)92226-2] [PMID: 2539141]

[238]   Lee WC, Lin KY, Chen CM, Chen ZT, Liu HJ, Lai YK. Induction of heat-shock response and alterations of protein phosphorylation by a novel topoisomerase II inhibitor, withangulatin A, in 9L rat brain tumor cells. J Cell Physiol 1991; 149(1): 66-76.
[http://dx.doi.org/10.1002/jcp.1041490110] [PMID: 1658010]

[239]   Gao C, Li R, Zhou M, Yang Y, Kong L, Luo J. Cytotoxic withanolides from *Physalis angulata.* Nat Prod Res 2018; 32(6): 676-81.
[http://dx.doi.org/10.1080/14786419.2017.1338281] [PMID: 28617049]

[240]   Kusumaningtyas R, Laily N, Limandha P. Potential of Ciplukan (Physalis angulata L) as Source of Functional Ingredient. 2nd Humboldt Kolleg in conjunction with International Conference on Natural Sciences 2014; 2015; pp. 367-72. Proc Chem

[241]   Ismail N, Alam M. A novel cytotoxic flavonoid glycoside from Physalis angulata. Fitoterapia 2001; 72(6): 676-9.
[http://dx.doi.org/10.1016/S0367-326X(01)00281-7] [PMID: 11543968]

[242] Gu J-Q, Li W, Kang Y-H, *et al.* Minor withanolides from Physalis philadelphica: structures, quinone reductase induction activities, and liquid chromatography (LC)-MS-MS investigation as artifacts. Chem Pharm Bull (Tokyo) 2003; 51(5): 530-9.
[http://dx.doi.org/10.1248/cpb.51.530] [PMID: 12736452]

[243] Xu Y-M, Liu MX, Grunow N, *et al.* Discovery of Potent 17β-Hydroxywithanolides for Castration-Resistant Prostate Cancer by High-Throughput Screening of a Natural Products Library for Androgen-Induced Gene Expression Inhibitors. J Med Chem 2015; 58(17): 6984-93.
[http://dx.doi.org/10.1021/acs.jmedchem.5b00867] [PMID: 26305181]

[244] Maldonado E, Torres FR, Martínez M, Pérez-Castorena AL. 18-Acetoxywithanolides from Physalis chenopodifolia1. Planta Med 2004; 70(1): 59-64.
[http://dx.doi.org/10.1055/s-2004-815457] [PMID: 14765295]

[245] Kawai M, Matsuura T, Kyuno S, Matsuki H, Takenaka M, Katsuoka T, *et al.* A New physalin from Physalis alkekengi: structure of physalin L. Phytochemistry 1987; 26(12): 3313-7.
[http://dx.doi.org/10.1016/S0031-9422(00)82495-4]

[246] Amir M, Kumar S. Possible industrial applications of genus Solanum in twentyfirst century-A review. J Sci Ind Res (India) 2004; 63(2): 116-24.

[247] Friedman M. Tomato glycoalkaloids: role in the plant and in the diet. J Agric Food Chem 2002; 50(21): 5751-80.
[http://dx.doi.org/10.1021/jf020560c] [PMID: 12358437]

[248] Weese TL, Bohs L. Eggplant origins: out of Africa, into the Orient. Taxon 2010; 59(1): 49-56.

[249] Prota Plant resources of tropical Africa database [Internet] 2007. [cited 2018 Feb 21] Available from: https://www.prota4u.org/database/searchresults.aspS

[250] Sękara A, Cebula S, Kunicki E. Cultivated eggplants–origin, breeding objectives and genetic resources, a review. Folia Hortic. 2007;19(1):97–114.

[251] Schmelzer GH. Medicinal plants. Vol. 1. Schmelzer GH, Gurib-Fakim A, Eds. Ede, Netherlands: Prota 2008. pp. 790.

[252] Facciola S. Cornucopia II: A Source Book of Edible Plants. 2nd ed. California: Kampong Publications 1998; p. 713.

[253] Yamashita T, Fujimura N, Yahara S, Nohara T, Kawanobu S, Fujieda K. Studies on constituents of Solanaceae.17. Structures of 3 new steroidal alkaloid glycosides, Solaverine-I, solaverine-II, solaverine-III from Solanum toxicarium and Solanum verbascifolium. Chem Pharm Bull (Tokyo) 1990; 38(3): 827-9.
[http://dx.doi.org/10.1248/cpb.38.827]

[254] da Silva LR, Silva B. Natural Bioactive Compounds from Fruits and Vegetables as Health Promoters Part I. United Arab Emirates: Bentham Science Publishers 2016.

[255] Milner SE, Brunton NP, Jones PW, O'Brien NM, Collins SG, Maguire AR. Bioactivities of glycoalkaloids and their aglycones from Solanum species. J Agric Food Chem 2011; 59(8): 3454-84.
[http://dx.doi.org/10.1021/jf200439q] [PMID: 21401040]

[256] Cao G, Sofic E, Prior RL. Antioxidant capacity of tea and common vegetables. J Agric Food Chem 1996; 44(11): 3426-31.
[http://dx.doi.org/10.1021/jf9602535]

[257] Namitha KK, Negi PS. Chemistry and biotechnology of carotenoids. Crit Rev Food Sci Nutr 2010; 50(8): 728-60.
[http://dx.doi.org/10.1080/10408398.2010.499811] [PMID: 20830634]

[258] Hernandez-Marin E, Galano A, Martínez A. Cis carotenoids: colorful molecules and free radical quenchers. J Phys Chem B 2013; 117(15): 4050-61.
[http://dx.doi.org/10.1021/jp401647n] [PMID: 23560647]

[259] George B, Kaur C, Khurdiya DS, Kapoor HC. Antioxidants in tomato (Lycopersium esculentum) as a function of genotype. Food Chem 2004; 84(1): 45-51.
[http://dx.doi.org/10.1016/S0308-8146(03)00165-1]

[260] Khachik F, Carvalho L, Bernstein PS, Muir GJ, Zhao D-Y, Katz NB. Chemistry, distribution, and metabolism of tomato carotenoids and their impact on human health. Exp Biol Med (Maywood) 2002; 227(10): 845-51.
[http://dx.doi.org/10.1177/153537020222701002] [PMID: 12424324]

[261] Shi J, Qu Q, Kakùda Y, Yeung D, Jiang Y. Stability and synergistic effect of antioxidative properties of lycopene and other active components. Crit Rev Food Sci Nutr 2004; 44(7-8): 559-73.
[http://dx.doi.org/10.1080/15417060490908962] [PMID: 15969328]

[262] Hanson PM, Yang R-Y, Tsou SCS, Ledesma D, Engle L, Lee T-C. Diversity in eggplant (*Solanum melongena*) for superoxide scavenging activity, total phenolics, and ascorbic acid. J Food Compos Anal 2006; 19(6): 594-600.
[http://dx.doi.org/10.1016/j.jfca.2006.03.001]

[263] Luthria DL, Mukhopadhyay S. Influence of sample preparation on assay of phenolic acids from eggplant. J Agric Food Chem 2006; 54(1): 41-7.
[http://dx.doi.org/10.1021/jf0522457] [PMID: 16390175]

[264] Whitaker BD, Stommel JR. Distribution of hydroxycinnamic acid conjugates in fruit of commercial eggplant (*Solanum melongena* L.) cultivars. J Agric Food Chem 2003; 51(11): 3448-54.
[http://dx.doi.org/10.1021/jf026250b] [PMID: 12744682]

[265] Davey MW, Van Montagu M, Inze D, Sanmartin M, Kanellis A, Smirnoff N, *et al.* Plant L-ascorbic acid: chemistry, function, metabolism, bioavailability and effects of processing. J Sci Food Agric 2000; 80(7): 825-60.
[http://dx.doi.org/10.1002/(SICI)1097-0010(20000515)80:7<825::AID-JSFA598>3.0.CO;2-6]

[266] Gould WA. Tomato production, processing and technology. The Netherlands: Elsevier 2013; p. 550.

[267] Burton GW, Traber MG. Vitamin E: antioxidant activity, biokinetics, and bioavailability. Annu Rev Nutr 1990; 10(1): 357-82.
[http://dx.doi.org/10.1146/annurev.nu.10.070190.002041] [PMID: 2200468]

[268] Seybold C, Fröhlich K, Bitsch R, Otto K, Böhm V. Changes in contents of carotenoids and vitamin E during tomato processing. J Agric Food Chem 2004; 52(23): 7005-10.
[http://dx.doi.org/10.1021/jf049169c] [PMID: 15537310]

[269] Asano N, Kato A, Matsui K, *et al.* The effects of calystegines isolated from edible fruits and vegetables on mammalian liver glycosidases. Glycobiology 1997; 7(8): 1085-8.
[http://dx.doi.org/10.1093/glycob/7.8.1085] [PMID: 9455909]

[270] Lee K-R, Kozukue N, Han J-S, *et al.* Glycoalkaloids and metabolites inhibit the growth of human colon (HT29) and liver (HepG2) cancer cells. J Agric Food Chem 2004; 52(10): 2832-9.
[http://dx.doi.org/10.1021/jf030526d] [PMID: 15137822]

[271] Kozukue N, Han J-S, Lee K-R, Friedman M. Dehydrotomatine and α-tomatine content in tomato fruits and vegetative plant tissues. J Agric Food Chem 2004; 52(7): 2079-83.
[http://dx.doi.org/10.1021/jf0306845] [PMID: 15053555]

[272] Friedman M, Levin CE. Dehydrotomatine content in tomatoes. J Agric Food Chem 1998; 46(11): 4571-6.
[http://dx.doi.org/10.1021/jf9804589]

[273] Friedman M. Anticarcinogenic, cardioprotective, and other health benefits of tomato compounds lycopene, α-tomatine, and tomatidine in pure form and in fresh and processed tomatoes. J Agric Food Chem 2013; 61(40): 9534-50.
[http://dx.doi.org/10.1021/jf402654e] [PMID: 24079774]

[274]   Friedman M. Chemistry and anticarcinogenic mechanisms of glycoalkaloids produced by eggplants, potatoes, and tomatoes. J Agric Food Chem 2015; 63(13): 3323-37.
[http://dx.doi.org/10.1021/acs.jafc.5b00818] [PMID: 25821990]

[275]   Velisek J. The Chemistry of Food. London: Wiley 2013; p. 1128.

[276]   Friedman M. Analysis of biologically active compounds in potatoes (Solanum tuberosum), tomatoes (Lycopersicon esculentum), and jimson weed (Datura stramonium) seeds. J Chromatogr A 2004; 1054(1-2): 143-55.
[http://dx.doi.org/10.1016/j.chroma.2004.04.049] [PMID: 15553139]

[277]   Nohara T, Ono M, Ikeda T, Fujiwara Y, El-Aasr M. The tomato saponin, esculeoside A. J Nat Prod 2010; 73(10): 1734-41.
[http://dx.doi.org/10.1021/np100311t] [PMID: 20853874]

[278]   Fujiwara Y, Takaki A, Uehara Y, Ikeda T, Okawa M, Yamauchi K, *et al.* Tomato steroidal alkaloid glycosides, esculeosides A and B, from ripe fruits. Tetrahedron 2004; 60(22): 4915-20.
[http://dx.doi.org/10.1016/j.tet.2004.03.088]

[279]   Katsumata A, Kimura M, Saigo H, *et al.* Changes in esculeoside A content in different regions of the tomato fruit during maturation and heat processing. J Agric Food Chem 2011; 59(8): 4104-10.
[http://dx.doi.org/10.1021/jf104025p] [PMID: 21395308]

[280]   Cataldi TRI, Lelario F, Bufo SA. Analysis of tomato glycoalkaloids by liquid chromatography coupled with electrospray ionization tandem mass spectrometry. Rapid Commun Mass Spectrom 2005; 19(21): 3103-10.
[http://dx.doi.org/10.1002/rcm.2176] [PMID: 16200652]

[281]   Yahara S, Uda N, Nohara T. Lycoperosides A-C, three stereoisomeric 23-acetoxyspirosolan-3β-ol β-lycotetraosides from Lycopersicon esculentum. Phytochemistry 1996; 42(1): 169-72.
[http://dx.doi.org/10.1016/0031-9422(95)00854-3]

[282]   Yahara S, Uda N, Yoshio E, Yae E. Steroidal alkaloid glycosides from tomato (Lycopersicon esculentum). J Nat Prod 2004; 67(3): 500-2.
[http://dx.doi.org/10.1021/np030382x] [PMID: 15043444]

[283]   Yamanaka T, Vincken J-P, de Waard P, Sanders M, Takada N, Gruppen H. Isolation, characterization, and surfactant properties of the major triterpenoid glycosides from unripe tomato fruits. J Agric Food Chem 2008; 56(23): 11432-40.
[http://dx.doi.org/10.1021/jf802351c] [PMID: 18998702]

[284]   Yoshizaki M, Matsushita S, Fujiwara Y, Ikeda T, Ono M, Nohara T. Tomato new sapogenols, isoesculeogenin A and esculeogenin B. Chem Pharm Bull (Tokyo) 2005; 53(7): 839-40.
[http://dx.doi.org/10.1248/cpb.53.839] [PMID: 15997148]

[285]   Ohno M, Ono M, Nohara T. New solanocapsine-type tomato glycoside from ripe fruit of *Solanum lycopersicum.* Chem Pharm Bull (Tokyo) 2011; 59(11): 1403-5.
[http://dx.doi.org/10.1248/cpb.59.1403] [PMID: 22041079]

[286]   Katambo M, Lawrence M. A systematic study of African Solanum L Wageningen University, The Netherlands2007.

[287]   Adam G, Huong HT, Khoi NH. The constituents of the Vietnamese drug plant *Solanum verbascifolium* L. Planta Med 1979; 36: 238-9.

[288]   Ichiyanagi T, Kashiwada Y, Shida Y, Ikeshiro Y, Kaneyuki T, Konishi T. Nasunin from eggplant consists of cis-trans isomers of delphinidin 3-[4-(p-coumaroyl)-L-rhamnosyl (1-->6)glucopyranosid-]-5-glucopyranoside. J Agric Food Chem 2005; 53(24): 9472-7.
[http://dx.doi.org/10.1021/jf051841y] [PMID: 16302764]

[289]   Azuma K, Ohyama A, Ippoushi K, *et al.* Structures and antioxidant activity of anthocyanins in many accessions of eggplant and its related species. J Agric Food Chem 2008; 56(21): 10154-9.

[http://dx.doi.org/10.1021/jf801322m] [PMID: 18831559]

[290] Singh AP, Luthria D, Wilson T, Vorsa N, Singh V, Banuelos GS, *et al.* Polyphenols content and antioxidant capacity of eggplant pulp. Food Chem 2009; 114(3): 955-61.
[http://dx.doi.org/10.1016/j.foodchem.2008.10.048]

[291] García-Salas P, Gómez-Caravaca AM, Morales-Soto A, Segura-Carretero A, Fernández-Gutiérrez A. Identification and quantification of phenolic compounds in diverse cultivars of eggplant grown in different seasons by high-performance liquid chromatography coupled to diode array detector and electrospray-quadrupole-time of flight-mass spectrometry. Food Res Int 2014; 57: 114-22.
[http://dx.doi.org/10.1016/j.foodres.2014.01.032]

[292] Long M, Millar DJ, Kimura Y, *et al.* Metabolite profiling of carotenoid and phenolic pathways in mutant and transgenic lines of tomato: identification of a high antioxidant fruit line. Phytochemistry 2006; 67(16): 1750-7.
[http://dx.doi.org/10.1016/j.phytochem.2006.02.022] [PMID: 16616263]

[293] Kalogeropoulos N, Chiou A, Pyriochou V, Peristeraki A, Karathanos VT. Bioactive phytochemicals in industrial tomatoes and their processing byproducts. Lebensm Wiss Technol 2012; 49(2): 213-6.
[http://dx.doi.org/10.1016/j.lwt.2011.12.036]

[294] Stommel JR, Whitaker BD. Phenolic acid content and composition of eggplant fruit in a germplasm core subset. J Am Soc Hortic Sci 2003; 128(5): 704-10.

[295] Sunseri F, Polignano GB, Alba V, Lotti C, Bisignano V, Mennella G, *et al.* Genetic diversity and characterization of African eggplant germplasm collection. Afr J Plant Sci 2010; 4(7): 231-41.

[296] Chou S-C, Huang T-J, Lin E-H, Huang C-H, Chou C-H. Antihepatitis B virus constituents of Solanum erianthum. Nat Prod Commun 2012; 7(2): 153-6.
[PMID: 22474941]

[297] Ohtsuki T, Miyagawa T, Koyano T, Kowithayakorn T, Ishibashi M. Isolation and structure elucidation of flavonoid glycosides from Solanum verbascifolium. Phytochem Lett 2010; 3(2): 88-92.
[http://dx.doi.org/10.1016/j.phytol.2010.02.002]

[298] Zhou LX, Ding Y. A cinnamide derivative from Solanum verbascifolium L. J Asian Nat Prod Res 2002; 4(3): 185-7.
[http://dx.doi.org/10.1080/10286020290011396] [PMID: 12118506]

[299] Domino EF, Hornbach E, Demana T. The nicotine content of common vegetables. N Engl J Med 1993; 329(6): 437.
[http://dx.doi.org/10.1056/NEJM199308053290619] [PMID: 8326992]

[300] Nielsen SS, Franklin GM, Longstreth WT, Swanson PD, Checkoway H. Nicotine from edible Solanaceae and risk of Parkinson disease. Ann Neurol 2013; 74(3): 472-7.
[http://dx.doi.org/10.1002/ana.23884] [PMID: 23661325]

[301] Ikeda T, Tsumagari H, Nohara T. Steroidal oligoglycosides from Solanum nigrum. Chem Pharm Bull (Tokyo) 2000; 48(7): 1062-4.
[http://dx.doi.org/10.1248/cpb.48.1062] [PMID: 10923841]

[302] Lee S-J, Oh P-S, Ko J-H, Lim K, Lim K-TA. A 150-kDa glycoprotein isolated from Solanum nigrum L. has cytotoxic and apoptotic effects by inhibiting the effects of protein kinase C alpha, nuclear factor-kappa B and inducible nitric oxide in HCT-116 cells. Cancer Chemother Pharmacol 2004; 54(6): 562-72.
[http://dx.doi.org/10.1007/s00280-004-0850-x] [PMID: 15349752]

[303] Adam G, Huong HT, Khoi NH. Solaverbascine—a new 22,26-epiminocholestane alkaloid from Solanum verbascifolium. Phytochemistry 1980; 19(5): 1002-3.
[http://dx.doi.org/10.1016/0031-9422(80)85168-5]

[304] Chen Y-C, Lee H-Z, Chen H-C, Wen C-L, Kuo Y-H, Wang G-J. Anti-inflammatory components from the root of Solanum erianthum. Int J Mol Sci 2013; 14(6): 12581-92.

[http://dx.doi.org/10.3390/ijms140612581] [PMID: 23771024]

[305] Abas F, Lajis NH, Israf DA, Khozirah S, Kalsom YU. Antioxidant and nitric oxide inhibition activities of selected Malay traditional vegetables. Food Chem 2006; 95(4): 566-73.
[http://dx.doi.org/10.1016/j.foodchem.2005.01.034]

[306] Lim K-T. Glycoprotein isolated from Solanum nigrum L. kills HT-29 cells through apoptosis. J Med Food 2005; 8(2): 215-26.
[http://dx.doi.org/10.1089/jmf.2005.8.215] [PMID: 16117614]

[307] Raju K, Anbuganapathi G, Gokulakrishnan V, Rajkapoor B, Jayakar B, Manian S. Effect of dried fruits of Solanum nigrum LINN against CCl4-induced hepatic damage in rats. Biol Pharm Bull 2003; 26(11): 1618-9.
[http://dx.doi.org/10.1248/bpb.26.1618] [PMID: 14600413]

[308] Prashanth Kumar V, Shashidhara S, Kumar MM, Sridhara BY. Cytoprotective role of Solanum nigrum against gentamicin-induced kidney cell (Vero cells) damage *in vitro*. Fitoterapia 2001; 72(5): 481-6.
[http://dx.doi.org/10.1016/S0367-326X(01)00266-0] [PMID: 11429239]

[309] Rani P, Khullar N. Antimicrobial evaluation of some medicinal plants for their anti-enteric potential against multi-drug resistant Salmonella typhi. Phytother Res 2004; 18(8): 670-3.
[http://dx.doi.org/10.1002/ptr.1522] [PMID: 15476301]

[310] Javed T, Ashfaq UA, Riaz S, Rehman S, Riazuddin S. In-vitro antiviral activity of Solanum nigrum against Hepatitis C Virus. Virol J 2011; 8(1): 26.
[http://dx.doi.org/10.1186/1743-422X-8-26] [PMID: 21247464]

[311] Jainu M, Devi CSS. Antiulcerogenic and ulcer healing effects of Solanum nigrum (L.) on experimental ulcer models: possible mechanism for the inhibition of acid formation. J Ethnopharmacol 2006; 104(1-2): 156-63.
[http://dx.doi.org/10.1016/j.jep.2005.08.064] [PMID: 16202548]

[312] Nirmal SA, Patel AP, Bhawar SB, Pattan SR. Antihistaminic and antiallergic actions of extracts of Solanum nigrum berries: possible role in the treatment of asthma. J Ethnopharmacol 2012; 142(1): 91-7.
[http://dx.doi.org/10.1016/j.jep.2012.04.019] [PMID: 22564816]

[313] Ahmed AH, Kamal IH, Ramzy RM. Studies on the molluscicidal and larvicidal properties of Solanum nigrum L. leaves ethanol extract. J Egypt Soc Parasitol 2001; 31(3): 843-52.
[PMID: 11775110]

[314] Burton-Freeman B, Reimers K. Tomato consumption and health: emerging benefits. Am J Lifestyle Med 2011; 5(2): 182-91.
[http://dx.doi.org/10.1177/1559827610387488]

[315] Figueiredo-González M, Valentão P, Andrade PB. Tomato plant leaves: From by-products to the management of enzymes in chronic diseases. Ind Crops Prod 2016; 94 (Suppl. C): 621-9.
[http://dx.doi.org/10.1016/j.indcrop.2016.09.036]

[316] Chang W, Li Y, Zhang M, Zheng S, Li Y, Lou H. Solasodine-3-O-beta-D-glucopyranoside kills Candida albicans by disrupting the intracellular vacuole. Food Chem Toxicol. 2017;106(A):139–46.

[317] Xie X, Zhu H, Zhang J, Wang M, Zhu L, Guo Z, *et al.* Solamargine inhibits the migration and invasion of HepG2 cells by blocking epithelial-to-mesenchymal transition 2017.
[http://dx.doi.org/10.3892/ol.2017.6147]

[318] Shiu LY, Chang LC, Liang CH, Huang YS, Sheu HM, Kuo KW. Solamargine induces apoptosis and sensitizes breast cancer cells to cisplatin. Food Chem Toxicol 2007; 45(11): 2155-64.
[http://dx.doi.org/10.1016/j.fct.2007.05.009] [PMID: 17619073]

[319] Ding X, Zhu FS, Li M, Gao SG. Induction of apoptosis in human hepatoma SMMC-7721 cells by solamargine from *Solanum nigrum* L. J Ethnopharmacol 2012; 139(2): 599-604.

[320] Sun L, Zhao Y, Li X, Yuan H, Cheng A, Lou H. A lysosomal–mitochondrial death pathway is induced by solamargine in human K562 leukemia cells. Toxicol Vitr 2010; 24(6): 1504-11.

[321] Chen Y, Li S, Sun F, *et al. in vivo* antimalarial activities of glycoalkaloids isolated from Solanaceae plants. Pharm Biol 2010; 48(9): 1018-24.
[http://dx.doi.org/10.3109/13880200903440211] [PMID: 20731554]

[322] Gu X-Y, Shen X-F, Wang L, *et al.* Bioactive steroidal alkaloids from the fruits of Solanum nigrum. Phytochemistry 2018; 147: 125-31.
[http://dx.doi.org/10.1016/j.phytochem.2017.12.020] [PMID: 29306798]

[323] Ikeda T, Ando J, Miyazono A, *et al.* Anti-herpes virus activity of Solanum steroidal glycosides. Biol Pharm Bull 2000; 23(3): 363-4.
[http://dx.doi.org/10.1248/bpb.23.363] [PMID: 10726897]

[324] Ding X, Zhu F, Yang Y, Li M. Purification, antitumor activity *in vitro* of steroidal glycoalkaloids from black nightshade (Solanum nigrum L.). Food Chem 2013; 141(2): 1181-6.
[http://dx.doi.org/10.1016/j.foodchem.2013.03.062] [PMID: 23790901]

[325] Fujiwara Y, Kiyota N, Hori M, *et al.* Esculeogenin A, a new tomato sapogenol, ameliorates hyperlipidemia and atherosclerosis in ApoE-deficient mice by inhibiting ACAT. Arterioscler Thromb Vasc Biol 2007; 27(11): 2400-6.
[http://dx.doi.org/10.1161/ATVBAHA.107.147405] [PMID: 17872457]

[326] Fujiwara Y, Yahara S, Ikeda T, Ono M, Nohara T. Cytotoxic major saponin from tomato fruits. Chem Pharm Bull (Tokyo) 2003; 51(2): 234-5.
[http://dx.doi.org/10.1248/cpb.51.234] [PMID: 12576668]

[327] Figueiredo-González M, Valentão P, Pereira DM, Andrade PB. Further insights on tomato plant: Cytotoxic and antioxidant activity of leaf extracts in human gastric cells. Food Chem Toxicol 2017; 109(Pt 1): 386-92.
[http://dx.doi.org/10.1016/j.fct.2017.09.018] [PMID: 28899774]

[328] Kim SP, Nam SH, Friedman M. The tomato glycoalkaloid α-tomatine induces caspase-independent cell death in mouse colon cancer CT-26 cells and transplanted tumors in mice. J Agric Food Chem 2015; 63(4): 1142-50.
[http://dx.doi.org/10.1021/jf5040288] [PMID: 25614934]

[329] Liu J, Kanetake S, Wu Y-H, *et al.* Antiprotozoal effects of the tomato tetrasaccharide glycoalkaloid tomatine and the aglycone tomatidine on mucosal trichomonads. J Agric Food Chem 2016; 64(46): 8806-10.
[http://dx.doi.org/10.1021/acs.jafc.6b04030] [PMID: 27934291]

[330] Lee S-T, Wong P-F, Cheah S-C, Mustafa MR. Alpha-tomatine induces apoptosis and inhibits nuclear factor-kappa B activation on human prostatic adenocarcinoma PC-3 cells. PLoS One 2011; 6(4): e18915.
[http://dx.doi.org/10.1371/journal.pone.0018915] [PMID: 21541327]

[331] Shieh J-M, Cheng T-H, Shi M-D, *et al.* α-Tomatine suppresses invasion and migration of human non-small cell lung cancer NCI-H460 cells through inactivating FAK/PI3K/Akt signaling pathway and reducing binding activity of NF-κB. Cell Biochem Biophys 2011; 60(3): 297-310.
[http://dx.doi.org/10.1007/s12013-011-9152-1] [PMID: 21264526]

[332] Ono H, Kozuka D, Chiba Y, Horigane A, Isshiki K. Structure and cytotoxicity of dehydrotomatine, a minor component of tomato glycoalkaloids. J Agric Food Chem 1997; 45(10): 3743-6.
[http://dx.doi.org/10.1021/jf970253k]

[333] Chagnon F, Guay I, Bonin M-A, *et al.* Unraveling the structure-activity relationship of tomatidine, a steroid alkaloid with unique antibiotic properties against persistent forms of Staphylococcus aureus. Eur J Med Chem 2014; 80: 605-20.
[http://dx.doi.org/10.1016/j.ejmech.2013.11.019] [PMID: 24877760]

[334] Mitchell G, Gattuso M, Grondin G, Marsault É, Bouarab K, Malouin F. Tomatidine inhibits replication of Staphylococcus aureus small-colony variants in cystic fibrosis airway epithelial cells. Antimicrob Agents Chemother 2011; 55(5): 1937-45.
[http://dx.doi.org/10.1128/AAC.01468-10] [PMID: 21357296]

[335] Yuan B, Byrnes D, Giurleo D, Villani T, Simon JE, Wu Q. Rapid screening of toxic glycoalkaloids and micronutrients in edible nightshades (Solanum spp.). Yao Wu Shi Pin Fen Xi In press
[http://dx.doi.org/10.1016/j.jfda.2017.10.005]

[336] Das S, Raychaudhuri U, Falchi M, Bertelli A, Braga PC, Das DK. Cardioprotective properties of raw and cooked eggplant (*Solanum melongena* L). Food Funct 2011; 2(7): 395-9.
[http://dx.doi.org/10.1039/c1fo10048c] [PMID: 21894326]

[337] Dos Santos R, Pimenta-Freire G, Dias-Souza MV. Carotenoids and flavonoids can impair the effectiveness of some antimicrobial drugs against clinical isolates of Escherichia coli and Staphylococcus aureus. Int Food Res J 2015; 22(5): 1772-82.

[338] Sun J, Huo H-X, Huang Z, Zhang J, Li J, Tu P-F. A new γ-alkylated-γ-butyrolactone from the roots of *Solanum melongena*. Chin J Nat Med 2015; 13(9): 699-703.
[http://dx.doi.org/10.1016/S1875-5364(15)30068-6] [PMID: 26412430]

[339] Nagase H, Sasaki K, Kito H, Haga A, Sato T. Inhibitory effect of delphinidin from *Solanum melongena* on human fibrosarcoma HT-1080 invasiveness *in vitro*. Planta Med 1998; 64(3): 216-9.
[http://dx.doi.org/10.1055/s-2006-957412] [PMID: 9581517]

[340] Nahmias Y, Goldwasser J, Casali M, *et al.* Apolipoprotein B-dependent hepatitis C virus secretion is inhibited by the grapefruit flavonoid naringenin. Hepatology 2008; 47(5): 1437-45.
[http://dx.doi.org/10.1002/hep.22197] [PMID: 18393287]

[341] Noda Y, Kneyuki T, Igarashi K, Mori A, Packer L. Antioxidant activity of nasunin, an anthocyanin in eggplant peels. Toxicology 2000; 148(2-3): 119-23.
[http://dx.doi.org/10.1016/S0300-483X(00)00202-X] [PMID: 10962130]

[342] Noda Y, Kaneyuki T, Igarashi K, Mori A, Packer L. Antioxidant activity of nasunin, an anthocyanin in eggplant. Res Commun Mol Pathol Pharmacol 1998; 102(2): 175-87.
[PMID: 10100509]

[343] Kayamori F, Igarashi K. Effects of dietary nasunin on the serum cholesterol level in rats. Biosci Biotechnol Biochem 1994; 58(3): 570-1.
[http://dx.doi.org/10.1271/bbb.58.570]

[344] Matsubara K, Kaneyuki T, Miyake T, Mori M. Antiangiogenic activity of nasunin, an antioxidant anthocyanin, in eggplant peels. J Agric Food Chem 2005; 53(16): 6272-5.
[http://dx.doi.org/10.1021/jf050796r] [PMID: 16076105]

[345] Shen G, Van Kiem P, Cai X-F, *et al.* Solanoflavone, a new biflavonol glycoside from *Solanum melongena*: seeking for anti-inflammatory components. Arch Pharm Res 2005; 28(6): 657-9.
[http://dx.doi.org/10.1007/BF02969354] [PMID: 16042073]

[346] Cavallini C, Trettene M, Degan M, *et al.* Anti-angiogenic effects of two cystine-knot miniproteins from tomato fruit. Br J Pharmacol 2011; 162(6): 1261-73.
[http://dx.doi.org/10.1111/j.1476-5381.2010.01154.x] [PMID: 21175567]

[347] Herbel V, Schäfer H, Wink M. Recombinant production of snakin-2 (an antimicrobial peptide from tomato) in E. coli and analysis of its bioactivity. Molecules 2015; 20(8): 14889-901.
[http://dx.doi.org/10.3390/molecules200814889] [PMID: 26287145]

[348] Manase MJ, Mitaine-Offer A-C, Pertuit D, *et al.* Solanum incanum and S. heteracanthum as sources of biologically active steroid glycosides: confirmation of their synonymy. Fitoterapia 2012; 83(6): 1115-9.
[http://dx.doi.org/10.1016/j.fitote.2012.04.024] [PMID: 22579841]

[349] Sun J, Gu Y-F, Su X-Q, *et al.* Anti-inflammatory lignanamides from the roots of *Solanum melongena* L. Fitoterapia 2014; 98: 110-6.
[http://dx.doi.org/10.1016/j.fitote.2014.07.012] [PMID: 25068200]

[350] Wang X, Zhang F, Yang L, *et al.* Ursolic acid inhibits proliferation and induces apoptosis of cancer cells *in vitro* and *in vivo*. J Biomed Biotechnol 2011; 2011: 419343.
[http://dx.doi.org/10.1155/2011/419343] [PMID: 21716649]

[351] Pinela J, Barros L, Carvalho AM, Ferreira ICFR. Nutritional composition and antioxidant activity of four tomato (Lycopersicon esculentum L.) farmer' varieties in Northeastern Portugal homegardens. Food Chem Toxicol 2012; 50(3-4): 829-34.
[http://dx.doi.org/10.1016/j.fct.2011.11.045] [PMID: 22154854]

[352] Stajčić S, Ćetković G, Čanadanović-Brunet J, Djilas S, Mandić A, Četojević-Simin D. Tomato waste: Carotenoids content, antioxidant and cell growth activities. Food Chem 2015; 172: 225-32.
[http://dx.doi.org/10.1016/j.foodchem.2014.09.069] [PMID: 25442547]

[353] Sudheesh S, Sandhya C, Sarah Koshy A, Vijayalakshmi NR. Antioxidant activity of flavonoids from *Solanum melongena*. Phytother Res 1999; 13(5): 393-6.
[http://dx.doi.org/10.1002/(SICI)1099-1573(199908/09)13:5<393::AID-PTR474>3.0.CO;2-8] [PMID: 10441778]

[354] Nisha P, Abdul Nazar P, Jayamurthy P. A comparative study on antioxidant activities of different varieties of *Solanum melongena*. Food Chem Toxicol 2009; 47(10): 2640-4.
[http://dx.doi.org/10.1016/j.fct.2009.07.026] [PMID: 19638291]

[355] Lo Scalzo R, Fibiani M, Mennella G, *et al.* Thermal treatment of eggplant (*Solanum melongena* L.) increases the antioxidant content and the inhibitory effect on human neutrophil burst. J Agric Food Chem 2010; 58(6): 3371-9.
[http://dx.doi.org/10.1021/jf903881s] [PMID: 20187646]

[356] Akanitapichat P, Phraibung K, Nuchklang K, Prompitakkul S. Antioxidant and hepatoprotective activities of five eggplant varieties. Food Chem Toxicol 2010; 48(10): 3017-21.
[http://dx.doi.org/10.1016/j.fct.2010.07.045] [PMID: 20691749]

[357] Shi J, Kakuda Y, Yeung D. Antioxidative properties of lycopene and other carotenoids from tomatoes: synergistic effects. Biofactors 2004; 21(1-4): 203-10.
[http://dx.doi.org/10.1002/biof.552210141] [PMID: 15630198]

[358] Han S-W, Tae J, Kim J-A, *et al.* The aqueous extract of *Solanum melongena* inhibits PAR2 agonist-induced inflammation. Clin Chim Acta 2003; 328(1-2): 39-44.
[http://dx.doi.org/10.1016/S0009-8981(02)00377-7] [PMID: 12559597]

[359] Ravi V, Saleem TSM, Maiti PP, Ramamurthy J. Phytochemical and pharmacological evaluation of Solanum nigrum Linn. Afr J Pharm Pharmacol 2009; 3(9): 454-7.

[360] Plazas M, Prohens J, Cuñat AN, *et al.* Reducing capacity, chlorogenic acid content and biological activity in a collection of scarlet (Solanum aethiopicum) and Gboma (S. macrocarpon) eggplants. Int J Mol Sci 2014; 15(10): 17221-41.
[http://dx.doi.org/10.3390/ijms151017221] [PMID: 25264739]

[361] Ravi V, Saleem TSM, Patel SS, Raamamurthy J, Gauthaman K. Anti-inflammatory effect of methanolic extract of Solanum nigrum Linn berries. Int J Appl Res Nat Prod 2009; 2(2): 33-6.

[362] Kaushik D, Jogpal V, Kaushik P, Lal S, Saneja A, Sharma C, *et al.* Evaluation of activities of Solanum nigrum fruit extract. Arch Appl Sci Res 2009; 1(1): 43-50.

[363] Kuo K-W, Hsu S-H, Li Y-P, *et al.* Anticancer activity evaluation of the solanum glycoalkaloid solamargine. Triggering apoptosis in human hepatoma cells. Biochem Pharmacol 2000; 60(12): 1865-73.
[http://dx.doi.org/10.1016/S0006-2952(00)00506-2] [PMID: 11108802]

[364] Liu L-F, Liang C-H, Shiu L-Y, Lin W-L, Lin C-C, Kuo K-W. Action of solamargine on human lung cancer cells--enhancement of the susceptibility of cancer cells to TNFs. FEBS Lett 2004; 577(1-2): 67-74.
[http://dx.doi.org/10.1016/j.febslet.2004.09.064] [PMID: 15527763]

[365] Liang C-H, Liu L-F, Shiu L-Y, Huang Y-S, Chang L-C, Kuo K-W. Action of solamargine on TNFs and cisplatin-resistant human lung cancer cells. Biochem Biophys Res Commun 2004; 322(3): 751-8.
[http://dx.doi.org/10.1016/j.bbrc.2004.07.183] [PMID: 15336528]

[366] Liang CH, Shiu LY, Chang LC, Sheu HM, Kuo KW. Solamargine upregulation of Fas, downregulation of HER2, and enhancement of cytotoxicity using epirubicin in NSCLC cells. Mol Nutr Food Res 2007; 51(8): 999-1005.
[http://dx.doi.org/10.1002/mnfr.200700044] [PMID: 17639997]

[367] Shiu LY, Chang LC, Liang CH, Huang YS, Sheu HM, Kuo KW. Solamargine induces apoptosis and sensitizes breast cancer cells to cisplatin. Food Chem Toxicol 2007; 45(11): 2155-64.
[http://dx.doi.org/10.1016/j.fct.2007.05.009] [PMID: 17619073]

[368] Shiu LY, Liang CH, Chang LC, Sheu HM, Tsai EM, Kuo KW. Solamargine induces apoptosis and enhances susceptibility to trastuzumab and epirubicin in breast cancer cells with low or high expression levels of HER2/neu. Biosci Rep 2009; 29(1): 35-45.
[http://dx.doi.org/10.1042/BSR20080028] [PMID: 18699774]

[369] Sun L, Zhao Y, Yuan H, Li X, Cheng A, Lou H. Solamargine, a steroidal alkaloid glycoside, induces oncosis in human K562 leukemia and squamous cell carcinoma KB cells. Cancer Chemother Pharmacol 2011; 67(4): 813-21.
[http://dx.doi.org/10.1007/s00280-010-1387-9] [PMID: 20563579]

[370] Shabana MM, Salama MM, Ezzat SM, Ismail LR. *in vitro* and *in vivo* anticancer activity of the fruit peels of *Solanum melongena* L. against hepatocellular carcinoma. J Carcinog Mutagen 2013; 4(3): 149-54.

[371] Friedman M, Levin CE, Lee S-U, *et al.* Tomatine-containing green tomato extracts inhibit growth of human breast, colon, liver, and stomach cancer cells. J Agric Food Chem 2009; 57(13): 5727-33.
[http://dx.doi.org/10.1021/jf900364j] [PMID: 19514731]

[372] Mamone L, Di Venosa G, Valla JJ, *et al.* Cytotoxic effects of Argentinean plant extracts on tumour and normal cell lines. Cell Mol Biol 2011; 57(2) (Suppl.): OL1487-99.
[PMID: 21624335]

[373] Brigelius-Flohé R, Kelly FJ, Salonen JT, Neuzil J, Zingg J-M, Azzi A. The European perspective on vitamin E: current knowledge and future research. Am J Clin Nutr 2002; 76(4): 703-16.
[http://dx.doi.org/10.1093/ajcn/76.4.703] [PMID: 12324281]

[374] Pesewu GA, Cutler RR, Humber DP. Antibacterial activity of plants used in traditional medicines of Ghana with particular reference to MRSA. J Ethnopharmacol 2008; 116(1): 102-11.
[http://dx.doi.org/10.1016/j.jep.2007.11.005] [PMID: 18096337]

[375] Fuentes E, Castro R, Astudillo L, *et al.* Bioassay-guided isolation and HPLC determination of bioactive compound that relate to the antiplatelet activity (adhesion, secretion, and aggregation) from Solanum lycopersicum. Evid Based Complement Alternat Med 2012; 2012: 147031.
[http://dx.doi.org/10.1155/2012/147031] [PMID: 23227097]

[376] Kwon Y-I, Apostolidis E, Shetty K. *in vitro* studies of eggplant (*Solanum melongena*) phenolics as inhibitors of key enzymes relevant for type 2 diabetes and hypertension. Bioresour Technol 2008; 99(8): 2981-8.
[http://dx.doi.org/10.1016/j.biortech.2007.06.035] [PMID: 17706416]

[377] Odetola AA, Iranloye YO, Akinloye O. Hypolipidaemic potentials of *Solanum melongena* and Solanum gilo on hypercholesterolemic rabbits. Pak J Nutr 2004; 3(3): 180-7.
[http://dx.doi.org/10.3923/pjn.2004.180.187]

[378] Umamageswari M, Maniyar YA. Evaluation of analgesic activity of aqueous extract of leaves of *Solanum melongena* Linn. In experimental animals. Asian J Pharm Clin Res 2015; 8(1): 327-30.

[379] Mutalik S, Paridhavi K, Rao CM, Udupa N. Antipyretic and analgesic effect of leaves of *Solanum melongena* Linn. in rodents. Indian J Pharmacol 2003; 35(5): 312-5.

[380] Bello SO, Muhammad B, Gammaniel KS, Aguye AI, Ahmed H, Njoku CH. Randomized double blind placebo controlled clinical trial of *Solanum melongena* L. fruit in moderate to severe asthmatics. J Med Sci 2004; 4: 263-9.
       [http://dx.doi.org/10.3923/jms.2004.263.269]

[381] Guimarães PR, Galvão AMP, Batista CM, *et al.* Eggplant (*Solanum melongena*) infusion has a modest and transitory effect on hypercholesterolemic subjects. Braz J Med Biol Res 2000; 33(9): 1027-36.
       [http://dx.doi.org/10.1590/S0100-879X2000000900006] [PMID: 10973133]

[382] Singh SP, Raghavendra K, Singh RK, Subbarao SK. Studies on larvicidal properties of leaf extract of Solanum nigrum Linn.(family Solanaceae). Curr Sci 2001; 81(12): 1529-30.

[383] Zakaria ZA, Gopalan HK, Zainal H, *et al.* Antinociceptive, anti-inflammatory and antipyretic effects of Solanum nigrum chloroform extract in animal models. Yakugaku Zasshi 2006; 126(11): 1171-8.
       [http://dx.doi.org/10.1248/yakushi.126.1171] [PMID: 17077618]

CHAPTER 10

Impact on Health of Artichoke and Cardoon Bioactive Compounds: Content, Bioaccessibility, Bioavailability, and Bioactivity

Isabella D'Antuono[1], Francesco Di Gioia[2,*], Vito Linsalata[1], Erin N. Rosskopf[3] and Angela Cardinali[1]

[1] *Institute of Sciences of Food Production, National Research Council, Bari, Italy*

[2] *Department of Plant Science, Pennsylvania State University, University Park, PA, USA*

[3] *USDA-ARS, U.S. Horticultural Research Laboratory, Fort Pierce, FL, USA*

Abstract: Artichoke, cultivated cardoon, and their common relative, the wild cardoon are botanical varieties of the species *Cynara cardunculus* L., a perennial plant native to the Mediterranean Basin and belonging to the *Asteraceae* family. While commonly used as food, leaf extracts of this plants have been traditionally used as a natural remedy in folk medicine. These plants are in fact a rich source of bioactive compounds such as polyphenols, inulin, and sesquiterpene lactones. Many studies demonstrated that these compounds and their metabolites are responsible for several beneficial properties attributed to the extracts of artichoke and cardoon. As we gain knowledge on the effects and mode of action of these compounds, artichoke and cardoon are considered 'functional food' and are increasingly used to extract bioactive compounds and for numerous pharmaceutical applications. In this chapter, after a brief introduction on the origin and the importance of these crops, each class of bioactive compounds is presented summarizing the specific chemical properties, the biosynthesis, and the concentration range in plant tissues. The third section discusses the main factors (plant portion, physiological stage, plant genotype, environment, pre-harvest agronomic practices, post-harvest handling and processing) influencing the concentration of bioactive compounds in artichoke and cardoon. The following section is focused on the physiological fate of the bioactive compounds, reviewing the results of the most recent *in vitro* and *in vivo* studies conducted to assess their bioaccessibility, bioavailability, and pharmacokinetics. Finally, in the last section the main health-promoting effects attributed to artichoke and cardoon polyphenols are reviewed.

Keywords: Anticancer, Anti-inflammatory, Antioxidant, Artichoke, Bioaccessibility, Bioavailability, Biological activity, *Cynara cardunculus*, Cardoon, Cynarin, Cynaropicrin, Flavonoids, Functional foods, Hepatoprotective,

* **Corresponding author Francesco Di Gioia:** Pennsylvania State University, Department of Plant Science, University Park, PA 16802, USA; Tel: +1 814 863 2195; E-mail: fxd92@psu.edu

Spyridon A. Petropoulos, Isabel C.F.R. Ferreira and Lillian Barros (Eds.)

Hypocholesterolemic, Hypoglycemic, Inulin, Pharmacokinetic, Polyphenols, Sesquiterpene lactones.

INTRODUCTION

Native to the Mediterranean Basin [1], the species *Cynara cardunculus* L. is a perennial plant belonging to the *Asteraceae* family, and includes three botanical varieties: artichoke [*Cynara cardunculus* var. *scòlymus* (L.) Fiori], cultivated cardoon [*Cynara cardunculus* var. *altilis* DC.], and their common relative, the wild cardoon [*Cynara cardunculus* var. *sylvestris* (Lamk) Fiori] [2]. Plants of wild cardoons are characterized by small green and thorny inflorescences and large spiny leaves and are distributed over the entire Mediterranean area. The process of domestication of the wild cardoon remains uncertain, however, it is likely that artichoke was domesticated in Southern Italy during the Roman Empire, by selecting for large non-spiny heads, while the cultivated cardoon was domesticated in a later period in Southern France and Spain by selecting for large stalked tender leaves [2 - 5]. Traditionally cultivated in Southern Europe and North Africa, today, being known for its nutritional value and therapeutic properties [5], artichoke is widely grown around the world. In Europe, Italy and Spain are the major producers of artichoke with cultivated areas of 41,299 and 15,002 ha, and an annual production of about 401,335 and 204,111 tones, respectively [6]. Within the Mediterranean region, the cultivation of artichoke is also common in France (7,692 ha), Greece (1,880 ha), Egypt (10,721 ha), Algeria (4,674 ha), Morocco (3,236 ha), Tunisia (2,436 ha), and Turkey (2,538 ha). Outside the Mediterranean region, the main artichoke producers are Peru (5,513 ha), Argentina (4,740 ha), Chile (1,342 ha), the United States (2,750 ha, mainly California), and China (11,552 ha), whose production emerged in the last two decades [4 - 6]. Artichoke represents an important component of the Mediterranean diet; its edible portion is composed of the immature inflorescences, called capitula or heads, including a flavorful receptacle and fleshy, tender external bracts. These components of artichoke are a rich source of bioactive phenolic compounds, inulin, fiber, vitamins, minerals, and sesquiterpene lactones [7, 8]. Artichoke heads contain 15-20% dry matter (DM), and on a fresh weight (FW) basis, 6.8% carbohydrates, 3.0% protein, 0.2% fat, and 1.0% ash [9]. Moreover, artichoke is considered a rich source of vitamin C (10 mg 100 g^{-1} FW), K and Ca (360 and 50 mg 100 g^{-1} FW, respectively), Fe (1.7–2.9 mg 100 g^{-1} FW), and Zn (0.5–1.17 mg 100 g^{-1} FW) [10 - 13]. Besides the heads, artichoke leaves were also used in ancient times as a source of beneficial and therapeutic compounds. Artichoke leaf extracts are widely used in herbal medicine as hepatoprotectors and choleretics [14]. Leaf extracts also have lipid-lowering, diuretic, anti-carcinogenic, anti-HIV, and antioxidative effects, together with antifungal and antibacterial properties [7, 15, 16]. For both heads and leaves, the described health-promoting characteristics are mainly attributable to the high levels of phenolic compounds. Mono- and di-caffeoylquinic acid derivatives such as: 5-O-caffeoylquinic acid (chlorogenic acid), 1,5-O- and 3,5-O-dicaffeoylquinic acids are the main phenolic compounds identified in artichoke. Moreover, flavonoids, such as apigenin and luteolin (both present as glucosides and rutinosides) and cyanidin (caffeoylglucoside derivatives), have been identified in artichoke tissues [17 - 20]. The cultivated or leafy cardoon, generally consumed in traditional preparation, is mainly produced in Spain, Italy, Greece, France and south Portugal; although its naturalization is now extended to Australia, and North and South Africa [21], and to a lesser extent it is also cultivated in the USA. Cardoon is well adapted to Mediterranean conditions and can produce large amounts of lignocellulosic biomass suitable for energy or paper pulp production; their seeds can be used for oil extraction; and the flowers in some countries, such as Italy, Spain, and Portugal, are employed as a substitute for rennet for cheese production [22, 23]. Similar to artichoke, the health promoting components of cardoon are represented by mono-caffeoylquinic acids, di-caffeoylquinic acids, and flavonoids [17, 24, 25]. Recent studies of cardoon heads have identified chlorogenic acid (with relative abundance ranging between 18-36%), di-caffeoylquinic acid (1,5-O-dicaffeoylquinic acid), and flavonoids (such as apigenin glucuronide, apigenin-7-O-glucoside, and apigenin-7-O-rutinoside). The cardoon leaves are richer in luteolin derivatives, all recognized compounds for their health-promoting activities [26, 27].

ARTICHOKE AND CARDOON BIOACTIVE COMPOUNDS

Chemistry and Biosynthesis of Polyphenols

Fruits and vegetables are recognized sources of secondary metabolites, such as phenolic molecules that serve important functions regarding sensory and nutritional quality of plants [28]. From a chemical point of view, polyphenolic compounds are constituted by one or two aromatic rings with one or more hydroxyl groups, including their functional derivatives (*e.g.* esters and glycosides). Polyphenols are present in nature in monomeric and polymeric forms, and therefore with low and high molecular mass [29]. High molecular weight phenols (*e.g.* proanthocyanidins) are defined as metabolic end-products and they are mainly accumulated in the tissues during the development and growth of the plants, whereas low molecular weight phenolics might undergo high turnover in living plant tissues [30]. Polyphenol biosynthesis *via* the shikimic-acid pathway begins with the amino acid phenylalanine, the availability of which controls the quantity of polyphenols produced. Conversely, polyphenol quality is influenced by the cinnamic acid level, another metabolite of the shikimate pathway [31]. Polyphenols play a key role in both protecting plants from external stress (UV rays, insects, fungi, bacteria) and lending color that characterizes the plant morphology. The polyphenols present in artichoke belong to two main classes: hydroxicinnamic acids (C3-C6 skeleton) and flavonoids (C6-C3-C6 skeleton). Among the hydroxycinnamic acids, chlorogenic, 3,5-O-dicaffeoylquinic and 1,5-O-dicaffeoylquinic acids are the most abundant polyphenols, followed by minor compounds [32]. Other mono- and di-caffeoylquinic derivatives occurring in artichoke extracts include 1-O-caffeoylquinic acid, 3-O-caffeoylquinic acid (neochlorogenic acid), 4-O-caffeoylquinic acid (cryptochlorogenic acid), 1,3-O-dicaffeoylquinic acid (cynarin), 1,4-O-dicaffeoylquinic acid, 3,4-O-dicaffeoylquinic acid, and 4,5-O-dicaffeoylquinic acid [7]. The main flavonoids identified in artichoke heads are apigenin-7-O-glucuronide, luteolin-7-O-glucuronide, apigenin-7-O-rutinoside, apigenin-7-O-glucoside, luteolin-7-O-glucoside, luteolin-7-O-rutinoside, naringenin-7-O-rutinoside, and naringenin-7-O-glucoside [33] (Fig. **1**). In addition, using different cultivars, Schutz and co-workers [34] characterized and quantified the artichoke anthocyanin profile. Cyanidin 3,5-diglucoside, cyanidin 3-glucoside, cyanidin 3,5-malonyldiglucoside, cyanidin 3-(3-malonyl)glucoside, and cyanidin 3-(6-malonyl)glucoside were identified, followed by several minor compounds as peonidin and delphinidin derivatives [34]. Wild and cultivated cardoon and artichoke have similar bioactive compounds. The wild cardoons contain the lowest content of both mono-, di-caffeoylquinc acids and flavonoids, with the latter mainly belonging to apigenin derivatives [12]. Ramos and co-workers [27], in their study on the determination of the phenolic composition of *Cynara cardunculus* L. var. *altilis* (DC), have identified twenty-eight phenolic compounds, with eriodictyol hexoside described for the first time in capitula florets of cultivated cardoon. In addition, other cardoon polyphenol components, such as 1,4-O-dicaffeoylquinic acid, naringenin-7-O-glucoside, naringenin-7-O-rutinoside, naringenin, luteolin acetyl-hexoside, and apigenin acetyl-hexoside were identified [27]. Like artichoke, cardoon leaves also contain cynarin, in addition to the other polyphenols that are not found in artichoke, such as: silymarin (0.9 to 2.7% of dry weight DW) a flavonolignans mixture common in other *Asteraceae* species [35], characterized by antioxidant and hepatoprotective properties [36]; and succinylcaffeoylquinic acid compounds [37] whose concentration was higher in wild cardoon (26.1-35.2%) than in cultivated cardoon (3.9-17.6%). Due to multiple beneficial effects of the polyphenols (antioxidant, anti-carcinogenic, cardio-protective, antimicrobial, anti-viral, and neuro-protective agents) numerous research studies are focused on the development of strategies to increase their levels in plant tissues, acting mainly on their biosynthetic pathways [38]. Three different biosynthetic routes have been proposed for the synthesis of chlorogenic acid in plants. The first route involves caffeoyl CoA and quinic acid, which, through the catalysis of hydroxycinnamoyl-

CoA quinate hydroxycinnamoyl transferase enzyme, are converted into chlorogenic acid. According to the second synthetic route hypothesis, the enzyme hydroxycinnamoyl D-glucose:quinate hydroxycinnamoyl transferase activates the synthesis of chlorogenic acid starting from caffeoyl glucoside and quinic acid as precursors. Finally, the third possible route, currently considered as the most likely pathway for chlorogenic acid synthesis, involves p-coumaroyl CoA and quinic acid to form p-coumaroyl quinate by acyl transferase activity. Further, the p-coumaroyl quinate is hydroxylated to form chlorogenic acid, by p-coumarate 3'-hydroxylase [39].

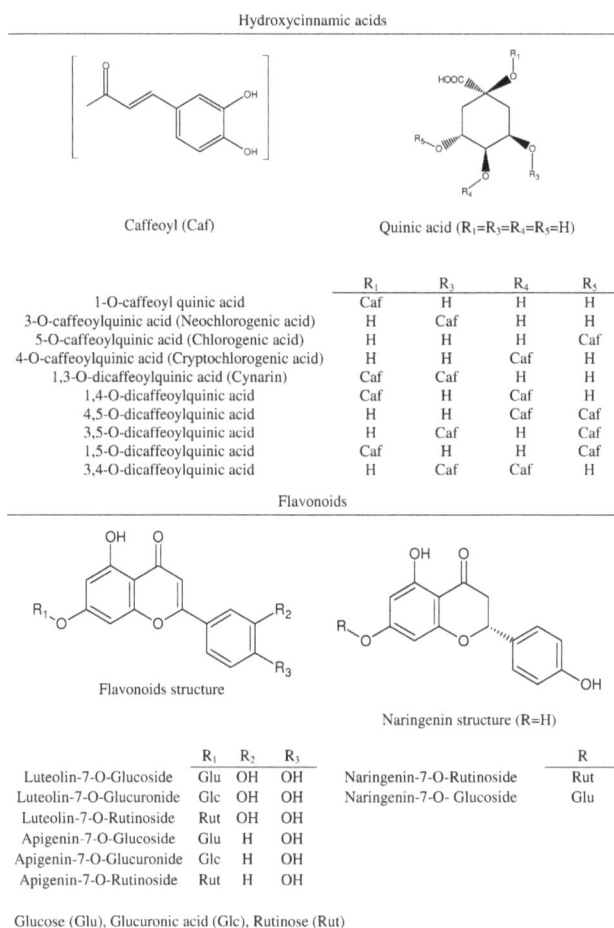

Hydroxycinnamic acids

Caffeoyl (Caf)   Quinic acid ($R_1=R_3=R_4=R_5=H$)

	R_1	R_3	R_4	R_5
1-O-caffeoyl quinic acid	Caf	H	H	H
3-O-caffeoylquinic acid (Neochlorogenic acid)	H	Caf	H	H
5-O-caffeoylquinic acid (Chlorogenic acid)	H	H	H	Caf
4-O-caffeoylquinic acid (Cryptochlorogenic acid)	H	H	Caf	H
1,3-O-dicaffeoylquinic acid (Cynarin)	Caf	Caf	H	H
1,4-O-dicaffeoylquinic acid	Caf	H	Caf	H
4,5-O-dicaffeoylquinic acid	H	H	Caf	Caf
3,5-O-dicaffeoylquinic acid	H	Caf	H	Caf
1,5-O-dicaffeoylquinic acid	Caf	H	H	Caf
3,4-O-dicaffeoylquinic acid	H	Caf	Caf	H

Flavonoids

Flavonoids structure

Naringenin structure (R=H)

	R_1	R_2	R_3		R
Luteolin-7-O-Glucoside	Glu	OH	OH	Naringenin-7-O-Rutinoside	Rut
Luteolin-7-O-Glucuronide	Glc	OH	OH	Naringenin-7-O- Glucoside	Glu
Luteolin-7-O-Rutinoside	Rut	OH	OH		
Apigenin-7-O-Glucoside	Glu	H	OH		
Apigenin-7-O-Glucuronide	Glc	H	OH		
Apigenin-7-O-Rutinoside	Rut	H	OH		

Glucose (Glu), Glucuronic acid (Glc), Rutinose (Rut)

Fig. (1). Chemical structure of polyphenols identified from *Cynara cardunculus* L. varieties (artichoke and cardoon).

Inulin

Another important component characteristic of artichoke and cardoon is inulin, a polymeric carbohydrate belonging to a class of highly water-soluble fibers called fructans. Chemically, fructans are composed of linear or branched fructose polymers with one or more β-linked fructose (Fig. **2**), and their interest as functional foods is increasing for their potential beneficial effect on human health. Inulin can be extracted from external bracts, artichoke roots, and flowers; the length of its chain is no more than 200 fructose units [40].

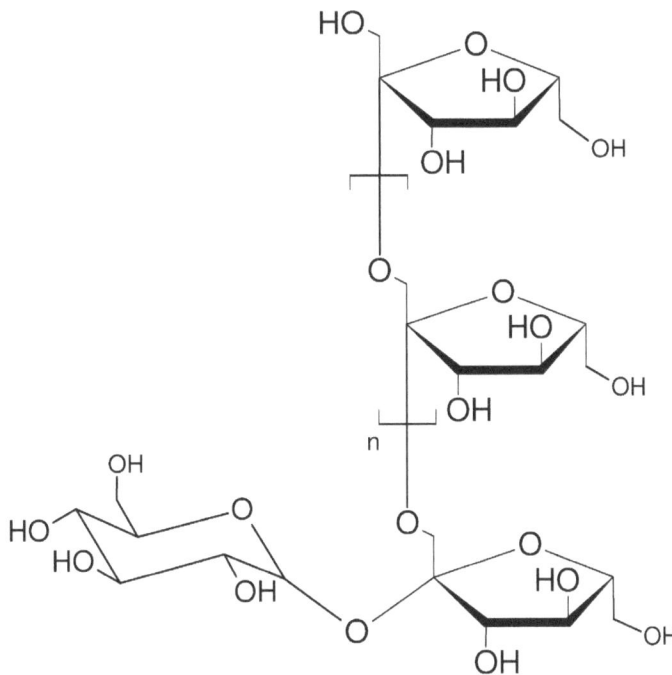

Fig. (2). Chemical structure of inulin polymer

Inulin is also present in the heads and roots of wild cardoon [41]. As a fructan polymer, inulin is characterized by β-2,1 bonds between the linear fructose and oligomers. These bonds prevent the digestion and absorption of inulin in the human small intestine as a dietary carbohydrate, even for the absence of enzymes that allow the fructans hydrolysis. Inulin reaches the colon without modification, where it is hydrolyzed and fermented by bacteria such as bifidobacteria and saccharolytic species, thus assuming prebiotic function [7, 42 - 44]. It is worth highlighting that inulin has low caloric value, positive influence on mineral absorption, on composition of blood lipid, and on colon cancer prevention. For the latter, the increased bifidobacteria component stimulates fecal short-chain fatty acid production and the reduction of cancerogenic substances like ammonia [45, 46]. Furthermore, inulin consumption is suggested for people with diabetes mellitus and can be used to produce low fat foods [7, 42, 47]. The external bracts of artichoke are rich in high molecular weight inulin with a polymerization degree 46 times higher than in other crops such as Jerusalem artichoke, dahlia, and chicory [7]. These differences in degree of polymerization are responsible for the different inulin functions [47]. In fact, long chain inulin can create microcrystals with a creamy texture giving a fat impression to the test. Moreover, inulin can be used to prevent the formation of ice crystals in frozen food [48] and, having a neutral taste while being not sugary, it does not add off-flavor or aftertaste, and can be used as a substitute for butter or margarine in bakery products, dairy products, frozen foods, and condiments [47]. Therefore, artichoke inulin may offer new opportunities to the food industry for developing novel and healthy foods [49].

Sesquiterpene Lactones

Sesquiterpene lactones (STLs) are an important and large group of bioactive terpenoid compounds characteristic of the *Asteraceae* family [50] (Fig. **3**). STLs are lipophilic compounds that are derived from the cyclization of farnesyl pyrophosphate (FPP) with subsequent oxidative modifications [51], and their biological activity is strictly linked to their structure. These secondary metabolites play a key role in plants as deterrents against herbivores and as anti-fungal, anti-bacterial, and allelopathic agents [52]. Moreover, STLs contribute to the bitter taste of many vegetables, including artichoke. In artichoke and cardoon plants, STLs accumulate primarily in the leaves (up to 87 g/kg), while are undetected in the inflorescence at harvest. The main STLs are cynaropicrin and at lower concentration grosheimin and its derivatives [8, 53]. Cynaropicrin and grosheimin are classified as guaianolide STLs [54]. Cynaropicrin is constituted by three isoprene C_5 units, which constitute the typical skeleton of the guaianolide, and it is synthesized *via* the mevalonate pathway. As for other STLs, cynaropicrin likely derives from the common precursor costunolide. The process starts with the cyclization of FPP catalyzed by (+)-germacrene A synthase (GAS) to form germacrene A, which is then subjected to three oxidation steps catalyzed by cytochrome P450 germacrene A oxidase (GAO) to form the corresponding acid. The acid is hydroxylated by (+)-costunolide synthase (COS) into an unstable intermediate that is subject to a non-enzymatic lactonization (addition of the lactone group) forming costunolide and then cynaropicrin [8, 55].Various studies demonstrated that cynaropicrin and many other STLs have several bioactive and medicinal properties, such as antioxidant, anti-photoaging, anti-inflammatory, anti-bacterial, anti-parasitic, antispasmodic, anti-hyperlipoproteinaemia, anti-hyperlipidaemia, anti-hepatitis C virus, anti-tumor, and cytotoxic activity against several types of cancer cells [56 - 60]. Alcohol extracts of artichoke leaves, traditionally used to produce bitter liqueurs, have also been used as a folk medicine remedy for digestion disorders [61]. While several clinical studies continue to explore the medicinal properties of STLs, there is increasing interest in the pharmaceutical industry regarding these compounds, and several studies are on-going aimed at improving the efficiency and the safety of current extraction methods as well the assessment of the bioaccessibility of these compounds [62, 63].

Sesquiterpens lactones

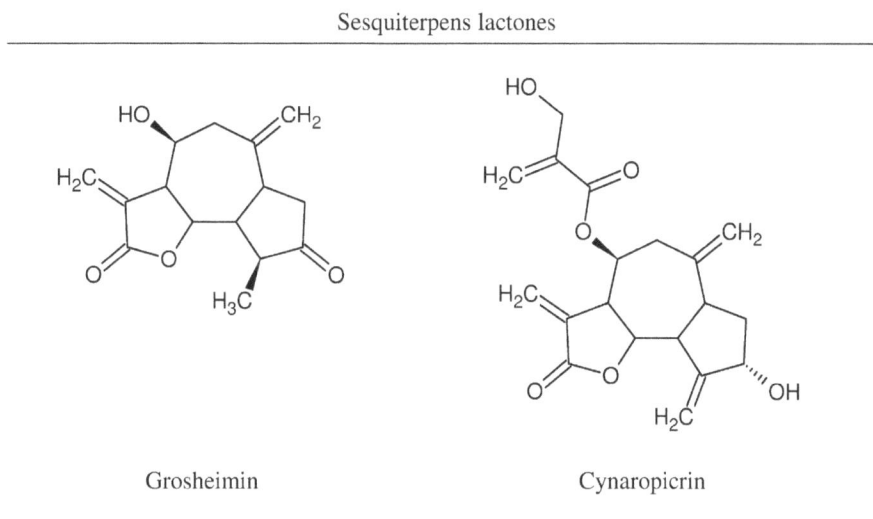

Grosheimin   Cynaropicrin

Fig. (3). Sesquiterpene lactone chemical structures

FACTORS INFLUENCING ARTICHOKE AND CARDOON BIOACTIVE COMPOUND CONTENT

The profile of bioactive compounds contained in artichoke and cardoon is variable and dependent upon several factors such as plant physiological stage, plant portion, genotype, environmental conditions, harvesting time, agronomic practices, postharvest handling, and processing.

Physiological Stage, Tissue, and Genotypic Variations

Polyphenols

Polyphenol distribution is influenced by physiological stage. Within the head, younger tissues contain higher levels of caffeoylquinic acids than senescent tissues, and their level increases from the external to the internal tissue: receptacle> inner bracts> intermediate bracts> outer bracts [7, 24, 64]. Analyzing the polyphenolic profile of different artichoke plant tissues, it was observed that the edible portion (head) contains lower polyphenols than leaves and their abundance could be related also to the genotypes analyzed. Analyzing different tissues (receptacle, inner bracts, outer bracts, floral stem) of nine artichoke genotypes, including local landraces, micropropagated clones, and seed-propagated hybrids, Lombardo and co-workers [24] observed large variations in phenolic profiles. In all cultivars analyzed, chlorogenic acid was mainly detected in the receptacle, floral stem, and inner bracts, with highest concentration in the inner bracts of *Violetto di Sicilia*. Cynarin was present mainly in the floral stem ranging from 21.5 to 90.2 mg/kg of DM in varieties *Concerto* and *Tema 2000*. *Violetto di Sicilia* had the highest content of 3,5- and 1,5-di-O-caffeoylquinic acids in all head portions analyzed, and the highest concentration of luteolin 7-O-rutinoside in the inner bracts. Conversely, *Romanesco* clone C3 and *Concerto* receptacles had the highest concentrations of apigenin 7-O-glucuronide, the predominant flavonoid compound found in all of the head portions analyzed [24]. The total phenol content of the edible heart was significantly lower in seed-propagated hybrids (*Tempo, Opal*, and *Madrigal*) with respect to the two standard vegetatively-propagated varieties (*Violetto di Provenza* and *Catanese*) [25]. Variations in the polyphenol profile were also observed among different clones of a single landrace, *Spinoso di Palermo* [65]. Examining the polyphenolic profiles of the leaves of several artichoke genotypes, different phenolic compounds were identified in leaf blades, petioles, and midribs [66, 67]. Luteolin glycoside derivatives prevailed in leaf blades; while hydroxycinnamic acid derivatives, prevailed in petioles and midribs [66]. The total phenolic content of leaf extracts from nineteen different artichoke cultivars ranged between 141.7 and 264.5 mg of gallic acid equivalents per 100 g FW (highest in *Campagnano, Grato 1*, and *Violetto di Provenza*) [67]. In the same study, leaf content of STLs was also examined. Grosheimin and its deoxydihydroxy derivative, cynaratriol, cynaropicrin and dihydroxy cynaropocrin were detected in the leaves of all the tested cultivars. The highest quantity of the identified STLs was recovered in *Blanca de Tudela* and *Cardena*, while *Romanesco clone C3, Italo*, and *Castellamare* had the lowest concentration [67]. Several authors have investigated the phenolic profiles of the three botanical varieties of *Cynara* species (artichoke, and cultivated and wild cardoon). Pandino and co-workers [22] found similar phenolic profiles in capitula of selected wild and cultivated cardoon accessions and the local artichoke selection *Cimiciusa di Mazzarino,.* All genotypes were characterized by a low content of caffeoylquinic acids and luteolin glycosides, and a high content of apigenin. In relation to the polyphenol concentration, the highest content of polyphenols was observed in the capitula of cultivated cardoon, followed by the wild cardoon *Sylvestris Kamaryna* and the artichoke *Cimiciusa di Mazzarin,* while the lowest concentration was observed in the artichoke *Tondo di Paestum* and in the wild cardoon *Sylvestris Creta.* In another study, analyzing the phenolic profile of leaves and floral stems of six genotypes of artichoke, two accessions of wild cardoon, and a selection of cultivated cardoon, Pandino and co-workers [23] found that luteolin derivates were the major flavonoids in the leaves of artichoke genotypes, while apigenin derivatives were the predominant flavonoids in the leaves of wild and cultivated cardoon. Caffeoylquinic acids were the major phenolic compounds in the floral stem (about 95%) and were highest in artichoke (*Violetto di Sicilia*) and in the wild cardoon (*Sylvestris Creta*). Other authors, comparing the phenolic profiles of two accessions of cultivated cardoon (*Nizza* and *Madrid 1*) and three accessions of wild cardoon (*Siena, Belgio*, and *Madrid 2*), found the di-caffeoylquinic acids as the major phenolic compounds in the leaves with the highest values in the accession *Belgio* [68]. Chlorogenic acid was the major phenolic compound in the buds, and its concentration was highest in the accessions *Madrid 1* and *Madrid 2*. The accession of cultivated cardoon *Nizza* had the highest concentration of luteolin-7-glucoside. Further, analyzing the phenolic profiles of seven genotypes of cultivated cardoons and one accession of wild cardoon, Ciancolini and co-workers [26] found that chlorogenic acid, 1,5-O-dicaffeoylquinic acid, and cynaroside were the most abundant phenolic compounds in both leaves and heads, while the content of apigenin derivatives was negligible. Based on the phenolic content, it is possible to classify artichoke genotypes with: i) high concentrations of polyphenols in the receptacle, optimal for fresh consumption; ii) high concentrations of polyphenols in leaves, floral stem, and outer bracts optimal for the extraction of phenolic compounds; and iii) lower polyphenol concentrations and consequent limited enzymatic oxidation by polyphenol oxidase (PPO) with reduced browning phenomena, optimal to produce processed food [7, 24, 69]. Similarly, the wide variability of phenolic profiles observed among cardoon genotypes suggest that there is great potential to select genotypes for biomass production, fresh consumption, and pharmaceutical purposes [26].

Inulin

Plant portion, physiological stage, species and cultivar also affect the content of inulin. Comparing fifteen genotypes (five for each species) of artichoke, cultivated, and wild cardoon, Raccuia and co-workers [70] observed that root inulin content ranged from 306 to 367 g/kg of dry weight (DW) in artichoke, between 189 and 326 g/kg DW in cultivated cardoon, and from 339 to 398 g/kg DW in wild cardoon. Inulin is accumulated in artichoke and cardoon roots as a reserve carbohydrate, and it is used during the stalk and capitula formation, or to compensate for drought stress conditions. During the growth cycle of artichoke, and cultivated and wild cardoon, inulin reaches the highest concentration in the roots at the end of the vegetative stage, and substantially decreases during the formation of the capitula and seed ripening [71, 72]. Inulin is also present in the receptacle, and its content is 75% of the total glucidic amount and can vary in relation to the artichoke cultivars. Comparing thirty-five cultivars of artichoke, inulin head content ranged between 1 g/100 g of FW in the cv *Hyerois* and 6 g/100 g of FW in *Centofoglie*, *Bayrampasa*, and *Mazzaferrata* [73]. Remarkable variation in inulin content were observed by Lattanzio and co-workers [7], who, comparing nine cultivars of artichoke, assessed that the inulin content of the capitula ranged from 18.9% up to 36.2% on DW basis in the cv *Violet Margot* and *Romanesco*, respectively. Such variation in inulin content could also be due to variations of the physiological stage of collected samples. Moreover, the same authors attributed the observed differences mainly to the morphological traits of the artichoke cultivars examined (Fig. **4**). Within the head, inulin is accumulated mostly in the receptacle, which explain why the cv *Romanesco*, characterized by a very large receptacle, had an inulin content of 36.2% of DW compared to only 21.5% of DW for that of *Violetto di Provenza* characterized by a smaller receptacle [74].

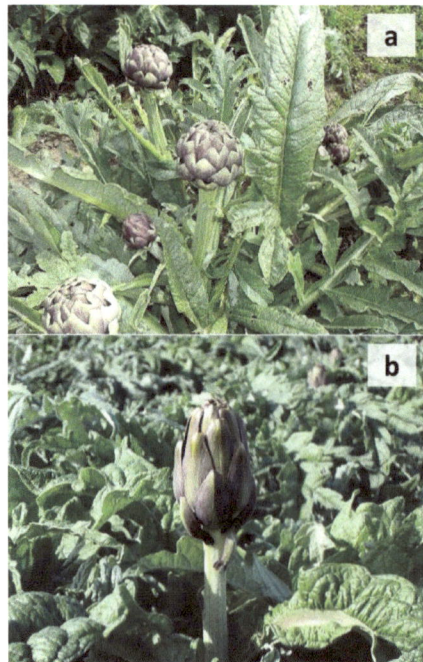

Fig. (4). Detail of (a) *Romanesco* and (b) *Violetto di Provenza* artichoke head morphology. Photos courtesy of Francesco Di Gioia.

Harvesting Time, Environmental, and Pre-harvest Agronomic Factors

The phenolic profile of a specific genotype is widely influenced by harvesting time, environmental conditions, and pre-harvest cultivation practices. Analyzing the phenolic profile of artichoke *Romanesco clone C3* grown in Sicily, Lombardo and co-workers [24] observed 16-fold increase in phenolic content in the floral stem when comparing winter to spring harvest. In the receptacle, the richest portion with regard to phenolic compounds, several compounds such as 1-O-caffeoylquinic acids, chlorogenic acid, caffeic acid, narirutin, and naringenin 7-O-glucoside were not detected in the winter harvest but were present in receptacles from the spring harvest. Testing seventeen different artichoke genotypes grown with a two year cycle, Lombardo and co-workers [75] observed a general decrease of total phenolic content from the first to the second year with a significant 'season × genotype' interaction. Genotypes such as *Tema 2000*, *Violetto di Sicilia clone 4/8*, and *Nobre* exhibited a 44%, 48%, and 78% decrease in the total phenol concentration from the first to the second year, respectively. Conversely, *Tempo* and *Camerys* showed no variation and a 150% increase in total polyphenols occurring from the first to the second year. These findings lead to the conclusion that content of phenolic compounds is influenced by seasonal weather conditions (solar radiation, temperature, rainfall) and harvesting time and that the crop response to these factors is mediated by the plant genotype [24, 75]. The content of polyphenols of both artichoke and cardoon is also influenced by environmental stress conditions. Moderate salinity stress (30 mM of NaCl) for instance, induced a reduction of plant biomass, but increased total polyphenols, chlorogenic acid, cynarin, and luteolin in the leaves of both artichoke and cultivated cardoon [76]. Polyphenol content of artichoke and cardoon leaves is also influenced by the source of salinity. Potassium chloride salinity stress (30 mM of KCl) enhanced total phenolic and flavonoid contents more than equivalent levels of NaCl and $CaCl_2$ salinity stress [77]. Environmental conditions also affect the accumulation of inulin. Comparing two growing locations Raccuia and co-workers [41] observed higher inulin concentration in the roots of wild cardoon in the location with lower precipitation and lower winter temperatures. The positive correlation between accumulation of inulin in the roots and drought stress conditions and low temperatures may be explained by its potential role in the regulation of the osmotic potential and as a cryoprotectant in plant cells [41]. From this perspective, irrigation management can significantly affect the accumulation of inulin, and higher irrigation rates can reduce root inulin concentration. Moreover, heads harvested during the winter season have higher inulin concentrations than those harvested in the spring [78]. Among the pre-harvest agronomic practices, artichoke and cardoon polyphenolic profiles may be influenced by planting time, plant density, nutrient and water availability, mycorrhizal associations, and microbial inoculants. Lombardo and co-workers [75] observed a linear increase of total polyphenol content with increasing plant density from 1.0 to 1.8 plant/m^2 and increased by 13% in plants established in twin-rows rather than in single rows. Planting in October rather than in September reduced the concentration of mono- and di-caffeoylquinic acids in the heads of seed-propagated artichoke *Romolo* and *Istar* [79]. Nevertheless, coating the seeds for the second planting time with a consortium of arbuscular mycorrhizal fungi and *Trichoderma atroviride* increased the polyphenol content of the primary heads, especially in *Romolo*. A similar increase of polyphenols was observed in plants of artichoke *Romanesco clone C3* inoculated at transplanting with other commercial inoculants containing *Glomus intraradices* and *G. massae* or *G. intraradices* and a mix of beneficial bacteria [80]. Artichoke head polyphenol content was also enhanced by foliar applications of the elicitor methyl jasmonate [81]. In another study conducted on seed-propagated artichoke and cardoon, increasing the nutrient solution concentration (4, 20, 36, 52, or 68 m$_{equiv}$/L) reduced total polyphenols, chlorogenic acid, caffeic acid, cynarin, and luteolin concentration in the leaves of both artichoke and cardoon [82].

Post-Harvest and Processing Factors

Bioactive compound concentration of artichoke heads is also influenced by post-harvest handling and storage conditions, as well as by processing practices. After storage for 8 days at 5 °C, total phenolic acid content of *Blanca de Tudela* artichoke heads increased, particularly in the inner bracts [10]. However, after storage, total phenolic acid content was influenced by the type of film used for the modified atmosphere packaging, and significantly higher concentrations of phenolic acids were observed in the internal bracts of artichoke packed with low density polyethylene (LDPE) and polyvinylchloride (PVC) compared to perforated polypropylene [83]. Palermo and co-workers [80] observed that phenolic compound concentration of artichoke heads increased even after cooking and was higher after steaming compared to microwaving. The storage temperature and time strongly influenced the inulin content in artichoke heads. Leroy and co-workers [84] observed that inulin depolymerization with consequent release of free fructose and sucrose, occurred with increased storage time and temperature. In the same study, inulin content was preserved, as consequence of a lower depolymerization rate, when artichokes were packed and stored in a micro-perforated polypropylene film.

PHYSIOLOGICAL FATE

Bioaccessibility

Before studying the mechanisms of action of artichoke polyphenols and their role in disease prevention, a thorough understanding of the factors that can influence their release from the vegetable matrix, their degree of absorption, and their fate in the human body, is fundamental [85]. Polyphenols bioactivity is firstly dependent on their digestive stability under gastrointestinal (GI) conditions, their release from the food matrix, and their availability for absorption, referred to as bioaccessibility. The bioaccessibility, in turn, can affect the polyphenols blood release and their capacity to reach target tissues and to have positive effects on human health [86]. To evaluate the bioaccessibility, different methods have been proposed, both static and dynamic, that simulate an *in vitro* GI digestion. These procedures generally reproduce oral, gastric, and small intestinal digestion, sometimes followed by Caco-2 cell uptake [87]. The *in vitro* methods are more rapid, less expensive, have high reproducibility, pose no ethical limitations, are conducted under controlled conditions, and overall are considered more suitable for mechanism of action studies. The simulated digestion, to reproduce the human physiological conditions of the GI tract, considers the chemical composition of digestive fluids, pH, and the residence time characteristics of each step [88]. One refined model is the computerized dynamic Dutch TNO gastrointestinal tract model 1 (TIM1), which simulates dynamic aspects of digestion, such as transport of digested matrix, variable enzyme concentrations, and pH changes over time [89]. This system has been recently complemented with a microbial fermentation step (TIM2) which simulates the large intestine and the colonic microbial activity [90]. Recently, other two dynamic systems have been proposed; NutraScan Artificial Gut tested by researchers at CSIRO's Division of Human and Animal Nutrition (Adelaide, South Australia), where they have developed an automated protocol for predicting glycemic index in cereal based foods, and the SHIME® and M-SHIME® systems (Ghent University, Belgium). SHIME® is a scientifically validated dynamic model to study physicochemical, enzymatic, and microbial parameters in the GI tract under *in vitro* controlled conditions, supplemented with a mucosal compartment integrated in the colonic regions, M-SHIME®. Besides the more sophisticated described dynamic systems, most models reported in the literature are static. The static digestive *in vitro* models have also been used to predict the artichoke polyphenol stability and possible modifications in the GI tract [91]. These authors have studied the composition, antioxidant activity, and stability of polyphenols present in artichoke infusions after GI digestion, reporting high stability of the identified flavonoids. A study performed by Garbetta and co-workers [92] evaluated the influence of GI digestion on the antioxidant effect of artichoke polyphenols, showing that *in vitro* digestion did not modify the antioxidant activity of polyphenols, with the exception of 1,5-O-dicaffeoylquinic acid that was less active. D'Antuono and co-workers [32] studied the bioaccessibility of artichoke polyphenols using an *in vitro* digestion model coupled with Caco-2 human intestinal cell line. They recovered 55.8% of total artichoke polyphenol bioaccessibility; chlorogenic acid was highly stable and bioaccessible (70.0%), while the two main di-caffeoylquinic acids, (3,5-O- and 1,5-O-) had lower stability and bioaccessibility (41.3% and 50.3%, respectively), highlighting their sensitivity to gastric and small intestinal digestive conditions. More recently, Juániz and co-workers [93] reported the bioaccessibility and metabolism of phenolic compounds identified in both raw and cooked cardoon, after a simulated GI digestion. The authors reported that the cooking process exerted a positive effect on the bioaccessibility of polyphenols. After *in vitro* GI digestion, only 2% of the total amount of polyphenolic compounds in raw cardoon was bioaccessible, while in cooked cardoon samples, between 60% and 67% of the total amount of polyphenolic compounds remained unmodified and thus bioaccessible. Exposing the samples to fecal fermentation after the simulated gastrointestinal digestion, an important microbial metabolic activity was observed at the expense of the caffeoylquinic acid, flavonoids and luteolin derivatives, with the production of several catabolites [93]. Using an *in vitro* digestion model coupled with high resolution mass spectrometry analysis, Colantuono and co-workers [62] tested the bioaccessibility of polyphenols and cynaropicrin from bread enriched with artichoke stem powder. They observed that 82% of totally bioaccessible polyphenols and 74% of cynaropicrin were released during the duodenal digestion, while 88% of caffeic acid was released in the colon step.

Bioavailability and Pharmacokinetics

The term "bioavailability" borrowed from the pharmacology field indicates the "rate and extent to which a drug reaches its site of action" [94]. This definition, from both physical and ethical points of view, is not attainable in humans, therefore the concept was modified for a more general definition such that "the fraction of an ingested nutrient or compound that reaches the systemic circulation and the specific sites where it can exert its biological action" [95]. The bioavailability varies highly among polyphenols and the most abundant compounds may not necessarily have the greatest bioavailability. Polyphenols bioavailability can be determined *in vitro*, using tissues or human cellular lines, and *in vivo*, using animal models and human studies. *In vivo* experiments, performed to explain the physiological mechanisms related to absorption and transport of polyphenols, generally are carried out with standard compounds that often do not reflect the food matrix composition and thus are not suitable for the understanding of produced effects. Concentrations generally used for *in vitro* models are not compatible with those recovered with *in vivo* plasma. Models *in vivo* cover a concentration range between μmolar and mmolar, while poly-phenol plasmatic concentration is in the order of nmolar. Nevertheless, the use of high concentrations may be functional to understand the biological mechanism of action exerted by polyphenols [94]. The *in vivo* animal studies provide results that are more compatible with the actual bioavailability levels in humans [96, 97], with the limitation related to the differences between the animal and human genome and microbiome. These differences can lead to some discrepancies in understanding the results and in the extrapolation to the human mechanism. In fact, some phenolic compounds, such as quercetin, are metabolized differently in rodents and humans [98, 99]. Concerning artichoke, *in vitro* studies using the Caco-2 model revealed very low small-intestinal accumulation (uptake) of polyphenols (0.16%), recovering apigenin-7-O-glucoside, caffeic acid, and coumaric acid derivatives as the only bioavailable compounds at a very low concentration (0.2%) [32]. The presence of the two hydroxycinnamic acid derivatives suggest the occurrence in the intestinal epithelium, simulated by Caco-2 cells, of metabolic activity on the chlorogenic acid [32]. Other *in vitro* Caco-2 experiments performed on chlorogenic acid, confirmed its low bioavailability in native form, although the detection of metabolic products in the basolateral site was neglected, with consequent underestimation of chlorogenic acid absorption [100]. *In vivo* studies on chlorogenic acid using animal models demonstrated that it could be absorbed directly in the stomach and it was identified in its intact form, in both gastric veins and aorta of rats [101]. Other research groups have assessed the influence of gut microflora on the absorption and metabolism of chlorogenic acid. Their results show that high amounts of chlorogenic acid reaches the colon almost unbroken and then it is hydrolyzed by the microbiota esterase activity with the production of caffeic acid and quinic acid. Further, caffeic acid is converted by microorganism to dihydroferulic and dihydrocaffeic acid, which can be easily absorbed through the colonic epithelium, reaching the blood stream at high concentrations [100 - 104]. Only a few human studies on the absorption and metabolism of artichoke polyphenols have been reported [105, 106]. A pilot study was performed to evaluate the absorption, metabolism, and pharmacokinetics of artichoke polyphenols after oral intake of cooked artichoke heads by human subjects [105]. The authors reported a plasma maximum chlorogenic acid concentration of 6.4 ng mL^{-1} after 1 h and its complete loss within 2 h. This rapid plasma detection of chlorogenic acid could be due to its direct absorption in the stomach as reported by Lafay and co-workers [101] in studies on rats. Further, a large amount of chlorogenic acid reaches the colon where it is metabolized by gastrointestinal microflora to caffeic acid and quinic acid. Caffeic acid reached a maximum plasma concentration of 19.5 ng/mL within 1 h, while the pharmacokinetics of absorption for ferulic acid showed a biphasic profile, with 6.4 ng/mL after 1 h and 8.4 ng/mL after 8 h. Simultaneously, after 8 h, the authors observed a significant increase in the total concentration of dihydrocaffeic and dihydroferulic acid; however, no flavonoids were detected in the blood stream [105]. Moreover, after oral administration of artichoke leaf extracts to humans, the polyphenol targets were not detected in plasma and urine samples, but their metabolites such as caffeic, ferulic, isoferulic, dihydrocaffeic, and dihydroferulic acids were [106]. Pharmacokinetic studies on dicaffoylquinic acids are focused on cynarin (1,3-O-dicaffoylquinic acid) and 1,5-O-dicaffoylquinic acid. The latter is one of the main phenolics present in artichoke heads and leaves and it is a potent HIV-1 integrase inhibitor [107]. These authors developed an analytical method to simultaneously determine 1,5-O-dicaffoylquinic acid and its two active metabolites, 1-caffeoyl-5-feruoylquinic acid and 1,5-O-diferuoylquinic acid in human plasma of five healthy volunteers. The recovered blood concentration, 2 h after a single oral administration of 600 mg 1,5-O-dicaffoylquinic acid, was 83.3 ng/mL. At the same time its metabolites, 1-caffeoyl-5-feruoylquinic acid and 1,5-O-diferuoylquinic acid, were detected at concentration of 5.3 ng/mL, and 62.9 ng/mL, respectively [107]. Recently, another *in vivo* study on rats assessed the blood concentrations of 1,3-O-dicaffeoylquinic acid and caffeic acid [108], suggesting that the two phenolics may be present in the peripheral tissues of rats, with a half-life ($t_{1/2}$) of elimination for caffeic acid lower than that of 1,3-O-dicaffeoylquinic acid. The results showed that caffeic acid, similar to its structural analog ferulic acid [109], is absorbed and eliminated faster than 1,3-O-dicaffeoylquinic acid.

HEALTH PROMOTING ACTIVITIES OF ARTICHOKE AND CARDOON POLYPHENOLS

Antioxidant Effect

The health-promoting properties of the polyphenols identified in artichoke and cardoon are mainly attributable to their antioxidant capacity. Polyphenols can scavenge different reactive radical oxygen species (ROS) such as superoxides, hydroxyls, and peroxyls and they are also able to chelate metal ions [110, 111]. *In vitro* studies performed by Gebhardt [112], demonstrated antioxidant capacities of polyphenols from artichoke leaves against oxidative stress induced by hydroperoxide in cultured rat hepatocytes. Jimenez-Escrig and co-workers [113] measured the antioxidant effect of aqueous extracts of artichoke heads using *in vitro* chemical methods: free radical DPPH˙ scavenging (DPPH) and ferric-reducing antioxidant power (FRAP). In this study, 1 g of artichoke dry matter, containing 31-58 mg of polyphenols, exerted an antioxidant capacity, assayed by the DPPH method, equivalent to 29.2 mg of vitamin C and 77.9 mg of vitamin E; using the FRAP assay the capacity was equivalent to 62.6 mg of vitamin C and 159 mg of vitamin E. Falè and co-workers [91], using the DPPH assay, demonstrated the high antioxidant power of artichoke infusion, while Pereira and co-workers [114] showed that the infusion had higher antioxidant capacity than artichoke dietary supplements. Other authors have also reported that the plant physiological state influences the antioxidant activity of artichoke, suggesting that baby artichokes exhibited higher scavenging capacity than mature ones [115]. The antioxidant capacity of artichoke polyphenols may also be influenced by processing practices. Ferracane and co-workers [116] found that the antioxidant activity of artichokes substantially increased after cooking, especially after steaming (up to 15-fold) and boiling (up to 8-fold). These results suggest that the cooking process can improve the extractability of bioactive polyphenols from the plant tissue. The antioxidant capacity of artichoke polyphenols can also be affected by their stability under GI conditions. Falé and co-workers [91] reported that the high stability and bioaccessibility of the main polyphenols identified in artichoke infusion during the *in vitro* digestion translated to high antioxidant activity during the digestion process. Other antioxidant assays, more physiologically related to the *in vivo* human condition, using different cell lines have been used to investigate the *in vitro* scavenging effect of bioactive compounds on reactive oxygen species (ROS). Garbetta and co-workers [92], using the HT-29 cell line, assessed the antioxidant power induced by artichoke head extract and by each of the single polyphenols identified. The results highlighted that artichoke extract was highly antioxidant compared to the single pure polyphenols. Comparing the single polyphenols, chlorogenic and 1,5-O-dicaffeoylquinic acid were more active than 3,5-O-dicaffeoylquinic acid. The same authors verified that *in vitro* digestion did not modify the antioxidant activity of artichoke polyphenols, except for 1,5-O-dicaffeoylquinic acid which became less active.

Anti-Inflammatory Effects

Currently, only a few studies have assessed the anti-inflammatory capacity of artichoke and cardoon. Most of the studies have focused attention on chlorogenic acid, the main phenolic compound found in artichoke, and its metabolite, caffeic acid. Recent studies reported the inhibitory effect exerted by chlorogenic acid on the secretion of interleukin-8 (IL-8) and on mRNA expression. Analog results were also reported for caffeic acid in Caco-2 cells and for chlorogenic acid and caffeic acid in a model of induced colitis by dextran sulfate sodium [117, 118]. Recently, the same authors proposed the mechanism of inhibition of IL-8 and PKD-IKKNFκB pathway as an effect of intracellular ROS scavenging [119], attributing the anti-inflammatory capacity of chlorogenic acid and caffeic acid to the presence of catecholic groups. The reported results also lead to the conclusion that all natural compounds with a catechol group can contribute to the prevention of inflammatory diseases. It is therefore possible to speculate that the same conclusions apply also in the case of artichoke and cardoon, although no *in vitro* or *in vivo* data are currently available for these crops. Recently, Ben Salem and co-workers [120] studied the anti-inflammatory capacity of artichoke leaf extract (ALE) using the *in vivo* carrageenan-induced paw edema (Carr), a well-known acute model of inflammation. The results obtained showed a dose-dependent protection of ALE against the Carr-induced acute inflammation through inhibition of inflammatory mediators, such as histamine, serotonin, and prostaglandins. Moreover, according to the authors, the inflammation induced by Carr generates ROS, which, as already suggested, can play an important role in various forms of inflammation. Finally, the authors conclude that the possible individual or synergistic anti-inflammatory effect of the leaf extract is attributable to chlorogenic acid and flavonoids [120]. Among artichoke and cardoon flavonoids, apigenin-7-O-glucoside had the capacity to block the enzymes involved in the inflammation process, such as lipoxygenases and cyclooxygenases [121] with consequent inhibition of NFkB proinflammatory molecules and the infiltration of neutrophils in tissues.

Hypocholesterolemic Effects

Hypocholesterolemic activity is attributable to cholesterol biosynthesis reduction and to Low Density Lipoproteins (LDL) oxidation inhibition. Several studies demonstrated the artichoke polyphenols affect lipid metabolism, through the reduction of cholesterol and endogenous triglyceride production. The effect of artichoke polyphenol extract was investigated with both *in vitro* and *in vivo* experiments. In particular, *in vitro* studies were performed on artichoke leaves and highlighted the properties of extracts in the support of cholesterol esterification in serum [122], and in the reduction of atherosclerotic plaques [123]. Gebhardt & Fausel [124] studied the possible influence of the polyphenolic components of artichoke leaves on cholesterol biosynthesis inhibition in rat hepatocytes. The authors demonstrated that the breakdown of the cholesterol synthesis is probably related to the enzymatic inhibition of HMG-CoA reductase from the flavonoidic components of artichoke leaves, such as luteolin and cynaroside, and from cynarin and caffeic acid, but to a lower extent [124]. Moreover, Preziosi and co-workers [125] tested the effects of cynarin and caffeic acid on rabbit serum, assessing the reduction of blood cholesterol levels for more than 72 hours initial treatment. In addition, Brown & Rice-Evans [126] have demonstrated a dose-dependent inhibition of LDL oxidation by flavonoids, determining that luteolin-7-O-glucoside was less effective than luteolin aglycon, assuming also a role of flavonoids in copper chelation. To highlight the structure-activity relationships, other *in vitro* experiments using human LDL have demonstrated the antioxidative properties of artichoke head hydroalcoholic extract compared with chlorogenic acid and caffeic acid [32]. The authors identified caffeic acid as the most active against LDL oxidation, followed by chlorogenic acid, and total artichoke extract. These results could be explained considering a different effect related to the chemical structures of single compounds and to the possible synergic/antagonistic effect occurring in the complex mixture of polyphenols in artichoke head extracts [32]. Comparable results were obtained by Coinu and co-workers [127] from artichoke leaves and outer bract extracts, with leaves more active in the inhibition of LDL oxidation than bracts, highlighting also the non-direct correlation with polyphenol content. Another *in vitro* study using cultured monocytes and endothelial cells demonstrated the beneficial characteristics of artichoke leaf extract in the prevention of LDL oxidation [128]. Further, Jimenez-Escrig and co-workers [113] indicated that artichoke head hydroalcholic extract had good activity in the inhibition of oxidation of *in vitro* human LDL. Recently, in clinical trials with patients affected by metabolic disease Rezazadeh and co-workers [129] showed the positive effect of artichoke leaf extract supplementation in the reduction of LDL oxidation and of triglyceride serum concentration.

Hypoglycemic Effect

Artichoke is also recognized for the capacity to reduce glycaemia; this effect could be attributed to the ability of polyphenols to modulate cellular defense systems acting both as an antioxidant and as an inhibitor of enzymatic pathways. Nazni and co-workers [130] have reported *in vivo* studies on type 2 diabetic individuals after intake of five bakery foods enriched with 6 g of artichoke powder. The results obtained on 30 patients recorded a significant reduction of post-prandial blood glucose when consuming biscuits enriched with artichoke. Recently, other authors have highlighted the hypoglycemic properties of artichoke leaves and heads using a rat model [131]. The authors working on type 2 diabetic rats, assessed the effect of both extract (leaves and heads) on the reduction of glucose in a dose-dependent manner, with the leaf extract being more active than head extract. The authors attributed this effect to the flavonoid component, and specifically to luteolin-7-O-glucoside, the most abundant flavonoid present in leaves.

Anticancer Effect

Among the biological activities attributed to polyphenols, anticancer properties have been most widely studied, highlighting their capacity to reduce the disease incidence. Some authors have evaluated the capacity of polyphenols to affect the cellular cycle, the proliferation, and the apoptosis on various human cancer cells and using a mice model. Studies have already been performed on the anti-proliferative and apoptotic capacity of artichoke and cardoon phenolic extracts on hepatoma, breast cancer, mesothelioma, and multiple myeloma cells lines [132 - 135]. All of the cited studies demonstrated that cell-growth and viability decreased using high doses of artichoke and/or cardoon extracts, while low concentrations of the artichoke extract caused premature senescence of breast cancer cells, most likely through activation of epigenetic and ROS-mediated mechanisms [133]. Further, Pulito and co-workers [134] have demonstrated the apoptotic effect of artichoke leaf extract on mesothelioma cell lines, and documented a simultaneous reduction of DNA damage in non-cancerous cells (mesothelial cell), highlighting its selective activity. The anti-hepatocellular carcinoma activity of artichoke leaf infusion was also studied by Pereira and co-workers [114] with a 50% net cell growth inhibition (GI$_{50}$ values) at a sample concentration of 52 μg/mL and a positive correlation with flavonoids and phenolic acids content. Other authors [136] described in *in vitro* and *in vivo* studies, that water-soluble polymer-encapsulated particles (nanoparticles) of luteolin exerted higher growth inhibition of lung cancer cells and squamous cell carcinoma of head and neck, than luteolin, the naturally occurring compound. Finally, *in vivo* experiments using the nude mouse model treated with xenografted mesothelioma tumors, reported that a 3-week feeding with artichoke leaf extract significantly decreased and in a dose-dependent manner, the growth of xenografted mesothelioma tumors. Further, the pretreatment of xenografted mesothelioma cells with artichoke leaf extract reduced cell engraftment in nude mouse, highlighting its therapeutic property against mesothelioma [134].

CONCLUDING REMARKS

Artichoke and cardoon are important ingredients of the Mediterranean diet and have been traditionally used for their medicinal properties. Both vegetable crops and their common wild relative are considered a rich source of health-promoting compounds, such as polyphenols, inulin, and STLs. As research studies continue to investigate and expand our knowledge of the beneficial properties and the mechanisms of action of these bioactive compounds, the interest for using these crops to develop novel functional food and pharmaceutical products continues to grow. From this perspective, further research investment is required to: i) select genotypes characterized by higher concentrations of specific bioactive compounds; ii) study pre-harvest agronomic techniques that can enhance the concentration of bioactive compounds; iii) define post-harvest and processing procedures aimed at preserving the content of bioactive compounds and/or enhancing their bioaccessibility and bioavailability; iv) exploit crop by-products more efficiently to extract bioactive compounds or develop novel food products and drugs; and v) continue to investigate and assess the beneficial health properties of single bioactive compounds, plant extracts, and novel artichoke-based food and pharmaceutical products.

CONSENT FOR PUBLICATION

Not applicable.

CONFLICT OF INTEREST

The author (editor) declares no conflict of interest, financial or otherwise.

ACKNOWLEDGEMENTS

Declare none.

REFERENCES

[1]   Rottenberg A, Zohary D. Wild genetic resources of cultivated artichoke. IV International Congress on Artichoke. 307-14.

[2]   Sonnante G, Pignone D, Hammer K. The domestication of artichoke and cardoon: from Roman times to the genomic age. Ann Bot 2007; 100(5): 1095-100.
[http://dx.doi.org/10.1093/aob/mcm127] [PMID: 17611191]

[3]   Foury C. Ressources génétiques et diversification de l'artichaut (*Cynara scolymus* L.). Acta Hortic 1989; (242): 155-66.
[http://dx.doi.org/10.17660/ActaHortic.1989.242.20]

[4]   Bianco VV. Present situation and future potential of artichoke in the Mediterranean basin. Acta Hortic 2005; (681): 39-55.
[http://dx.doi.org/10.17660/ActaHortic.2005.681.2]

[5]   Marzi V, Lattanzio V, Vanadia S. Il Carciofo Pianta Medicinale. Palo, Bari: Ed. Liantonio 1975.

[6]   Food FAO. Trade Statistics. FAO, & Italy, 2015; last update December 2017 January 17; 2018 January 17; access

[7]   Lattanzio V, Kroon PA, Linsalata V, Cardinali A. Globe artichoke: A functional food and source of nutraceutical ingredients. J Funct Foods 2009; 1: 131-44.
[http://dx.doi.org/10.1016/j.jff.2009.01.002]

[8]   Menin B, Comino C, Portis E, *et al.* Genetic mapping and characterization of the globe artichoke (+)-germacrene A synthase gene, encoding the first dedicated enzyme for biosynthesis of the bitter sesquiterpene lactone cynaropicrin. Plant Sci 2012; 190: 1-8.
[http://dx.doi.org/10.1016/j.plantsci.2012.03.006] [PMID: 22608514]

[9]   Lattanzio V, Lafiandra D, Morone-Fortunato I. Composizione chimica e valore nutritivo del carciofo (Cynara scolymus L.).Studi sul Carciofo. Industrie Grafiche Laterza 1981; pp. 117-208.

[10]   Gil-Izquierdo A, Gil MI, Conesa MA, Ferreres F. The effect of storage temperatures on vitamin C and phenolics content of artichoke (*Cynara scolymus* L.) heads. Innov Food Sci Emerg Technol 2001; 2: 199-202.
[http://dx.doi.org/10.1016/S1466-8564(01)00018-2]

[11]   Romani A, Pinelli P, Cantini C, Cimato A, Heimler D. Characterization of Violetto di Toscana, a typical Italian variety of artichoke (*Cynara scolymus* L.). Food Chem 2006; 95: 221-5.
[http://dx.doi.org/10.1016/j.foodchem.2005.01.013]

[12]   Pandino G, Lombardo S, Williamson G, Mauromicale G. Polyphenol Profile and Content in Wild and Cultivated *Cynara Cardunculus* L. Ital J Agron 2012; 7(e35): 254-61.

[13]   Petropoulos SA, Pereira C, Ntatsi G, Danalatos N, Barros L, Ferreira ICFR. Nutritional value and chemical composition of Greek artichoke genotypes. Food Chem 2018; 267: 296-302.
[http://dx.doi.org/10.1016/j.foodchem.2017.01.159] [PMID: 29934171]

[14]   Saénz Rodriguez T, García Giménez D, de la Puerta Vázquez R. Choleretic activity and biliary elimination of lipids and bile acids induced by an artichoke leaf extract in rats. Phytomedicine 2002; 9(8): 687-93.

[http://dx.doi.org/10.1078/094471102321621278] [PMID: 12587687]

[15]   Gebhardt R. Anticholestatic activity of flavonoids from artichoke (*Cynara scolymus* L.) and of their metabolites. Med Sci Monit 2001; 7 (Suppl. 1): 316-20.
[PMID: 12211745]

[16]   Thompson Coon JS, Ernst E. Herbs for serum cholesterol reduction: a systematic view. J Fam Pract 2003; 52(6): 468-78.
[PMID: 12791229]

[17]   Fratianni F, Tucci M, De Palma M, Pepe R, Nazzaro F. Polyphenolic composition in different parts of some cultivars of globe artichoke (*Cynara cardunculus* var. *scolymus* (L.) Fiori). Food Chem 2007; 104: 1282-6.
[http://dx.doi.org/10.1016/j.foodchem.2007.01.044]

[18]   Lattanzio V, Cardinali A, Di Venere D, Linsalata V, Palmieri S. Browning phenomena in stored artichoke (*Cynara scolymus* L.) heads: enzymic or chemical reactions? Food Chem 1994; 50: 1-7.
[http://dx.doi.org/10.1016/0308-8146(94)90083-3]

[19]   Llorach R, Espín JC, Tomás-Barberán FA, Ferreres F. Artichoke (*Cynara scolymus* L.) byproducts as a potential source of health-promoting antioxidant phenolics. J Agric Food Chem 2002; 50(12): 3458-64.
[http://dx.doi.org/10.1021/jf0200570] [PMID: 12033811]

[20]   Orlovskaya TV, Luneva IL, Chelombitko VA. Chemical composition of *Cynara scolymus* leaves. Chem Nat Compd 2007; 43: 239-40.
[http://dx.doi.org/10.1007/s10600-007-0093-2]

[21]   Gatto A, De Paola D, Bagnoli F, Vendramin GG, Sonnante G. Population structure of *Cynara cardunculus* complex and the origin of the conspecific crops artichoke and cardoon. Ann Bot 2013; 112(5): 855-65.
[http://dx.doi.org/10.1093/aob/mct150] [PMID: 23877076]

[22]   Pandino G, Courts FL, Lombardo S, Mauromicale G, Williamson G. Caffeoylquinic acids and flavonoids in the immature inflorescence of globe artichoke, wild cardoon, and cultivated cardoon. J Agric Food Chem 2010; 58(2): 1026-31.
[http://dx.doi.org/10.1021/jf903311j] [PMID: 20028012]

[23]   Pandino G, Lombardo S, Mauromicale G, Williamson G. Phenolic acids and flavonoids in leaf and floral stem of cultivated and wild *Cynara cardunculus* L. genotypes. Food Chem 2011; 126: 417-22.
[http://dx.doi.org/10.1016/j.foodchem.2010.11.001]

[24]   Lombardo S, Pandino G, Mauromicale G, *et al.* Influence of genotype, harvest time and plant part on polyphenolic compositionof globe artichoke [*Cynara cardunculus* L. var. *scolymus* (L.) Fiori]. Food Chem 2010; 119: 1175-81.
[http://dx.doi.org/10.1016/j.foodchem.2009.08.033]

[25]   Bonasia A, Conversa G, Lazzizera C, Gambacorta G, Elia A. Morphological and qualitative characterisation of globe artichoke head from new seed-propagated cultivars. J Sci Food Agric 2010; 90(15): 2689-93.
[http://dx.doi.org/10.1002/jsfa.4141] [PMID: 20853305]

[26]   Ciancolini A, Alignan M, Pagnotta MA, Vilarem G, Crinò P. Selection of Italian cardoon genotypes as industrial crop for biomass and polyphenol production. Ind Crops Prod 2013; 51: 145-51.
[http://dx.doi.org/10.1016/j.indcrop.2013.08.069]

[27]   Ramos PA, Santos SA, Guerra ÂR, *et al.* Phenolic composition and antioxidant activity of different morphological parts of *Cynara cardunculus* L. var. *altilis* (DC). Ind Crops Prod 2014; 61: 460-71.
[http://dx.doi.org/10.1016/j.indcrop.2014.07.042]

[28]   Ignat I, Volf I, Popa VI. A critical review of methods for characterisation of polyphenolic compounds in fruits and vegetables. Food Chem 2011; 126(4): 1821-35.

[http://dx.doi.org/10.1016/j.foodchem.2010.12.026] [PMID: 25213963]

[29] Manach C, Scalbert A, Morand C, Rémésy C, Jiménez L. Polyphenols: food sources and bioavailability. Am J Clin Nutr 2004; 79(5): 727-47.
[http://dx.doi.org/10.1093/ajcn/79.5.727] [PMID: 15113710]

[30] Hättenschwiler S, Vitousek PM. The role of polyphenols in terrestrial ecosystem nutrient cycling. Trends Ecol Evol (Amst) 2000; 15(6): 238-43.
[http://dx.doi.org/10.1016/S0169-5347(00)01861-9] [PMID: 10802549]

[31] Jones CG, Hartley SE. A protein competition model of phenolic allocation. Oikos 1999; 86: 27-44.
[http://dx.doi.org/10.2307/3546567]

[32] D'Antuono I, Garbetta A, Linsalata V, Minervini F, Cardinali A. Polyphenols from artichoke heads (*Cynara cardunculus* (L.) subsp. *scolymus* Hayek): *in vitro* bio-accessibility, intestinal uptake and bioavailability. Food Funct 2015; 6(4): 1268-77.
[http://dx.doi.org/10.1039/C5FO00137D] [PMID: 25758164]

[33] Schütz K, Kammerer D, Carle R, Schieber A. Identification and quantification of caffeoylquinic acids and flavonoids from artichoke (*Cynara scolymus* L.) heads, juice, and pomace by HPLC-DAD-ESI/MS(n). J Agric Food Chem 2004; 52(13): 4090-6.
[http://dx.doi.org/10.1021/jf049625x] [PMID: 15212452]

[34] Schütz K, Persike M, Carle R, Schieber A. Characterization and quantification of anthocyanins in selected artichoke (*Cynara scolymus* L.) cultivars by HPLC-DAD-ESI-MSn. Anal Bioanal Chem 2006; 384(7-8): 1511-7.
[http://dx.doi.org/10.1007/s00216-006-0316-6] [PMID: 16534575]

[35] Fernández J, Curt MD, Aguado PL. Industrial applications of *Cynara cardunculus* L. for energy and other uses. Ind Crops Prod 2006; 24(3): 222-9.
[http://dx.doi.org/10.1016/j.indcrop.2006.06.010]

[36] Wellington K, Jarvis B. Silymarin: a review of its clinical properties in the management of hepatic disorders. BioDrugs 2001; 15(7): 465-89.
[http://dx.doi.org/10.2165/00063030-200115070-00005] [PMID: 11520257]

[37] Pinelli P, Agostini F, Comino C, Lanteri S, Portis E, Romani A. Simultaneous quantification of caffeoyl esters and flavonoids in wild and cultivated cardoon leaves. Food Chem 2007; 105(4): 1695-701.
[http://dx.doi.org/10.1016/j.foodchem.2007.05.014]

[38] Lattanzio V, Cardinali A, Ruta C, *et al.* Relationship of secondary metabolism to growth in oregano (*Origanum vulgare* L.) shoot cultures under nutritional stress. Environ Exp Bot 2009; 65: 54-62.
[http://dx.doi.org/10.1016/j.envexpbot.2008.09.002]

[39] Niggeweg R, Michael AJ, Martin C. Engineering plants with increased levels of the antioxidant chlorogenic acid. Nat Biotechnol 2004; 22(6): 746-54.
[http://dx.doi.org/10.1038/nbt966] [PMID: 15107863]

[40] Hellwege EM, Czapla S, Jahnke A, Willmitzer L, Heyer AG. Transgenic potato (*Solanum tuberosum*) tubers synthesize the full spectrum of inulin molecules naturally occurring in globe artichoke (*Cynara scolymus*) roots. Proc Natl Acad Sci USA 2000; 97(15): 8699-704.
[http://dx.doi.org/10.1073/pnas.150043797] [PMID: 10890908]

[41] Raccuia SA, Melilli MG, Tringali S. Genetic and environmental influence on inulin yield in wild cardoon (*Cynara cardunculus* L. var. *sylvestris* Lam.). Acta Hortic 2004; (660): 47-53.
[http://dx.doi.org/10.17660/ActaHortic.2004.660.4]

[42] Frutos MJ, Guilabert-Antón L, Tomás-Bellido A, Hernández-Herrero JA. Effect of artichoke (*Cynara scolymus* L.) fiber on textural and sensory qualities of wheat bread. Food Sci Technol 2008; 14: 49-55.
[http://dx.doi.org/10.1177/1082013208094582]

[43] Kelly G. Inulin-type prebiotics-a review: part 1. Altern Med Rev 2008; 13(4): 315-29.

[PMID: 19152479]

[44]   Franck A, Bosscher A. Inulin.Fiber ingredients: Food applications and health benefits Taylor and Francis Group, NW. USA: CRC Press 2009.
[http://dx.doi.org/10.1201/9781420043853-c4]

[45]   Gallaher DD, Stallings WH, Blessing LL, Busta FF, Brady LJ. Probiotics, cecal microflora, and aberrant crypts in the rat colon. J Nutr 1996; 126(5): 1362-71.
[http://dx.doi.org/10.1093/jn/126.5.1362] [PMID: 8618132]

[46]   Heyer AG, Lloyd JR, Kossmann J. Production of modified polymeric carbohydrates. Curr Opin Biotechnol 1999; 10(2): 169-74.
[http://dx.doi.org/10.1016/S0958-1669(99)80030-5] [PMID: 10209134]

[47]   López-Molina D, Navarro-Martínez MD, Rojas Melgarejo F, Hiner AN, Chazarra S, Rodríguez-López JN. Molecular properties and prebiotic effect of inulin obtained from artichoke (*Cynara scolymus* L.). Phytochemistry 2005; 66(12): 1476-84.
[http://dx.doi.org/10.1016/j.phytochem.2005.04.003] [PMID: 15960982]

[48]   Costabile A, Kolida S, Klinder A, *et al.* A double-blind, placebo-controlled, cross-over study to establish the bifidogenic effect of a very-long-chain inulin extracted from globe artichoke (*Cynara scolymus*) in healthy human subjects. Br J Nutr 2010; 104(7): 1007-17.
[http://dx.doi.org/10.1017/S0007114510001571] [PMID: 20591206]

[49]   Christaki E, Bonos EL, Florou-Paneri P. Nutritional and functional properties of *Cynara* crops (globe artichoke and cardoon) and their potential application: a review. Int J Appl Sci Technol 2012; 2(2): 64-70.

[50]   Seaman FC. Sesquiterpene lactones as taxonomic characteristic in the *Asteraceae*. Bot Rev 1982; 48: 121-594.
[http://dx.doi.org/10.1007/BF02919190]

[51]   Rodriguez E, Towers GH, Mitchell JC. Biological activities of sesquiterpene lactones. Phytochemistry 1976; 15(11): 1573-80.
[http://dx.doi.org/10.1016/S0031-9422(00)97430-2]

[52]   Picman AK. Biological activities of sesquiterpene lactones. Biochem Syst Ecol 1986; 14(3): 255-81.
[http://dx.doi.org/10.1016/0305-1978(86)90101-8]

[53]   Ramos PAB, Guerra AR, Guerreiro O, *et al.* Lipophilic extracts of *Cynara cardunculus* L. var. *altilis* (DC): a source of valuable bioactive terpenic compounds. J Agric Food Chem 2013; 61(35): 8420-9.
[http://dx.doi.org/10.1021/jf402253a] [PMID: 23915287]

[54]   Suchy M, Herout V, Šorm F. Terpenes. CXVI. Structure of cynaropicrin. Collect Czech Chem Commun 1960; 25: 2777-82.
[http://dx.doi.org/10.1135/cccc19602777]

[55]   Eljounaidi K, Comino C, Moglia A, *et al.* Accumulation of cynaropicrin in globe artichoke and localization of enzymes involved in its biosynthesis. Plant Sci 2015; 239: 128-36.
[http://dx.doi.org/10.1016/j.plantsci.2015.07.020] [PMID: 26398797]

[56]   Cho JY, Baik KU, Jung JH, Park MH. *in vitro* anti-inflammatory effects of cynaropicrin, a sesquiterpene lactone, from *Saussurea lappa*. Eur J Pharmacol 2000; 398(3): 399-407.
[http://dx.doi.org/10.1016/S0014-2999(00)00337-X] [PMID: 10862830]

[57]   Cho JY, Kim AR, Jung JH, Chun T, Rhee MH, Yoo ES. Cytotoxic and pro-apoptotic activities of cynaropicrin, a sesquiterpene lactone, on the viability of leukocyte cancer cell lines. Eur J Pharmacol 2004; 492(2-3): 85-94.
[http://dx.doi.org/10.1016/j.ejphar.2004.03.027] [PMID: 15178350]

[58]   Elsebai MF, Mocan A, Atanasov AG. Cynaropicrin: A comprehensive research review and therapeutic potential as an anti-hepatitis C virus agent. Front Pharmacol 2016; 7: 472.
[http://dx.doi.org/10.3389/fphar.2016.00472] [PMID: 28008316]

[59] Englisch W, Beckers C, Unkauf M, Ruepp M, Zinserling V. Efficacy of Artichoke dry extract in patients with hyperlipoproteinemia. Arzneimittelforschung 2000; 50(3): 260-5.
[PMID: 10758778]

[60] Shimoda H, Ninomiya K, Nishida N, *et al.* Anti-hyperlipidemic sesquiterpenes and new sesquiterpene glycosides from the leaves of artichoke (*Cynara scolymus* L.): structure requirement and mode of action. Bioorg Med Chem Lett 2003; 13(2): 223-8.
[http://dx.doi.org/10.1016/S0960-894X(02)00889-2] [PMID: 12482428]

[61] Cravotto G, Nano GM, Binello A, Spagliardi P, Seu G. Chemical and biological modification of cynaropicrin and grosheimin: a structure–bitterness relationship study. J Sci Food Agric 2005; 85(10): 1757-64.
[http://dx.doi.org/10.1002/jsfa.2180]

[62] Colantuono A, Ferracane R, Vitaglione P. Potential bioaccessibility and functionality of polyphenols and cynaropicrin from breads enriched with artichoke stem. Food Chem 2018; 245: 838-44.
[http://dx.doi.org/10.1016/j.foodchem.2017.11.099] [PMID: 29287449]

[63] de Faria ELP, Gomes MV, Cláudio AFM, Freire CSR, Silvestre AJD, Freire MG. Extraction and recovery processes for cynaropicrin from *Cynara cardunculus* L. using aqueous solutions of surface-active ionic liquids. Biophys Rev 2018; 10(3): 915-25.
[http://dx.doi.org/10.1007/s12551-017-0387-y] [PMID: 29294260]

[64] Negro D, Montesano V, Grieco S, *et al.* Polyphenol compounds in artichoke plant tissues and varieties. J Food Sci 2012; 77(2): C244-52.
[http://dx.doi.org/10.1111/j.1750-3841.2011.02531.x] [PMID: 22251096]

[65] Pandino G, Lombardo S, Mauro RP, Mauromicale G. Variation in polyphenol profile and head morphology among clones of globe artichoke selected from a landrace. Sci Hortic (Amsterdam) 2012; 138: 259-65.
[http://dx.doi.org/10.1016/j.scienta.2012.02.032]

[66] Petropoulos SA, Pereira C, Barros L, Ferreira ICFR. Leaf parts from Greek artichoke genotypes as a good source of bioactive compounds and antioxidants. Food Funct 2017; 8(5): 2022-9.
[http://dx.doi.org/10.1039/C7FO00356K] [PMID: 28492621]

[67] Rouphael Y, Bernardi J, Cardarelli M, *et al.* Phenolic compounds and sesquiterpene lactones profile in leaves of nineteen artichoke cultivars. J Agric Food Chem 2016; 64(45): 8540-8.
[http://dx.doi.org/10.1021/acs.jafc.6b03856] [PMID: 27792334]

[68] Di Venere D, Linsalata V, Calabrese N, Cardinali A, Sergio L. Biochemical characterization of wild and cultivated cardoon accessions. Acta Hortic 2005; (681): 523-8.
[http://dx.doi.org/10.17660/ActaHortic.2005.681.73]

[69] Cefola M, D'Antuono I, Pace B, *et al.* Biochemical relationships and browning index for assessing the storage suitability of artichoke genotypes. Food Res Int 2012; 48: 397-403.
[http://dx.doi.org/10.1016/j.foodres.2012.04.012]

[70] Raccuia SA, Melilli MG. *Cynara cardunculus* L., a potential source of inulin in the Mediterranean environment: screening of genetic variability. Aust J Agric Res 2004; 55(6): 693-8.
[http://dx.doi.org/10.1071/AR03038]

[71] Melilli MG, Raccuia SA. Inulin and water-soluble-sugars variations in Cynara roots during the biological cycle. Acta Hortic 2007; (730): 475-81.
[http://dx.doi.org/10.17660/ActaHortic.2007.730.62]

[72] Raccuia SA, Melilli MG. Seasonal dynamics of biomass, inulin, and water-soluble sugars in roots of *Cynara cardunculus* L. Field Crops Res 2010; 116(1-2): 147-53.
[http://dx.doi.org/10.1016/j.fcr.2009.12.005]

[73] Di Venere D, Linsalata V, Pace B, Bianco VV, Perrino P. Polyphenol and inulin content in a collection of artichoke. Acta Hortic 2005; (681): 453-9.

[http://dx.doi.org/10.17660/ActaHortic.2005.681.63]

[74]   Lattanzio V, Cicco N, Terzano R, *et al.* Potenziale utilizzo di sottoprodotti derivanti dalla lavorazione industriale del carciofo: Antiossidanti di natura fenolica ed inulina. Tipografia IiritiAtti XIX Convegno SICA. Reggio Calabria 2002; pp. 251-8.

[75]   Lombardo S, Pandino G, Mauro R, Mauromicale G. Variation of phenolic content in globe artichoke in relation to biological, technical and environmental factors. Ital J Agron 2009; 4(4): 181-90.
[http://dx.doi.org/10.4081/ija.2009.4.181]

[76]   Colla G, Rouphael Y, Cardarelli M, Svecova E, Rea E, Lucini L. Effects of saline stress on mineral composition, phenolic acids and flavonoids in leaves of artichoke and cardoon genotypes grown in floating system. J Sci Food Agric 2013; 93(5): 1119-27.
[http://dx.doi.org/10.1002/jsfa.5861] [PMID: 22936423]

[77]   Borgognone D, Cardarelli M, Rea E, Lucini L, Colla G. Salinity source-induced changes in yield, mineral composition, phenolic acids and flavonoids in leaves of artichoke and cardoon grown in floating system. J Sci Food Agric 2014; 94(6): 1231-7.
[http://dx.doi.org/10.1002/jsfa.6403] [PMID: 24105819]

[78]   Almela L, Muñoz JA, Roca MJ, Fernández-López JA. Appraisal of oxidative enzymatic activities and inulin content during artichoke growth. Acta Hortic 2000; (681): 529-36.
[http://dx.doi.org/10.17660/ActaHortic.2005.681.74]

[79]   Rouphael Y, Colla G, Graziani G, Ritieni A, Cardarelli M, De Pascale S. Phenolic composition, antioxidant activity and mineral profile in two seed-propagated artichoke cultivars as affected by microbial inoculants and planting time. Food Chem 2017; 234: 10-9.
[http://dx.doi.org/10.1016/j.foodchem.2017.04.175] [PMID: 28551211]

[80]   Palermo M, Colla G, Barbieri G, Fogliano V. Polyphenol metabolite profile of artichoke is modulated by agronomical practices and cooking method. J Agric Food Chem 2013; 61(33): 7960-8.
[http://dx.doi.org/10.1021/jf401468s] [PMID: 23865390]

[81]   Martínez-Esplá A, Valero D, Martínez-Romero D, *et al.* Preharvest application of methyl jasmonate as an elicitor improves the yield and phenolic content of artichoke. J Agric Food Chem 2017; 65(42): 9247-54.
[http://dx.doi.org/10.1021/acs.jafc.7b03447] [PMID: 28960971]

[82]   Rouphael Y, Cardarelli M, Lucini L, Rea E, Colla G. Nutrient solution concentration affects growth, mineral composition, phenolic acids, and flavonoids in leaves of artichoke and cardoon. HortScience 2012; 47(10): 1424-9.

[83]   Gil-Izquierdo A, Conesa M, Ferreres F, Gil M. Influence of modified atmosphere packaging on quality, vitamin C and phenolic content of artichokes (*Cynara scolymus* L.). Eur Food Res Technol 2002; 215(1): 21-7.
[http://dx.doi.org/10.1007/s00217-002-0492-3]

[84]   Leroy G, Grongnet JF, Mabeau S, Corre DL, Baty-Julien C. Changes in inulin and soluble sugar concentration in artichokes (*Cynara scolymus* L.) during storage. J Sci Food Agric 2010; 90(7): 1203-9.
[http://dx.doi.org/10.1002/jsfa.3948] [PMID: 20394002]

[85]   Cadenas E, Packer L, Eds. Handbook of Antioxidants. New York: Marcel Dekker Press 2002.

[86]   Manach C, Williamson G, Morand C, Scalbert A, Rémésy C. Bioavailability and bioefficacy of polyphenols in humans. I. Review of 97 bioavailability studies. Am J Clin Nutr 2005; 81(1) (Suppl.): 230S-42S.
[http://dx.doi.org/10.1093/ajcn/81.1.230S] [PMID: 15640486]

[87]   Courraud J, Berger J, Cristol JP, Avallone S. Stability and bioaccessibility of different forms of carotenoids and vitamin A during *in vitro* digestion. Food Chem 2013; 136(2): 871-7.
[http://dx.doi.org/10.1016/j.foodchem.2012.08.076] [PMID: 23122139]

[88]   D'Antuono I, Garbetta A, Ciasca B, *et al.* Biophenols from table olive cv Bella di Cerignola: chemical characterization, bioaccessibility, and intestinal absorption. J Agric Food Chem 2016; 64(28): 5671-8.
[http://dx.doi.org/10.1021/acs.jafc.6b01642] [PMID: 27355793]

[89]   Minekus M. The TNO Gastro-Intestinal Model (TIM).The impact of food bioactives on health. Cham: Springer 2015; pp. 37-46.

[90]   Verwei M, Minekus M, Zeijdner E, Schilderink R, Havenaar R. Evaluation of two dynamic *in vitro* models simulating fasted and fed state conditions in the upper gastrointestinal tract (TIM-1 and tiny-TIM) for investigating the bioaccessibility of pharmaceutical compounds from oral dosage forms. Int J Pharm 2016; 498(1-2): 178-86.
[http://dx.doi.org/10.1016/j.ijpharm.2015.11.048] [PMID: 26688035]

[91]   Falé PL, Ferreira C, Rodrigues AM, *et al.* Antioxidant and anti-acetylcholinesterase activity of commercially available medicinal infusions after *in vitro* gastrointestinal digestion. J Med Plants Res 2013; 7: 1370-8.
[http://dx.doi.org/10.5897/JMPR13.4438]

[92]   Garbetta A, Capotorto I, Cardinali A, *et al.* Antioxidant activity induced by main polyphenols present in edible artichoke heads: influence of *in vitro* gastro-intestinal digestion. J Funct Foods 2014; 10: 456-64.
[http://dx.doi.org/10.1016/j.jff.2014.07.019]

[93]   Juániz I, Ludwig IA, Bresciani L, *et al.* Bioaccessibility of (poly) phenolic compounds of raw and cooked cardoon (*Cynara cardunculus* L.) after simulated gastrointestinal digestion and fermentation by human colonic microbiota. J Funct Foods 2017; 32: 195-207.
[http://dx.doi.org/10.1016/j.jff.2017.02.033]

[94]   D'Archivio M, Filesi C, Varì R, Scazzocchio B, Masella R. Bioavailability of the polyphenols: status and controversies. Int J Mol Sci 2010; 11(4): 1321-42.
[http://dx.doi.org/10.3390/ijms11041321] [PMID: 20480022]

[95]   Porrini M, Riso P. Factors influencing the bioavailability of antioxidants in foods: a critical appraisal. Nutr Metab Cardiovasc Dis 2008; 18(10): 647-50.
[http://dx.doi.org/10.1016/j.numecd.2008.08.004] [PMID: 18996686]

[96]   Gladine C, Rock E, Morand C, Bauchart D, Durand D. Bioavailability and antioxidant capacity of plant extracts rich in polyphenols, given as a single acute dose, in sheep made highly susceptible to lipoperoxidation. Br J Nutr 2007; 98(4): 691-701.
[http://dx.doi.org/10.1017/S0007114507742666] [PMID: 17475083]

[97]   Zaripheh S, Erdman JW Jr. The biodistribution of a single oral dose of [14C]-lycopene in rats prefed either a control or lycopene-enriched diet. J Nutr 2005; 135(9): 2212-8.
[http://dx.doi.org/10.1093/jn/135.9.2212] [PMID: 16140900]

[98]   de Boer VC, Dihal AA, van der Woude H, *et al.* Tissue distribution of quercetin in rats and pigs. J Nutr 2005; 135(7): 1718-25.
[http://dx.doi.org/10.1093/jn/135.7.1718] [PMID: 15987855]

[99]   Mullen W, Edwards CA, Crozier A. Absorption, excretion and metabolite profiling of methyl-, glucuronyl-, glucosyl- and sulpho-conjugates of quercetin in human plasma and urine after ingestion of onions. Br J Nutr 2006; 96(1): 107-16.
[http://dx.doi.org/10.1079/BJN20061809] [PMID: 16869998]

[100]   Dupas C, Marsset Baglieri A, Ordonaud C, Tomé D, Maillard MN. Chlorogenic acid is poorly absorbed, independently of the food matrix: A Caco-2 cells and rat chronic absorption study. Mol Nutr Food Res 2006; 50(11): 1053-60.
[http://dx.doi.org/10.1002/mnfr.200600034] [PMID: 17054098]

[101]   Lafay S, Gil-Izquierdo A, Manach C, Morand C, Besson C, Scalbert A. Chlorogenic acid is absorbed in its intact form in the stomach of rats. J Nutr 2006; 136(5): 1192-7.

[http://dx.doi.org/10.1093/jn/136.5.1192] [PMID: 16614403]

[102] Plumb GW, Garcia Conesa MT, Kroon PA, Rhodes M, Ridley S, Williamson G. Metabolism of chlorogenic acid by human plasma, liver, intestine and gut microflora. J Sci Food Agric 1999; 79: 390-2.
[http://dx.doi.org/10.1002/(SICI)1097-0010(19990301)79:3<390::AID-JSFA258>3.0.CO;2-0]

[103] Gonthier MP, Verny MA, Besson C, Rémésy C, Scalbert A. Chlorogenic acid bioavailability largely depends on its metabolism by the gut microflora in rats. J Nutr 2003; 133(6): 1853-9.
[http://dx.doi.org/10.1093/jn/133.6.1853] [PMID: 12771329]

[104] Williamson G, Clifford MN. Role of the small intestine, colon and microbiota in determining the metabolic fate of polyphenols. Biochem Pharmacol 2017; 139: 24-39.
[http://dx.doi.org/10.1016/j.bcp.2017.03.012] [PMID: 28322745]

[105] Azzini E, Bugianesi R, Romano F, *et al.* Absorption and metabolism of bioactive molecules after oral consumption of cooked edible heads of *Cynara scolymus* L. (cultivar Violetto di Provenza) in human subjects: a pilot study. Br J Nutr 2007; 97(5): 963-9.
[http://dx.doi.org/10.1017/S0007114507617218] [PMID: 17408528]

[106] Wittemer SM, Ploch M, Windeck T, *et al.* Bioavailability and pharmacokinetics of caffeoylquinic acids and flavonoids after oral administration of Artichoke leaf extracts in humans. Phytomedicine 2005; 12(1-2): 28-38.
[http://dx.doi.org/10.1016/j.phymed.2003.11.002] [PMID: 15693705]

[107] Gu R, Dou G, Wang J, Dong J, Meng Z. Simultaneous determination of 1,5-dicaffeoylquinic acid and its active metabolites in human plasma by liquid chromatography-tandem mass spectrometry for pharmacokinetic studies. J Chromatogr B 2007; 852(1-2): 85-91.
[http://dx.doi.org/10.1016/j.jchromb.2006.12.055] [PMID: 17267301]

[108] Dai G, Ma S, Sun B, *et al.* Simultaneous determination of 1, 3-dicaffeoylquinic acid and caffeic acid in rat plasma by liquid chromatography/tandem mass spectrometry and its application to a pharmacokinetic study. Anal Methods 2015; 7: 3587-92.
[http://dx.doi.org/10.1039/C4AY02968B]

[109] Du Y, He B, Li Q, He J, Wang D, Bi K. Simultaneous determination of multiple active components in rat plasma using ultra-fast liquid chromatography with tandem mass spectrometry and application to a comparative pharmacokinetic study after oral administration of Suan-Zao-Ren decoction and Suan-Zao-Ren granule. J Sep Sci 2017; 40(10): 2097-106.
[http://dx.doi.org/10.1002/jssc.201601383] [PMID: 28345817]

[110] Singh M, Ramassamy C. Beneficial effects of phenolic compounds from fruit and vegetables in neurodegenerative disease.Improving health-promoting properties of fruit and vegetable compound products. Cambridge, UK: Woodhead Publishing Limited 2008; pp. 145-81.
[http://dx.doi.org/10.1533/9781845694289.2.145]

[111] Cardinali A, Pati S, Minervini F, D'Antuono I, Linsalata V, Lattanzio V. Verbascoside, isoverbascoside, and their derivatives recovered from olive mill wastewater as possible food antioxidants. J Agric Food Chem 2012; 60(7): 1822-9.
[http://dx.doi.org/10.1021/jf204001p] [PMID: 22268549]

[112] Gebhardt R. Antioxidative and protective properties of extracts from leaves of the artichoke (*Cynara scolymus* L.) against hydroperoxide-induced oxidative stress in cultured rat hepatocytes. Toxicol Appl Pharmacol 1997; 144(2): 279-86.
[http://dx.doi.org/10.1006/taap.1997.8130] [PMID: 9194411]

[113] Jiménez-Escrig A, Dragsted LO, Daneshvar B, Pulido R, Saura-Calixto F. *in vitro* antioxidant activities of edible artichoke (*Cynara scolymus* L.) and effect on biomarkers of antioxidants in rats. J Agric Food Chem 2003; 51(18): 5540-5.
[http://dx.doi.org/10.1021/jf030047e] [PMID: 12926911]

[114] Pereira C, Calhelha RC, Barros L, Ferreira IC. Antioxidant properties, anti-hepatocellular carcinoma

activity and hepatotoxicity of artichoke, milk thistle and borututu. Ind Crops Prod 2013; 49: 61-5.
[http://dx.doi.org/10.1016/j.indcrop.2013.04.032]

[115]   Lutz M, Henríquez C, Escobar M. Chemical composition and antioxidant properties of mature and baby artichokes (*Cynara scolymus* L.), raw and cooked. J Food Compos Anal 2011; 4: 49-54.
[http://dx.doi.org/10.1016/j.jfca.2010.06.001]

[116]   Ferracane R, Pellegrini N, Visconti A, *et al.* Effects of different cooking methods on antioxidant profile, antioxidant capacity, and physical characteristics of artichoke. J Agric Food Chem 2008; 56(18): 8601-8.
[http://dx.doi.org/10.1021/jf800408w] [PMID: 18759447]

[117]   Zhao Z, Shin HS, Satsu H, Totsuka M, Shimizu M. 5-caffeoylquinic acid and caffeic acid down-regulate the oxidative stress- and TNF-α-induced secretion of interleukin-8 from Caco-2 cells. J Agric Food Chem 2008; 56(10): 3863-8.
[http://dx.doi.org/10.1021/jf073168d] [PMID: 18444659]

[118]   Shin HS, Satsu H, Bae MJ, *et al.* Anti-inflammatory effect of chlorogenic acid on the IL-8 production in Caco-2 cells and the dextran sulphate sodium-induced colitis symptoms in C57BL/6 mice. Food Chem 2015; 168: 167-75.
[http://dx.doi.org/10.1016/j.foodchem.2014.06.100] [PMID: 25172696]

[119]   Shin HS, Satsu H, Bae MJ, Totsuka M, Shimizu M. Catechol groups enable reactive oxygen species scavenging-mediated suppression of PKD-NFkappaB-IL-8 signaling pathway by chlorogenic and caffeic acids in human intestinal cells. Nutrients 2017; 9(2): 165.
[http://dx.doi.org/10.3390/nu9020165] [PMID: 28230729]

[120]   Ben Salem M, Affes H, Athmouni K, *et al.* Chemicals compositions, antioxidant and anti-inflammatory activity of Cynara scolymus leaves extracts, and analysis of major bioactive polyphenols by HPLC. Evid-Based Compl Alt 2017.Evid-Based Compl Alt 2017. Article ID 4951937

[121]   Baumann J, von Bruchhausen F, Wurm G. Flavonoids and related compounds as inhibition of arachidonic acid peroxidation. Prostaglandins 1980; 20(4): 627-39.
[http://dx.doi.org/10.1016/0090-6980(80)90103-3] [PMID: 6781013]

[122]   Del Vecchio A. Action of an extract of artichoke (*Cynara scolymus*) on cholesterolesterase *in vitro*. Boll Soc Ital Biol Sper 1953; 29(1): 48-50.
[PMID: 13066651]

[123]   Rondanelli M, Monteferrario F, Perna S, Faliva MA, Opizzi A. Health-promoting properties of artichoke in preventing cardiovascular disease by its lipidic and glycemic-reducing action. Monaldi Arch Chest Dis 2013; 80(1): 17-26.
[PMID: 23923586]

[124]   Gebhardt R, Fausel M. Antioxidant and hepatoprotective effects of artichoke extracts and constituents in cultured rat hepatocytes. Toxicol in vitro 1997; 11(5): 669-72.
[http://dx.doi.org/10.1016/S0887-2333(97)00078-7] [PMID: 20654368]

[125]   Preziosi P, Loscalzo B. Pharmacological properties of 1, 4 dicaffeylquinic acid, the active principle of Cynara scolimus. Arch Int Pharmacodyn Ther 1958; 117(1-2): 63-80.
[PMID: 13606913]

[126]   Brown JE, Rice-Evans CA. Luteolin-rich artichoke extract protects low density lipoprotein from oxidation *in vitro*. Free Radic Res 1998; 29(3): 247-55.
[http://dx.doi.org/10.1080/10715769800300281] [PMID: 9802556]

[127]   Coinu R, Carta S, Urgeghe PP, *et al.* Dose-effect study on the antioxidant properties of leaves and outer bracts of extracts obtained from Violetto di Toscana artichoke. Food Chem 2007; 101: 524-31.
[http://dx.doi.org/10.1016/j.foodchem.2006.02.009]

[128]   Zapolska-Downar D, Zapolski-Downar A, Naruszewicz M, Siennicka A, Krasnodębska B, Kołdziej B. Protective properties of artichoke (*Cynara scolymus*) against oxidative stress induced in cultured

endothelial cells and monocytes. Life Sci 2002; 71(24): 2897-08.
[http://dx.doi.org/10.1016/S0024-3205(02)02136-7] [PMID: 12377270]

[129] Rezazadeh K, Aliashrafi S, Asghari-Jafarabadi M, Ebrahimi-Mameghani M. Antioxidant response to artichoke leaf extract supplementation in metabolic syndrome: A double-blind placebo-controlled randomized clinical trial. Clin Nutr 2018; 37(3): 790-6.
[http://dx.doi.org/10.1016/j.clnu.2017.03.017] [PMID: 28410922]

[130] Nazni P, Poongodi Vijayakumar P, Alagianambi P, Amirthaveni M. Hypoglycemic and hypolipidemic effect of *Cynara Scolymus* among selected type 2 diabetic individuals. Pak J Nutr 2006; 5: 147-51.
[http://dx.doi.org/10.3923/pjn.2006.147.151]

[131] Magied MMA, Hussien SED, Zaki SM, Said RME. Artichoke (*Cynara scolymus* L.) leaves and heads extracts as hypoglycemic and hypocholesterolemic in rats. J Food Nutr Res 2016; 4: 60-8.

[132] Miccadei S, Di Venere D, Cardinali A, *et al.* Antioxidative and apoptotic properties of polyphenolic extracts from edible part of artichoke (*Cynara scolymus* L.) on cultured rat hepatocytes and on human hepatoma cells. Nutr Cancer 2008; 60(2): 276-83.
[http://dx.doi.org/10.1080/01635580801891583] [PMID: 18444161]

[133] Mileo AM, Di Venere D, Linsalata V, Fraioli R, Miccadei S. Artichoke polyphenols induce apoptosis and decrease the invasive potential of the human breast cancer cell line MDA-MB231. J Cell Physiol 2012; 227(9): 3301-9.
[http://dx.doi.org/10.1002/jcp.24029] [PMID: 22170094]

[134] Pulito C, Mori F, Sacconi A, *et al. Cynara scolymus* affects malignant pleural mesothelioma by promoting apoptosis and restraining invasion. Oncotarget 2015; 6(20): 18134-50.
[http://dx.doi.org/10.18632/oncotarget.4017] [PMID: 26136339]

[135] Genovese C, Brundo MV, Toscano V, Tibullo D, Puglisi F, Raccuia SA. Effect of Cynara extracts on multiple myeloma cell lines. Acta Hortic 2015; (1147): 113-8.
[http://dx.doi.org/10.17660/ActaHortic.2016.1147.16]

[136] Majumdar D, Jung KH, Zhang H, *et al.* Luteolin nanoparticle in chemoprevention: *in vitro* and *in vivo* anticancer activity. Cancer Prev Res (Phila) 2014; 7(1): 65-73.
[http://dx.doi.org/10.1158/1940-6207.CAPR-13-0230] [PMID: 24403290]

CHAPTER 11

Phytochemicals Content and Health Effects of *Abelmoschus esculentus* (Okra)

Spyridon A. Petropoulos[1,*], **Sofia Plexida**[1], **Nikolaos Tzortzakis**[2] and **Isabel C.F.R. Ferreira**[3]

[1] *University of Thessaly, Department of Agriculture, Crop Production and Rural Environment, 38446 N. Ionia, Magnissia, Greece*

[2] *Cyprus University of Technology, Department of Agricultural Sciences, Biotechnology and Food Science, 3036, Lemesos, Cyprus*

[3] *Centro de Investigação de Montanha (CIMO), Instituto Politécnico de Bragança, Campus de Santa Apolónia, 5300-253 Bragança, Portugal*

Abstract: Okra (*Abelmoschus esculentus* L. Moench.) or lady finger, is a tropical vegetable of the Malvaceae family, which is usually consumed for its immature fresh or dried pods, while other parts of the plant such as leaves and seeds are also edible or have alternative uses. In addition, it is a versatile species and considered as a multipurpose crop, since plant tissues contain many chemical compounds that find applications in the food and pharmaceutical industry, as well as in other not widely known industrial uses (*e.g.* making of ropes, sacks, fishing lines, paper, biofuel, blood plasma replacement, stabilize foams). This chapter will describe chemical composition and uses of okra plant tissues, including pharmaceutical and industrial uses of the species. Moreover, special focus will be given on the health effects of the various plant parts and products (fruit, seeds, leaves, roots, flours, mucilage) and the mechanisms involved, while the most recent research results from both *in vivo* and *in vitro* models will be presented in order to establish the health effects of okra products and by-products. In conclusion, considering that okra is an underutilized plant for most parts of the world, the potential of further exploiting the species and the future perspectives will be highlighted.

Keywords: *Abelmoschus esculentus*, Antioxidant activity, Anti-diabetic, Anti-inflammatory, Bioactive compounds, Bepatoproctive, Flavonoids, Lady's finger, Lectins, Mucilage, Okra, Okra gum, Pectins, Quercetin, Proteins, Seed oil.

INTRODUCTION

Okra, *Abelmoschus esculentus* L. (Syn. *Hibiscus esculentus* L.) is a member of the

* **Corresponding author Spyridon A. Petropoulos:** University of Thessaly, School of Agricultural Sciences, Volos, Greece; Tel: +302421093196; E-mail: spetropoulos@uth.gr

Malvaceae family. The center of origin for the species is considered Africa and Ethiopia in particular, while other researchers suggest that it originates from Asia or West Africa [1, 2]. There are historical reports that okra was cultivated by the ancient Egyptians since the 12th century B.C., while according to Nikolai Vavilov the plant had been cultivated in tropical regions of Africa in Ethiopia; hence this region is considered the center of origin for okra and its wild relatives [2]. Nowadays, there are many cultivars and heirloom varieties of common okra, including commercial cultivars which are widely used during the last decades (*e.g.* 'Clemson Spineless', 'Indiana', 'Emerald', 'Pusa Sawani *etc.*) and vary in pod characteristics, plant growth and earliness in pod formation [2]. However, many local landraces and farmer's varieties also exist throughout the world with special quality features and adaptation to the specific edaphoclimatic conditions which make the conservation of this valuable genetic material of special importance [3].

As a vegetable crop, okra is widely cultivated throughout the world, as long as the climate requirements of the species (regions with tropical-subtropical climate and warm temperatures) are met [4, 5]. Apart from vegetable, okra is considered a multipurpose crop due to the various uses of all the plant tissues (*e.g.* fresh leaves, buds, flowers, immature pods, stems, roots and seeds) [6]. The most commonly used plant part is immature pods, consumed in fresh or dried form, which have a high nutritional value [7]. Nutritionally, it is considered a rich source of dietary fibers, vitamin C, folates, minerals such as Fe, Zn, Mn, Ni, Ca, P and I and low fat content [1, 6, 8, 9]. Fresh pods contain mucilage or okra gum, a viscous hydrocolloid material with important industrial uses, which constitutes mainly of pectin and polysaccharides, such as galactose, rhamnose and arabinose [10, 11], while Kpodo *et al.* [12] have reported that structural properties and chemical composition of fruit pectins are genotype dependent features. Leaves, flower buds and seeds of okra plants are also edible and find many culinary uses in various traditional dishes throughout the world [6, 13].

The species is also highly appreciated for its medicinal value, since many recent studies confirm the medicinal properties and ethno-pharmacological uses of okra plant parts throughout the centuries. According to the literature, okra plant tissues show significant bioactivities, including antioxidative, immunomodulatory, hepatoprotective, anti-tumor, anti-microbial, anti-diabetic, antihyperlipidemic, anti-inflammatory, anti-ulcer and hypoglycemic, as well as activities against gastric peristalsis, goiter and constipation among others [14 - 20]. Other bioactive properties include its effect against genito-urinary disorders, spermatorrhoea, chronic dysentery and hemorrhoids, as well as its use as blood plasma replacement or blood volume expander [21].

Okra is not the only species from the genus *Abelmoschus* which is commercially cultivated, since there are also other species with similar uses to okra. Among them, *A. manihot* is a species with plants of small height, distributed in Asia, Africa and South America, which is usually consumed as a leafy vegetable; *A. moschatus* is used for its aromatic seeds for culinary purposes; *A. caillei* is used for its edible fruits and leaves, while seeds and flower buds are also used in various dishes [2, 22, 23]. However, since this chapter is focused on chemical composition and bioactive properties of *A. esculentus*, only limited information regarding the related species will be presented.

TAXONOMY OF THE SPECIES

Okra is well-known with various common names throughout the world. It is also known as lady fingers and okra plant in English language, while in Asian countries, there are several names for the same species, such as bhindi, krajiab kheaw, ochro, okoro, quimgombo and so forth. In Middle East and the Mediterranean basin where okra is a very popular vegetable, it is also known as bamia, bamya or bamieh and gumbo [1].

Confusion among the species of the *Abelmoschus* genus makes it very important to describe plant morphology differences of the most important species, especially when considering that literature references on okra may also include information related to other species, as well as the similar uses that related species may have. Therefore, plants of *A. caillei* and *A. moschatus* are different from *A. esculentus* in several morphological features; however, flower and pod morphology differences are the most reliable features for discrimination between these species [24]. According to Kyriakopoulou *et al.* [25] and Saifullah and Rabbani [7] phenotypic variation is very common within the species and plant and pod morphology varies significantly among the numerous cultivars and local landraces.

CHEMICAL COMPOSITION

A significant variation in the number of chromosomes number and ploidy levels between the species of the *Abelmoschus* genus has been described so far, which indicates great genetic variation and consequently variant chemical composition and phytochemical content. Moreover, considering the multipurpose character of the species with various plant parts finding edible and/or industrial uses, it is of major concern to elucidate chemical composition of different plant parts. As mentioned before, there are several species in this genus with same edible uses as okra, including cultivated and wild species such as *A. angulosus*, *A. manihot*, *A.manihot* var. *tetraphyllus*, *A. caillei*, *A. tuberculatus* and *A. ficulneus*, which however differ in terms of chemical composition Table **1**). This variation in chemical composition is becoming even bigger when considering the various subspecies within each of these subspecies that have been described so far [26], as well as the variation in ploidy levels and number of chromosomes [27] and the great number of existing accessions of okra and related species [28]. Significant differences between okra and okra related species have been observed not only in edible parts but also in seed oils content and composition [28] (Table **2**).

Table 1. Composition of edible parts of okra and related species (mean values expressed in 100 g fresh weight for fruit and leaves and 100g dry weight for seeds).

Species	Water (%)	Fat (g)	Proteins (g)	Carbohydrates (g)	Fibre (g)	Energy (Kcal)
A. caillei (L)	n/a	n/a	10.16**	n/a	n/a	n/a
A. esculentus (F)*	88.6	0.2	2.1	8.2	1.7	36
A. esculentus (L)	81.5-83.91	0.2-2.3	1.3-4.8	4.8-11.3	1.13-2.5	42-56
A. esculentus (S)	6.5-11.3	26-29	37.4-40.7	25.3-31.3	23.43	502-523
A. manihot	88.4	1.77	4.9	2.2	n/a	n/a
	Ash (g)	P (mg)	Ca (mg)	Fe (mg)	K (mg)	Mg (mg)
A. caillei	n/a	0.46**	1.01**	n/a	2.73**	0.54**
A. esculentus (F)	n/a	90	84	1.2	n/a	n/a
A. esculentus (L)	1.91-3.23	65-70	428-532	0.7	368	13
A. esculentus (S)	4.8-5.7		681-3052	5.83-7.52	1800-4000	428-600
A. manihot (L)	1.61	197-635	197-635	0.8-8.7	265-630	79-264
	β-carotene (μg)	Riboflavin (mg)	Thiamin (mg)	Niacin (mg)	ascorbic acid (mg)	
A. caillei	9.07	n/a	n/a	n/a	n/a	n/a
A. esculentus (F)	185	0.08	0.04	0.6	47	n/a
A. esculentus (L)	232-385	2.8	0.25	0.2	8.4-59	n/a
A. esculentus (S)	90-350	n/a	n/a	n/a	n/a	n/a
A. manihot (L)	315	n/a	n/a	n/a	n/a	n/a

*F: fruit; L: leaves; S: Seeds ** Values expressed as % on a dry weight basis. n/a: not available data Sources [29 - 36].

Table 2. Seed oil content (%) and the main fatty acids composition (relative percentage %) of *Abelmoschus* sp. (okra and related species).

Species	Oil content	Palmitic acid (16:0)	Oleic acid (18:1)	Linoleic acid (18:2)
A. caillei	12.6	30.4	21.1	37.8
A. esculentus	16.6	30.0	25.6	31.6
A. ficulneus	14.0	24.4	18.7	48.5

(Table 2) cont.....

Species	Oil content	Palmitic acid (16:0)	Oleic acid (18:1)	Linoleic acid (18:2)
A. manihot	11.6	27.0	31.9	27.1
A. moschatus	14.5	19.8	30.4	40.2
A. tuberculatus	15.7	21.7	21.2	49.6

Modified from Jarret *et al.* [28].

There are many studies regarding the qualitative phytochemicals screening of okra plant parts which report the presence and/or absence of alkaloids, cardiac glycosides, flavonoids, tannins, anthraquinones, saponins, volatile oils, cyanogenic glycosides, coumarins, sterols and/or triterpenes [18, 23, 37]. The basic composition of the species has been studied intensively and the reports show that the various plant parts contain valuable nutritional ingredients and bioactive compounds, such as proteins, polysaccharides, fatty acids, flavonoids, amino acids, minerals, and vitamins [14, 38 - 40]. For example, Lin *et al.* [38] reported three flavonoids, all being identified as quercetin derivatives (quercetin -3- O-gentiobiopyranoside, quercetin-3 -O- [β-D- xyl-(1 → 2)]- β-D- glucopyranoside and quercetin -4″ -O- methy- 3-O- β- D- glucopyranoside), where contents differed significantly between plant parts of the various *A. esculentus* cultivars. Moreover, chemical composition of okra parts has been reported to vary significantly and being dependent not only on genotypes but also on growing conditions and harvest stage [42]. In a recent study of Petropoulos *et al.* [31], a significant variation between seeds of various okra genotypes was reported in terms of antioxidant activity and chemical composition.

Pods

Immature fruit (pods) are the most commonly used part of the plant that is consumed either as a fresh or frozen vegetable or after processing as canned or sun-dried product. Okra pods are the basic ingredient in many dishes throughout the world and they are considered a valuable source of various nutrients and phytochemicals.

Okra pods are a good source of carbohydrates, fibers, Ca, K, Mg, vitamins and phenolic compounds [1, 42, 43], while El-Qudah [44] have reported the presence of various carotenoids (namely neoxanthin, violaxanthin, lutein, chlorophylls, β-carotene, 9- cis-β-carotene, lycopene and 13-cis-β-carotene), with high amounts of lutein and β-carotene and high ratios of 9-cis/all trans β-carotene being noted. According to Wang *et al.* [45] who performed a target-guided isolation of antioxidant compounds in okra extracts, the main compounds with antioxidant properties were identified as L-tryptophan, quercetin-3-*O*-sophoroside, 5,7,3′,4′-tetrahydroxyflavonol-3-*O*-[β-D-rhamnopyranosil-(1→2)]-β-D-glucopyranoside, and quercetin-3-*O*-glucoside (Fig. **1**).

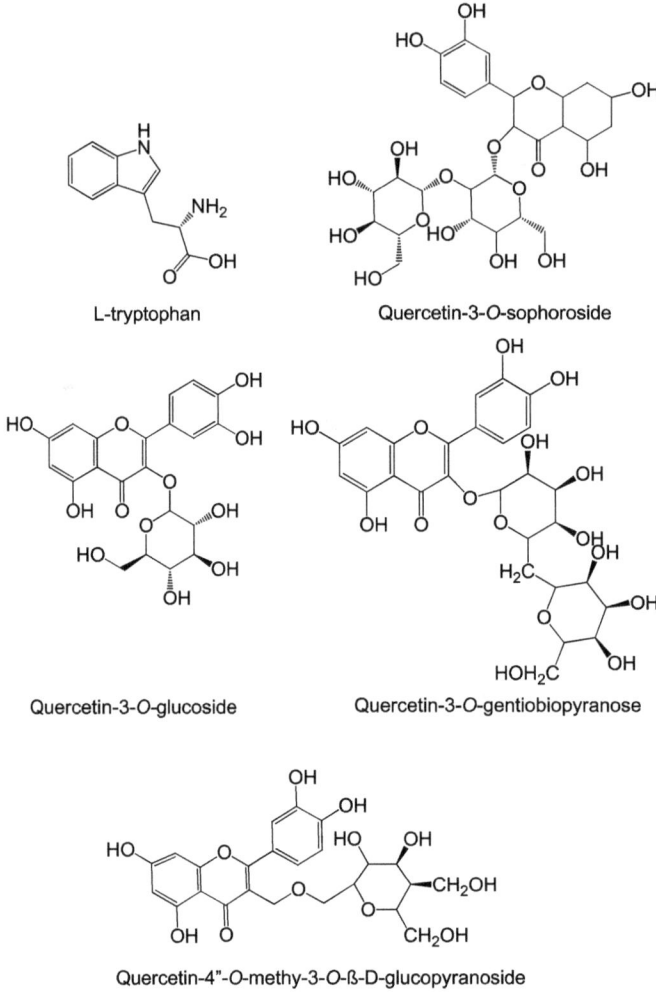

L-tryptophan

Quercetin-3-*O*-sophoroside

Quercetin-3-*O*-glucoside

Quercetin-3-*O*-gentiobiopyranose

Quercetin-4"-*O*-methy-3-*O*-ß-D-glucopyranoside

Fig. (1). The chemical structure of the main compounds in okra pods.

Regarding phenolics composition [5], reported a sectional distribution of phenolic compounds in pods, with skins being abundant in hydroxycinnamic derivatives, while seeds were rich in oligomeric catechins and flavonol derivatives (Fig. **2**). The main detected phenolic compounds in pod skins were quercetin-3-*O*-diglucose, quercetin-3-*O*-glucose, sinapoyl-hexose and a catechin derivative, while in seeds the most abundant compounds were also quercetin-3-*O*-diglucose and quercetin-3-*O*-glucose [5].

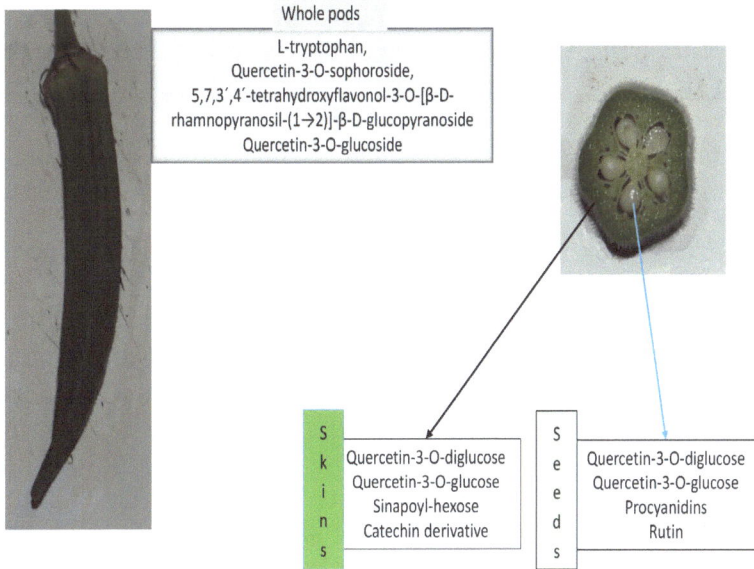

Fig. (2). Phenolic compounds composition of okra pod parts (Photos provided from Dr. Petropoulos personal record).

In contrast, in a later study [46], the main phenolic compounds in fruit were identified as catechin and epicatechin, with lower amounts of procyanidin B2 and rutin being detected, while Shui and Peng [47] identified as major phenolics quercetin derivatives and epigallocatechin. Moreover, in the same study of Shui and Peng [47] four new quercetin derivatives were identified in okra fruit for the first time, namely quercetin 3-*O*-xylosyl (1‴→2″) glucoside, quercetin 3-*O*-glucosyl (1″→6″) glucoside, quercetin 3-*O*-glucoside and quercetin 3-*O*-(6″-O-malonyl)-glucoside. Liao *et al.* [40] detected two flavonoids in pods of *A. esculentus*, one of which was newly found and characterized as 5,7,3′,4′-tetrahydroxy-4″-*O*-methyl flavonol-3-*O*-β-D- glucopyranoside, while the second one was identified as 5,7,3′,4′-tetrahydroxy flavonol -3-*O*-[β-D-glucopyrano-yl-(1→6)]-β-D-glucopyranoside and was reported in okra for the first time.

Moreover, Yang *et al.* [48] identified twelve compounds with antiproliferative activity against cancer cells by using high-speed countercurrent chromatographic separation (HSCCS), with N-trans-coumaroyltyramine, trans-feruloydopamine and 4′-Hydroxy phenethyl trans-ferulate being the most abundant compounds, while carolignan was the most potent compound of all the detected compounds. Other recently found compounds were two new pentacyclic triterpenes, namely (3β,21β)-19,21-epoxylup-20(29)-en-3-yl acetate and (3β)- 9,18-dihydroxyolean-

12-en-3-yl acetate, which were reported in the recent study of Zhou *et al*. [50]. Similarly, Ahiakpa *et al*. [41] evaluated phytochemical compounds of twenty-five accessions of okra collected from different locations in Ghana. The results of the study showed that total flavonoid and phenolic compounds contents (TFC and TPC, respectively), as well as total antioxidant activity of ethanolic and aqueous extracts of okra pods are genotype dependent, while growth conditions have also a significant impact on these parameters.

Other valuable compounds present in okra are fatty acids and amino acids. In the study of Sami *et al*. [50], samples of sun-dried okra pods grown in four different locations were analyzed to determine fatty acid and amino acid compositions in relation to growing conditions. Significant differences were observed between the studied locations with total lipid content ranging from 4.34% to 4.52%, while the main okra fatty acids were palmitic acid (29.18–43.26%) and linoleic acid (32.22–43.07%), followed by alpha-linolenic acid (6.79–12.34%), stearic acid (6.36–7.73%), oleic acid (4.31–6.98%), arachidic acid (0–3.48%), margaric acid (1.44–2.16%), pentadecylic acid (0.63–0.92%), and myristic acid (0.21–0.49%). Moreover, Berry [52] and Rakhimov *et al*. [53] have reported significant differences in total amount of lipids depending on harvest stage, with green fruit containing more lipids than ripe fruit (2.0% and 1.1%, respectively), whereas linolenic acid reduced with maturity stage (40.3% and 18.2% for green and ripe fruit, respectively). Regarding amino acids content, aspartic acid, proline, and glutamic acids were the most abundant amino acids, while cysteine and tyrosine were detected in lower amounts. Recently, Petropoulos *et al*. [3] have studied the effect of fruit size on chemical composition of various local landraces and cultivars cultivated in the Mediterranean basin and reported that both genotype and fruit size may have a significant effect on chemical composition of okra pods. The results of the study showed that in most cases small fruit size had a higher nutritional value than large fruit size. However, large fruit of specific genotypes may be considered for other uses apart from raw consumption, since they have a higher antioxidant activity than small fruit. Especially for the case of processed okra food products, processing treatments and domestic cooking practices may also affect chemical composition, since heating application has an impact on both physical and chemical quality of pods and vitamin C and β-carotene in particular [53].

Seeds

Chemical composition of seeds has gained the interest of researchers and the food industry since many years ago by the imperative challenges for alternative protein sources and food security. Proteins play a particularly important role in human nutrition and deficiency of proteins is one of the main malnutrition reasons throughout the world, especially in the undeveloped countries [54]. In the early study of Karakoltsidis and Constantinides [4], it has been reported that okra seeds may be used as a good protein source in human diet and animal feed considering the amount and quality of their proteins content, as well as their high content in unsaturated fatty acids (mainly oleic and linoleic acids) [56, 57]. As reported, okra seeds contain about 20% proteins and 20% of oil, amounts that are comparable to many cereals and oil seed crops [57 - 60], while *A. esculentus* seeds are the richest in oil content among the various related species of the genus [28]. Apart from proteins and oil, okra seeds extracts contain alkaloids, carbohydrates, phenols, terpenoids, flavonoids, tannins and sterols, whereas anti-nutrient compounds such as cardiac glycosides and saponins were not detected [17]. According to Khomsug *et al.* [46], the main phenolic compounds in seeds were two procyanidins (B2 and B1), followed by quercetin and rutin which were detected in lower amounts, while Thanakosai and Phuwapraisirisan [61] identified isoquercetin and quercetin-3-O-β-glucopyranosyl-(1"→6")-glucoside. Okra seeds are commonly used for their oil, which is mostly consisted of linoleic acid (67.5%), making it a rich source of unsaturated fatty acids [63, 64], although András *et al.* [64], Lee *et al.* [65] and Topkafa *et al.* [66] report different fatty acids profiles with lower amounts of linoleic acid (up to 47.5%), followed by palmitic and oleic acid (up to 36.1% and 19.4%, respectively), as well as significant amounts of β-sitosterol and alpha- and gamma-tocopherols. Regarding the triacylglycerol (TAG) profile of seed oil, the main identified TAGs were palmitodilinolein (PLL), dipalmitolinolein (PPL) and palmitolinoleo-olein/palmito-oleo-linolein (PLO/POL) in amounts that contribute up to 53.3% of total TAGs [65]. According to Rakhimov *et al.* [52], oil composition and content of seeds varies with harvest age (2.1% and 19.5% for seeds from green and ripe fruit, respectively), whereas although palmitic and oleic acid were detected in similar amounts, stearic, linoleic and alpha-linolenic acid differed significantly between seeds of different stages and different extraction methods (supercritical carbon dioxide, screw press and solvent extraction) [56]. Moreover, okra seeds from genotypes cultivated mainly in Greece showed a great potential for oil production, with contents varying between 15-20%, depending on the extraction method [64, 67], while Anwar *et al.* [68] and Bryant and Montecalvo [32] have suggested the use of seeds for biofuel production and as a new protein source, respectively. However, apart from energy production uses, spectroscopic and thermoxidative analyses have shown that okra seeds oil has also a great potential for human consumption as an alternative vegetable oil due to high oxidation temperature and low peroxide contents [55, 59].

The amino acid content and composition of fruit and vegetables, and their digestibility by humans characterize the protein's biological value. Al-Wandawi [57] evaluated amino acids composition in seeds of two okra cultivars and identified the high potential of using these seeds as high-protein source due to their high lysine content which is usually the most limiting amino acid in cereal-based diets. Moreover, anti-nutrients such as gossypol and cyclopropenoids that could impede the safe use of seeds in the food industry were detected in very low amounts and significantly lower contents than other related cultivated species [69, 70]. According to Bryant and Montecalvo [32], the amino acid composition of okra seed proteins is comparable to that of soybean and includes, tryptophan and sulfur-containing amino acids, while the protein efficiency ratio (PER) and net protein utilization (NPU) of okra is higher than that of soybean and other legumes, which usually have a high content in anti-nutrient factors that decrease protein digestibility [71 - 73]. These features make it very common to use okra seeds for flour and baked products preparation enriched with alternative sources of plant-based protein, especially in African countries where okra is widely cultivated [1, 62, 74].

Seeds are also a good source of phenolic compounds, where according to Arapitsas [5] they contain significantly higher amounts of flavonols and oligomeric catechins comparing to pod skins (3387.12 and 2518.44 µg/g comparing to 147.73 and 320.14 µg/g), although hydroxycinammic derivatives were detected only in skins. In contrast, Xia *et al.* [75] have reported that total flavonoids and isoquercitrin and quercetin-3-O-gentiobiose in particular, were detected only in seeds. Similarly, Hu *et al.* [76] identified quercetin 3-O-glucosyl glucoside and quercetin 3-O-glucoside as the major phenolic compounds, whereas in the earlier study of Atawodi *et al.* [77] only quercetin glucoside and an unidentified flavonoid were detected.

Leaves

Okra leaves are a good source of carbohydrates, coumarins, terpenoids, fibers, Ca, β- carotene and vitamins [1, 78], whereas they contain low amounts of anti-nutrients such as saponins, tannins, alkaloids and lectins [37, 79]. Carbohydrates consist of galactose and rhamnose and are detected mainly in mucilage, which also contains high amounts of galacturonic acid and amino acids [27, 80, 81]. According to Li *et al.* [82] aqueous extracts of leaves contain two polysaccharides that consist of arabinose, xylose, glucose, mannose, and galactose in different ratios. Moreover, Taroreh *et al.* [83] studied the chemical composition of methanolic extracts from gedi leaves (*A. manihot* L.) and based on UV, IR, 1H-NMR and 13C-NMR spectra identified a new flavonoid (3,3',5' trihydroxy-4' methoxy flavone) which exhibited significant antioxidant activities against the 2,2-diphenyl-1-picrylhydrazyl (DPPH) assay.

According to Caluête *et al.* [79], although processing treatments (bleaching and cooking) did not affect proteins, fibers and carbohydrates, they may affect macronutrients contents, with decreased levels of Ca, Mn and K compared to fresh okra leaves. Freeze drying processing accelerated the composition and levels of minerals of okra leaves comparing to fresh tissue. Phytochemicals content of leaves, especially antinutrients and tannins in particular, showed an increase after freeze-drying processing, while cooking decreased its content.

Roots

Tomoda *et al.* [84] isolated from plant roots a polysaccharide which consists of rhamnose, galactose, galacturonic acid and glucuronic acid, while it contains various amino acids including glutamic and aspartic acid, alanine and methionine.

Volatile Compounds of Okra Parts

Okra contains volatile compounds which can be found in leaves, flowers, broths, as well as in fruit and seeds [64, 85 - 87]. According to Jiang *et al.* [13], leaves contain a significantly higher amount of volatile compounds comparing to flowers and seeds, which are mainly esters, acids and terpenoids, while recorded IC_{50} values for antioxidant activity of leaves were between flowers and seeds. In the same study, 43 different compounds were identified in flowers, 20 compounds in leaves, and 29 compounds in seeds, with only 11 compounds being common among the three plant parts. Most of the detected compounds were phenols, alcohols, aldehydes and hydrocarbons, while the most abundant compound was n-Hexadecanoic acid (44.3% and 23.4% in seeds and flowers, respectively), followed by butylated hydroxytoluene (35.1% in seeds) and 2,3-dihydroxypropylester,(Z,Z,Z)-9,12,15-octadecarienoicacid (7.1% in flowers).

Similarly, Ames and Macleod [86] analyzed the essential oils of okra pods and detected 148 volatile compounds with terpenes being the most abundant chemical class (26.9% of total compounds). The most abundant compound was 2-methoxy-4-vinylphenol, while various citronel esters were detected in lower amounts.

A few years later, Camciuc *et al.* [85] studied the volatile constituents of seed coats and identified 72 different compounds with three different extraction techniques (n-hexane for solvent extraction followed by steam distillation and head space technique). All of the compounds were classified as esters or sesquiterpenes and their corresponding derivatives, while the most abundant compound was 2-methylbutyl 2-methylbutanoate. Moreover, the fractionation of ethanolic extracts of seeds revealed more compounds that are responsible for the distinctive aroma of seeds, depending on the fraction [88]. In a recent study, Molfetta *et al.* [87] evaluated volatile compounds of seed oils from various accessions of *A. esculentus* and *A. moschatus*. Significant differences in volatile compounds profile were observed not only between the two species, but also between the accessions of *A. esculentus*. For seed samples of *A. esculentus* accessions, the most abundant compounds were isopentyl 2-methyl butanoate and heptanoic acid 2-methylbutyl ester from a total of 70 compounds, while n-tridecane, isopentyl 2-methyl butanoate and decanal were the major compounds in *A. moschatus* accession from a total of 93 compounds. Moreover, Du *et al.* [89] identified 58 nitrogen-containing compounds in *A. moschatus* CO_2 seed extracts, including four new natural compounds [1-(6-ethyl-3- hydroxypyridin-2-yl)ethanone, 1-(3-hydroxy-5,6-dimethylpyridin-2-yl)ethanone, 1-(3-hydroxy- 6-methylpyridin-2-yl)ethanone, and 1-(3-hydroxy-5-methylpyridin-2-yl)ethanone].

Anti-Nutrients

Apart from nutrient compounds, okra plant parts may also include anti-nutrients that affect digestibility and bioavailability of other nutrients. Phytate, tannins, saponins, trypsin, hemagglutinin and oxalates are the most common anti-nutrients found in okra pods and seeds [42, 90]. However according to Gemede *et al.* [91] and Adetuyi and Osagie [42] there is a significant variation between okra accessions and varieties which indicates a genotype-dependent effect on these compounds content that could be valorized in breeding programs for selection of elite genotypes with low amounts of anti-nutrients.

Postharvest Preservation and Innovative Techniques

Okra is widely consumed as a fresh vegetable and in order to increase its availability in the market, it is essential to extent the shelf-life of fresh pods beyond the production periods. The major changes taking place during its senescence include degradation of chlorophyll, loss of turgidity, off-odor development, loss of nutritional value, and blackening of the cap, ridges, and tips, which contribute to quality loss and consumers' loathing. Several means in postharvest technology examined the storability of raw okra pods. Aderiye [92] showed that when fresh okra pods were dipped in ascorbic acid aqueous solution maintained texture, aroma and greenness whereas deterioration symptoms showed only after 20 days of storage at 2 °C. Moreover, Saleh [93] reported that ascorbic acid and cysteine dipping solution were the best anti-coloring agents for okra blackening inhibition at 5 °C, due to their strong ability to inhibit polyphenoloxidase (PPO) and to react with the resulted colored quinones to give colorless products. Similarly, okra dipped in $CaCl_2$ had increased cell membrane integrity leading to improved texture, minimized weight loss, decreased microbial load, and reduced blackness *via* prevention of PPO activity from contacting its phenolic substrates [93].

Storage at low temperatures delayed fruit senescence and extended postharvest life of okra pods, although the beneficial effects may be limited by the development of chilling injury-associated disorders, including skin browning, translucency or pitting [94]. Therefore, inhibition or reduction of chilling injury development is an effective way to extend storage life and maintain quality of okra pods at low temperatures. Methyl jasmonate (MeJA) is a natural plant growth regulator and plays important roles in plant growth and development, fruit ripening, and responses to environmental stress. Boonyaritthongchai *et al.* [95] found that MeJA treated okra can prevent chilling injury symptoms and maintain the quality of okra (lower weight loss, electrolyte leakage, and cellular damage index though lipid peroxidation) at 4 °C. Application of 1-methylcyclopropene (1-MCP) maintained okra quality and was also effective against chilling injury disorders (browning and pitting), associated with lower membrane permeability and lipid peroxidation, following storage for 18 d at 7 °C [96].

Modified atmosphere packaging (MAP) is an efficient tool for fresh produce preservation, with Rai and Balasubramanian [97] showing that MAP stored okra in 300 and 400 g packages were much superior in terms of maintenance of chlorophyll (green color) in the pericarp, β-carotene and ascorbic acid in the seeds as well as other qualitative parameters.

Novel processing techniques have shown promising results in okra pods products. Pulsed electric field (PEF) treatment, for example, has been applied to treat many plant materials to facilitate their further processing. Saetung *et al.* [98] applied low-voltage direct current electricity to increase phenolics content during postharvest storage of okra. In that study, total phenolics content significantly increased when using a current of 0.30 A, application time of 900–3600 s, with brine as the electrolytes. Nevertheless, the cost of a PEF instrument is rather high and may not be affordable by a small or medium food processing industry.

Storage of fresh okra pods may also affect chemical composition and quality of the final product. Analysis of metabolite profiles by 1H high resolution nuclear magnetic resonance spectroscopy (NMR) was a good way to investigate the chemical changes of okra stored for 19 days at 25 °C. Liu *et al.* [99] identified 17 metabolites among which were isoleucine, fatty acids, γ-aminobutyrate, glutamine, asparagine, unsaturated lipids, choline, phosphocholine and cinnamic acid. Moreover, the decrease in quality attributes after storage was related to the decreases of glucose and sucrose levels, while accumulation of cinnamic acid was involved in the lignification of okra tissues in the late storage period resulting in firmer pods. Extended storage affected protein content due to amino acids and γ-aminobutyrate levels increase, which indicate the degradation of proteins.

HEALTH EFFECTS AND BIOACTIVE PROPERTIES

Okra has been attributed with significant bioactive properties, such as antidiabetic, antibacterial, anticancer, antihyperlipidemic, antiinflammatory and so forth, which are mainly correlated with various antioxidant compounds [9, 100 - 102]. Plant parts (leaves, seeds, pods, roots) are considered effective against genito-urinary disorders, spermatorrhoea, chronic dysentery, ulcers, urinary calculi and hemorrhoids, and are commonly used since ancient times in traditional medicine for therapeutic purposes [103]. Okra pods in particular are considered to be stomachic stimulant, diuretic, antispasmodic and nervine and effective against constipation, leucorrhea, spermatorrhea, diabetes and jaundice [16, 104, 105]. Several bioactive compounds have been identified in okra parts so far, including alkaloids, coumarins, flavonoids, glycosides, saponins, steroids, tannins and terpenoids [37], with various plant parts presenting different chemical composition and bioactive properties [27, 90]. The health effects of okra plant parts are summarized in (Table **3**).

Antioxidant Activity

Fruit and vegetables are considered the main source of antioxidant and bioactive compounds, therefore their incorporation in human diet is of fundamental importance towards a well-balanced nutrition and well-being in general [42, 106]. Antioxidants are substances within food matrices that are able to prevent oxidative processes and consequently lead to food quality deterioration, while after ingestion they help towards inhibition of degenerative diseases in human body [107]. Many phytochemicals have been attributed with antioxidant activities, including vitamins C and E, carotenoids and phenolic compounds (flavonoids and phenolic acids) and so forth [43, 47, 102, 108]. Determination of total phenolic compounds in food matrices and plant products is a safe indicator of antioxidant activity since its content is positively correlated with antioxidant activity [109, 110]. According to Cai *et al.* [106], plants contain several compounds with radical scavenging properties and rich antioxidant activity, while there are numerous assays for antioxidant activity determination with different chemical principles and reactions involved [110]. Therefore, it is essential to apply different methods for each matrix to ensure that results are representative of real food and biological systems [111]. The assessment of flavonoids and phenolics composition in okra, as well as total antioxidant capacity helps to associate the functional properties of the species with specific compounds. Regarding flavonoids, antioxidant activity is positively correlated with the presence of hydroxyl groups which react with free radicals, hence the higher the number of hydroxyl groups the higher the antioxidant activity [112]. According to Liao *et al.* [39, 40], phenolic compounds (total phenolics and total flavonoids) were correlated with antioxidant activity as determined by DPPH and ferric reducing/antioxidant power (FRAP) assays in methanolic extracts from various okra parts (flowers, leaves and seeds), while two newly detected flavonoids showed strong antioxidant properties.

Antioxidant activity in okra plant parts has been determined with various assays so far, including DPPH, oxygen radical absorbance capacity (ORAC) and trolox equivalent antioxidant capacity (TEAC) assays in pods [41, 113], DPPH, superoxide radical, hydroxyl radical and reducing power in seeds [74, 76], DPPH, FRAP and 2,2'-Azino bis-3-ethylbenzothiazoline-6-sulfonic acid (ABTS) in okra snacks [114], DPPH, FRAP, β-carotene-linoleic acid, iron chelation and free radical scavenging activity (FRSA) in okra leaves, flowers, pods and seeds [13, 17, 39, 115], and DPPH, reducing power, nitric oxide-scavenging activity, metal chelating activity and ferric thiocyanate method (FTC) in leaves [43, 116]. Okra leaves are usually consumed as leafy vegetables and possess significant bioactive properties. According to Ebrahimzadeh *et al.* [116] who studied antioxidant properties of leaves methanolic extracts, okra leaves showed good DPPH radical-scavenging activity, Fe^{+2} chelating ability and reducing power, whereas nitric oxide scavenging activity and β-carotene/linoleic acid EC_{50} values were weak. Li *et al.* [82] have also reported strong radical-scavenging activities (DPPH) and good reducing power of aqueous extracts of leaves which were attributed to polysaccharides content, while crude polysaccharides exhibited stronger antioxidant efficacy than purified ones. Moreover, flowers ethanolic extracts exhibited good antioxidant properties determined with DPPH and reducing power assays, which could be attributed to the high amounts of detected phenolic compounds [117].

Okra seed oil has also been attributed with significant antioxidant properties which depend on extraction method, since according to Dong *et al.* [56] screw press extraction resulted in higher antioxidant activity comparing to supercritical carbon dioxide and solvent extraction.

Hepatoprotective Effects

According to Alqasoumi *et al.* [16], oral administration of ethanolic extracts of fresh okra pods at 250 and 500 mg/kg body weight were effective against carbon tetrachloride-induced liver injury in rodents. The administration of okra extracts reduced significantly the hepatic markers enzymes such as SGOT, SGPT, ALP, GGT, and cholesterol, triglycerides and malondialdehyde in serum, while non-protein sulfhydryls and total protein levels in liver increased comparing to carbon tetrachloride treated test group. Moreover, the results were comparable to silymarin which is the commonly used standard compound for the evaluation of hepatoprotective effects. Okra seeds have also beneficial effects, since according to Hu *et al.* [76] who studied the hepatoprotective effects of methanolic seeds extracts against oxidative stress induced by carbon tetrachloride in rat hepatocytes, a significant protective effect of total phenolic extracts and quercetin 3-O-glucoside was observed. Hepatoprotective effects have been also attributed to okra roots, since according to Sunilson *et al.* [118] oral administration of root extracts in doses of 250 and 500 mg/kg body weight reduced hepatoxicity in rats treated with carbon tetrachloride, while *in vitro* studies with HepG2 cells incubated with carbon tetrachloride showed similar results [119].

Apart from okra, *A. manihot* L. Medic has significant hepatoprotective properties and is being used in traditional Chinese medicine for the treatment of various types of hepatitis. Yan *et al.* [120] have studied the effect of decoctions from flowers of the species and suggested that the main phytochemicals were flavonoid compounds. Moreover, they evaluated the hepatoprotective effects of flower decoctions *in vitro* against carbon tetrachloride, as well as *in vitro* against α-naphthylisothiocyanate (ANIT)-induced cholestatic liver injury in rats. The results of this study showed the hepatoprotective properties of flavonoids, mostly due to their antioxidative and anti-inflammatory effects, as well as the regulation of hepatic transporters expression. Similar results have been reported by Ai *et al.* [121] who also confirm the hepatoprotective role of *A. manihot* flavonoids through antioxidant and anti-inflammatory effects.

Anticancer Activities

Ying *et al.* [48] used High-speed Countercurrent Chromatographic Separation (HSCCC) with ethyl-acetate–n-butanol as a gradient solvent system in order to identify compounds with anti-cancer activity from okra. The eluted compounds were tested for antiproliferative activity against cancer cell lines (A594, HL-60, MCF-7 and HO8910) and carolignan, although not detected in significant amounts, exhibited the strongest antiproliferative activity against all the tested cancer cell lines. In addition, 4′-Hydroxy phenethyl trans-ferulate which was detected in significant amounts showed the second highest antiproliferative activity. Moreover, Ren and Chen [122] tested the efficacy of polysaccharides extracted from okra fruit against different human cancer cell lines (*e.g.* OVCAR-3, MCF-7, Hela and MCG-803) and reported that raw polysaccharides and purified extracts showed inhibitory effects against proliferation of all the tested cell lines in a dose-dependent manner. Mollick *et al.* [123] used okra fruit pulp extracts in order to synthesize silver nanoparticles as a means of eco-friendly and green technique, while they also observed significant anticancer and antimicrobial activities.

Apart from fruit, seeds have been also attributed with anticancer activities. In particular, Monte *et al.* [124] tested the *in vitro* potency of a newly-discovered lectin against human breast cancer (MCF7) and skin fibroblast (CCD-1059 sk) cell lines and observed a significant inhibition in growth of breast cancer cells by up to 63% through the expression of pro-apopoptic genes and the consequent increase of Bax/Bcl ratio. However, no cytotoxic effects against skin fibroblast cells were observed, indicating a selective anti-tumor activity of okra seeds lectin.

Antihypoxic, Antifatigue Activities

Seed extracts have been reported to possess antioxidant and antihypoxic properties, where according to Ebrahimzadeh *et al.* [125] the administration *via* intraperitoneal injection in mice showed antihypoxic activity against both haemic and circulatory hypoxia in a dose-dependent manner, as well as good reducing power and nitric oxide scavenging activity. Moreover, okra seeds have been attributed with anti-fatigue and antioxidant properties due to their high content in total flavonoids, and isoquercitrin and quercetin-3-O-gentiobiose in particular, which were proposed as the main active compounds [75, 126]. In the study of Xia *et al.* [75], it was suggested that the mechanism of this anti-fatigue activity was the result of *in vitro* blood lactic acid and nitrogen urea reduction, which concomitantly increased storage of hypatic glycogen and antioxidant ability in male mice. In contrast, in the study of Li *et al.* [127] the *in vitro* antifatigue effects of okra fruit ethanolic extracts were attributed to their high content in polysaccharides.

Antimicrobial Activities

Polysaccharides of plant origin have been used in traditional medicine for dermatological therapies and skin treatments probably due to the proliferation of keratinocytes [128], while okra gum has been used in drug formulation for drug release control in chronotherapeutic drugs [11]. Okra pods contain significant amounts of pectin (okra gum) which has been attributed with various health effects. It is associated with lowering serum glucose and blood cholesterol levels, as well inhibitory activity against melanoma and colon cancer cells and bacterial adhesion of *Helicobacter pylori* and other bacteria [129 - 131]. Moreover, Lengsfeld *et al.* [132] evaluated the effect of crude and purified carbohydrate containing extracts from fresh okra pods and reported a complete inhibition of *in situ* adhesion of *H. pylori* on human gastric mucosa sections when fresh pods juice was applied, while this effect lessened for crude and purified extracts' application. These antiadhesive effects could be attributed to glycoproteins and highly acidic sugar compounds that are present in high amounts in fresh pod juice. Antibacterial properties of okra pods extracts have been also confirmed by De Carvalho *et al.* [133] who observed significant bactericidal effects against *Rhodococcus opacus*, *Mycobacterium* sp. and *M. aurum*, *Staphylococcus aureus*, *Escherichia coli*, and *Xanthobacter* Py2, attributed to the lipid fraction of okra gum and palmitic and stearic acids in particular. However, Messing *et al.* [134] report that under *in vitro* conditions *H. pylori* is located in or on the gastric mucus phase where okra polysaccharides have no affinity and therefore okra extracts could only interact with bacteria floating in stomach liquids.

In a recent study of Petropoulos *et al.* [31], the antimicrobial properties of okra seeds against *Listeria monocytogenes, Salmonella enteritidis* and *S. typhimurium, Aspergillus versicolor, Cladosporium cladosporioides* and *Penicillium funiculosum* have been reported, while Singh [135] indicated the efficiency of seed extracts against *Bacillus subtilis, S. aureus, Klebsiella pneumoniae, E. coli* and *Pseudomonas fluorescens.* Moreover, Nwaiwu *et al.* [136] reported the efficacy of aqueous okra and roselle (*Hibiscus sabdariffa*) seed extracts against various enterobacteria, which could be further exploited as natural antimicrobials with drinking water treatments, especially in regions of the world where infections of such bacteria are very common due to lack of sanitation reasons. Seed aqueous extracts have been used in the synthesis of gold nanoparticles which acted as fungicidal agents, since they exhibited significant antifungal activity against *Puccinia graminis tritci, Aspergillus flavus, Aspergillus niger* and *Candida albicans* [137]. Antimicrobial effects against various bacteria have been also reported for gold nanoparticles synthesized with okra pulp extracts by Mollick *et al.* [138] who also observed significant *in vitro* anticancer activities.

Antidiabetic Properties

Okada *et al.* [139] carried out a screening of various vegetable seeds extracts for potential antidiabetic activity by evaluating their *in vitro* inhibitory activity against tumor necrosis factor-alpha (TNF-α) and the production of adiponectin (ACRP30). TNF-α activity has been correlated with a decrease in insulin sensitivity and disruption of lipid and glucose metabolism, while ACRP30 is secreted by the adipocytes and has anti-diabetic effects, therefore any effect of the tested seed extracts on these compounds could have beneficial anti-diabetic activity. The *in vitro* results of the study showed that although okra seeds extracts were responsible for adiponectin production, TNF-α levels were not reduced when higher concentration extracts (4 μL comparing to 2 μL) were applied and further studies to elucidate the exact inhibitory mechanism are required.

Sabitha *et al.* [140], Ramachandran *et al.* [141] and Saha *et al.* [142] have suggested the antidiabetic and antihyperlipidemic effects of okra seeds, pod peels and fruit powder in streptozotocin-induced and alloxan-induced diabetic rats. According to this study, the oral administration of 100 and 200 mg/kg of powder reduced significantly blood glucose level, glycosylated hemoglobin (HbA1c), and serum glutamate-pyruvate transferase (SGPT), while it increased body weight, total proteins and hemoglobin (Hb). Moreover, a reduction in lipids to normal levels was also observed, which indicates concurrent antihyperlipidemic effects.

Similar properties have been observed by Fan *et al.* [100] who tested the effect of okra ethanolic extracts, as well as its major flavonoid compounds (isoquercitrin and quercetin 3-O-gentiobioside) on blood glucose and serum insulin in obese mice. Moreover, the oral administration of okra extracts and isoquercetin improved liver morphology, while the properties of okra extracts seems to be correlated with the inhibition of PPARγ transcription factor expression. Amin [143] and Mishra *et al.* [144] have also confirmed hypoglycemic effects of okra fruit extracts in streptozotocin-induced and alloxan-induced diabetic rats, respectively, while similar results have been observed by Subrahmanyam *et al.* [145] in alloxan-induced diabetic rabbits. Antidiabetic effects have been also attributed to okra polysaccharides, since according to Fan *et al.* [146] the crude extract of fresh okra pods reduced body weight and blood glucose levels, while it improved glucose tolerance, and decreased total cholesterol levels in serum of obese mice through the inhibition of liver X and peroxisome proliferator-activated receptors expression in liver. Huang *et al.* [147] and Peng *et al.* [148] used a step-wise fractionation of okra extracts in order to elucidate the effects and chemical composition of each subfraction. According to their results, the first two subfractions (F1 and F2, respectively) were rich in quercetin glucosides and pentacyclic triterpene esters, and carbohydrates and polysaccharides respectively, and exhibited significant effects against palmitate-induced β-cells apoptosis which is associated with diabetic nephropathy. Moreover, Karim *et al.* [149] have suggested the regular consumption of okra fruit as a means towards maintaining blood sugar in normal levels in diabetic patients, since even after domestic processing (cooking, blanching, baking) okra retains its anti-amylolytic properties which is essential for retaining low blood sugar levels. The *in vitro* inhibitory effects of okra extracts against α-amylase and α-glucosidase have been also reported by Ahmed and Kumar [150] who studied the antidiabetic properties of ethanol, methanol and aqueous extracts of okra fruit, while Lu *et al.* [151] and Sabitha *et al.* [152] reported similar effects (for okra seeds and seed and fruit peels extracts, respectively) and attributed these effects to their (epi)gallocatechins and (epi)catechins contents.

In the study of Tian *et al.* [101], the potential effects of okra extracts against gestational diabetes mellitus (GDM) were evaluated and the results indicate that the oral administration of 200 mg/kg/d okra extract in pregnant rats significantly improved blood glucose levels, probably by inhibition of oxidative stress and increase of insulin resistance which are common in pregnant women with GDM. Administration of water-soluble fractions of fruit extracts improved glycemic control in alloxan-induced diabetic rats by reducing glucose oral absorption [153], while aqueous and ethanolic fruit extracts reduced serum glucose levels and the activity of alkaline phosphatase (ALP), aspartate aminotransferase (AST) and alanine aminotransferase (ALT) [154]. Moreover, oral administration of leaves crude extracts in alloxan-induced diabetic rats resulted in an increase of albumin and a decrease of billrubin levels, which both indicate antidiabetic efficacy since hyperglycemia induces protein breakdown through glycation and gluconeogenesis and consequently albumin content increases and billrubin decreases [155]. In addition, okra gum from pods and roots has been confirmed with anticomplementary and hypoglycemic activities in mice [156].

Zhou *et al.* [157] have suggested the pretreatment of streptozotocin-induced diabetic rats with flavone glycosides from *A. manihot* against diabetic nephropathy, while hyperoside (the major flavonoid of the species) mitigated cultured podocyte apoptosis which forebodes diabetic nephropathy.

Anticholesterolemic-Antihyperlipidemic Effects

Okra fruit have been reported to bind bile acids *in vitro* [158] more efficiently than other vegetables, showing a binding capacity of 15.9% of positive control (cholysteramine). The higher binding capacity of okra fruit comparing to other vegetables has been attributed to dietary fibers content which was the highest among the vegetables tested [158]. Hypolipidemic effects of fruit and plant tissues dichloromethane and ethanol extracts in tyloxapol-induced hiperlipidemic rats were investigated by Ngoc *et al.* [159] who reported that plant tissues extracts showed better results than fruit extracts in terms of blood cholesterol and triglycerides levels. The antihyperlipidemic effect of okra extracts could be attributed to down regulation of nicotinamide adenine dinucleotide phosphate (NADPH) and nicotinamide adenine dinucleotide (NADH) cofactors or the oxidation of NADPH in fat metabolism, as previously reported by Al-Dosari *et al.* [160] and Fernandes *et al.* [161] for *Beta vulgaris* and *Mormodica charantia* extracts, respectively.

Okra seed oil consumption has been associated with hypo-cholesterolemic properties, since according to Rao *et al.* [162], 10% of seed oil in a casein based diet resulted in higher reduction of total lipids and triglycerides in animal models comparing to groundnut oil.

Immunomodulatory ffects

Flower extracts of okra have exhibited immunomodulating activity against the proliferation of HepG-2 cells and induced phagocytic ability, production of NO and TNF-α and IL-1β secretion of RAW264.7 cells [163]. According to the authors, these properties were attributed to a water-soluble polysaccharide (OFPS11) which contains mainly galactose and rhamnose. Sheu and Lai [164] investigated *in vitro* immunomodulating effects of crude and hydrolyzed fruit extracts on function and maturation of dendritic cells from bone marrow hematopoietic cells of rats and reported that polysaccharides from okra extracts activated dendritic cells, reduced endocytosis and induced the secretion of T_H1 cytokines.

Neuroprotective Effects

Tongjaroenbuangam *et al.* [165] have reported that ethanolic extracts of okra fruit, as well as rutin and quercetin exhibited protective properties to neuronal function in mice subjected to dexamethasone treatments, while they also improved learning and memory deficits due to protection of CA3 region against morphological changes. Furthermore, okra fruit extracts could be beneficial towards decreasing the risk of developing Alzheimer's (AD) and other neurogenerative diseases, since according to Mairuae *et al.* [166] *in vitro* studies with SH-SY5Y cells treated with okra extracts showed decreasing trends of oxidative stress, tau phosphorylation and intracellular iron levels, all of which are associated with AD cellular processes. Moreover, in a recent study of Mairuae *et al.* [167] okra extracts have shown inhibitory effects against lipopolysaccharide-induced pro-inflammatory mediators in BV2 microglial cells.

Other Health Effects

Doreddula *et al.* [7]studied the *in vitro* and *in vitro* potency of ethanolic and aqueous seed extracts of okra against scopolamine-induced cognitive impairment in mice and suggested that oral administration of 200 mg/kg okra extracts improved nootropic activities, while at the same time blood glucose, corticosterone, cholesterol, and triglyceride levels were reduced after stress inducing conditions. Moreover, administration of okra extracts showed no toxicity symptoms at levels up to 2000 mg/kg. Ebrahimzadeh *et al.* [116] suggested that methanolic extracts of okra leaves showed significant potency against *in vitro* erythrocyte hemolysis in rats in a dose-dependent manner. Güne *et al.* [168] who carried out an ethnopharmacological survey in Turkey have also recorded the traditional use of okra flowers and seeds decoctions and fomentations against bronchitis as well as in abscess healing. Seed lectin has been attributed with antiinflammatory (*in vitro* determination in rats using paw oedema model), antinociceptive (*in vitro* determination with writhing test in rats) and hemagglutinating activities (*in vitro* determination in human and rabbit erythrocytes) [169], while similar antiinflammatory, analgesic and antinociceptive effects have been reported for ethanolic and methanolic fruit peel extracts [170]. Ingestion of fresh okra fruit has been traditionally used in folk medicine in Turkey for gastroprotective purposes against ulcers and gastric pain, while Gürbüz *et al.* [19] have confirmed these properties *in vitro* by administrating orally methanolic pod extracts in rats with ethanol induced gastric lesions. Histopathological tests showed significant inhibition of ulcers (95.8% inhibition comparing to control) and mild edema, bleeding and degenerative changes in gastric tissues. Moreover, Ribeiro *et al.* [171] attributed the gastroprotective effects of okra fruit extracts to lectins through the involvement of opioids and Alpha-2 adrenergic receptors. Lectins from seed powder have shown antiinflammatory activities against zymosan-induced temporomandibular joint inflammations through inhibiting TNF-α/IL-1β expression and promoting HO-1 overexpression [172].

Abelmoschus Manihot is a basic ingredient in Chinese traditional medicine with various therapeutic properties. Ethanolic extracts from flowers of *A. manihot* have shown anticonvulsant and antidepressant-like effects, which could be attributed to the presence of flavonoids, including isoquercitrin, hyperoside, hibifolin, quercetin-3′-O-glucoside and their derivatives [173]. Huangkui, a commercial product based on *A. manihot* extracts has shown antiinflammatory effects against renal inflammation and glomerular injury in adriamycin-induced nephropathy rats [174].

Table 3. Health effects and bioactivity mechanisms of okra plant parts.

Health effect	Mechanism of action	Plant part	Administration	Reference
Antioxidant	-Radical scavenging activity	-Fruit, seeds, leaves and flower extracts -Seed oil	-*in vitro* assays (DPPH, FRAP, ORAC, TEAC, ABTS, β-carotene-linoleic acid, FRSA	[13, 41, 56, 74, 113, 125]
Hepatoprotective	- Reduction of hepatic markers enzymes such as SGOT, SGPT, ALP, GGT	-Ethanolic extracts of fruit -Methanolic seeds extracts - Ethanolic root extracts	-Oral administration of 250 and 500 mg fruit extracts per kg in rodents and rats -*in vitro* assays with 50-200 mg fruit extract per kg in rats - Oral administration of 250 and 500 mg root extracts per kg	[16, 76, 118, 119]
Anticancer	-Antiproliferative activity against cancer cell lines - Expression of pro-apopoptic genes and increase of Bax/Bcl ratio	-Ethanolic fruit extracts -Seed extracts	-*in vitro* assays against cancer cells	[48, 122, 124, 138]

(Table 3) cont.....

Health effect	Mechanism of action	Plant part	Administration	Reference
Antihypoxic Antifatigue	-*in vitro* blood lactic acid and nitrogen urea reduction -Decrease of serum lactic acid and blood urea nitrogen levels -Enhance of hepatic glycogen storage - Lowering of malondialdehyde (MDA) levels -increase of superoxide dismutase (SOD) and glutathione peroxidase (GSH-PX) levels	-Ethanolic extracts of fruit -Methanolic extracts of seeds -Methanolic extracts of fruit Aqueous extracts of okra pods and pod parts (seeds and skins)	-Oral administration of 0.8-3.2 g extract per kg - Intraperitoneal injection of 0-1000 mg/kg in mice - Oral administration of 25-75 mg/kg of quercetin-3-O-gentiobiose -Oral administration of pod extracts	[75, 125 - 127]
Antimicrobial	-Inhibitory activity against bacterial adhesion of *Helicobacter pylori* and other bacteria -Bastericidal effects -Antifungal activities	-Aqueous fruit extracts -Ethanolic seed extracts -Seed aqueous extracts	-*in vitro* assays with human gastric epithelial AGS cells and human gastric mucosa -*in vitro* assays on agar plates -Synthesis of gold nanoparticles	[31, 131 - 133, 137]
Antidiabetic	-Adiponectin production -Reduction of blood glucose level, glycosylated hemoglobin (HbA1c), and serum glutamate-pyruvate transferase (SGPT) - Inhibition of PPARγ transcription factor expression - Inhibition of liver X and peroxisome proliferator-activated receptors expression in liver - Inhibition of palmitate-induced β-cells apoptosis - Inhibition of oxidative stress and increase of insulin resistance - Increase of albumin and a decrease of billrubin levels	-Ethanolic extracts of seeds -Aqueous extracts of pod skins and seeds	-*in vitro* assays with 2 and 4 μL of extracts in mouse 3T3-L1 cells -Oral administration of 100 and 200 mg extract per kg -Oral administration of 200-800 mg leaves extracts per kg	[139 - 142, 147, 148, 154]

(Table 3) cont.....

Health effect	Mechanism of action	Plant part	Administration	Reference
Anticholesterolemic-antihyperlipidemic	-Binding of bile acids - Down regulation of nicotinamide adenine dinucleotide phosphate (NADPH) and nicotinamide adenine dinucleotide (NADH) cofactors - Oxidation of NADPH in fat metabolism	-Okra fruit -Dichloromethane and ethanol fruit extracts -Seed oil	-*in vitro* bile acid binding - Oral administration of 30 g of dry extract per kg -Oral administration of seed oil (10% of feed)	[158, 159, 162]
Immunomodulatory	- Inhibition of HepG-2 cells proliferation - Increase of the phagocytic, elevation of NO production, TNF-α and IL-1β secretion of RAW264.7 cells -Activation of dendritic cells, reduction of endocytosis and induction the secretion of T_H1 cytokines	-Aqueous flower extracts - Crude and hydrolyzed fruit extracts	-*in vitro* assays with 0-1000 μg/mL in human hepatocellular carcinoma HepG-2 and mouse macrophage cell line RAW264.7 cells -*in vitro* assays with 100 μg/mL in dendritic cells from rat bone marrow hematopoietic cells	[163, 164]
Neuroprotective	-Protection of CA3 region against morphological changes -Decrease of oxidative stress, tau phosphorylation and intracellular iron levels -Inhibition of pro-inflammatory mediators in BV2 microglial cells	-Ethanolic extracts of okra fruit	-Oral administration of 60 mg kg^{-2} in mice -*in vitro* assays with 50-200 μg mL in murine BV2 cell cultures -*in vitro* assays with 0-50 μg mL in human neuroblastoma SH-SY5Y cells	165–167]
Against cognitive impairment	Inhibition of acetylcholinesterase	- Aqueous and methanolic extracts of seeds	-Oral administration of 200 mg/kg in mice	[17]
Against erythrocyte hemolysis		-Seed lectin -Methanolic extracts of leaves	-Intravenous administration of 0.01-1 mg/kg in rats and *in vitro* studies in rabbit and human cells -*in vitro* assays in rats erythrocytes	[116, 169]
Antinociceptive				
Hemagglutinating activities				

(Table 3) cont.....

Health effect	Mechanism of action	Plant part	Administration	Reference
Gastroprotective	-Mediation by alpha-2 adrenergic and opioid receptors activation	-Seed lectin -Methanolic ectracts of fruit	-1 mg/kg in mice -5ml/kg of methanlic extracts in rats	[19, 171]
Antiinflammatory	Inhibition of TNF-α/IL-1β expression and promotion of HO-1 overexpression	-Seed powder extracts -Ethanolic and methanolic fruit peel extracts	-0.1-1 mg/kg to male Wistar rats	[170, 172]

OTHER USES

Pod extracts are low-cost and naturally available biopolymers with low toxicity that find many uses in industrial applications. The predominant compounds in these extracts are mainly pectins, which are described as polysaccharides with three segments, *e.g.* homogalacturonan (HG), rhamnogalacturonan I (RG-I) and rhamnogalacturonan II (RG-II) regions [175]. There is great diversity in their structure not only between the various plant species but also within the same species and between the different tissues and cells of the same plant [175].

The roots and stems of okra plants have been used as clarification agents in sugar production [1], while okra gum has been suggested for the preparation of artificial saliva [176].

Alternative used of okra seeds have been proposed, such as poultry feeding or as a coffee substitute [50], while Bryant and Montecalvo [32] have suggested the use of okra seeds for flour, protein concentrate and protein isolate due to their high protein solubility properties. According toAdelakun *et al.* [14, 76], seed flour is even richer in protein (up to 46%) after pretreatments such as oil extraction and roasting which makes its use more interesting for the food industry. Moreover, pretreatments such as soaking and roasting improve digestibility and antioxidant activity of seed flours, while the most of antioxidant properties are exhibited in the intestinal phase within the gastrointestinal tract [14, 74]. Additionally, okra seed flour has been suggested as an ingredient in weaning and adult foods with improved nutritional value, with fortified "ogi" being an example of such products [177]. Other uses include biodiesel derived from okra seeds, since according to Anwar *et al.* [68] fuel properties such as acid value, cetane number, cold flow properties, density, flash point, kinematic viscosity, lubricity, oxidative stability and sulfur contents comply with ASTM and EU specifications for biodiesel fuels.

Ghori *et al.* [178] examined the potential use of okra extracts in pharmaceutical and food formulations by evaluating swelling and dissolution behavior and emulsification capacity and suggested that okra polysaccharides may be used in hydrophilic matrix tablets or as emulsification agents and stabilizers in drug formulations. The high amounts of polysaccharides in okra extracts make them suitable for controlled-release tablet formulations as a hydrophilic polymer or a drug carrier [179, 180]. Moreover, Sinha *et al.* [181, 182] have reported that mixture of okra gum with other alginates (carbohydrate hydrocolloids) has a great potential as mucoadhesive agent in oral sustained delivery of pharmaceuticals and drugs with controlled-release properties, while Alba *et al.* [183] and Tamenouga *et al.* [184] have suggested the use of okra extracts as emulsifiers in acidic conditions. Blends of okra gum with synthetic polymers such as carbopol 71G increased significantly $t_{70\%}$ (time needed for release of 70% of drug) and drug release kinetics in matrix tablets of zidovudine [180]. Similar uses of okra gum have been reported for other controlled release matrix tablets, such as povidone [185], furosemide and diclofenac sodium [186], sulphaguanidine [187] and paracetamole [188]. Moreover, Bakre and Jaiyeoba [189] reported the use of okra powder in metronidazole tablet formulation and further suggested that powder from sun-dried pods showed better disintegration and dissolution properties than oven-dried pods. According to Sharma *et al.* [190, 191], polysaccharides from okra fruit can be used for the formation of a mucoadhesive gel and microspheres suitable for nasal drug delivery of rizatriptan benzoate, since the permeation properties of okra-based gel were superior to synthetic polymers.

Okra pod skins have been proposed as sources of hydrocolloids and hydroxyapatite that could find industrial applications in the development of bone prosthetics and cartilage engineering [192].

Okra by-products, namely stalks, core and bast contain amounts of alpha-cellulose and lignin which are comparable to wood and other resources used in the paper-making industry, rendering them appropriate for use in paper production [67, 193]. Moreover, according to Alam and Khan [194] the high cellulose content of okra bast, as well as its dye exhaustion properties, tensile strength and low percentage in lignin make it useable in cellulose based and textile fiber industries.

Okra seeds could be incorporated in food products for biofortification purposes. According to Ray *et al.* [195], fortification of dhokla (an Indian food that contains chickpea and rice flour) with different amounts of okra seeds flour (1-7%) resulted in an increase in folic acid, vitamin C, β-carotene, phenolics and flavonoids. Moreover, water holding properties and antioxidant activity of biofortified dokhla increased, comparing to control treatment (no okra seeds flour added).

Okra mucilage (okra gum) has also culinary uses as a thickener in soups, while extraction procedure may affect its chemical composition [196]. Moreover, it can be used for pharmaceutical purposes as blood volume expander, plasma replacement and excipient in tablet formulations, as well as a spreading agent in the paper industry [2, 6, 15]. Okra gum in dried form has been suggested to sustain the release of propranolol hydrochloride in tablet formulations due to its viscous properties [179].

CONCLUDING REMARKS AND FUTURE PROSPECTS

Okra is a valuable vegetable with high nutritional value and bioactive properties. Its multidimensional character renders it a species with high economic potential, since all plant parts may find uses in the food and related industries. However, although it is a popular crop in some regions of the world, especially in the tropics where it is usually found, it is considered a minor crop and an underutilized species and its further exploitation is of major importance. So far, its uses include fruit consumption in fresh or processed form, however the interesting results regarding the bioactive properties of the species indicate that alternative uses could find application in the food industry. Considering the recent trends for "healthy" foods and/or food products with enhanced bioactivities, okra products such as seed oils and flours or okra-based products could be very promising for the food and nutraceuticals industry.

Future research need to focus on unravelling the mechanisms involved in the therapeutic and pharmaceutical properties of okra plants in order to identify specific compounds or classes of compounds which are responsible for the health effects of the species. Further clinical studies are also needed to confirm the health effects of okra products that are well-known in traditional and folk medicine and have been already verified in *in vitro* or *in vitro* studies with animal models. The identification of specific compounds with high bioactivity potential would help towards tailor-made breeding programs, where apart from agronomical features, qualitative features could be also considered in order to select elite cultivars with high nutritional value and enhanced health effects.

CONSENT FOR PUBLICATION

Not applicable.

CONFLICT OF INTEREST

The author (editor) declares no conflict of interest, financial or otherwise.

ACKNOWLEDGEMENT

Declare none.

REFERENCES

[1]   Singh P, Chauhan V, Tiwari BK, Chauhan SS, Simon S, Bilal S, *et al.* An overview on okra (*Abelmoschus esculentus*) and it's importance as a nutritive vegetable in the world. Int J Pharma Bio Sci 2014; 4(2): 227-33.

[2]   Lamont WJ, Wall A. Okra — A Versatile Vegetable Crop. Horttechnology 1999; 9(2): 179-84.

[3]   Petropoulos S, Fernandes Â, Barros L, Ferreira ICFR. Chemical composition, nutritional value and antioxidant properties of Mediterranean okra genotypes in relation to harvest stage. Food Chem 2018; 242: 466-74.
[http://dx.doi.org/10.1016/j.foodchem.2017.09.082] [PMID: 29037716]

[4]   Karakoltsidis PA, Constantinides SM. Okra seeds: a new protein source. J Agric Food Chem 1975; 23(6): 1204-7.
[http://dx.doi.org/10.1021/jf60202a041] [PMID: 1238449]

[5]   Arapitsas P. Identification and quantification of polyphenolic compounds from okra seeds and skins. Food Chem 2008; 110(4): 1041-5.
[http://dx.doi.org/10.1016/j.foodchem.2008.03.014] [PMID: 26047300]

[6]   Gemede HF, Ratta N, Haki GD, Woldegiorgis AZ. Nutritional Quality and Health Benefits of Okra (*Abelmoschus esculentus*): A Review. Food Sci Qual Manag 2014; 33: 87-97.

[7]   Saifullah M, Rabbani MG. Evaluation and characterizatialon of okra (*Abelmoschus esculentus* L. Moench.) genotypes. SAARC J Agric 2009; 7(1): 92-9.

[8]   Moyin-Jesu EI. Use of plant residues for improving soil fertility, pod nutrients, root growth and pod weight of okra (Abelmoschus esculentum L). Bioresour Technol 2007; 98(11): 2057-64.
[http://dx.doi.org/10.1016/j.biortech.2006.03.007] [PMID: 17336057]

[9]   Sobukola O. Effect of Pre-Treatment on the Drying Characteristics and Kinetics of Okra (Abelmoschus esculetus (L.) Moench) Slices. Int J Food Eng 2009; 5(2): 1-20.
[http://dx.doi.org/10.2202/1556-3758.1191]

[10]   Alba K, Laws AP, Kontogiorgos V. Isolation and characterization of acetylated LM-pectins extracted from okra pods. Food Hydrocoll 2015; 43: 726-35.
[http://dx.doi.org/10.1016/j.foodhyd.2014.08.003]

[11]   Newton AMJ, Indana VL, Kumar J. Chronotherapeutic drug delivery of Tamarind gum, Chitosan and Okra gum controlled release colon targeted directly compressed Propranolol HCl matrix tablets and in-vitro evaluation. Int J Biol Macromol 2015; 79: 290-9.
[http://dx.doi.org/10.1016/j.ijbiomac.2015.03.031] [PMID: 25936283]

[12]   Kpodo FM, Agbenorhevi JK, Alba K, Bingham RJ, Oduro IN, Morris GA, *et al.* Pectin isolation and characterization from six okra genotypes. Food Hydrocoll 2017; 72: 323-30.
[http://dx.doi.org/10.1016/j.foodhyd.2017.06.014]

[13]   Jiang DQ, Xu D, Yuan K. Chemical composition and antioxidant activities of essential oil of *Abelmoschus esculentus* L. from different parts. 2011 IEEE Int Symp IT. Med Educ 2011; 1: 165-8.

[14]   Adelakun OE, Ade-Omowaye BIO, Adeyemi IA, Van de Venter M. Mineral composition and the functional attributes of Nigerian okra seed (*Abelmoschus esculentus* Moench) flour. Food Res Int 2012; 47(2): 348-52.
[http://dx.doi.org/10.1016/j.foodres.2011.08.003]

[15]   Ameena K, Dilip C, Saraswathi R, Pn K, Sankar C, Sp S. Isolation of the mucilages from Hibiscus rosasinensis linn. and Okra (*Abelmoschus esculentus* linn.) and studies of the binding effects of the

mucilages. Asian Pac J Trop Med 2010; 3(7): 539-43.
[http://dx.doi.org/10.1016/S1995-7645(10)60130-7]

[16]   Alqasoumi SI. 'Okra' *Hibiscus esculentus* L.: A study of its hepatoprotective activity. Saudi Pharm J 2012; 20(2): 135-41.
[http://dx.doi.org/10.1016/j.jsps.2011.10.002] [PMID: 23960784]

[17]   Doreddula SK, Bonam SR, Gaddam DP, Desu BS, Ramarao N, Pandy V. Phytochemical analysis, antioxidant, antistress, and nootropic activities of aqueous and methanolic seed extracts of ladies finger (*Abelmoschus esculentus* L.) in mice. Sci World J 2014; 2014: 519848.
[http://dx.doi.org/10.1155/2014/519848] [PMID: 25401145]

[18]   Kamalesh P, Subrata D, Asraf AK, Pranabesh C. Phytochemical investigation and hypoglycaemic effect of *Abelmoschus esculentus*. Res J Pharm Technol 2016; 9(2): 162-4.
[http://dx.doi.org/10.5958/0974-360X.2016.00028.7]

[19]   Gürbüz I, Ustün O, Yesilada E, Sezik E, Kutsal O. Anti-ulcerogenic activity of some plants used as folk remedy in Turkey. J Ethnopharmacol 2003; 88(1): 93-7.
[http://dx.doi.org/10.1016/S0378-8741(03)00174-0] [PMID: 12902057]

[20]   Takahashi J, Toshima G, Matsumoto Y, *et al. in vitro* screening for antihyperlipidemic activities in foodstuffs by evaluating lipoprotein profiles secreted from human hepatoma cells. J Nat Med 2011; 65(3-4): 670-4.
[http://dx.doi.org/10.1007/s11418-011-0542-x] [PMID: 21562909]

[21]   Singh MK, Pandey V, Singh S. Package of practices for organic farming in okra. Rashtriyakrishi 2014; 9(1): 71-2. [*Abelmoschus esculentus* (L.) Moench].

[22]   Siemonsma JS. West african okra - Morphological and cytogenetical indications for the existence of a natural amphidiploid of *Abelmoschus esculentus* (L.) Moench and A. manihot (L.) Medikus. Euphytica 1982; 31(1): 241-52.
[http://dx.doi.org/10.1007/BF00028327]

[23]   Onakpa MM. Ethnomedicinal, phytochemical and pharmacological profile of genus Abelmoschus. Phytopharmacology 2013; 4(3): 648-63.

[24]   Kumar DS, Tony DE, Kumar AP, Kumar KA, Rao DBS, Nadendla R. 2013.

[25]   Kyriakopoulou OG, Arens P, Pelgrom KTB, Karapanos I, Bebeli P, Passam HC. Genetic and morphological diversity of okra (*Abelmoschus esculentus* [L.] Moench.) genotypes and their possible relationships, with particular reference to Greek landraces. Sci Hortic (Amsterdam) 2014; 171: 58-70.
[http://dx.doi.org/10.1016/j.scienta.2014.03.029]

[26]   Charrier A. Genetic resources of Abelmoschus (okra) Rome, Italy. 1984.

[27]   Benchasri S. Okra (*Abelmoschus esculentus* (L.) Moench) as a Valuable Vegetable of the World. Ratar Povrt 2012; 49: 105-12.
[http://dx.doi.org/10.5937/ratpov49-1172]

[28]   Jarret RL, Wang ML, Levy IJ. Seed oil and fatty acid content in okra (*Abelmoschus esculentus*) and related species. J Agric Food Chem 2011; 59(8): 4019-24.
[http://dx.doi.org/10.1021/jf104590u] [PMID: 21413797]

[29]   www.prota.org2017.

[30]   Zoro AF, Zoue LT, Kra SAK, Yepie AE, Niamke SL. An Overview of Nutritive Potential of Leafy Vegetables Consumed in Western Côte d'Ivoire. Pak J Nutr 2013; 12(10): 949-56.
[http://dx.doi.org/10.3923/pjn.2013.949.956]

[31]   Petropoulos S, Fernandes Â, Barros L, Ciric A, Sokovic M, Ferreira ICFR. The chemical composition, nutritional value and antimicrobial properties of *Abelmoschus esculentus* seeds. Food Funct 2017; 8(12): 4733-43.
[http://dx.doi.org/10.1039/C7FO01446E] [PMID: 29165457]

[32]   Bryant LA, Montecalvo J. Functional, and Nutritional Seed Products Properties of Okra. J Food Sci 1988; 53(3): 810-6.
[http://dx.doi.org/10.1111/j.1365-2621.1988.tb08960.x]

[33]   Nwachukwu EC, Nulit R, Go R. Nutritional and biochemical properties of Malaysian okra variety. Adv Med Plant Res 2014; 2(1): 16-9.

[34]   Roy A, Shrivastava SL, Mandal SM. Functional properties of Okra *Abelmoschus esculentus* L. (Moench): traditional claims and scientific evidences. Plant Sci Today 2014; 1(3): 121-30.
[http://dx.doi.org/10.14719/pst.2014.1.3.63]

[35]   Rubiang-Yalambing L, Arcot J, Greenfield H, Holford P. Aibika (Abelmoschus manihot L.): Genetic variation, morphology and relationships to micronutrient composition. Food Chem 2016; 193: 62-8.
[http://dx.doi.org/10.1016/j.foodchem.2014.08.058] [PMID: 26433288]

[36]   Rao KS, Dominic TR, Singh K, Kaluwin C, Rivett DE, Jones GP, *et al.* Lipid, Fatty Acid, Amino Acid, and Mineral Compositions of Five Edible Plant Leaves. J Agric Food Chem 1990; 38: 2137-9.
[http://dx.doi.org/10.1021/jf00102a007]

[37]   Ayushi T, Prachee D, Gupta SK, Geeta W. Screened phytochemicals of A. esculentus leaves and their therapeutic role as an antioxidant. Int J Pharmacogn Phytochem Res 2016; 8(9): 1509-15.

[38]   Lin Y, Lu M, Liao H, Li Y, Han W, Yuan K. Content determination of the flavonoids in the different parts and different species of *Abelmoschus esculentus* L by reversed phase-high performance liquid chromatograph and colorimetric method Pharmacogn Mag 2014; 10(39): 278-84.

[39]   Liao H, Dong W, Shi X, Liu H, Yuan K. Analysis and comparison of the active components and antioxidant activities of extracts from *Abelmoschus esculentus* L. Pharmacogn Mag 2012; 8(30): 156-61.
[http://dx.doi.org/10.4103/0973-1296.96570] [PMID: 22701290]

[40]   Liao H, Liu H, Yuan K. A new flavonol glycoside from the *Abelmoschus esculentus* Linn. Pharmacogn Mag 2012; 8(29): 12-5.
[http://dx.doi.org/10.4103/0973-1296.93303] [PMID: 22438657]

[41]   Ahiakpa K, Kwatei E, Energy A, Achel G, Commission E, Achoribo E, *et al.* Total flavonoid, phenolic contents and antioxidant scavenging activity in 25 accessions of okra (Abelmoschus spp L.). Afr J Food Sci Technol 2014; 4(5): 129-35.

[42]   Adetuyi F, Osagie A. Nutrient, antinutrient, mineral and zinc bioavailability of okra *Abelmoschus esculentus* (L) Moench variety. Am J Food Nutr 2011; 1(2): 49-54.
[http://dx.doi.org/10.5251/ajfn.2011.1.2.49.54]

[43]   Ademoyegun OT, Akin-Idowu PE, Ibitoye DO, Adewuyi GO. Phenolic contents and free radical scavenging activity in some leafy vegetables. Int J Veg Sci 2013; 19: 126-37.
[http://dx.doi.org/10.1080/19315260.2012.677943]

[44]   El-Qudah JM. Identification and quantification of major carotenoids in some vegetables. Am J Appl Sci 2009; 6(3): 492-7.
[http://dx.doi.org/10.3844/ajassp.2009.492.497]

[45]   Wang R, Liu Q, Wu Z, Wang M, Chen X. Target-guided isolation of polar antioxidants from *Abelmoschus esculentus* (L). Moench by high-speed counter-current chromatography method coupled with wavelength switching and extrusion elution mode. J Sep Sci 2016; 39(20): 3983-9.
[http://dx.doi.org/10.1002/jssc.201600617] [PMID: 27542571]

[46]   Khomsug P, Thongjaroenbuangam W, Pakdeenarong N, Suttajit M, Chantiratikul P. Antioxidative Activities and Phenolic Content of Extracts from Okra (*Abelmoschus esculentus* L.). Res J Biol Sci 2010; 5(4): 310-3.
[http://dx.doi.org/10.3923/rjbsci.2010.310.313]

[47]   Shui G, Peng LL. An improved method for the analysis of major antioxidants of *Hibiscus esculentus*

Linn. J Chromatogr A 2004; 1048(1): 17-24.
[http://dx.doi.org/10.1016/S0021-9673(04)01187-2] [PMID: 15453414]

[48] Ying H, Jiang H, Liu H, Chen F, Du Q. Ethyl acetate-n-butanol gradient solvent system for high-speed countercurrent chromatography to screen bioactive substances in okra. J Chromatogr A 2014; 1359: 117-23.
[http://dx.doi.org/10.1016/j.chroma.2014.07.029] [PMID: 25069743]

[49] Zhou Y, Jia X, Shi J, Xu Y, Jing L, Jia L. Two New Pentacyclic Triterpenes from *Abelmoschus esculentus*. Helv Chim Acta 2013; 96: 533-7.
[http://dx.doi.org/10.1002/hlca.201200275]

[50] Sami R, Lianzhou J, Yang L, Ma Y, Jing J. Evaluation of Fatty Acid and Amino Acid Compositions in Okra (*Abelmoschus esculentus*) Grown in Different Geographical Locations Biomed Res Int 2013 Article ID 574283

[51] Berry SK. The fatty acid composition and cyclopropene fatty acid content of the maturing okra (*Hibiscus esculentus* L.) fruits. Pertanika 1980; 3(2): 82-6.

[52] Rakhimov DA, Chernenko TV, Asia M. Lipid-carbohydrate composition of Hibiscus esculenthus. Chem Nat Compd 2003; 39(3): 246-8.
[http://dx.doi.org/10.1023/A:1025462216825]

[53] Arlai A, Nakkong R, Samjamin N, Sitthipaisarnkun B. The effects of heating on physical and chemical constitutes of organic and conventional okra. Procedia Eng 2012; 32: 38-44.
[http://dx.doi.org/10.1016/j.proeng.2012.01.1234]

[54] Asgar MA, Fazilah A, Huda N, Bhat R, Karim AA. Nonmeat Protein Alternatives as Meat Extenders and Meat Analogs. Compr Rev Food Sci Food Saf 2010; 9(5): 513-29.
[http://dx.doi.org/10.1111/j.1541-4337.2010.00124.x]

[55] de Sousa Ferreira Soares G, Gomes VdeM, Dos Reis Albuquerque A, *et al.* Spectroscopic and thermooxidative analysis of organic okra oil and seeds from *Abelmoschus esculentus*. Sci World J 2012; 2012: 847471.
[http://dx.doi.org/10.1100/2012/847471] [PMID: 22645459]

[56] Dong Z, Zhang JG, Tian KW, Pan WJ, Wei ZJ. The fatty oil from okra seed: Supercritical carbon dioxide extraction, composition and antioxidant activity. Curr Top Nutraceutical Res 2014; 12(3): 75-84.

[57] Al-Wandawi H. Chemical composition of seeds of two okra cultivars. J Agric Food Chem 1983; 31(6): 1355-8.
[http://dx.doi.org/10.1021/jf00120a051] [PMID: 6655144]

[58] Kumar S, Dagnoko S, Haougui A, Ratnadass A, Pasternak D, Kouame C. Okra (Abelmoschus spp.) in West and Central Africa: Potential and progress on its improvement. Afr J Agric Res 2010; 5(25): 3590-8.

[59] Anwar F, Rashid U, Mahmood Z, Iqbal T, Sherazi TH. Inter-varietal variation in the composition of okra (*Hibiscus esculentus* L.) seed oil. Pak J Bot 2011; 43(1): 271-80.

[60] Çalişir S, Özcan M, Haciseferoğullari H, Yildiz MU. A study on some physico-chemical properties of Turkey okra (Hibiscus esculenta L.) seeds. J Food Eng 2005; 68(1): 73-8.
[http://dx.doi.org/10.1016/j.jfoodeng.2004.05.023]

[61] Thanakosai W, Phuwapraisirisan P. First identification of α-glucosidase inhibitors from okra (*Abelmoschus esculentus*) seeds. Nat Prod Commun 2013; 8(8): 1085-8.
[PMID: 24079173]

[62] Savello PA, Martin FW, Hill JM. Nutritional composition of okra seed meal. J Agric Food Chem 1980; 28(6): 1163-6.
[http://dx.doi.org/10.1021/jf60232a021] [PMID: 7451742]

[63]   Gebhardt SE, Thomas RG. Nutritive Value of Foods. United States Department of Agriculture 2002.

[64]   András CD, Simándi B, Örsi F, Lambrou C, Missopolinou-Tatala D, Panayiotou C, *et al.* Supercritical carbon dioxide extraction of okra (*Hibiscus esculentus* L) seeds. J Sci Food Agric 2005; 85(8): 1415-9.
[http://dx.doi.org/10.1002/jsfa.2130]

[65]   Lee ST, Radu S, Ariffin A, Ghazali HM. Physico-Chemical Characterization of Oils Extracted from Noni, Spinach, Lady's Finger, Bitter Gourd and Mustard Seeds, and Copra. Int J Food Prop 2015; 18(11): 2508-27.
[http://dx.doi.org/10.1080/10942912.2014.986577]

[66]   Topkafa M. Evaluation of chemical properties of cold pressed onion, okra, rosehip, safflower and carrot seed oils: triglyceride, fatty acid and tocol compositions. Anal Methods 2016; 8(21): 4220-5.
[http://dx.doi.org/10.1039/C6AY00709K]

[67]   Camciuc M, Deplagne M, Vilarem G, Gaset A. Okra - *Abelmoschus esculentus* L. (Moench.) a crop with economic potential for set aside acreage in France. Ind Crops Prod 1998; 7(2–3): 257-64.
[http://dx.doi.org/10.1016/S0926-6690(97)00056-3]

[68]   Anwar F, Rashid U, Ashraf M, Nadeem M. Okra (*Hibiscus esculentus*) seed oil for biodiesel production. Appl Energy 2010; 87(3): 779-85.
[http://dx.doi.org/10.1016/j.apenergy.2009.09.020]

[69]   Martin FW, Rhodes AM. Seed characteristics of okra and related *Abelmoschus* species. Qual Plant Plant Foods Hum Nutr 1983; 33(1): 41-9.
[http://dx.doi.org/10.1007/BF01093736]

[70]   Martin FW, Telek L, Ruberté R, Santiago AG. Protein, Oil and Gossypol Contents of a Vegetable Curd Made From Okra Seeds. J Food Sci 1979; 44(5): 1517-9.
[http://dx.doi.org/10.1111/j.1365-2621.1979.tb06476.x]

[71]   Muzquiz M, Varela A, Burbano C, Cuadrado C, Guillamón E, Pedrosa MM. Bioactive compounds in legumes: Pronutritive and antinutritive actions. implications for nutrition and health. Phytochem Rev 2012; 11(2–3): 227-44.
[http://dx.doi.org/10.1007/s11101-012-9233-9]

[72]   Duranti M, Gius C. Legume seeds : protein content and nutritional value. Field Crops Res 1997; 53: 31-45.
[http://dx.doi.org/10.1016/S0378-4290(97)00021-X]

[73]   Gutiérrez-Uribe JA, Guajardo-Flores D, López-Barrios L. Legumes in the Diet. ¹. The Netherlands: Elsevier LtdEncyclopedia of Food and Health 2016; 3: pp. 539-43.Encyclopedia of Food and Health 2016; 3: pp.
[http://dx.doi.org/10.1016/B978-0-12-384947-2.00420-7]

[74]   Adelakun OE, Oyelade OJ, Ade-Omowaye BIO, Adeyemi IA, Van de Venter M, Koekemoer TC. Influence of pre-treatment on yield chemical and antioxidant properties of a Nigerian okra seed (*Abelmoschus esculentus* moench) flour. Food Chem Toxicol 2009; 47(3): 657-61.
[http://dx.doi.org/10.1016/j.fct.2008.12.023] [PMID: 19146911]

[75]   Xia F, Zhong Y, Li M, *et al.* Antioxidant and Anti-Fatigue Constituents of Okra. Nutrients 2015; 7(10): 8846-58.
[http://dx.doi.org/10.3390/nu7105435] [PMID: 26516905]

[76]   Hu L, Yu W, Li Y, Prasad N, Tang Z. Antioxidant activity of extract and its major constituents from okra seed on rat hepatocytes injured by carbon tetrachloride. BioMed Res Int 2014; 2014: 341291.
[http://dx.doi.org/10.1155/2014/341291] [PMID: 24719856]

[77]   Atawodi SE, Atawodi JC, Idakwo GA, *et al.* Polyphenol composition and antioxidant potential of *Hibiscus esculentus* L. fruit cultivated in Nigeria. J Med Food 2009; 12(6): 1316-20.
[http://dx.doi.org/10.1089/jmf.2008.0211] [PMID: 20041787]

[78] Gopalan C, Rama Sastri B, Balasubramanian S. Proximate principles: Common foods Nutritive value of Indian foods. Hyderabad, India: National Institute of Nutrition 2012; p. 52.

[79] Caluête EME, de Souza LMP, Ferreira E dos S, de França AP, Gadelha CADA, Aquino JDS, *et al.* Nutritional, antinutritional and phytochemical status of okra leaves (*Abelmoschus esculentus*) subjected to different processes. Afr J Biotechnol 2015; 14(8): 683-7.
[http://dx.doi.org/10.5897/AJB2014.14356]

[80] Kumar R, Patil MB, Patil SR, Paschapur MS. Evaluation of *Abelmoschus esculentus* mucilage as suspending agent in paracetamol suspension. Int J Pharm Tech Res 2009; 1(3): 658-65.

[81] Liu I-M, Liou S-S, Lan T-W, Hsu F-L, Cheng J-T. Myricetin as the active principle of *Abelmoschus moschatus* to lower plasma glucose in streptozotocin-induced diabetic rats. Planta Med 2005; 71(7): 617-21.
[http://dx.doi.org/10.1055/s-2005-871266] [PMID: 16041646]

[82] Li Q, Zhao T, Bai SQ, Mao GH, Zou Y, Feng WW, *et al.* Water-Soluble Polysaccharides from Leaves of *Abelmoschus esculentus*: Purification, Characterization, and Antioxidant Activity. Chem Nat Compd 2017; 53(3): 412-6.
[http://dx.doi.org/10.1007/s10600-017-2011-6]

[83] Taroreh MIR, Widiyantoro A, Murdiati A, Hastuti P, Raharjo S. Identification of flavonoid from leaves of gedi (Abelmoschus manihot L.) and its antioxidant activity. AIP Conf Proc 2016; 1755(1)
[http://dx.doi.org/10.1063/1.4958518]

[84] Tomoda M, Shimizu N, Gonda R. Isolation and characterization of a mucilage, "Okra-Mucilage R", from the roots of *Abelmoschus esculentus*. Jpn J Crop Sci 1981; 50(4): 476-80.
[http://dx.doi.org/10.1626/jcs.50.476]

[85] Camciuc M, Marie J, Re B, Vilarem R, Gaset A, Bessière JM, *et al.* Volatile components in okra seed coat. Phytochemistry 1998; 48(2): 311-5.
[http://dx.doi.org/10.1016/S0031-9422(97)01127-8]

[86] Ames JM, Macleod G. Volatile component of okra. Phytochemistry 1990; 29(4): 1201-7.
[http://dx.doi.org/10.1016/0031-9422(90)85429-J]

[87] Molfetta I, Ceccarini L, Macchia M, Flamini G, Cioni PL. *Abelmoschus esculentus* (L.) Moench. and *Abelmoschus* moschatus Medik: seeds production and analysis of the volatile compounds. Food Chem 2013; 141(1): 34-40.
[http://dx.doi.org/10.1016/j.foodchem.2013.02.030] [PMID: 23768323]

[88] Camciuc M, Vilarem G, Gaset A, Bessière JM. Volatile constituents of the seed teguments of *Abelmoschus esculentus* (L.) moench. J Essent Oil Res 1999; 11(5): 545-52.
[http://dx.doi.org/10.1080/10412905.1999.9701211]

[89] Du Z, Clery RA, Hammond CJ. Volatile organic nitrogen-containing constituents in ambrette seed Abelmoschus moschatus Medik (Malvaceae). J Agric Food Chem 2008; 56(16): 7388-92.
[http://dx.doi.org/10.1021/jf800958d] [PMID: 18656937]

[90] Rao PU. Chemical composition and biological evaluation of okra (*Hibiscus esculentus*) seeds and their kernels. Qual Plant 1985; 35: 389-96.
[http://dx.doi.org/10.1007/BF01091784]

[91] Gemede HF, Haki GD, Beyene F, Woldegiorgis AZ, Rakshit SK. Proximate, mineral, and antinutrient compositions of indigenous Okra (*Abelmoschus esculentus*) pod accessions: implications for mineral bioavailability. Food Sci Nutr 2015; 4(2): 223-33.
[http://dx.doi.org/10.1002/fsn3.282] [PMID: 27004112]

[92] Aderiye BI. Effects of ascorbic acid and pre-packaging on shelf-life and quality of raw and cooked okra (*Hibiscus esculentus*). Food Chem 1985; 16: 69-77.
[http://dx.doi.org/10.1016/0308-8146(85)90020-2]

[93] Saleh MA. Effects of anti-coloring agents on blackening inhibition and maintaining physical and chemical quality of fresh-cut okra during storage. Ann Agric Sci 2013; 58(2): 239-45.
[http://dx.doi.org/10.1016/j.aoas.2013.07.008]

[94] Boontongto N, Srilaong V, Uthairatanakij A, Wongs-Aree C, Aryusuk K. Effect of methyl jasmonate on chilling injury of okra pod. Acta Hortic 2007; (746): 323-8.
[http://dx.doi.org/10.17660/ActaHortic.2007.746.37]

[95] Boonyaritthongchai P, Srilaong V, Wongs-Aree C, Techavuthiporn C. Effect of methyl jasmonate on reducing chilling injury symptom and maintaining postharvest quality of okra (Hisbiscus esculentus L.). Acta Hortic 2013; (1012): 1119-24.
[http://dx.doi.org/10.17660/ActaHortic.2013.1012.151]

[96] Huang S, Li T, Jiang G, Xie W, Chang S, Jiang Y, *et al.* 1-Methylcyclopropene reduces chilling injury of harvested okra (*Hibiscus esculentus* L.) pods. Sci Hortic (Amsterdam) 2012; 141: 42-6.
[http://dx.doi.org/10.1016/j.scienta.2012.04.016]

[97] Rai DR, Balasubramanian S. Qualitative and Textural Changes in Fresh Okra Pods (*Hibiscus esculentus* L.) under Modified Atmosphere Packaging in Perforated Film Packages. Rev Agaroquim Tecnol Aliment 2009; 15(2): 131-8.

[98] Saetung T, Devahastin S, Chiewchan N. Use of low-voltage direct current electricity treatment to increase phenolics content of postharvest okra: effects of some treatment parameters. Int J Food Sci Technol 2017; 53(2): 1-8.

[99] Liu J, Yuan Y, Wu Q, Zhao Y, Jiang Y, John A, *et al.* Analyses of quality and metabolites levels of okra during postharvest senescence by1H-high resolution NMR. Postharvest Biol Technol 2017; 132: 171-8.
[http://dx.doi.org/10.1016/j.postharvbio.2017.07.002]

[100] Fan S, Zhang Y, Sun Q, *et al.* Extract of okra lowers blood glucose and serum lipids in high-fat diet-induced obese C57BL/6 mice. J Nutr Biochem 2014; 25(7): 702-9.
[http://dx.doi.org/10.1016/j.jnutbio.2014.02.010] [PMID: 24746837]

[101] Tian ZH, Miao FT, Zhang X, Wang QH, Lei N, Guo LC. Therapeutic effect of okra extract on gestational diabetes mellitus rats induced by streptozotocin. Asian Pac J Trop Med 2015; 8(12): 1038-42.
[http://dx.doi.org/10.1016/j.apjtm.2015.11.002] [PMID: 26706676]

[102] Valko M, Leibfritz D, Moncol J, Cronin MTD, Mazur M, Telser J. Free radicals and antioxidants in normal physiological functions and human disease. Int J Biochem Cell Biol 2007; 39(1): 44-84.
[http://dx.doi.org/10.1016/j.biocel.2006.07.001] [PMID: 16978905]

[103] Purkait K, Das S, Ali KA, Chakraborty P. Phytochemical Investigation and Hypoglycaemic Effect of *Abelmoschus esculentus*. Res J Pharm Technol 2016; 9(2): 162-4.
[http://dx.doi.org/10.5958/0974-360X.2016.00028.7]

[104] Maramag RP. Diuretic Potential of Capsicum Frutescens Linn., Corchorus Oliturius Linn., and *Abelmoschus esculentus* Linn. Asian J Nat Appl Sci 2013; 2(1): 60-9.

[105] Prommakool A, Sajjaanantakul T, Janjarasskul T, Krochta JM. Whey protein-okra polysaccharide fraction blend edible films: tensile properties, water vapor permeability and oxygen permeability. J Sci Food Agric 2011; 91(2): 362-9.
[http://dx.doi.org/10.1002/jsfa.4194] [PMID: 20960421]

[106] Cai Y, Luo Q, Sun M, Corke H. Antioxidant activity and phenolic compounds of 112 traditional Chinese medicinal plants associated with anticancer. Life Sci 2004; 74(17): 2157-84.
[http://dx.doi.org/10.1016/j.lfs.2003.09.047] [PMID: 14969719]

[107] Liu B, Zhu Y. Extraction of flavonoids from flavonoid-rich parts in tartary buckwheat and identification of the main flavonoids. J Food Eng 2007; 78: 584-7.
[http://dx.doi.org/10.1016/j.jfoodeng.2005.11.001]

[108] Prasad K, Singh J, Chandra D. Quantification of antioxidant phytochemical in fresh vegetables using High Performance Liquid Chromatography. Res J Phytochem 2014; 8(4): 162-7.
[http://dx.doi.org/10.3923/rjphyto.2014.162.167]

[109] Kaur C, Kapoor HC. Antioxidant activity and total phenolic content of some Asian vegetables. Int J Food Sci Technol 2002; 37: 153-61.
[http://dx.doi.org/10.1046/j.1365-2621.2002.00552.x]

[110] Huang D, Ou B, Prior RL. The chemistry behind antioxidant capacity assays. J Agric Food Chem 2005; 53(6): 1841-56.
[http://dx.doi.org/10.1021/jf030723c] [PMID: 15769103]

[111] Karadag A, Ozcelik B, Saner S. Review of Methods to Determine Antioxidant Capacities. Food Anal Methods 2009; 2: 41-60.
[http://dx.doi.org/10.1007/s12161-008-9067-7]

[112] Shahidi F, Naczk M, Eds. Phenolics in Food and Nutraceuticals. Boka Raton, FL: CRC Press 2004; p. 557.

[113] Chao PY, Lin SY, Lin KH, *et al.* Antioxidant activity in extracts of 27 indigenous Taiwanese vegetables. Nutrients 2014; 6(5): 2115-30.
[http://dx.doi.org/10.3390/nu6052115] [PMID: 24858497]

[114] Jiang N, Liu C, Li D, Zhang Z, Liu C, Wang D, *et al.* Evaluation of freeze drying combined with microwave vacuum drying for functional okra snacks: Antioxidant properties, sensory quality, and energy consumption. Lebensm Wiss Technol 2017; 82: 216-26.
[http://dx.doi.org/10.1016/j.lwt.2017.04.015]

[115] Ansari NM, Houlihan L, Hussain B, Pieroni A. Antioxidant activity of five vegetables traditionally consumed by South-Asian migrants in Bradford, Yorkshire, UK. Phytother Res 2005; 19(10): 907-11.
[http://dx.doi.org/10.1002/ptr.1756] [PMID: 16261524]

[116] Ebrahimzadeh AM, Nabavi FS, Nabavi MS. Antihemolytic and antioxidant activity of *Hibiscus esculentus* leaves. Pharmacologyonline 2009; 2: 1097-105.

[117] Geng S, Liu Y, Ma H, Chen C. Extraction and Antioxidant Activity of Phenolic Compounds from Okra Flowers. Trop J Pharm Res 2015; 14: 807-14.
[http://dx.doi.org/10.4314/tjpr.v14i5.10]

[118] Sunilson JAJ, Jayaraj P, Syam MM, Kumari AAG, Varatharajan R. Antioxidant and hepatoprotective effect of the roots of *Hibiscus esculentus* Linn. Int J Green Pharm 2008; 2(4): 200-3.
[http://dx.doi.org/10.4103/0973-8258.44731]

[119] Saravanan S, Pandikumar P, Pazhanivel N, Paulraj MG, Ignacimuthu S. Hepatoprotective role of *Abelmoschus esculentus* (Linn.) Moench., on carbon tetrachloride-induced liver injury. Toxicol Mech Methods 2013; 23(7): 528-36.
[http://dx.doi.org/10.3109/15376516.2013.796032] [PMID: 23581558]

[120] Yan JY, Ai G, Zhang XJ, Xu HJ, Huang ZM. Investigations of the total flavonoids extracted from flowers of Abelmoschus manihot (L.) Medic against α-naphthylisothiocyanate-induced cholestatic liver injury in rats. J Ethnopharmacol 2015; 172: 202-13.
[http://dx.doi.org/10.1016/j.jep.2015.06.044] [PMID: 26133062]

[121] Ai G, Liu Q, Hua W, Huang Z, Wang D. Hepatoprotective evaluation of the total flavonoids extracted from flowers of Abelmoschus manihot (L.) Medic: *in vitro* and *in vivo* studies. J Ethnopharmacol 2013; 146(3): 794-802.
[http://dx.doi.org/10.1016/j.jep.2013.02.005] [PMID: 23422335]

[122] Ren D, Chen G. Inhibition Effect of Okra Polysaccharides on Proliferation of Human Cancer Cell Lines. Food Sci 2010; •••: 21.

[123] Mollick MMR, Rana D, Dash SK, Chattopadhyay S, Bhowmick B, Maity D, *et al.* Studies on green

synthesized silver nanoparticles using *Abelmoschus esculentus* (L.) pulp extract having anticancer (*in vitro*) and antimicrobial applications. Arab J Chem J Med Chem in press
[http://dx.doi.org/10.1016/j.arabjc.2015.04.033]

[124]   Monte LG, Santi-Gadelha T, Reis LB, *et al*. Lectin of *Abelmoschus esculentus* (okra) promotes selective antitumor effects in human breast cancer cells. Biotechnol Lett 2014; 36(3): 461-9.
[http://dx.doi.org/10.1007/s10529-013-1382-4] [PMID: 24129958]

[125]   Ebrahimzadeh AM, Nabavi FS, Nabavi MS, Eslami B. Antihypoxic and antioxidant activity of *Hibiscus esculentus* seeds. Crasas Aceites 2010; 61(1): 30-6.
[http://dx.doi.org/10.3989/gya.053809]

[126]   Lin Y, Liu HL, Fang J, Yu CH, Xiong YK, Yuan K. Anti-fatigue and vasoprotective effects of quercetin-3-O-gentiobiose on oxidative stress and vascular endothelial dysfunction induced by endurance swimming in rats. Food Chem Toxicol 2014; 68: 290-6.
[http://dx.doi.org/10.1016/j.fct.2014.03.026] [PMID: 24685824]

[127]   Li Y-X, Yang Z-H, Lin Y, Han W, Jia S-S, Yuan K. Antifatigue effects of ethanol extracts and polysaccharides isolated from *Abelmoschus esculentus*. Pharmacogn Mag 2016; 12(47): 219-24.
[http://dx.doi.org/10.4103/0973-1296.186341] [PMID: 27601853]

[128]   Deters AM, Lengsfeld C, Hensel A. Oligo- and polysaccharides exhibit a structure-dependent bioactivity on human keratinocytes *in vitro*. J Ethnopharmacol 2005; 102(3): 391-9.
[http://dx.doi.org/10.1016/j.jep.2005.06.038] [PMID: 16111846]

[129]   Vayssade M, Sengkhamparn N, Verhoef R, *et al*. Antiproliferative and proapoptotic actions of okra pectin on B16F10 melanoma cells. Phytother Res 2010; 24(7): 982-9.
[PMID: 20013817]

[130]   Jackson CL, Dreaden TM, Theobald LK, *et al*. Pectin induces apoptosis in human prostate cancer cells: correlation of apoptotic function with pectin structure. Glycobiology 2007; 17(8): 805-19.
[http://dx.doi.org/10.1093/glycob/cwm054] [PMID: 17513886]

[131]   Thöle C, Brandt S, Ahmed N, Hensel A. Acetylated Rhamnogalacturonans from Immature Fruits of *Abelmoschus esculentus* Inhibit the Adhesion of Helicobacter pylori to Human Gastric Cells by Interaction with Outer Membrane Proteins. Molecules 2015; 20(9): 16770-87.
[http://dx.doi.org/10.3390/molecules200916770] [PMID: 26389872]

[132]   Lengsfeld C, Titgemeyer F, Faller G, Hensel A. Glycosylated compounds from okra inhibit adhesion of Helicobacter pylori to human gastric mucosa. J Agric Food Chem 2004; 52(6): 1495-503.
[http://dx.doi.org/10.1021/jf030666n] [PMID: 15030201]

[133]   De Carvalho CCCR, Cruz PA, Da Fonseca MMR, Xavier-Filho L. Antibacterial properties of the extract of *Abelmoschus esculentus*. Biotechnol Bioprocess Eng; BBE 2011; 16(5): 971-7.
[http://dx.doi.org/10.1007/s12257-011-0050-6]

[134]   Messing J, Thöle C, Niehues M, *et al*. Antiadhesive properties of *Abelmoschus esculentus* (Okra) immature fruit extract against Helicobacter pylori adhesion. PLoS One 2014; 9(1): e84836.
[http://dx.doi.org/10.1371/journal.pone.0084836] [PMID: 24416297]

[135]   Singh K. Phytochemical determination and antibacterial activity of Trichosanthes dioica Roxb (patal), Cucurbita maxima (pumpkin) and Abelmoschus esculentus Moench (okra) plant seeds. India: National Institute of Technology 2012.

[136]   Nwaiwu N, Mshelia F, Raufu I. Antimicrobial activities of crude extracts of Moringa oleifera, Hibiscus sabdariffa and *Hibiscus esculentus* seeds against some Enterobacteria. J Appl Phytotechnology Environ Sanit 2012; 1: 11-6.

[137]   Jayaseelan C, Ramkumar R, Rahuman AA, Perumal P. Green synthesis of gold nanoparticles using seed aqueous extract of *Abelmoschus esculentus* and its antifungal activity. Ind Crops Prod 2013; 45: 423-9.
[http://dx.doi.org/10.1016/j.indcrop.2012.12.019]

[138] Mollick MMR, Bhowmick B, Mondal D, Maity D, Rana D, Dash SK, *et al.* Anticancer (*in vitro*) and antimicrobial effect of gold nanoparticles synthesized using *Abelmoschus esculentus* (L.) pulp extract *via* a green route. RSC Advances 2014; 4: 37838-48.
[http://dx.doi.org/10.1039/C4RA07285E]

[139] Okada Y, Okada M, Sagesaka Y. Screening of dried plant seed extracts for adiponectin production activity and tumor necrosis factor-alpha inhibitory activity on 3T3-L1 adipocytes. Plant Foods Hum Nutr 2010; 65(3): 225-32.
[http://dx.doi.org/10.1007/s11130-010-0184-2] [PMID: 20717728]

[140] Sabitha V, Ramachandran S, Naveen KR, Panneerselvam K. Antidiabetic and antihyperlipidemic potential of *Abelmoschus esculentus* (L.) Moench. in streptozotocin-induced diabetic rats. J Pharm Bioallied Sci 2011; 3(3): 397-402.
[http://dx.doi.org/10.4103/0975-7406.84447] [PMID: 21966160]

[141] Ramachandran S, Sandeep VS, Srinivas NK, Dhanaraju MD. Anti-diabetic activity of *Abelmoschus esculentus* Linn. on alloxan–induced diabetic rats. Res Rev Biosci 2010; 4(3): 18-20.

[142] Saha D, Jain B, Jain VK. Phytochemical evaluation and characterization of hypoglycemic activity of various extracts of *Abelmoschus esculentus* Linn. Fruit. Int J Pharm Pharm Sci 2011; 3(2): 183-5.

[143] Amin IM. Hypoglyclemic Effects in Response to Abelmoshus Esculentus Treatment: A Research Framework using STZ-Induced Diabetic Rats. Int J Biosci Biochem Bioinform 2011; 1(1): 63-7.
[http://dx.doi.org/10.7763/IJBBB.2011.V1.12]

[144] Mishra N, Kumar D, Rizvi SI. Protective effect of *Abelmoschus esculentus* against alloxan-induced diabetes in Wistar strain rats. J Diet Suppl 2016; 13(6): 634-46.
[http://dx.doi.org/10.3109/19390211.2016.1164787] [PMID: 27065051]

[145] Subrahmanyam GV, Sushma M, Alekya A, Neeraja C, Harsha HSS, Ravindra J. Antidiabetic Activity of *Abelmoschus esculentus* Fruit Extact. Int J Res Pharm Chem 2011; 1(1): 17-20.

[146] Fan S, Guo L, Zhang Y, Sun Q, Yang B, Huang C. Okra polysaccharide improves metabolic disorders in high-fat diet-induced obese C57BL/6 mice. Mol Nutr Food Res 2013; 57(11): 2075-8.
[http://dx.doi.org/10.1002/mnfr.201300054] [PMID: 23894043]

[147] Huang CN, Wang CJ, Lee YJ, Peng CH. Active subfractions of *Abelmoschus esculentus* substantially prevent free fatty acid-induced ß cell apoptosis *via* inhibiting dipeptidyl peptidase-4. PLoS One 2017; 12(7): 1-16.
[http://dx.doi.org/10.1371/journal.pone.0180285]

[148] Peng C-H, Chyau C-C, Wang C-J, Lin H-T, Huang C-N, Ker Y-B. *Abelmoschus esculentus* fractions potently inhibited the pathogenic targets associated with diabetic renal epithelial to mesenchymal transition. Food Funct 2016; 7(2): 728-40.
[http://dx.doi.org/10.1039/C5FO01214G] [PMID: 26787242]

[149] Karim MR, Islam MS, Sarkar SM, Murugan AC, Makky EA, Rashid SS, *et al.* Anti-amylolytic activity of fresh and cooked okra (*Hibiscus esculentus* L.) pod extract. Biocatal Agric Biotechnol 2014; 3(4): 373-7.
[http://dx.doi.org/10.1016/j.bcab.2014.07.006]

[150] Ahmed BT, Kumar SA. Antioxidant and antidiabetic properties of *Abelmoschus esculentus* extract - An *in vitro* assay. Res J Biotechnol 2016; 11(3): 34-41.

[151] Lu Y, Demleitner MF, Song L, Rychlik M, Huang D. Oligomeric proanthocyanidins are the active compounds in *Abelmoschus esculentus* Moench for its α-amylase and α-glucosidase inhibition activity. J Funct Foods 2016; 20: 463-71.
[http://dx.doi.org/10.1016/j.jff.2015.10.037]

[152] Sabitha V, Panneerselvam K, Ramachandran S. *in vitro* α-glucosidase and α-amylase enzyme inhibitory effects in aqueous extracts of Abelmoscus esculentus (L.) Moench. Asian Pac J Trop Biomed 2012; 2(1): S162-4.

[http://dx.doi.org/10.1016/S2221-1691(12)60150-6]

[153] Khatun H, Rahman A, Biswas M, Islam AU. Water-soluble Fraction of *Abelmoschus esculentus* L Interacts with Glucose and Metformin Hydrochloride and Alters Their Absorption Kinetics after Coadministration in Rats. ISRN Pharm 2011; 2011: 260537.
[http://dx.doi.org/10.5402/2011/260537] [PMID: 22389848]

[154] Uraku AJ, Onuoha SC, Offor CE, Ogbanshi ME, Ndidi US. The Effects of *Abelmoschus esculentus* Fruits on ALP, AST and ALT of diabetic albino rats. Int J Sci Nat 2011; 2(3): 582-6.

[155] Uraku AJ, Ajah PM, Okaka ANC, Ibiam UA, Onu PN. Effects of Crude Extracts of *Abelmoschus esculentus* on Albumin and Total Bilirubin of Diabetic Albino Rats. Int J Sci Nat 2001; 1(1): 38-41.

[156] Tomoda M, Shimizu N, Gonda R, Kanari M, Yamada H, Hikino H. Anticomplementary and hypoglycemic activity of okra and hibiscus mucilages. Carbohydr Res 1989; 190(2): 323-8.
[http://dx.doi.org/10.1016/0008-6215(89)84136-9] [PMID: 2805010]

[157] Zhou L, An X-F, Teng S-C, *et al.* Pretreatment with the total flavone glycosides of Flos *Abelmoschus manihot* and hyperoside prevents glomerular podocyte apoptosis in streptozotocin-induced diabetic nephropathy. J Med Food 2012; 15(5): 461-8.
[http://dx.doi.org/10.1089/jmf.2011.1921] [PMID: 22439874]

[158] Kahlon TS, Chapman MH, Smith GE. *in vitro* binding of bile acids by okra, beets, asparagus, eggplant, turnips, green beans, carrots, and cauliflower. Food Chem 2007; 103: 676-80.
[http://dx.doi.org/10.1016/j.foodchem.2006.07.056]

[159] Ngoc TH, Ngoc QN. Van a TT. Hypolipidemic Effect of Extracts from *Abelmoschus esculentus* L. (Malvaceae) on Tyloxapol-Induced Hyperlipidemia in Mice. Mahidol Univ J Pharm Sci 2008; 35(1–4): 42-6.

[160] Al-Dosari M, Alqasoumi S, Ahmad M, Al-Yahya M, Ansari MN, Rafatullah S. Effect of Beta vulgaris L. on cholesterol rich diet-induced hypercholesterolemia in rats. Farmacia 2014; 59(5): 669-78.

[161] Fernandes NP, Lagishetty CV, Panda VS, Naik SR. An experimental evaluation of the antidiabetic and antilipidemic properties of a standardized Momordica charantia fruit extract. BMC Complement Altern Med 2007; 7(29): 29.
[http://dx.doi.org/10.1186/1472-6882-7-29] [PMID: 17892543]

[162] Rao S, Udayasekhara Rao P, Sesikeran B. Serum cholesterol, triglycerides and total lipid fatty acids of rats in response to okra(hibiscus esciuentus) seed oil. J Am Oil Chem Soc 1991; 68(6): 433-5.
[http://dx.doi.org/10.1007/BF02663762]

[163] Zheng W, Zhao T, Feng W, *et al.* Purification, characterization and immunomodulating activity of a polysaccharide from flowers of *Abelmoschus esculentus*. Carbohydr Polym 2014; 106(1): 335-42.
[http://dx.doi.org/10.1016/j.carbpol.2014.02.079] [PMID: 24721087]

[164] Sheu S-C, Lai M-H. Composition analysis and immuno-modulatory effect of okra (*Abelmoschus esculentus* L.) extract. Food Chem 2012; 134(4): 1906-11.
[http://dx.doi.org/10.1016/j.foodchem.2012.03.110] [PMID: 23442637]

[165] Tongjaroenbuangam W, Ruksee N, Chantiratikul P, Pakdeenarong N, Kongbuntad W, Govitrapong P. Neuroprotective effects of quercetin, rutin and okra (*Abelmoschus esculentus* Linn.) in dexamethasone-treated mice. Neurochem Int 2011; 59(5): 677-85.
[http://dx.doi.org/10.1016/j.neuint.2011.06.014] [PMID: 21740943]

[166] Mairuae N, Connor JR, Lee SY, Cheepsunthorn P, Tongjaroenbuangam W. The effects of okra (*Abelmoschus esculentus* Linn.) on the cellular events associated with Alzheimer's disease in a stably expressed HFE neuroblastoma SH-SY5Y cell line. Neurosci Lett 2015; 603: 6-11.
[http://dx.doi.org/10.1016/j.neulet.2015.07.011] [PMID: 26170247]

[167] Mairuae N, Cheepsunthorn P, Cheepsunthorn CL, Tongjaroenbuangam W. Okra (*Abelmoschus esculentus* Linn) inhibits lipopolysaccharide-induced inflammatory mediators in BV2 microglial cells. Trop J Pharm Res 2017; 16(6): 1285-92.

[http://dx.doi.org/10.4314/tjpr.v16i6.11]

[168] Güne S, Savran A, Yavuz M, Ko M. Ethnopharmacological survey of medicinal plants in Karaisali and its surrounding (Adana-Turkey). J Herb Med 2017; 8: 68-75.
[http://dx.doi.org/10.1016/j.hermed.2017.04.002]

[169] de Sousa Ferreira Soares G, Assreuy AMS, de Almeida Gadelha CA, *et al.* Purification and biological activities of *Abelmoschus esculentus* seed lectin. Protein J 2012; 31(8): 674-80.
[http://dx.doi.org/10.1007/s10930-012-9447-0] [PMID: 22965555]

[170] Naim Z, Billah MM, Ibrahim M, Debnath D, Masud Rana SM, Arefin P, *et al.* Anti-Inflammatory, Analgesic and Anti-Nociceptive Efficacy of Peel of *Abelmoschus esculentus* Fruits in Laboratory Animal. Curr Drug Ther 2015; 10(2): 113-21.
[http://dx.doi.org/10.2174/1574885510021512221614090]

[171] Ribeiro KA, Chaves HV, Filho SMP, *et al.* Alpha-2 Adrenergic and Opioids Receptors Participation in Mice Gastroprotection of *Abelmoschus esculentus* Lectin. Curr Pharm Des 2016; 22(30): 4736-42.
[http://dx.doi.org/10.2174/1381612822666160201152438] [PMID: 26831461]

[172] Freitas RS, do Val DR, Fernandes ME, *et al.* Lectin from *Abelmoschus esculentus* reduces zymosan-induced temporomandibular joint inflammatory hypernociception in rats *via* heme oxygenase-1 pathway integrity and tnf-α and il-1β suppression. Int Immunopharmacol 2016; 38: 313-23.
[http://dx.doi.org/10.1016/j.intimp.2016.06.012] [PMID: 27344040]

[173] Guo J, Xue C, Duan JA, Qian D, Tang Y, You Y. Anticonvulsant, antidepressant-like activity of Abelmoschus manihot ethanol extract and its potential active components *in vivo*. Phytomedicine 2011; 18(14): 1250-4.
[http://dx.doi.org/10.1016/j.phymed.2011.06.012] [PMID: 21784623]

[174] Tu Y, Sun W, Wan YG, *et al.* Huangkui capsule, an extract from Abelmoschus manihot (L.) medic, ameliorates adriamycin-induced renal inflammation and glomerular injury *via* inhibiting p38MAPK signaling pathway activity in rats. J Ethnopharmacol 2013; 147(2): 311-20.
[http://dx.doi.org/10.1016/j.jep.2013.03.006] [PMID: 23518420]

[175] Willats WGT, Knox JP, Mikkelsen JD. Pectin: New insights into an old polymer are starting to gel. Trends Food Sci Technol 2006; 17(3): 97-104.
[http://dx.doi.org/10.1016/j.tifs.2005.10.008]

[176] Manosroi A, Pattamapun K, Khositsuntiwong N, *et al.* Physicochemical properties and biological activities of Thai plant mucilages for artificial saliva preparation. Pharm Biol 2015; 53(11): 1653-60.
[http://dx.doi.org/10.3109/13880209.2014.1001402] [PMID: 25853963]

[177] Aderonke A, Moronkeji A, Vivian I, Chinwe O, Rotimi S, Henry O, *et al.* Dietary Fortification of Ogi (Maize slurry) with Okra seed flour and its Nutritional value. Sch J Agric Sci 2014; 4(4): 213-7.

[178] Ghori MU, Alba K, Smith AM, Conway BR, Kontogiorgos V. Okra extracts in pharmaceutical and food applications. Food Hydrocoll 2014; 42(P3): 342-7.
[http://dx.doi.org/10.1016/j.foodhyd.2014.04.024]

[179] Zaharuddin ND, Noordin MI, Kadivar A. The Use of *Hibiscus esculentus* (Okra) Gum in Sustaining the Release of Propranolol Hydrochloride in a Solid Oral Dosage Form Biomed Res Int. 2014. Article ID 735891

[180] Emeje M, Olaleye O, Isimi C, *et al.* Oral sustained release tablets of zidovudine using binary blends of natural and synthetic polymers. Biol Pharm Bull 2010; 33(9): 1561-7.
[http://dx.doi.org/10.1248/bpb.33.1561] [PMID: 20823575]

[181] Sinha P, Ubaidulla U, Hasnain MS, Nayak AK, Rama B. Alginate-okra gum blend beads of diclofenac sodium from aqueous template using ZnSO4 as a cross-linker. Int J Biol Macromol 2015; 79: 555-63.
[http://dx.doi.org/10.1016/j.ijbiomac.2015.04.067] [PMID: 25987461]

[182] Sinha P, Ubaidulla U, Nayak AK. Okra (*Hibiscus esculentus*) gum-alginate blend mucoadhesive beads for controlled glibenclamide release. Int J Biol Macromol 2015; 72: 1069-75.

[http://dx.doi.org/10.1016/j.ijbiomac.2014.10.002] [PMID: 25312603]

[183]   Alba K, Ritzoulis C, Georgiadis N, Kontogiorgos V. Okra extracts as emulsifiers for acidic emulsions. Food Res Int J 2013; 54: 1730-7.
[http://dx.doi.org/10.1016/j.foodres.2013.09.051]

[184]   Temenouga V, Charitidis T, Avgidou M, Karayannakidis PD, Dimopoulou M. Novel emulsifiers as products from internal Maillard reactions in okra hydrocolloid mucilage. Food Hydrocoll 2016; 52: 972-81.
[http://dx.doi.org/10.1016/j.foodhyd.2015.08.026]

[185]   Ahad HA, Mallapu RE, Gangadhar P, Suma PB, Lavanya G. Fabrication and Evaluation of Glipizide Abelmonschus esculentus Fruit Mucilage Povidone Controlled Release Matrix Tablets. Int J Res Ayurveda Pharm 2011; 2(2): 592-6.

[186]   Ofoefule SI, Chukwu A. Application of *Abelmoschus esculentus* gum as a mini matrix for furosemide and diclofenac sodium tablets. Indian J Pharm Sci 2001; 63(6): 532-5.

[187]   Ofoefule SI, Chukwu A, Anyakoha N, Ebebe IM. Application Of *Abelmoschus esculentus* In Solid Dosage Formulation 1 : Use As A Binder For A Poorly Water Soluble Drug. Indian J Pharm Sci 2001; 63(3): 234-8.

[188]   Biswal B, Karna N, Patel R. Okra mucilage act as a potential binder for the preparation of tablet formulation. Der Pharm Lett 2014; 6(3): 31-9.

[189]   Bakre LG, Jaiyeoba KT. Effects of drying methods on the physicochemical and compressional characteristics of Okra powder and the release properties of its metronidazole tablet formulation. Arch Pharm Res 2009; 32(2): 259-67.
[http://dx.doi.org/10.1007/s12272-009-1231-0] [PMID: 19280157]

[190]   Sharma N, Kulkarni GT, Sharma A. Development of *Abelmoschus esculentus* (Okra) -Based Mucoadhesive Gel for Nasal Delivery of Rizatriptan. Trop J Pharm Res 2013; 12: 149-53.

[191]   Sharma N, Kulkarni GT, Sharma A, Bhatnagar A, Kumar N. Natural mucoadhesive microspheres of *Abelmoschus esculentus* polysaccharide as a new carrier for nasal drug delivery. J Microencapsul 2013; 30(6): 589-98.
[http://dx.doi.org/10.3109/02652048.2013.764941] [PMID: 23379506]

[192]   Dimopoulou M, Ritzoulis C, Papastergiadis ES. Composite materials based on okra hydrocolloids and hydroxyapatite. Food Hydrocoll 2014; 42: 348-54.
[http://dx.doi.org/10.1016/j.foodhyd.2014.04.015]

[193]   Duldulao M, Watanabe H, Kamaya Y, Suzuki K. Papermaking potential of okra stalks. J Packag Sci Technol 2010; 19(4): 305-15.

[194]   Alam S, Khan GMA. Chemical Analysis of Okra Bast Fiber (*Abelmoschus esculentus*) and Its Physicochemical Properties. J Text App Technol Manag 2007; 5(4): 1-9.

[195]   Ray S, Kumar S, Utpal S, Runu R. Preparation of okra-incorporated dhokla and subsequent analysis of nutrition, antioxidant, color, moisture and sensory profile. J Food Meas Charact 2017; 11(2): 639-50.
[http://dx.doi.org/10.1007/s11694-016-9433-x]

[196]   Ndjouenkeu R, Goycoolea FM, Morrisav ER, Akingbala JO. Rheology of okra (Hllbiscus esculentus L.) and dika nut (Irvingia gabonensis) polysaccharides. Carbohydr Polym 1996; 29: 263-9.
[http://dx.doi.org/10.1016/0144-8617(96)00016-1]

CHAPTER 12

Phytochemical, Nutritional and Pharmacological Properties of Unconventional Native Fruits and Vegetables from Brazil

Maria Fernanda Frankelin, Tatiane Francielli Vieira, Jessica Amanda Andrade Garcia, Rubia Carvalho Gomes Correa, Antonio Roberto Giriboni Monteiro, Adelar Bracht and Rosane Marina Peralta[*]

Post-graduate Program of Food Science, Universidade Estadual de Maringá, Maringá, Paraná, Brazil

Abstract: Diversity of the Brazilian flora is very prominent. The land possesses over 50,000 different flowering plants species among native and non-native. Approximately 5,000 of them are edible. This richness, however, is far from being optimally utilized. The term unconventional vegetables is applied to those that are consumed by a small part of the population, mostly by communities restricted to highly specific areas. These vegetables are frequently of high nutritional value and have good taste. Globalization and the consequently increased use of processed foods, however, is having a negative influence on the consumption and cultivation of these vegetables in all regions. This trend is manifesting in both rural and urban areas. The final consequence is a diminution of the consumption of foods from local and regional sources and, consequently, a modification of eating habits of the Brazilians of all social classes. In this chapter, attention was given to six unconventional native vegetables from Brazil that are regionally consumed as food (baru, cubiu, jambu, ora-pro-nobis, pinhão, and pequi), emphasizing on their nutritional characteristics and the health benefits resulting from their consumption.

Keywords: Baru, Brazilian Biomes, Cubiu, Jambu, Ora-Pro-Nobis, Pequi, Pinhão, Phytochemicals, Unconventional Vegetal.

INTRODUCTION

The great biodiversity of edible plant species in Brazil is due to several factors. They include the large territorial extension (8.5 millions km^2), the diversity of climate, which varies considerably from the tropical north to the sub-tropical and temperate regions located in the south of the Tropic of Capricorn, and the exis-

[*] **Corresponding author Marina Peralta:** Universidade Estadual de Maringá, Maringá, Paraná, Brazil; Tel: +44 3011-4715; E-mails: rosanemperalta@gmail.com, rmperalta@uem.br

Spyridon A. Petropoulos, Isabel C.F.R. Ferreira and Lillian Barros (Eds.)

tence of different biomes. Brazil has its territory occupied by six terrestrial biomes: the tropical rainforest "Amazonia"; the tropical scrub forest "Caatinga", the tropical grassland and savannah "Cerrado"; the tropical deciduous forest "Mata Atlântica"; the flooded grassland "Pantanal"; and the subtropical prairies or grasslands "Pampas" (Fig. **1**). In Brazil, it is estimated that from over 50,000 flowering plant species, among native and non-native ones, around 5,000 of them are edible [1]. However, only a few dozens make up the basis of the common Brazilian diet [2].

In recent years, there has been a growing interest in the Brazilian native species. Good examples are the already widely consumed Brazil nuts (*Bertholletia excelsa*) [3], jaboticaba (*Plinia cauliflora*) [4 - 6], açaí (*Euterpe oleraceae*) [7], juçara (*Euterpe edulis*) [8], and pitanga (*Eugenia uniflora*) [9, 10]. Unconventional, neglected and underutilised species have also been the subject of worldwide interest. This group is formed by species that are consumed by a limited part of the population, mostly by communities restricted to highly specific areas. Considerable efforts have been made to evaluate the potential of using these species against chronic diseases and several studies revealed that they may be important sources of bioactive substances.

In this chapter, the scientific literature on six unconventional native vegetables from Brazil, regionally consumed as food, is reviewed (Fig. **1**; Table **1**): two belongs to the Amazonic Forest biome, cubiu and jambu; two belongs to the Cerrado biome, pequi and baru; and two belongs to the Atlantic Forest, ora-pro-nobis and pinhão.

Three factors were considered in the selection of these species: their routine use in regional dishes and in folk medicine and the existence of scientific studies evaluating their phytochemical, nutritional and pharmacological properties. Information on recent studies was obtained from the Web of Science, PubMed and Scopus. The website http://floradobrasil.jbrj.gov.br/2010/ was used to confirm the mentioned species as being native of Brazil.

Table 1. Six unconventional native vegetables from Brazil.

Families and species	Vernacular names	Brazil region biomes	Edible parts	Plant parts used in folk medicine
Solanaceae *Solanum sessiliflorum*	Cubiu Maná-cubiu	North, Amazonic Forest	Fruits	Fruits, leaves roots
Asteraceae *Acmella oleracea*	Jambu	North, Amazonic Forest	Leaves, flowers	Leaves, flowers

(Table 1) cont.....

Families and species	Vernacular names	Brazil region biomes	Edible parts	Plant parts used in folk medicine
Caryocaraceae *Caryocar brasiliense* *Caryocar coriaceum* *Caryocar villosum*	Pequi	Central-West, Cerrado Northeast, Caatinga North, Amazonic Forest	Cooked fruits	Fruits, leaves, seeds
Fabaceae *Dipteryx alata*	Baru	Central-West, Cerrado	Nuts	Leaves, Nuts
Cactaceae *Pereskia aculeata* *Periskia grandfolia*	Ora-pro-nobis	Southeast, South, Northeast, Atlantic Forest	Mainly cooked leaves, but also flowers and fruits	Fruits, leaves, flowers
Araucariaceae *Araucaria angustifolia*	Pinhão, Pinion	South, Atlantic forest	Cooked seeds	Leaves, bark seeds coat

Fig. (1). Biomes and six unconventional native vegetables from Brazil.

Cubiu (*Solanum sessiliflorum* Dunal)

One of the most important sources of vegetables and of these species is the Solanaceae family. Several species are widely known and constitute a great part of the human diet worldwide. The list includes potatoes, tomatoes and eggplants, for example [11]. In spite of the great potential of the latter species, however, there are other plants from this family that remain unexplored.

Commonly found in the Amazon Forest, *Solanum sessiliflorum* is a shrub whose consumption as food is basically restricted to local communities [12 - 15]. The species was domesticated by indigenous populations [13 - 15]. The fruit of this plant, which possess an unusual flavor, are called cubiu of maná-cubiu. They are used mainly in juices and salads and for making jellies and cakes [12].

Due to its rusticity and abundant fruit production, this plant has potentiality in the modern agricultural industry [14, 15]. In recent years, there has been an increase in the appreciation of this fruit by the Brazilian consumers [15], which resulted in an expansion of its distribution to other places including *Zona da Mata of Pernambuco* and some locations in the southeast and southern regions [16].

Nutritional Characteristics of Cubiu

Cubiu fruits have a histology similar to tomato (*Solanum lycopersicum* L., Solanaceae), *i.e*, thin peel (exocarp), fleshy pulp (mesocarp), and axial placenta (endocarp); they are often described as the Indian tomato [17] (Fig. **2**). The color of the peel starts from green, when the fruit is unripe, becomes yellow-orange when the fruit is getting mature, and finally turns red at full maturity. The presence of carotenoids gives to the pulp a light yellow color.

The fruit can be consumed at all ripening stages and once manually harvested, carefully washed, and dried, they may be stored at ambient temperatures (*e.g.*, ± 29 °C) and remain suitable for consumption for approximately five days [18, 19]. The fruit weight varies between 20 and 450 g and contain between 200 and 500 oval flat seeds (Fig. **2**). The fruits have the most diverse forms, being spherical, oval or cylindrical. The cubiu fruit are 5−6 cm in diameter, and approximately 91% (w/w) of their total fresh weight is edible. The water content of the fruit corresponds to around 90% water (w/w). In chemical terms, carbohydrates are important components (32 g/100 g of dry weight), glucose and fructose being predominant. Another important component is citric acid (14 g/100 g of dry weight) [12].

There are a few reports of studies aiming to characterize the main bioactive molecules of cubiu. The pulp of cubiu is rich in iron, selenium, manganese, zinc, carotenoids and phenolic compounds [16, 17, 20]. Hydro-ethanolic extraction of the edible part of the fruit revealed the presence of coumarins, phenols, alkaloids and flavonoid glycosides. Aqueous extraction, on the other hand, revelead volatile and fixed acids, anthocyanins, tannins, gums and mucilages [21].

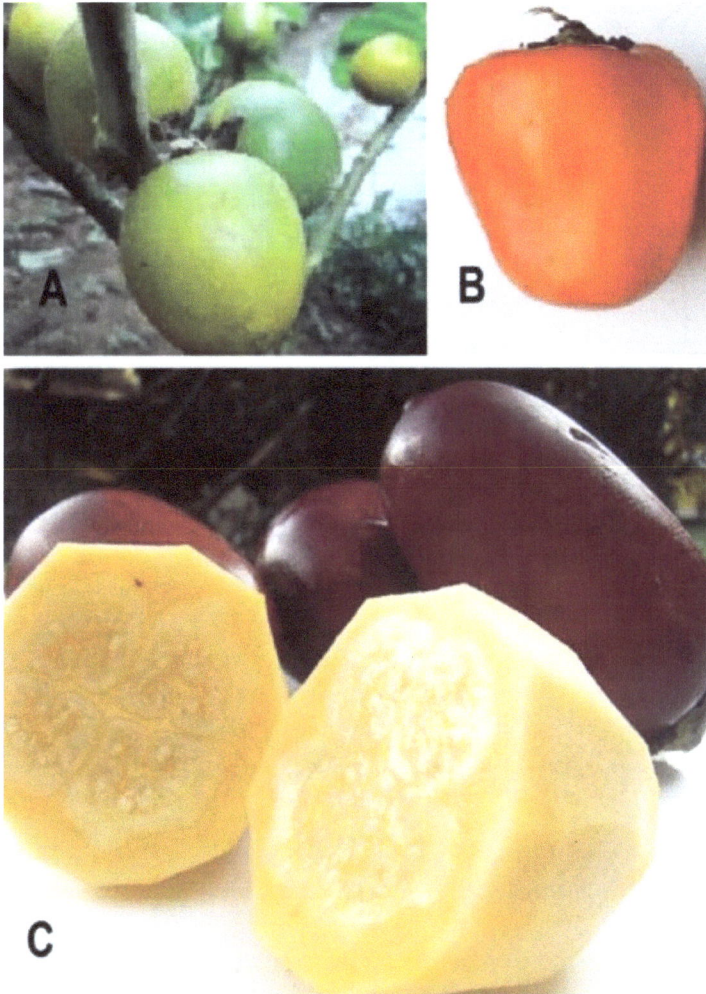

Fig. (2). Cubiu in two different stages of rippining: green (A); orange (B); pulp of mature fruit (C).

In a hydrophillic extract of mana-cubiu three phenolic compounds were identified in addition to seventeen carotenoids [17]. Among the latter the most abundant were (all-E)-lutein (2.41 µg/g of dry weight) and (all-E)-β-carotene (7.15 µg/g of dry weight). Among the phenolics, 5-caffeoylquinic acid (1351 µg/g of dry weight) represented more than 78% (w/w) of the total phenolic compounds content (Fig. **3**). Dihydrocaffeoyl spermidines were also found. The hydrophilic extract presented a pronounced capacity of scavenging HOCl and H_2O_2. This capacity was attributed to the conjugates of spermidine caffeic acid [17].

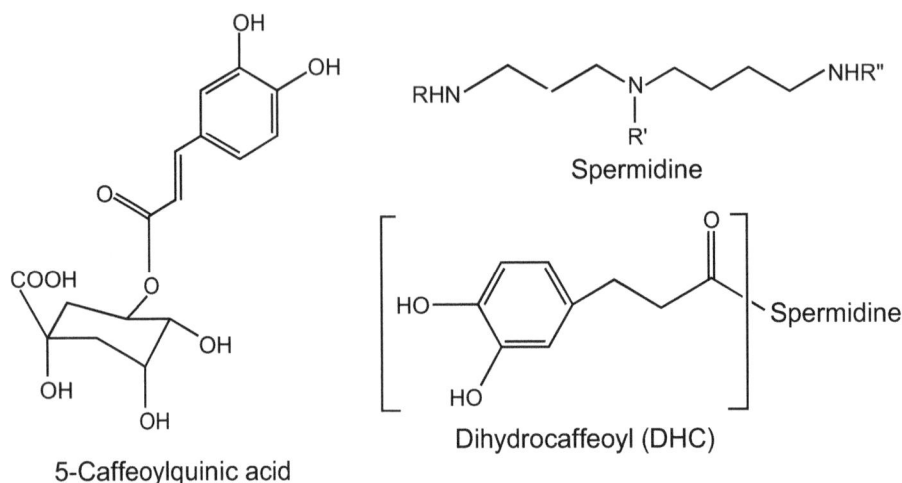

Compound	R	R'	R"
Spermidine	-H	-H	-H
N^1,N^5-bis(dihydrocaffeoyl)spermidine	-DHC	-DHC	-H
N^5,N^{10}-bis(dihydrocaffeoyl)spermidine	-H	-DHC	-DHC
N^1,N^5,N^{10}-trisdihydrocaffeoyl)spermidine	-DHC	-DHC	-DHC

Fig. (3). Main phytochemicals identified in mana-cubiu fruit (modified from reference [17]).

Polysaccharides from cubiu fruits have also been recently studied [22]. Their hemicelluloses are apparently composed by arabinogalactoxyloglucans, xylans and galactoglucomannans. A great part of the pectic fractions presented low degrees of methyl-esterification and their contents in uronic acid were above 65%. The polysaccharides from the peel are characterized by the presence of a mixture of homogalacturonans (degree of methyl-esterification is 56.9%) plus a small amount of rhamnogalacturonan I branched by type I and type II arabinogalactans. These observations suggest that cubiu could be regarded as a possible source of pectins with high uronic acid content and extractable by low-cost and environmental friendly procedures.

Health Effects and Ethnopharmacological Studies on Cubiu

Both fruit and leaves of cubiu have been used popularly for treating skin infections, snake bites, and scorpion stings [24]. Their use as hypoglycemic and hypocholesterolemic agents is also common in traditional medicine [24]. However, despite the pharmacological potentials of Cubiu, rigorous evaluations of its functional and toxicological properties are scarce. In one study, the genotoxic and/or antigenotoxic potential of the maná-cubiu pulp was evaluated [25]. These experiments were carried out using Wistar rats in which the micronucleus test and the alkaline single-cell comet assay were performed. Additionally, oxidative stress levels in heart and liver cells were assessed by measuring gluthatione contents (GSH) and thiobarbituric acid reactive substances (TBARS) [25]. The study detected neither significant cytotoxicity to bone marrow cells nor genotoxic effects. The conclusion was that Cubiu is safe for human consumption at least with reference to the parameters that were investigated.

Biological activities of cubiu extracts include antioxidant and antimicrobial activities as well as wound healing activity [26]. A hydrophilic extract of mana-cubiu revealed ability of scavenging reactive oxygen and nitrogen species, as for example, $ONOO^-$, $HOCl$, $ROO\bullet$,$HO\bullet$and H_2O_2, in a dose-dependent way [17]. Cubiu extracts were also able to inhibit cholesterol degradation under heating conditions suggesting a potential use in food products containing highly polyunsaturated lipids [27]. The only documented scientific study conducted on humans showed that the administration of cubiu extract (40 mL/day) for three days decreased the levels of low density lipoprotein (LDL), serum triglycerides, cholesterol and glucose and raised the concentrations of serum high density lipoprotein (HDL) in 100 volunteers [24].

Jambu (*Acmella oleracea*)

Acmella oleracea (L.) R.K. Jansen, Asteraceae ocurs in tropical regions near the Equator line especially in Asia and South America. It is regarded as a native Amazonian plant. In Brazil, the species is known as jambu and agrião-do-pará. It is an herbaceous, perennial plant, 20-40 cm of height, semi-straight, almost flat, with cylindrical, fleshy and decumbent branches. The leaves are simple, opposing, membranaceous, petiolate, with petioles 20-60 mm long, flat, with furrows on the surface, slightly winged and slightly hairy. The flowers are small, yellowish, with distinct purple areas in the chalice palle (Fig. **4**).

Nutritional Characteristics of Jambu

Flowers and especially leaves of jambu are usually consumed as food preparations typical of the Amazon Region, such as duck in tucupi, tacacá, chicken in tucupi, and fish in tucupi. The leaves consist of 90% water (w/w), 5.3% carbohydrates, 3.1% proteins, and 0.25% lipids. Calcium, iron, phosphorus, vitamins B_1, B_2, B_3 and C are the main microelements of jambu [28].

Fig. (4). Flowers and leaves of jambu (*Acmella oleracea*).

Health Effects and Ethnopharmacological Studies of Jambu

The plant is popularly used for the treatment of toothaches, skin problems, dysentery, bladder calculus, pertussis, infestation, stomach upset, weakness, dog bite, snake bites, cough, pulmonary tuberculosis, allergies, candidiasis, herpes simplex and gingivitis. In addition, it acts as a purgative and local anesthetic [29]. The plant is also recommended against anemia, asthma, malaria, fever, scurvy, urinary tract diseases, sore throats, cavities and canker sores, in addition to improving digestion and stimulating gastric secretion [1]. Among its active ingredients are essential oils, saponins, espilantin, affinin, espilantol, filosterin, choline and triterpenoids. It has localized anaesthetic, antifungal, antiseptic, antiviral, diuretic and immune system stimulative properties [30]. Flower′s aqueous extracts were considered poor in phenolics and flavonoids and presented low antioxidant activity [31]. On the other hand, a polysaccharide (rhamnogalacturonan) from the leaves of jambu has been found to be an inhibitor of gastric lesions induced by ethanol. This phenomenon suggest the use of this polysaccharide as a gastroprotective agent [32].

Spilanthol ($C_{14}H_{23}NO$, 221.339 g/mol; Fig. (**5**) is the main N-alkylamide found in

Acmella oleracea [33]. This compound has also been reported to occur in several plants that are employed worldwide in traditional medicines, as for example *Heliopsis longipes* and other *Acmella* species. The biological and pharmacological effects of spilanthol were recently reviewed and include insecticidal, analgesic, anticancer, anti-inflammatory, antilarvicidal, anti-molluscicidas, antimalarial, antimicrobial, antinoceptive, neuroprotective, antimutagenic and antioxidant activities [34]. Other therapeutic properties attributed to spilanthol are anticonvulsant, aphrodisiac, antimicrobial, diuretic, vasorelaxant and pancreatic lipase inhibitory activities. Activity against the human immunodeficiency virus has also been reported [35]. The chemical groups in the spilanthol molecule that seem to be essential for its analgesic activity, tingling and mouth-watering effects (pharmacophores) are the amide and the unsaturated (alkenyl) fatty acyl [34].

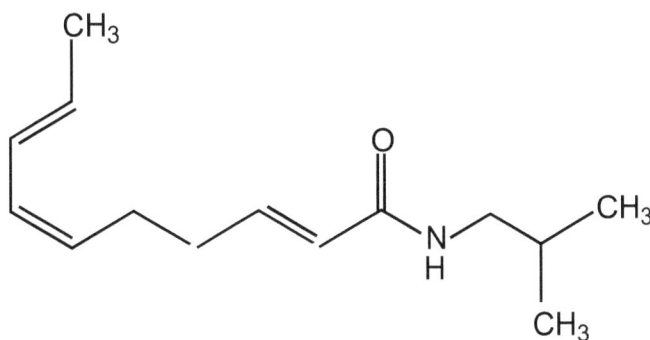

Fig. (5). Spilanthol, the main phytochemical found in *Acmella oleraceae.*

Recently an anaesthetic mucoadhesive film that contains an *Acmella oleracea* (jambu) extract was proposed for topical use in the oral mucosa [36]. A dichloromethane fraction of jambu (containing 100% spilanthol) inhibited the tyrosinase enzyme, and is potentially useful in formulations to prevent and/or reduce skin hyperpigmentation processes. Spilanthol can also be useful in the controlling the enzymatic browning in fruits and vegetables [37].

Pequi (*Caryocar* sp.)

Caryocar brasiliense Camb, *Caryocar coriaceum* Wittm and *Caryocar villosum* received the Brazilian-Portuguese names of pequi or piqui. These plants belong to the Caryocaraceae family, which is amply distributed over the tropical forests of Central and South America. The main representative of the Caryocaraceae family in Brazil is *C. brasiliense*. This species is abundant in the Brazilian Cerrado where it is also amply cultivated [38]. The tree can reach 15 m in height (Fig. **6A**). Its fruits are globose drupes with a green pericarp that envelopes from one to four pyrenes, which in the common language are known as stones (Fig. **6B**). The woody, thorny endocarp containing a white kernel or seed is surrounded by a mesocarp which consists of an external portion (leathery fleshy) and an internal one (yellow, fleshy edible part) Figs. (**6C** and **6D**).

Caryocar coriaceum Wittm, is found in the Brazilian Northeast, more precisely it its northernmost part. It exerts a significant economic and social role in the vicinities of the Araripe Plateau and in the states of Piauí, Pernambuco and Ceará [39]. A third species, *Caryocar villosum* (Aubl. Pers.), is found in the Amazon Region [40 - 42].

Fig. (6). *Caryocar brasiliensis.* Tree (A), flower (B), green and mature fruit (C, D) and closed and opened mesocarp (E, F).

Nutritional Characteristics of Pequi

The pequi fruit (*C. brasiliense*) is amply used in the cuisine of the Goiás State and in other Central-West States of Brazil. It is always consumed after cooking, never in natura, and is added preferably to savory dishes rather than to sweet ones. Two highly appreciated popular dishes containing pequi are pequi rice and chicken with pequi. The fruit pulps from the three species are rich sources of energy, lipids (20.0-33.0%), fibers (10.0-15.0%), minerals, and vitamins, especially vitamin C and β-carotene [43]. Essential for the nutritive value of pequi is its lipid fraction. Its main fatty acids are oleic acid (54.0%) and palmitic acid (39.0%) [38]. Significant amounts of the carotenoids zeaxantin, α- and β-carotenes and β-criptoxantin were equally found in the pequi fruit [44]. The pequi oil is used as a butter substitute and in the cosmetics industry [45].

In the epicarp and external mesocarp of the pequi fruits several phytosterols and phenolic compounds were detected. The following secondary metabolites were identified in aqueous extracts of pequi (*C. brasiliense*): *p*-coumaric acid, β-carotene, gallic acid, ellagic acid, zeaxanthin, and ferulic acid [46]. Ethanolic extracts of pequi (*C. brasiliense*) present high antioxidant activities and the main components responsible for this activity are possibly quercetin 3-*O*-arabinose, quercetin, quinic acid, and gallic acid [47]. The pectins in the pequi peel revealed a high esterification degree. At least, when microwave-assisted extraction was performed the yields were found to be comparable to those of other important sources of pectin [48]. The characteristic aroma of the pequi fruit seems to be given firstly by ethyl hexanoate, followed by ethyl octanoate, tetrahydrofur-furylalcohol, ethyl butanoate, butyl palmitate, isobutyl stearate, 3-methylvaleric acid and phenylace phenyl-acetaldehyde [49].

Several triterpenes and phenolic glycosides were identified in the leaves, fruits and steam barks of *C. brasiliense* [50 - 52]. The essential oils of seeds and leaves of *C. brasiliense* are composed mainly of ethyl hexanoate, while octacosane, heptadecane and hexadecanol were also found in the leaf oil [53].

Concerning the chemical constituents of *C. villosum*, seeds, pulps and shells were extracted with ethanol alone and with a mixture of ethanol and water (8:2) [54]. These procedures lead to the identification of ellagic acid, ellagic deoxyhexoside, and gallic acid. The extracts presented high antioxidant and anti-inflammatory activities, were only moderately toxic to human fibroblasts, but highly toxic to tumor strains [54].

Hydroethanolic and aqueous extracts of *C. villosum* presented high antioxidant activities and can be used as sources of natural antioxidants [40, 41]. Seventeen phenolics were identified, being ellagic acid deoxyhexoside, ellagic acid, gallic acid and monogalloyl hexoside the most abundant ones. The extracts of *C. villosum* revealed to possess free-radical scavenging capacities against all tested reactive oxygen species and reactive nitrogen species. These capacities correlated well with the content in phenolic compounds [40, 41].

Health Effects and Ethnopharmacological Studies on Pequi

Both fruit oil and leaves of *C. brasiliense, C. coriaceum* and *C. villosum* are used in the folk medicine for treating burns, edema, bronchitis, cold and cough [55]. The seeds, in turn, have been considered as aphrodisiacs [55].

In the context of natural medicine, the Pequi oil has been widely used. These are reports of balm-like effects on rheumatism in addition to anti-inflammatory and healing effects when treating respiratory diseases, gastric ulcers, muscular and rheumatic pains [56]. Pharmacologic studies suggest chemopreventive, anti-inflammatory, antigenotoxic, anticlastogenic, and hypocholesterolemic activities [57, 58]. A carotenoid-rich oil from Pequi has been reported to be an effective tissue-injury reducer in runners, especially in women. Furthermore, it has been claimed that the same formulation diminishes DNA damage in both genders. For these reasons, this type of oil would be adequate for being used as an antiaging supplement and in general as an antioxidant agent [59, 60]. It has also been claimed that the antioxidant activity of the pequi oil might confer to it a tumour-growth inhibitory ability associated to a capacity of increasing the lymphocyte-dependent immunity and of reducing the adverse effects associated to the doxorubicin-induced oxidative damage to normal cells [61]. In addition to its anti-inflammatory activities the pequi oil has also been described as possessing anti-microbial activities [47, 62, 63].

With respect to the *Caryocar brasiliense* oil, there are a series of novel results published recently:

1. The toxicological potential of pequi was evaluated by means of acute and subchronic toxicity tests in Wistar rats [58]. The results demonstrated a low acute and subchronic toxicity. The maternal, embryotoxic and teratogenic effects of the pequi oil were evaluated in pregnant Wistar female rats. These investigations revealed that the pequi oil affects neither the ponderal evolution of the pregnant females nor their behavior and hematological, biochemical and histopathological parameters. All these results indicate absence of maternal toxicity.
2. Pequi oil had a positive effect on the healing process of skin lesions in rats by promoting quicker tissue repair [56].
3. Pequi oil and extracts protected against DNA damage induced by urethane in mice. An association between this protective action and the modulation of lipid peroxidation, reduction of nitric oxide synthase expression, and improvement in the gene expression of antioxidant enzymes was proposed [64].
4. The actions of a prolongued intake of pequi oil on cardiovascular risk factors and on the *ex vivo* cardiac function of rats were investigated. The preparation improved the *ex vivo* cardiac function in consequence of an increased cardiac contractility and relaxation. The systemic cardiovascular risk factors, however, were not affected by the pequi oil. These effects of the pequi oil were attributed to its high content in carotenoids and oleic plus palmitic acid [65].
5. The antioxidant and anti-inflammatory properties of the Pequi almond oil were considered to be the main causes for its attenuating actions on acute hepatic damage induced by carbon tetrachloride in rats. Cold pressed oil and high doses of handmade oil were more effective [66].

6. The anti-inflammatory actions of the fruit oil as well as its possible actions on arterial blood pressure and postprandial lipidemia in female and male athletes were investigated. A general trend for reduced arterial pressure was found, an observation that suggests hypotensive effects [60].

Other recent studies have been done with the *Caryocar coriaceum* oil:

1. *Caryocar coriaceum*'s oil presented anti-inflammatory activity and hypolipidemic effects in healthy and dyslipidemic mice (tyloxapol-induced). These results support the traditional use of the oil in folk medicine and suggest a potential cardioprotective effect [67].
2. The action of pequi oil on topical inflammation and its healing effects on cutaneous wounds were studied in mice. The pequi oil inhibited the ear edema and showed a significant reduction of the unhealed wound area, with increased percentage of wound contraction. The pequi oil inhibited the topical inflammation and accelerated cutaneous wound repair [68].

Baru (*Dipteryx lata*)

The baru nut is the seed of the baruzeiro plant fruit (*Dipteryx alata* Vog.). This plant belongs to the Leguminosae family which is native to the Cerrado biomes. Blooming occurs from November to May and the fruits develop from July to October. The fruit consists of a light brown drupe and the edible seed, usually called almond, has a brown color and an elliptic form Fig. (**7**).

Nutritional Characteristics of Baru

In regional terms the baru seed is very important and its nutritional composition has been the object of a number of recent publications. The baru nuts present great nutritional value. They are rich in high-quality proteins (23.9 to 29.9 g/100 g) and lipids (38.2 to 41.9 g/100 g). Among the latter, unsaturated fatty acids (approximately 81.2%) predominate. The most common are oleic, linolenic, gadoleic, and erucic acids [69]. The baru nuts also contain high concentrations zinc, calcium and iron, as well as vitamin E, phytate and tannins [70].

In a study with the baru seed oil in which gas chromatography–tandem mass spectrometry (GC–MS) was used, eleven compounds were identified [71]. The list includes steroids, mono and sesquiterpenes and tocopherol derivatives. Several limonenes (β-elemene, γ-elemene, α-caryophyllene, β-caryophyllene) and sterols (campesterol, stigmasterol, β-sitosterol and cycloartenol) were also identified.

Fig. (7). Baru (*Dipteryx alata*). Tree (A); fruits (B); opened fruits (C); edible seed (D).

Health Effects and Ethnopharmacological Studies of Baru

The main justification for the popular medicinal use of the baru seed oil is based on its putative anti-rheumatic properties [72]. In addition to these properties, there are also claims about possible sudorific, tonic and menstrual regulatory activities [72]. The use of the baru seed as a medicinal agent has also benefited from its high content in unsaturated fatty acids, mainly because health effects such as cholesterol reduction and atherosclerosis prevention are usually attributed to these compounds [73 - 75]. In this respect, experiments with 20 mildly hypercholesterolemic subjects were conducted [76]. The study consisted of a randomized, crossover, placebo-controlled study in which two treatment periods of 6 weeks were separated by a 4-week washout period. In these alternated periods the subjects were randomly allocated. The treatment consisted of dietary supplementation with baru almonds or placebo at daily doses of 20 g. Compared to placebo, supplementation of baru almonds reduced total cholesterol, low-density lipoprotein cholesterol and non high-density lipoprotein. In other words, supplementation of mildly hypercholesterolemic subjects with baru almonds improved serum lipid parameters. This observation recommends the inclusion of baru seeds in diets aiming at reducing risks of cardiovascular disease [76].

Another series of experiments was conducted with aqueous and ethyl acetate extracts of the baru nut [77]. Consumption of these extracts by rats supplemented orally with iron diminished the iron-induced oxidative stress. This is an activity that might be exerted by the phytic acid content of the baru nut, but the involvement of phenolic compounds is equally possible [77].

Ora-Pro-Nobis (*Pereskia* sp.)

The genus *Pereskia* includes 17 species which occur only in mesic or slightly arid regions. The species are widely distributed in the Caribbean and Central as well as South America in a range of drier forest habitats [78]. In Brazil, two species (*P. aculeata* and *P. grandifolia*) are native to the Atlantic Forest and are designated popularly as ora-pro-nobis (Fig. **8**). When not in the flowering stage, *P. aculeata and P. grandifolia* can be easily confused (Fig. **8A**). Distinction is possible, however, by the flower and fruit colors. *P. aculeata* has yellow flowers and the mature fruits are also yellow (Fig. **8B,D**), whereas in *P. grandfolia* both flowers and mature fruits are purple (Figs. **8C,E**).

The leaves of both plants are used as food by natives in some regions of Brazil. For example, in Diamantina, Minas Gerais State, seventy-seven per cent of the population eats this nutritious vegetable [79]. However, *P. grandifolia* is appreciated mainly as an ornamental plant and for landscaping purposes [80]. Raw and cooked leaves, fruits and flowers can be used in several culinary preparations. Leaves can be used in salads, soups, omelettes or pies, as well as for enriching breads, cakes, and pastas. Its mucilage can substitute eggs in these preparations. The fruits can be used for juices, jellies, mousses and liquors. The seeds can be germinated to produce shoots. The young flowers can be used in salads, with meat and omelette.

Nutritional Characteristics of Ora-Pro-Nobis

Ora-pro-nobis leaves are considered rich in proteins (28.4 g/100 g, dry wet) with high tryptophan content (5.52 g/100 g protein) [81]. For this reason, they are nicknamed *meat of the poor*. The leaves have also high amounts of dietary fibers (39.1 g/100 g), minerals (iron, zinc, manganese, magnesium and calcium) and vitamins. The monosaccharides arabinose, galactose, rhamnose and glucose, and the polysaccharides arabinogalactan and galacto-mannans were also found in *P. grandifolia* and *P. aculeata* leaves [82]. These biopolymers have the ability to combine with different ions including Fe^{2+}, Co^{2+} and Ni^{2+}.

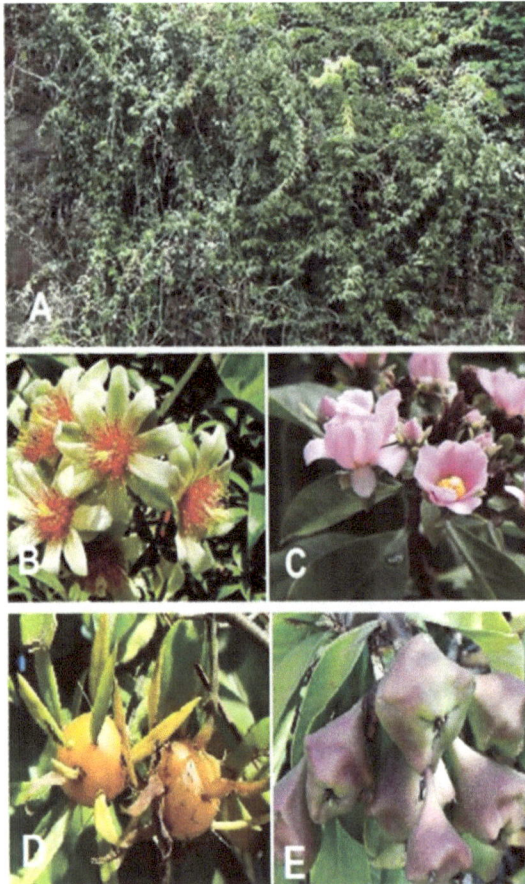

Fig. (8). *Pereskia* sp. Tree (A); flowers (B) and fruits (D) of *P. aculeata*; flowers (C) and fruit (E) of *P. grandifolia*.

Health Effects and Ethnopharmacological Studies of Ora-Pro-Nobis

Pereskia sp. finds application in popular medicine, its leaves being used to make an emollient known for mitigating inflammation and for healing burns [83 - 85]. The flour of the leaves is also used for preventing and/or treating chronic diseases such as diabetes mellitus, dyslipidaemia and obesity [80, 86]. The fruits of both species are used in popular medicine as expectorants and antisyphilitics [97]. Carotenoids and phenolics have been identified in leaves and flowers of *P. aculeata* [87 - 90] and *P. grandifolia* [91]. Such molecules could be responsible for several of the biological effects popularly attributed to ora-pro-nobis, such as antifungal, antibacterial and anti-inflammatory actions. Studies have related wound healing by leaves to their high mucilage content [87, 92, 93]. A study was conducted in models of acute and chronic ear dermatitis in mice using a hexane fraction of *P. aculeate* leaves [94]. The hexane extract contains appreciable amounts of phytosterols and triterpenes and it reduced the inflammatory process caused by several irritating agents.

Ora-pro-nobis is undoutbly a rich source of nutrients for human consumption, but it has also been suggested that its use could contribute for the improvement of a series of clinical parameters such as serum levels of glucose and triacyl-glycerols. The intake of ora-pro-nobis has also been associated to the prevention of anaemia caused by iron deficiency, cancer, osteoporosis and constipation [80].

Apparently confirming traditional medicine, a recent scientific publication attributes diuretic and hypotensive properties to the leaves of *P. grandifolia* [95]. Similarly, antioxidant and cytotoxic activities against cancer cell proliferation were demonstrated for the leaves of *P. aculeata* [96]. No such activity was found in normal cells [96]. However, modest cytotoxic actions against human breast adenocarcinoma cells (MCF-7 cell line) and human promyelocytic leukemia cells (HL60 cell line) were reported for various fractions of a methanolic extract of *P. aculeata* leaves [96].

Pinhão (Brazilian Pine, *Araucaria angustifolia*)

The Brazilian pine (*Araucaria angustifolia* (Bert.) O. Kuntze, also known as Paraná pine, is the sole native gymnosperm of the tropical deciduous forest "Mata Atlântica". It has great cultural, social and economic importance [97]. The Brazilian pine is a tall and lush tree of 20-50 m height, with an upright trunk 90-180 cm in diameter (Fig. **9A**). The flowering process of *A. angustifolia* produces a fruit, which is called pinecone (Fig. **9B**). Each of these has a diameter of 10-25 cm, containing around 150 seeds (Fig. **9C**) [98]. A typical seed, currently known as pinhão, is about 3-8 cm in length, and 1-2.5 cm wide and weights in average 8.7 g (Fig. **9C**). In Brazil, the pinhões are found in greater amounts from April to June [99].

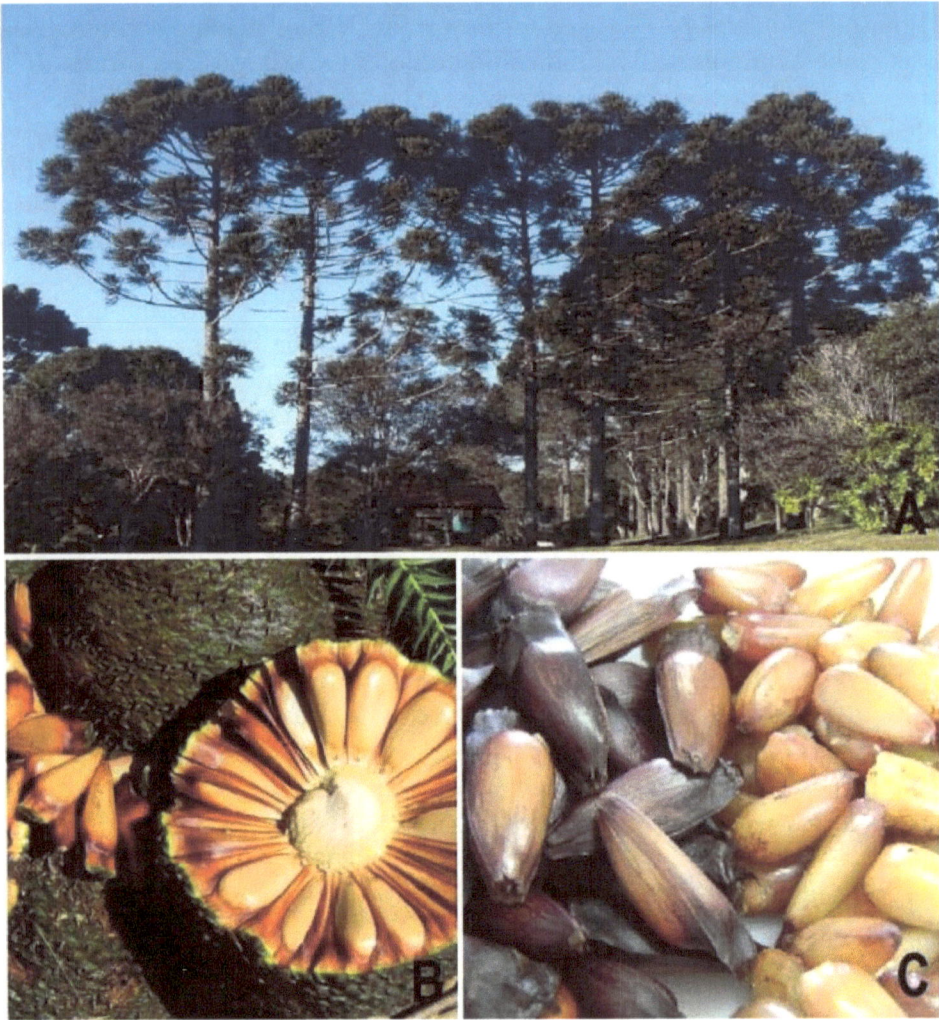

Fig. (9). *Araucaria angustifolia.* Tree (A); pinhão (B); edible seed with and without coat (C).

Nutritional Characteristics of Pinhão

For human consumption, the seeds are usually cooked in water or roasted. Flours of raw or cooked pinhão are used in the preparation of regional dishes, including cookies, cakes and breads. A common practice in Southern Brazil is to cook the seeds in water, followed by preservation in salt and vinegar [100]. The edible part of the seeds is an energetic food due to its high starch content. Pinhão is considered as an unconventional source of starch (resistant starch), with tougher texture when raw and soft after cooking. The pinhão has low levels of lipids and sugars, contains magnesium and copper, and as a gluten-free food it is suitable for consumption by sufferers of celiac disease [101, 102].

The content of soluble sugars (glucose, fructose and sucrose) in the *A. angustifolia* seed is relatively low, especially after cooking [102]. The main fatty acids are linoleic acid (18:2n-6), oleic acid (18:1n-9) and palmitic acid (16:0) [103]. The seed is also rich in tocopherols and phytosterols.

Health Effects and Ethnopharmacological Studies of Pinhão

Different parts of the Brazilian pine *A. angustifolia* have been used in folk medicine [104]. Tinctures of nodes are used topically or orally for the treatment of rheumatism. Infusions of nodes are used in the treatment of renal and sexually transmitted diseases. Infusions of barks are used in the topical treatment of muscle strains and varices. Infusions of leaves are used in the treatment of anaemia, fatigue and scrofula. Tinctures of leaves are used against dried skin, wounds and shingles. Finally, the syrup produced from the resin is used for the treatment of respiratory tract infections [105, 106].

Two lectins were purified from the pinhão [107, 108]. More recent studies have demonstrated that the pinhão lectins exert interesting biological activities such as agglutinating activity against rabbit erythrocytes [109], anti-inflammatory effects [110], and also depressant activity on the central nervous system [111].

Pinhão coat extracts being rich in condensed tannins they present elevated antioxidant activity and have been explored for possible use in cosmetic formulations [112]. The *in vivo* and *in vitro* antioxidant activity, the capability to protect MRC5 cells against H_2O_2-induced mortality and oxidative damage to lipids, proteins, and DNA, as well as the antigenotoxic activity of the bracts (undeveloped seeds) of the pines have been attributed to their high contents in phenolics, especially apigenin, catechin, epicatechin, quercetin, and rutin [113, 114].

Different extraction solvents, water, ethanol and ethanol plus water, revealed different profiles of phenolic compounds. The most abundant phenolic compound identified in aqueous extracts was protocatechuic acid, but in hydroethanolic extracts the most abundant were (+)-catechin and an (epi) catechin dimer Fig. (**10**) [115].

Zein films formulated with the cooking water of the coated pinhão seeds presented potential as a mechanical properties enhancer since a strong and tough film was obtained [115]. Extracts rich in tannin obtained from the pinhão coat inhibited in a concentration dependent manner both pancreatic α-amylase [116] and pancreatic lipase [117]. These dose-dependent inhibition activities, attenuated the postprandial hyperglycemic peak after a starchy meal and reduced the absorption of triglycerides. These activities suggest that pinhão coat extracts could be useful as alternative treatment for diabetes and obesity.

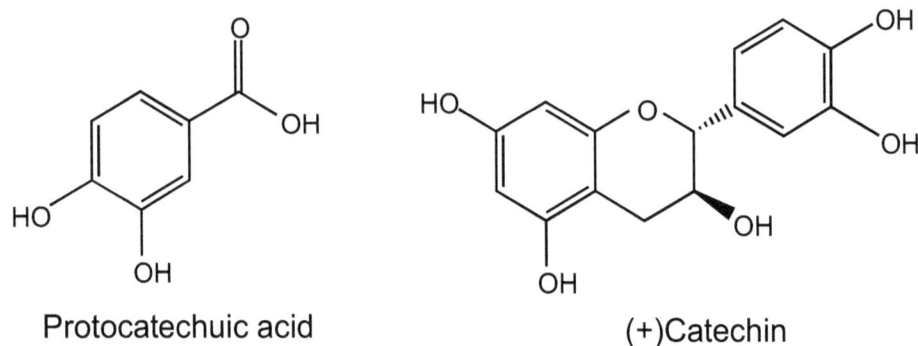

Protocatechuic acid   (+)Catechin

Fig. (10). Main phenolic compounds identified in aqueous and hydroethanolic pinhão coat extracts.

CONCLUDING REMARKS

It can be concluded from the scientific literature about the six unconventional fruit and vegetables native to Brazil that all these species present high potential for improving the nutritional quality of the Brazilian diet. This is especially true for the feeding habits at the most popular level, where shortage of financial resources often plays a highly crucial role. For this and other reasons, recent efforts in popularizing these and other species and in making them more readily available to the general population are highly welcome. It is also clear that more studies must be done in order to elucidate the chemical and biological potential of these species.

CONSENT FOR PUBLICATION

Not applicable.

CONFLICT OF INTEREST

The author (editor) declares no conflict of interest, financial or otherwise.

ACKNOWLEDGEMENT

This work was financially supported by grants from the Conselho Nacional de Desenvolvimento Científico e Tecnológico (CNPq, Proc. 302615/2011-3 and Proc. 307944/2015-8) and Fundação Araucária (Proc. 215/2014). Rúbia CG Corrêa were post-doc fellowship recipients of the CNPq (Proc.405329/2017-2).

REFERENCES

[1]   Kinupp VF, Lorenzi H. Plantas alimentícias não convencionais (PANC) no Brasil: Guia de identificação, aspectos nutricionais e receitas Instituto Plantarum de estudos da Flora. 2014. São Paulo

[2]   Koehnlein EA, Bracht A, Nishida VS, Peralta RM. Total antioxidant capacity and phenolic content of the Brazilian diet: a real scenario. Int J Food Sci Nutr 2014; 65(3): 293-8.
[http://dx.doi.org/10.3109/09637486.2013.879285] [PMID: 24490825]

[3]   Cardoso BR, Duarte GBS, Reis BZ, Cozzolino SMF. Brazil nuts: Nutritional composition, health benefits and safety aspects. Food Res Int 2017; 100(Pt 2): 9-18.
[http://dx.doi.org/10.1016/j.foodres.2017.08.036] [PMID: 28888463]

[4]   Abe LT, Lajolo FM, Genovese MI. Potential dietary sources of ellagic acid and other antioxidants among fruits consumed in Brazil: jabuticaba (*Myrciaria jaboticaba* (Vell.) Berg). J Sci Food Agric 2012; 92(8): 1679-87.
[http://dx.doi.org/10.1002/jsfa.5531] [PMID: 22173652]

[5]   Sá LZCM, Castro PFS, Lino FMA, Bernardes MJC, Viegas JCJ, Dinis TCP, *et al.* Antioxidant potential and vasodilatory activity of fermented beverages of jabuticaba berry (*Myrciaria jaboticaba*). J Funct Foods 2014; 8: 169-79.
[http://dx.doi.org/10.1016/j.jff.2014.03.009]

[6]   Wu SB, Long C, Kennelly EJ. Phytochemistry and health benefits of jaboticaba, an emerging fruit crop from Brazil. Food Res Int 2013; 54: 148-59.
[http://dx.doi.org/10.1016/j.foodres.2013.06.021]

[7]   Ulbricht C, Brigham A, Burke D, *et al.* An evidence-based systematic review of acai (*Euterpe oleracea*) by the Natural Standard Research Collaboration. J Diet Suppl 2012; 9(2): 128-47.
[http://dx.doi.org/10.3109/19390211.2012.686347] [PMID: 22607647]

[8]   Schulz M, Biluca FC, Gonzaga LV, *et al.* Bioaccessibility of bioactive compounds and antioxidant potential of juçara fruits (*Euterpe edulis* Martius) subjected to *in vitro* gastrointestinal digestion. Food Chem 2017; 228: 447-54.
[http://dx.doi.org/10.1016/j.foodchem.2017.02.038] [PMID: 28317748]

[9]   Josino Soares D, Walker J, Pignitter M, *et al.* Pitanga (*Eugenia uniflora* L.) fruit juice and two major constituents thereof exhibit anti-inflammatory properties in human gingival and oral gum epithelial cells. Food Funct 2014; 5(11): 2981-8.
[http://dx.doi.org/10.1039/C4FO00509K] [PMID: 25228206]

[10]   Martinez-Correa HA, Magalhães PM, Queiroga CL, Peixoto CA, Oliveira AL, Cabra FA. Extracts from pitanga (*Eugenia uniflora* L.) leaves: Influence of extraction process on antioxidant properties and yield of phenolic compounds. J Supercrit Fluids 2011; 55: 998-1006.
[http://dx.doi.org/10.1016/j.supflu.2010.09.001]

[11]   Samuels J. Biodiversity of food species of the Solanaceae family: a preliminary taxonomic inventory of subfamily Solanoideae. Resources 2015; 4: 277-322.
[http://dx.doi.org/10.3390/resources4020277]

[12]   Marx F, Andrade EHA, Maia JG. Chemical composition of the fruit of *Solanum sessiliflorum*. Z Lebensm-Unters Forsch A 2015; Article ID 364185: 8 pages.
[http://dx.doi.org/10.1155/2015/364185]

[13]   Lopes JC, Pereira MD. Germinação de sementes de cubiu em diferentes substratos e temperaturas. Rev Bras Sementes 2005; 27: 146-50.
[http://dx.doi.org/10.1590/S0101-31222005000200021]

[14]   Silva Filho DF, Andrade JS, Clement CR, Machado FM, Noda H. Correlações fenotípicas, genética e ambientais entre descritores morfológicos e químicos em frutos de cubiu (*Solanum sessiliflorum* Dunal) da Amazônia. Acta Amazon 1999; 29: 503-11.
[http://dx.doi.org/10.1590/1809-43921999294511]

[15]   Silva Filho DF. Yuyama LKO, Aguiar J P L, Oliveira MC, Martins LHP. Caracterização e avaliação do potencial agronômico e nutricional de etnovariedades de cubiu (*Solanum sessiliflorum* Dunal) da Amazônia. Acta Amazon 2005; 35: 399-406.
[http://dx.doi.org/10.1590/S0044-59672005000400003]

[16]   Pires AMB, Silva OS, Nardelli PM, Gomes JC, Ramos AM. Caracterização e processamento de cubiu (*Solanum sessiliflorum*). Rev Ceres Viçosa 2006; 53: 309-16.

[17]   Rodrigues E, Mariutti LR, Mercadante AZ. Carotenoids and phenolic compounds from *Solanum sessiliflorum*, an unexploited Amazonian fruit, and their scavenging capacities against reactive oxygen and nitrogen species. J Agric Food Chem 2013; 61(12): 3022-9.
[http://dx.doi.org/10.1021/jf3054214] [PMID: 23432472]

[18]   Andrade MC Jr, Andrade JS, Costa SS, Leite EAS. Nutrients of cubiu fruits (*Solanum sessiliflorum* Dunal, Solanaceae) as a function of tissues and ripening stages. J Food Nutr Res 2017; 5: 674-83.
[http://dx.doi.org/10.12691/jfnr-5-9-7]

[19]   Andrade MC Jr, Andrade JS, Costa SS. Biochemical changes of cubiu fruits (*Solanum sessiliflorum* Dunal, Solanaceae) according to different tissue portions and ripening stages. Food Nutr Sci 2016; 7: 1191-219.
[http://dx.doi.org/10.4236/fns.2016.712111]

[20]   Yuyama LKO, Macedo SHM, Aguiar JP, Filho DS, Yuyama K, Fávaro DIT, *et al.* Macro and micro nutrients quantification of some cubiu ethnovarieties (*Solanum sessiliflorum* Dunal). Acta Amazon 2007; 37: 425-9.
[http://dx.doi.org/10.1590/S0044-59672007000300014]

[21]   Mascato DRLH, Monteiro JB, Passarinho MM, Galeno DML, Cruz RJ, Ortiz C, *et al.* Evaluation of antioxidant capacity of Solanum sessiliflorum (Cubiu) extract: An *in vitro* assay. An in vitro assay J Nut Metab. 2015. Article ID 364185; 8 pages
[http://dx.doi.org/10.1155/2015/36418]

[22]   Colodel C, Bagatin RMDG, Tavares TM, Petkowicz CLO. Cell wall polysaccharides from pulp and peel of cubiu: A pectin-rich fruit. Carbohydr Polym 2017; 174: 226-34.
[http://dx.doi.org/10.1016/j.carbpol.2017.06.052] [PMID: 28821062]

[23]   Silva Filho DF, Noda H, Yuyama K, Yuyama LKO, Aguiar JPL, Machado FM. Cubiu (*Solanum sessiliflorum*, Dunal): A medicinal plant from Amazonia in the process of selection for cultivation in Manaus, Amazonas, Brasil. Rev Bras Plantas Med 2003; 5: 65-70.

[24]   Pardo MA. Efecto de *Solanum sessiliflorum* Dunal sobre el metabolismo lipidico y de la glucosa. Cien Inv 2004; 7: 43-8.

[25]   Hernandes LC, Aissa AF, Almeida MR, Darin JDC, Rodrigues E, Batista BL, *et al. In vivo* assessment of the cytotoxic, genotoxic and antigenotoxic potential of maná-cubiu (*Solanum sessiliflorum* Dunal) fruit. Food Res Int 2014; 62: 121-7.
[http://dx.doi.org/10.1016/j.foodres.2014.02.036]

[26]   Gonçalves KM, Soldati PP, Silva AF, Venâncio RP, Chaves MGAM, Raposo NRB. Biological activities of *Solanum sessiliflorum* Dunal. Biosci J 2013; 29: 1028-37.

[27]   Barriuso B, Mariutti LRB, Ansorena D, Astiasar I, Bragagnolo N. *Solanum sessiliflorum* (mana-cubiu) antioxidant protective effect toward cholesterol oxidation: Influence of docosahexaenoic acid. Eur J Lipid Sci Technol 2016; 118: 1125-31.
[http://dx.doi.org/10.1002/ejlt.201500285]

[28]   Aguiar JPL, Yuyama LKO, Souza FCA, Pessoa A. Biodisponibilidade do ferro do jambu (*Spilanthes oleracea* L.): estudo em murinos. Rev Pan-Amaz Saude 2014; 5: 19-24.
[http://dx.doi.org/10.5123/S2176-62232014000100002]

[29]   Rizzo JA, Campos IFP, Jaime MC, Munhoz G, Morgado WF. Utilização de plantas medicinais nas cidade de Goiás e Pirenópolis, Estado de Goiás. Rev Bras Cienc Farm 1999; 20: 431-47.

[30]   Favoreto R, Gilbert B. *Acmella oleracea* (L.) R. K. Jansen (Asteraceae) – Jambu. Rev Fitos 2010; 5: 83-91.

[31]   Romão NF, Silva FC, Viana RN, Ferraz ABF. Phytochemical analyses and antioxidant potential of *Spilanthes acmella* flowers extract. South Am J Bas Edu Tech Tecnol 2015; 2: 23-32.

[32]   Nascimento AM, de Souza LM, Baggio CH, *et al.* Gastroprotective effect and structure of a rhamnogalacturonan from *Acmella oleracea.* Phytochemistry 2013; 85: 137-42.
[http://dx.doi.org/10.1016/j.phytochem.2012.08.024] [PMID: 23014505]

[33]   Cheng Y-B, Liu RH, Ho M-C, *et al.* Alkylamides of *Acmella oleracea.* Molecules 2015; 20(4): 6970-7.
[http://dx.doi.org/10.3390/molecules20046970] [PMID: 25913934]

[34]   Barbosa AF, Carvalho MG, Smith RE, Sabaa-Srur AUO. Spilanthol: occurrence, extraction, chemistry and biological activities. Rev Bras Farmacogn 2016; 26: 128-33.
[http://dx.doi.org/10.1016/j.bjp.2015.07.024]

[35]   Dubey S, Maity S, Singh M, Saraf SA, Saha S. Phytochemistry, pharmacology and toxicology of *Spilanthes acmella*: a review. Adv Pharmacol Sci 2013; 2013: 423750.
[http://dx.doi.org/10.1155/2013/423750] [PMID: 24371437]

[36]   Santana de Freitas-Blanco V, Franz-Montan M, Groppo FC, *et al.* Development and evaluation of a novel mucoadhesive film containing *Acmella oleracea* extract for oral mucosa topical anesthesia. PLoS One 2016; 11(9): e0162850.
[http://dx.doi.org/10.1371/journal.pone.0162850] [PMID: 27626796]

[37]   Barbosa AF, Silva KCB, Oliveira MCC, Carvalho MG, Srur AUOS. Effects of *Acmella oleracea* methanolic extract and fractions on the tyrosinase enzyme. Rev Bras Farmacogn 2016; 26: 321-5.
[http://dx.doi.org/10.1016/j.bjp.2016.01.004]

[38]   Lima A, Oliveira e Silva AM, Trindade RA, Torres RP, Mancini-Filho J. Composição química e compostos bioativos presentes na polpa e na amêndoa do Pequi (*Caryocar brasiliense*, Camb.). Rev Bras Frutic 2007; 29: 695-8.
[http://dx.doi.org/10.1590/S0100-29452007000300052]

[39]   Silva RRV, Gomes LJ, Albuquerque UP. Plant extractivism in light of game theory: a case study in northeastern Brazil. J Ethnobiol Ethnomed 2015; 11: 6-12.
[http://dx.doi.org/10.1186/1746-4269-11-6] [PMID: 25971348]

[40]   Chisté RC, Freitas M, Mercadante AZ, Fernandes E. The potential of extracts of *Caryocar villosum* pulp to scavenge reactive oxygen and nitrogen species. Food Chem 2012; 135(3): 1740-9.
[http://dx.doi.org/10.1016/j.foodchem.2012.06.027] [PMID: 22953916]

[41]   Chisté RC, Mercadante AZ. Identification and quantification, by HPLC-DAD-MS/MS, of carotenoids and phenolic compounds from the Amazonian fruit *Caryocar villosum.* J Agric Food Chem 2012; 60(23): 5884-92.
[http://dx.doi.org/10.1021/jf301904f] [PMID: 22612541]

[42]   Chisté RC, Benassic MT, Mercadante AZ. Efficiency of different solvents on the extraction of bioactive compounds from the amazonian fruit *Caryocar villosum* and the effect on its antioxidant and colour properties. Phytochem Anal 2014; 25: 364-72.
[http://dx.doi.org/10.1002/pca.2489]

[43]   Rodrigues LJ, Paula NFR, Pinto DM, Vilas Boas EVB. Growth and maturation of pequi fruit of the Brazilian cerrado. Food Sci Technol (Campinas) 2015; 35: 11-7.
[http://dx.doi.org/10.1590/1678-457X.6378]

[44]   Ribeiro DM, Fernandes DC, Alves AM, Naves MMV. Carotenoids are related to the colour and lipid content of the pequi (*Caryocar brasiliense* Camb.) pulp from the Brazilian Savannah. Food Sci Technol 2014; 34: 507-12.
[http://dx.doi.org/10.1590/1678-457x.6369]

[45]   Pianovski AR, Vilela AFG, Silva AAS, Lima CG, Silva KK, Carvalho VFM, *et al.* Use of pequi oil (*Caryocar brasiliense*) in cosmetics emulsions: development and evaluate of physical stability. Braz J Pharm Sci 2008; 44: 249-59.

[46]   Machado MTC, Mello BCBS, Hubinger MD. Evaluation of pequi (*Caryocar brasiliense* Camb.) aqueous extract quality processed by membranes. Food Bioprod Process 2015; 95: 304-12.
[http://dx.doi.org/10.1016/j.fbp.2014.10.013]

[47]   Roesler R, Catharino RR, Malta LG, Eberlin MN, Pastore G. Antioxidant activity of *Caryocar brasiliense* (pequi) and characterization of components by electrospray ionization mass spectrometry. Food Chem 2008; 110: 711-7.
[http://dx.doi.org/10.1016/j.foodchem.2008.02.048]

[48]   Leão DP, Botelho BG, Oliveira LS, Franca AS. Potential of pequi (*Caryocar brasiliense* Camb.) peels as sources of highly esterified pectins obtained by microwave assisted extraction. Lebensm Wiss Technol 2018; 87: 575-80.
[http://dx.doi.org/10.1016/j.lwt.2017.09.037]

[49]   Maia JGS, Andrade EHA, Silva MHL. Aroma volatiles of pequi fruit (*Caryocar brasiliense* Camb.). J Food Compos Anal 2008; 21: 574-6.
[http://dx.doi.org/10.1016/j.jfca.2008.05.006]

[50]   Alabdul Magid A, Voutquenne-Nazabadioko L, Renimel I, Harakat D, Moretti C, Lavaud C. Triterpenoid saponins from the stem bark of *Caryocar villosum*. Phytochemistry 2006; 67(19): 2096-102.
[http://dx.doi.org/10.1016/j.phytochem.2006.07.009] [PMID: 16930644]

[51]   Alabdul Magid A, Voutquenne L, Harakat D, *et al.* Triterpenoid saponins from the fruits of *Caryocar villosum*. J Nat Prod 2006; 69(6): 919-26.
[http://dx.doi.org/10.1021/np060097o] [PMID: 16792411]

[52]   Magid AA, Voutquenne-Nazabadioko L, Harakat D, Moretti C, Lavaud C. Phenolic glycosides from the stem bark of *Caryocar villosum* and *C. glabrum*. J Nat Prod 2008; 71(5): 914-7.
[http://dx.doi.org/10.1021/np800015p] [PMID: 18412393]

[53]   Passos XS, Castro ACM, Pires JS, Garcia ACF, Campos FC, Fernandes OFL, *et al.* Composition and antifungal activity of the essential oils of *Caryocar brasiliensis.* Pharm Biol 2003; 41: 319-24.
[http://dx.doi.org/10.1076/phbi.41.5.319.15936]

[54]   Yamaguchi KKL, Lamarão CV, Aranha ESP, Souza ROS, Oliveira PDA, Vasconcellos MC, *et al.* HPLC-DAD profile of phenolic compounds, cytotoxicity, antioxidant and anti-inflammatory activities of the Amazon fruit *Caryocar villosum*. Quim Nova 2017; 40: 483-90.

[55]   Vieira RF, Martins MVM. Recursos genéticos de plantas medicinais do cerrado: uma compilação de dados. Rev Bras Plantas Med 2000; 3: 13-36.

[56]   Bezerra NKMS, Barros TL, Coelho NPMF. A ação do óleo de pequi (*Caryocar brasiliense*) no processo cicatricial de lesões cutâneas em ratos. Rev Bras Pl Med 2015; 17: 875-80.
[http://dx.doi.org/10.1590/1983-084X/14_061]

[57]   Almeida MR, Darin JDC, Hernandes LC, *et al.* Antigenotoxic effects of piquiá (*Caryocar villosum*) in multiple rat organs. Plant Foods Hum Nutr 2012; 67(2): 171-7.
[http://dx.doi.org/10.1007/s11130-012-0291-3] [PMID: 22562095]

[58]   Traesel GK, Menegati SELT, Dos Santos AC, *et al.* Oral acute and subchronic toxicity studies of the oil extracted from pequi (*Caryocar brasiliense*, Camb.) pulp in rats. Food Chem Toxicol 2016; 97: 224-31.
[http://dx.doi.org/10.1016/j.fct.2016.09.018] [PMID: 27639543]

[59]   Miranda-Vilela AL, Akimoto AK, Alves PC, *et al.* Dietary carotenoid-rich pequi oil reduces plasma lipid peroxidation and DNA damage in runners and evidence for an association with MnSOD genetic variant -Val9Ala. Genet Mol Res 2009; 8(4): 1481-95.

[http://dx.doi.org/10.4238/vol8-4gmr684] [PMID: 20082261]

[60] Miranda-Vilela AL, Pereira LC, Gonçalves CA, Grisolia CK. Pequi fruit (*Caryocar brasiliense* Camb.) pulp oil reduces exercise-induced inflammatory markers and blood pressure of male and female runners. Nutr Res 2009; 29(12): 850-8.
[http://dx.doi.org/10.1016/j.nutres.2009.10.022] [PMID: 19963158]

[61] Miranda-Vilela AL, Grisolia CK, Longo JPF, *et al.* Oil rich in carotenoids instead of vitamins C and E as a better option to reduce doxorubicin-induced damage to normal cells of Ehrlich tumor-bearing mice: hematological, toxicological and histopathological evaluations. J Nutr Biochem 2014; 25(11): 1161-76.
[http://dx.doi.org/10.1016/j.jnutbio.2014.06.005] [PMID: 25127291]

[62] Araruna MKA, Santos KKA, Costa JGM, Coutinho HDM, Boligon AA, Stefanello ST, *et al.* Phenolic composition and *in vitro* activity of the Brazilian fruit tree *Caryocar coriaceum* Wittm. Eur J Integr Med 2013; 5: 178-83.
[http://dx.doi.org/10.1016/j.eujim.2012.11.007]

[63] Oliveira Campos LZ, Albuquerque UP, Peroni N, Araújo EL. Do socioeconomic characteristics explain the knowledge and use of native food plants in semiarid environments in Northeastern Brazil? J Arid Environ 2015; 115: 53-61.
[http://dx.doi.org/10.1016/j.jaridenv.2015.01.002]

[64] Colombo NBR, Rangel MP, Martins V, *et al. Caryocar brasiliense* camb protects against genomic and oxidative damage in urethane-induced lung carcinogenesis. Braz J Med Biol Res 2015; 48(9): 852-62.
[http://dx.doi.org/10.1590/1414-431X20154467] [PMID: 26200231]

[65] Oliveira LG, Moreno LG, Melo DS, *et al. Caryocar brasiliense* oil improves cardiac function by increasing Serca2a/PLB ratio despite no significant changes in cardiovascular risk factors in rats. Lipids Health Dis 2017; 16(1): 37-44.
[http://dx.doi.org/10.1186/s12944-017-0422-9] [PMID: 28179001]

[66] Torres LRO, Santana FC, Torres-Leal FL, *et al.* Pequi (*Caryocar brasiliense* Camb.) almond oil attenuates carbon tetrachloride-induced acute hepatic injury in rats: Antioxidant and anti-inflammatory effects. Food Chem Toxicol 2016; 97: 205-16.
[http://dx.doi.org/10.1016/j.fct.2016.09.009] [PMID: 27623180]

[67] de Figueiredo PRL, Oliveira IB, Neto JBS, *et al. Caryocar coriaceum* Wittm. (Pequi) fixed oil presents hypolipemic and anti-inflammatory effects *in vitro* and *in vitro*. J Ethnopharmacol 2016; 191: 87-94.
[http://dx.doi.org/10.1016/j.jep.2016.06.038] [PMID: 27321275]

[68] de Oliveira ML, Nunes-Pinheiro DCS, Tomé AR, *et al. In vivo* topical anti-inflammatory and wound healing activities of the fixed oil of *Caryocar coriaceum* Wittm. seeds. J Ethnopharmacol 2010; 129(2): 214-9.
[http://dx.doi.org/10.1016/j.jep.2010.03.014] [PMID: 20332017]

[69] Takemoto E, Okada IA, Garbelotti ML, Tavares M, Aued-Pimentel S. Composição química da semente e do óleo de baru (*Dipteryx alata* Vog.) nativo do Município de Pirenópolis, Estado de Goiás. Rev Inst Adolfo Lutz 2001; 60: 113-7.

[70] Marin AM, Siqueira EM, Arruda SF. Minerals, phytic acid and tannin contents of 18 fruits from the Brazilian savanna. Int J Food Sci Nutr 2009; 60 (Suppl. 7): 180-90.
[http://dx.doi.org/10.1080/09637480902789342] [PMID: 19353365]

[71] Marques FG, Oliveira Neto JR, Cunha LC, Paula JR, Bara MTF. Identification of terpenes and phytosterols in *Dipteryx alata* (baru) oil seeds obtained through pressing. Rev Bras Farmacogn 2015; 25: 522-5.
[http://dx.doi.org/10.1016/j.bjp.2015.07.019]

[72] Sano SM, Ribeiro JF, Brito MA. Baru: biologia e uso.biologia e uso In: Documentos 116 EMBRAPA, Planaltina, Brasil. 2004; pp. 11-20.

[73]   Ausman LM, Rong N, Nicolosi RJ. Hypocholesterolemic effect of physically refined rice bran oil: studies of cholesterol metabolism and early atherosclerosis in hypercholesterolemic hamsters. J Nutr Biochem 2005; 16(9): 521-9.
[http://dx.doi.org/10.1016/j.jnutbio.2005.01.012] [PMID: 16115540]

[74]   Gromadzka J, Wardencki W. Trends in edible vegetable oils analysis. Part A. Determination of different components of edible oils – a review. J Food Nut Sci 2011; 61: 33-43.

[75]   Plat J, Mensink RP. Vegetable oil based *versus* wood based stanol ester mixtures: effects on serum lipids and hemostatic factors in non-hypercholesterolemic subjects. Atherosclerosis 2000; 148(1): 101-12.
[http://dx.doi.org/10.1016/S0021-9150(99)00261-0] [PMID: 10580176]

[76]   Bento AP, Cominetti C, Simões Filho A, Naves MM. Baru almond improves lipid profile in mildly hypercholesterolemic subjects: a randomized, controlled, crossover study. Nutr Metab Cardiovasc Dis 2014; 24(12): 1330-6.
[http://dx.doi.org/10.1016/j.numecd.2014.07.002] [PMID: 25149894]

[77]   Siqueira EMA, Marin AMF, da Cunha MSB, Fustinoni AM, de Sant'Ana LP, Arruda SF. Consumption of baru seeds (*Dipteryx alata* Vog.), a Brazilian savanna nut, prevents iron-induced oxidative stress in rats. Food Res Int 2012; 45: 427-33.
[http://dx.doi.org/10.1016/j.foodres.2011.11.005]

[78]   Edwards EJ, Nyffeler R, Donoghue MJ. Basal cactus phylogeny: implications of *Pereskia* (Cactaceae) paraphyly for the transition to the cactus life form. Am J Bot 2005; 92(7): 1177-88.
[http://dx.doi.org/10.3732/ajb.92.7.1177] [PMID: 21646140]

[79]   Dias ACP, Pinto NAVD, Yamada LTP, Mendes KL, Fernandes AG. Avaliação do consumo de hortaliças não convencionais pelos usuários das unidades do Programa Saúde da Família (PSF) de Diamantina – MG. Aliment Nutr 2005; 16: 279-84.

[80]   Almeida MEF, Corrêa AD. Utilization of cacti of the genus *Pereskia* in the human diet in a municipality of Minas Gerais. Cienc Rural 2012; 42: 751-6.
[http://dx.doi.org/10.1590/S0103-84782012000400029]

[81]   Takeiti CY, Antonio GC, Motta EM, Collares-Queiroz FP, Park KJ. Nutritive evaluation of a non-conventional leafy vegetable (*Pereskia aculeata* Miller). Int J Food Sci Nutr 2009; 60 (Suppl. 1): 148-60.
[http://dx.doi.org/10.1080/09637480802534509] [PMID: 19468927]

[82]   Sierakowski MR. Location of *O*-acetyl groups in the heteropolysaccharide of the cactus *Pereskia aculeata*. Carbohydr Res 1990; 201: 277-84.
[http://dx.doi.org/10.1016/0008-6215(90)84243-N]

[83]   Duarte MR, Hayashi SS. Estudo anatômico de folha e caule de *Pereskia aculeata* Mill. (Cactaceae). Rev Bras Farmacogn 2005; 15: 103-9.
[http://dx.doi.org/10.1590/S0102-695X2005000200006]

[84]   Farago PV, Takeda IJM, Budel JM, Duarte MR. Análise morfo-anatômica de folhas de *Pereskia grandifolia* Haw, Cactaceae. Lat Am J Pharm 2004; 23: 323-7.

[85]   Sartor CFP, Amaral V, Guimarães HET, Barros KN, Felipe DF, Cortez IER, *et al.* Estudo da ação cicatrizante de folhas de *Pereskia aculeata*. Saúde Pesqui 2010; 3: 149-54.

[86]   de Almeida ME, Simão AA, Corrêa AD, de Barros Fernandes RV. Improvement of physiological parameters of rats subjected to hypercaloric diet, with the use of *Pereskia grandifolia* (Cactaceae) leaf flour. Obes Res Clin Pract 2016; 10(6): 701-9.
[http://dx.doi.org/10.1016/j.orcp.2015.10.011] [PMID: 26616446]

[87]   Pinto NdeC, Scio E. The biological activities and chemical composition of *Pereskia* species (Cactaceae)--a review. Plant Foods Hum Nutr 2014; 69(3): 189-95.
[http://dx.doi.org/10.1007/s11130-014-0423-z] [PMID: 24862084]

[88]   Agostini-Costa TS, Pêssoa GKA, Silva DB, Gomes IS, Silva JP. Carotenoid composition of berries and leaves from a Cactaceae – *Pereskia* sp. J Funct Foods 2014; 11: 178-84.
[http://dx.doi.org/10.1016/j.jff.2014.09.015]

[89]   Agostini-Costa TS, Wondraceck DC, Rocha WS, Silva DB. Carotenoids profile and total polyphenols in fruits of *Pereskia aculeata* Miller. Rev Bras Frutic 2012; 34: 234-8.
[http://dx.doi.org/10.1590/S0100-29452012000100031]

[90]   Souza LF, Caputo L, Inchausti De Barros IB, Fratianni F, Nazzaro F, De Feo V. *Pereskia aculeata* Muller (Cactaceae) leaves: chemical composition and biological activities. Int J Mol Sci 2016; 17(9): 1478-60.
[http://dx.doi.org/10.3390/ijms17091478] [PMID: 27598154]

[91]   Nurestri MAS, Sim KS, Norhanom AW. Phytochemical and cytotoxic investigations of *Pereskia grandifolia* Haw. (Cactaceae) leaves. J Biol Sci 2009; 9: 488-93.
[http://dx.doi.org/10.3923/jbs.2009.488.493]

[92]   Carvalho EG, Soares CP, Blaua L, Menegon RF, Joaquim WM. Wound healing properties and mucilage content of *Pereskia aculeata* from different substrates. Rev Bras Farmacogn 2014; 24: 677-82.
[http://dx.doi.org/10.1016/j.bjp.2014.11.008]

[93]   Pinto NCC, Cassini-Vieira P, Souza-Fagundes EM, Barcelos LS, Castañon MCMN, Scio E. *Pereskia aculeata* Miller leaves accelerate excisional wound healing in mice. J Ethnopharmacol 2016; 194: 131-6.
[http://dx.doi.org/10.1016/j.jep.2016.09.005] [PMID: 27599609]

[94]   Pinto NdeC, Machado DC, da Silva JM, *et al.* *Pereskia aculeata* Miller leaves present *in vitro* topical anti-inflammatory activity in models of acute and chronic dermatitis. J Ethnopharmacol 2015; 173: 330-7.
[http://dx.doi.org/10.1016/j.jep.2015.07.032] [PMID: 26226436]

[95]   Kazama CC, Uchida DT, Canzi KN, *et al.* Involvement of arginine-vasopressin in the diuretic and hypotensive effects of *Pereskia grandifolia* Haw. (Cactaceae). J Ethnopharmacol 2012; 144(1): 86-93.
[http://dx.doi.org/10.1016/j.jep.2012.08.034] [PMID: 22960548]

[96]   Pinto NCC, Santos RC, Machado DC, Florêncio JR, Fagundes EMZ, Antinarelli LMR, *et al.* Cytotoxic and antioxidant activity of *Pereskia aculeate* Miller. Pharmacologyonline 2012; 3: 63-9.

[97]   Peralta RM, Koehnlein EA, Oliveira RF, Correa VG, Corrêa RCG, Bertonha L, *et al.* Biological activities and chemical constituents of *Araucaria angustifolia*: An effort to recover a species threatened by extinction. Trends Food Sci Technol 2016; 54: 85-93.
[http://dx.doi.org/10.1016/j.tifs.2016.05.013]

[98]   Lima EC, Royer B, Vaghetti JCP, *et al.* Adsorption of Cu(II) on *Araucaria angustifolia* wastes: determination of the optimal conditions by statistic design of experiments. J Hazard Mater 2007; 140(1-2): 211-20.
[http://dx.doi.org/10.1016/j.jhazmat.2006.06.073] [PMID: 16876938]

[99]   Amarante CVT, Mota CS, Megguer CA, Ide GM. Conservação pós colheita de pinhões sementes de *Araucaria angustifolia* (Bertoloni) Otto Kuntze armazenados em diferentes temperaturas. Cienc Rural 2007; 37: 346-51.
[http://dx.doi.org/10.1590/S0103-84782007000200008]

[100] Leite DMC, Jong EV, Noren CPZ, Brandelli A. Nutritional evaluation of *Araucaria angustifolia* seed flour as a protein complement for growing rats. J Sci Food Agric 2008; 88: 1166-71.
[http://dx.doi.org/10.1002/jsfa.3192]

[101] Conforti PA, Lupano CE. Comparative study of the starch digestibility of *Araucaria angustifolia* and *Araucaria araucana* seed flour. Starke 2008; 60: 192-8.
[http://dx.doi.org/10.1002/star.200700671]

[102] Cordenunsi BR, De Menezes Wenzel E, Genovese MIS, Colli C, De Souza Gonçalves A, Lajolo FM. Chemical composition and glycemic index of Brazilian pine (*Araucaria angustifolia*) seeds. J Agric Food Chem 2004; 52(11): 3412-6.
[http://dx.doi.org/10.1021/jf034814l] [PMID: 15161207]

[103] Silva CM, Zanqui AB, Souza AHP, Gohara AK, Gomes STMG, da Silva EA, *et al*. Extraction of oil and bioactive compounds from *Araucaria angustifolia* (Bertol.) Kunze using subcritical n-propane and organic solvents. J Supercrit Fluids 2016; 112: 14-21.
[http://dx.doi.org/10.1016/j.supflu.2016.02.003]

[104] Aslam MS, Choudhary BA, Uzair M, Ijaz AS. Phytochemical and ethno-pharmacological review of the genus *Araucaria*: a Review. Trop J Pharm Res 2013; 12: 651-9.

[105] Freitas AM, Almeida MTR, Andrighetti-Fröhner CR, *et al*. Antiviral activity-guided fractionation from *Araucaria angustifolia* leaves extract. J Ethnopharmacol 2009; 126(3): 512-7.
[http://dx.doi.org/10.1016/j.jep.2009.09.005] [PMID: 19761825]

[106] Gontijo VS, Dos Santos MH, Viegas C Jr. Biological and chemical aspects of natural biflavonoids from plants: a brief review. Mini Rev Med Chem 2017; 17(10): 834-62.
[http://dx.doi.org/10.2174/1389557517666161104130026] [PMID: 27823559]

[107] Datta PK, Figueroa MO, Lajolo FM. Purification and characterization of two major lectins from *Araucaria brasiliensis* syn. *Araucaria angustifolia* seeds (Pinhão). Plant Physiol 1991; 97(3): 856-62.
[http://dx.doi.org/10.1104/pp.97.3.856] [PMID: 16668523]

[108] Datta PK, Figueiroa MODCR, Lajolo FMJ. Chemical modification and sugar binding properties of two major lectins from pinhão (*Araucaria brasiliensis*) seeds. J Agric Food Chem 1993; 41: 1851-5.
[http://dx.doi.org/10.1021/jf00035a009]

[109] Santi-Gadelha T, de Almeida Gadelha CA, Aragão KS, *et al*. Purification and biological effects of *Araucaria angustifolia* (Araucariaceae) seed lectin. Biochem Biophys Res Commun 2006; 350(4): 1050-5.
[http://dx.doi.org/10.1016/j.bbrc.2006.09.149] [PMID: 17045568]

[110] Mota MRL, Criddle DN, Alencar NMN, *et al*. Modulation of acute inflammation by a chitin-binding lectin from *Araucaria angustifolia* seeds via mast cells. Naunyn Schmiedebergs Arch Pharmacol 2006; 374(1): 1-10.
[http://dx.doi.org/10.1007/s00210-006-0097-7] [PMID: 16957941]

[111] Vasconcelos SM, Lima SR, Soares PM, *et al*. Central action of *Araucaria angustifolia* seed lectin in mice. Epilepsy Behav 2009; 15(3): 291-3.
[http://dx.doi.org/10.1016/j.yebeh.2009.05.002] [PMID: 19446042]

[112] Mota GST, Arantes AB, Sacchetti G, Spagnoletti A, Ziosi P, Scalambra E, *et al*. Antioxidant activity of cosmetic formulations based on novel extracts from seeds of Brazilian *Araucaria* (Bertol) Kuntze. J Cosm Dermatol Sci Appl 2014; 4: 190-202.

[113] Michelon F, Branco CS, Calloni C, *et al*. *Araucaria angustifolia*: a potential nutraceutical with antioxidant and antimutagenic activities. Curr Nutr Food Sci 2012; 8: 155-8.
[http://dx.doi.org/10.2174/157340112802651103]

[114] Souza MO, Branco CS, Sene J, *et al*. Antioxidant and antigenotoxic activities of the Brazilian pine *Araucaria angustifolia* (Bert.). Antioxidants 2014; 3(1): 24-37.
[http://dx.doi.org/10.3390/antiox3010024] [PMID: 26784661]

[115] Freitas TB, Santos CHK, Silva MV, Shirai MA, Dias MI, Barros L, *et al*. Antioxidants extraction from Pinhão (*Araucaria angustifolia* (Bertol.) Kuntze) coats and application to zein films. Food Packag Shelf Life in press
[http://dx.doi.org/10.1016/j.fpsl.2017.10.006]

[116] Silva SM, Koehnlein EA, Bracht A, Castoldi R, de Morais GR, Baesso ML, *et al*. Inhibition of salivary and pancreatic a-amylases by a pinhão coat (*Araucaria angustifolia*) extract rich in condensed

tannin. Food Res Int 2014; 56: 1-8.
[http://dx.doi.org/10.1016/j.foodres.2013.12.004]

[117] Oliveira RF, Gonçalves GA, Inácio FD, *et al.* Inhibition of pancreatic lipase and triacylglycerol intestinal absorption by a pinhão coat (*Araucaria angustifolia*) extract rich in condensed tannin. Nutrients 2015; 7(7): 5601-14.
[http://dx.doi.org/10.3390/nu7075242] [PMID: 26184295]

AUTHOR INDEX

SUBJECT INDEX